A Primer on the Physics of the Cosmic Microwave Background

A Primer on the Physics of the Cosmic Microwave Background

Massimo Giovannini
Centro Enrico Fermi, Italy
CERN, Switzerland

NEW JERSEY · LONDON · SINGAPORE · BEIJING · SHANGHAI · HONG KONG · TAIPEI · CHENNAI

Published by

World Scientific Publishing Co. Pte. Ltd.

5 Toh Tuck Link, Singapore 596224

USA office: 27 Warren Street, Suite 401-402, Hackensack, NJ 07601

UK office: 57 Shelton Street, Covent Garden, London WC2H 9HE

British Library Cataloguing-in-Publication Data
A catalogue record for this book is available from the British Library.

A PRIMER ON THE PHYSICS OF THE COSMIC MICROWAVE BACKGROUND

Copyright © 2008 by World Scientific Publishing Co. Pte. Ltd.

All rights reserved. This book, or parts thereof, may not be reproduced in any form or by any means, electronic or mechanical, including photocopying, recording or any information storage and retrieval system now known or to be invented, without written permission from the Publisher.

For photocopying of material in this volume, please pay a copying fee through the Copyright Clearance Center, Inc., 222 Rosewood Drive, Danvers, MA 01923, USA. In this case permission to photocopy is not required from the publisher.

ISBN-13 978-981-279-142-9
ISBN-10 981-279-142-6

Printed in Singapore.

To Sergio Fubini

Preface

More than three score years ago, high-energy physicists were driven to scrutinize the properties of the cosmic radiation then available (i.e. cosmic rays). Today the same situation is realized not only with cosmic rays but also with different cosmological data: most notably, with the Cosmic Microwave Background (CMB in what follows). While I am writing this preface, European science is at the forefront of the developments in high-energy physics and cosmology thanks to the Large Hadron Collider program and thanks also to the Planck explorer mission. Today laboratory physics and celestial physics give us contradictory indications: it seems that all the matter accessible to terrestrial laboratory experiments contributes only 5% to the total energy budget of the Universe.

Cosmologists and astrophysicists today cannot ignore the knowledge of the micro-world provided by high-energy physics. In similar terms, high-energy physicists cannot avoid being exposed to some of the key concepts in modern gravitation and cosmology. While grand unifications of all fundamental forces are one of the intriguing hopes suggested by current theoretical speculations, the opportunity of a small unification lies already before us in the years to come: the construction of a common language which will allow, in the near future, a more effective exchange of information and ideas between contiguous branches of the physics community. The present book seeks to be a modest contribution to this mighty endeavor.

This book grew through the last decade because of various series of lectures that were either directly or indirectly connected to CMB physics and, more generally, to gravitation. In the last couple of years I came to the conclusion that an effective way of presenting a cosmology course (either for last year undergraduate or for PhD students) is to use CMB as a guiding theme. While lecturing to PhD students I have been confronted with the problem of giving a sufficiently accurate and updated information to an audience that was, very often, rather composite. Not all PhD students were exposed to General Relativity or field theory in their undergraduate courses. Similarly, not all PhD students have a preliminary knowledge of astrophysics. I have tried, therefore, to present the material in a reasonably self-contained manner also in view of the time limitations imposed by a PhD course.

My warm acknowledgment goes to the Enrico Fermi center for a senior research grand entitled *From the Planck Scale to the Hubble Radius*. Without this support my efforts would have been forlorn. In commencing this script I wish also to express my very special and sincere gratitude to Prof. G. Cocconi and Prof. E. Picasso. I am indebted to G. Cocconi for his advice in the preparation of the first section. I am indebted to E. Picasso for delightful discussions which have been extremely relevant both for the selection of topics and for the overall quality of the manuscript.

Massimo Giovannini

Contents

Preface   vii

Part I   1

1. Why CMB Physics?   3
   - 1.1 The blackbody spectrum and its physical implications .   8
   - 1.2 A bit of history of CMB observations   12
   - 1.3 The entropy of the CMB and its implications   14
   - 1.4 The time evolution of the CMB temperature   16
   - 1.5 A quick glance to the Sunyaev-Zeldovich effect   18
   - 1.6 Cosmological parameters   22

2. From CMB to the Standard Cosmological Model   29
   - 2.1 The Standard Cosmological Model (SCM)   30
     - 2.1.1 Homogeneity and isotropy   31
     - 2.1.2 Perfect barotropic fluids   32
     - 2.1.3 General Relativity   37
   - 2.2 Friedmann-Lemaître equations   40
   - 2.3 Matter content of the SCM   44
   - 2.4 The future of the Universe   47
   - 2.5 The past of the Universe   51
     - 2.5.1 Hydrogen recombination   54
     - 2.5.2 Coulomb scattering: the baryon-electron fluid . .   59
     - 2.5.3 Thompson scattering: the baryon-photon fluid .   60
   - 2.6 Simplified numerical estimates   64

Part II 67

3. Problems with the SCM — 69
   - 3.1 The horizon problem — 70
   - 3.2 The spatial curvature problem — 74
   - 3.3 The entropy problem — 75
   - 3.4 The structure formation problem — 76
   - 3.5 The singularity problem — 79

4. SCM and Beyond — 81
   - 4.1 The horizon and the flatness problems — 81
   - 4.2 Classical and quantum fluctuations — 88
   - 4.3 The entropy problem — 93
   - 4.4 The problem of geodesic incompleteness — 97

5. Essentials of Inflationary Dynamics — 101
   - 5.1 Fully inhomogeneous Friedmann-Lemaître equations — 101
   - 5.2 Homogeneous evolution of a scalar field — 108
   - 5.3 Classification(s) of inflationary backgrounds — 112
   - 5.4 Exact inflationary backgrounds — 116
   - 5.5 Slow-roll dynamics — 120
   - 5.6 Slow-roll parameters — 123

Part III 127

6. Inhomogeneities in FRW Models — 129
   - 6.1 Decomposition of inhomogeneities in FRW Universes — 131
   - 6.2 Gauge issues for the scalar modes — 133
   - 6.3 Super-adiabatic amplification — 136
   - 6.4 Quantum mechanical description of the tensor modes — 141
   - 6.5 Spectra of relic gravitons — 150
   - 6.6 Quantum state of cosmological perturbations — 151
   - 6.7 Digression on different vacua — 156
   - 6.8 Numerical estimates of the mixing coefficients — 164

7. The First Lap in CMB Anisotropies — 171
   - 7.1 Tensor Sachs-Wolfe effect — 172

	7.2	Scalar Sachs-Wolfe effect	175
	7.3	Scalar modes in the pre-decoupling phase	179
		7.3.1 Scale crossing and CMB initial conditions	182
	7.4	CDM-radiation system	183
	7.5	Adiabatic and non-adiabatic modes: an example	188
	7.6	Sachs-Wolfe plateau: mixture of initial conditions	196
8.	Improved Fluid Description of Pre-Decoupling Physics		205
	8.1	The general plasma with four components	206
	8.2	CDM component	208
	8.3	Tight-coupling between photons and baryons	216
	8.4	Shear viscosity and silk damping	218
	8.5	The adiabatic solution	220
	8.6	Pre-equality non-adiabatic initial conditions	223
		8.6.1 The CDM-radiation mode	224
		8.6.2 The baryon-entropy mode	228
		8.6.3 The neutrino-entropy mode	229
	8.7	Numerics in the tight-coupling approximation	232
		8.7.1 Interpretation of the numerical results	237
		8.7.2 Numerical estimates of diffusion damping	242
9.	Kinetic Hierarchies		247
	9.1	Collisionless Boltzmann equation	248
	9.2	Boltzmann hierarchy for massless neutrinos	251
	9.3	Brightness perturbations of the radiation field	255
	9.4	Evolution equations for the brightness perturbations	257
		9.4.1 Visibility function	259
	9.5	Line of sight integrals	261
		9.5.1 Angular power spectrum and observables	267
	9.6	Tight-coupling expansion	270
	9.7	Zeroth order in tight-coupling: acoustic oscillations	272
		9.7.1 Solutions of the evolution of monopole and dipole	274
		9.7.2 Estimate of the sound horizon at decoupling	276
	9.8	First order in tight-coupling: polarization	278
		9.8.1 Improved estimates of polarization	278
		9.8.2 Polarization power spectra	281

9.9	Second order in tight-coupling: diffusion damping	287
9.10	Semi-analytical approach to Doppler oscillations	292

10. **Early Initial Conditions?** — 305

10.1	Minimally coupled scalar field	307
	10.1.1 Gauge-invariant description	308
	10.1.2 Curvature perturbations and scalar normal modes	310
10.2	Spectral relations	312
	10.2.1 Some slow-roll algebra	312
	10.2.2 Tensor power spectra	315
	10.2.3 Scalar power spectra	318
	10.2.4 Consistency relation	319
10.3	Curvature perturbations and density contrasts	320
10.4	Hamiltonians for the scalar problem	322
10.5	Trans-Planckian problems?	324
	10.5.1 Minimization of canonically related Hamiltonians	327
	10.5.2 Back-reaction effects	331
10.6	How many adiabatic modes?	334

Part IV — 339

11. **Surfing on the Gauges** — 341

	11.0.1 Generalities on scalar gauge transformations	342
11.1	The longitudinal gauge	346
	11.1.1 Gauge-invariant generalizations	347
11.2	The synchronous gauge	349
	11.2.1 Evolution equations in the synchronous gauge	352
	11.2.2 The adiabatic mode in the synchronous gauge	357
	11.2.3 Entropic modes in the synchronous gauge	359
11.3	Comoving orthogonal hypersurfaces	362
11.4	Uniform density hypersurfaces	364
11.5	The off-diagonal gauge	367
	11.5.1 Evolution equations in the off-diagonal gauge	369
11.6	Mixed gauge-invariant treatments	372

12.	Interacting Fluids	377
	12.1 Interacting fluids with bulk viscous stresses	379
	12.2 Evolution equations for the entropy fluctuations	381
	12.3 Specific physical limits	385
	12.4 Mixing between entropy and curvature perturbations	386
13.	Spectator Fields	391
	13.1 Spectator fields in a fluid background	393
	13.2 Unconventional inflationary models	400
	13.3 Conventional inflationary models	403

Appendix A	The Concept of Distance in Cosmology	409
	A.1 The proper coordinate distance	409
	A.2 The redshift	411
	A.3 The distance measure	413
	A.4 Angular diameter distance	415
	A.5 Luminosity distance	416
	A.6 Horizon distances	417
	A.7 Few simple applications	417

Appendix B	Kinetic Description of Hot Plasmas	421
	B.1 Generalities on thermodynamic systems	421
	B.2 Fermions and bosons	423
	B.3 Thermal, kinetic and chemical equilibrium	425
	B.4 An example of primordial plasma	427
	B.5 Electron-positron annihilation and neutrino decoupling	429
	B.6 Big-bang nucleosynthesis (BBN)	431

Appendix C	Scalar Modes of the Geometry	435
	C.1 Fluctuations of the Einstein tensor	435
	C.2 Fluctuations of the energy-momentum tensor(s)	437
	C.3 Fluctuations of the covariant conservation equations	439
	C.4 Some algebra with the scalar modes	441

Appendix D	Metric Fluctuations: Gauge-Independent Treatment	445
	D.1 The scalar problem	446
	D.2 The vector problem	447

D.3	The tensor problem	448
D.4	Inhomogeneities of the sources	449

Bibliography   455

Index   467

Part I

Cosmic Microwave Background Physics and the Formulation of the Standard Cosmological Model

Chapter 1

Why CMB Physics?

When approaching a new subject of study, especially within the realm of empirical sciences, the relevant question to ask is always the same: why should we learn about this? So, why should we learn about CMB physics? To answer this kind of questions one might be tempted to invoke either historical or subjective arguments. For instance one could say that, historically, blackbody emission is rather interesting in itself since it represented, at the dawn of the century, one of the fragile bridges that allowed us to pass from a classical description of macroscopic phenomena to the quantum mechanical language which is today the most appropriate for the discussion of microscopic physics. On a more aesthetic (and hence subjective) level, one could also affirm that blackbody emission is beautiful since it depends only upon one crucial parameter, i.e. the temperature. Subjectivity in science is very important since it drives the enthusiasm of researchers towards new and exciting fields of investigation. At the same time any subjective self-excitation should be gauged by more objective elements of judgment. Objectivity, for natural scientists, rhymes with testability. The quest for objectivity does not imply the lack of fantasy but, on the contrary, it just focuses our theoretical endeavor.

In this introductory chapter the theme will be to stress that there are objective elements that make CMB physics one of the most attractive and promising frameworks for gathering indirect informations on the early stages of the life of our own Universe. After a general introduction to blackbody emission, the motivations of this script will be spelled out. The bottom line will be that, indeed, the CMB is cosmological and represents the dominant component of the detected extra-galactic emission.

The whole observable Universe will therefore be approached, in the first approximation, as a system emitting electromagnetic radiation. The topics to be treated in the present chapter are therefore the following:

- electromagnetic emission of the Universe;
- the blackbody spectrum;
- a bit of history of the CMB observations;
- the entropy of the CMB and its implications;
- the time evolution of the CMB temperature;
- a quick glance at the Sunyaev-Zeldovich effect.

All along this script the natural system of units will be adopted. In this system

$$\hbar = c = k_\mathrm{B} = 1, \tag{1.1}$$

where $\hbar = h/2\pi$, c is the speed of light and k_B is the Boltzmann constant. In order to pass from one system of units to the other it is useful to recall that

- $\hbar c = 197.327\,\mathrm{MeV\,fm}$;
- $\mathrm{K} = 8.617 \times 10^{-5}\,\mathrm{eV}$;
- $(\hbar c)^2 = 0.389\,\mathrm{GeV}^2\,\mathrm{mbarn}$;
- $c = 2.99792 \times 10^{10}\,\mathrm{cm/sec}$.

In Fig. 1.1 a rather intriguing plot summarizes the electromagnetic emission of our own Universe. Only the extra-galactic emissions are reported.[a] On the horizontal axis we have the logarithm of the energy of the photons (expressed in eV). On the vertical axis we reported the logarithm (to base 10) of $\Omega_\gamma(E)$ which is defined as

$$\Omega_\gamma(E) = \frac{1}{\rho_\mathrm{crit}} \frac{d\rho_\gamma}{d\ln E}. \tag{1.2}$$

The specific form of $\Omega_\gamma(E)$ in the case of the CMB branch of the spectrum will be discussed in the following section (see, for instance, Eq. (1.12)). For the moment it suffices to note that $\Omega_\gamma(E)$ measures the energy density of the emitted radiation in critical units. The critical energy density ρ_crit can be understood, grossly speaking, as the mean energy density of the Universe, i.e. for the current values of the cosmological parameters, the energy density equivalent to about six proton masses per cubic meter (see, for instance, Eq. (1.11)).

---

[a]By extra-galactic emissions we mean radiation coming from the outside of our galaxy. Of course, as stressed later on, it must be borne in mind that our own galaxy is also an efficient emitter of electromagnetic radiation.

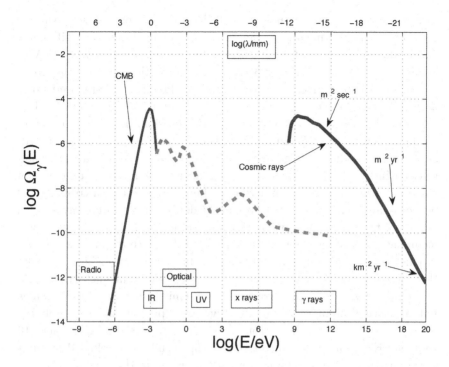

Fig. 1.1 In this cartoon the (extragalactic) electromagnetic emission is sketched. On the vertical axis the logarithm (to base 10) of the emitted energy density is reported in units of $\rho_{\rm crit}$ (see Eq. (1.9)). The logarithm of energy of the photons is instead reported on the horizontal axis. The wavelength scale is inserted at the top of the plot. The cosmic ray spectrum is included for comparison and in the same units used to describe the electromagnetic contribution.

For comparison also the associated wavelength of the emitted radiation is illustrated (see the top of the figure) in units of mm. Figure 1.1 motivates the choice of studying accurately the properties of CMB.

In Fig. 1.1 the maximum of $\Omega_\gamma(E)$ is located for a wavelength of the mm (see the scale of wavelengths at the top of Fig. 1.1) corresponding to typical energies of the order[b] of 10^{-3} eV. In the optical and ultraviolet range of wavelengths the energy density drops almost two orders of magnitude. In the x-rays (i.e. 10^{-6} mm $< \lambda < 10^{-9}$ mm) the energy density of the emitted radiation drops more than three orders of magnitude in comparison with the maximum. The x-ray range corresponds to photon energies $E >$ keV.

---

[b]In natural units $\hbar = c = k_{\rm B} = 1$ we have $E_k = \hbar\omega = \hbar c k$ and that $k = 2\pi/\lambda$. So $E_k = k$ and $\omega = 2\pi\nu$.

In the γ-rays (i.e. 10^{-9} mm $< \lambda < 10^{-12}$ mm) the spectral amplitude is roughly 5 orders of magnitude smaller than in the case of the millimeter maximum. The range of γ-rays occurs for photon energies $E >$ GeV. While the CMB represents 0.93 of the extragalactic emission, the infrared and visible part give, respectively, 0.05 and 0.02. The x-ray and γ-ray branches contribute, respectively, by 2.5×10^{-4} and 2.5×10^{-5}. The CMB is therefore the 93% of the total extragalactic emission. The CMB spectrum has been discovered by Penzias and Wilson [1] (see also [2]) and predicted, on the basis of the hot big-bang model, by Gamow, Alpher and Herman (see, for instance, [3]). Wavelengths as large as $\lambda \sim$ m lead to an emission which is highly anisotropic and will not be treated here as a cosmological probe. In any case, for $\lambda \geq$ m we are in the domain of the radio-waves. This branch of the spectrum is of upmost importance for a variety of problems including, for instance, large scale magnetic fields (both in galaxies and in clusters) [4] and pulsar astronomy [5]. In fact it should be mentioned that our own galaxy is also an efficient emitter of electromagnetic radiation. Since our galaxy possess a magnetic field, it emits synchrotron radiation as well as thermal bremsstrahlung. A very daring project that will probably be at the forefront of radio-astronomical investigations during the next 10 years is SKA (Square Kilometer Array) [6]. While the technical features of the instrument cannot be thoroughly discussed in the present script, it suffices to note that the collecting area of the instrument, as the name suggests will be of $10^6 \, \text{m}^2$. The design of SKA will probably allow full sky surveys of Faraday rotation and better understanding of galactic emission.[c]

In Fig. 1.1 the spectrum of the cosmic rays is also reported, for comparison. This inclusion is somehow arbitrary since the cosmic rays of moderate energy are known to come from within the galaxy. It is however useful to report also the energy spectrum of cosmic rays and to compare it, in the same units, with the energy spectrum of CMB photons. The energy density of the cosmic rays is, roughly, of the same order of the energy density of the CMB. For energies smaller than 10^{15} eV the rate is approximately of one particle per m^2 and per second. For energies larger than 10^{15} eV the rate is approximately of one particle per m^2 and per year. The difference in

---

[c]We will not enter here in the vast subject of CMB foregrounds. It suffices to appreciate that while the spectrum of synchrotron increases with frequency, for wavelengths shorter than the mm the emission is dominated by thermal dust emission whose typical spectrum decreases with frequency. It is opinion of the author that a better understanding of the spectral slope of the synchrotron would really be needed (not only from extrapolation). This seems important especially in the light of forthcoming satellite missions.

these two rates corresponds to a slightly different spectral behaviour of the cosmic ray spectrum, the so-called knee. Finally, for energies larger than 10^{18} eV, the rate of the so-called ultra-high-energy cosmic rays (UHECR) is even smaller and of the order of one particle per km^2 and per year. The sudden drop in the flux corresponds to another small change in the spectral behaviour, the so-called ankle. In the parametrization chosen in Fig. 1.1 the cosmic ray spectrum does not decrease as E^{-3} but rather as E^{-2}. The rationale for this difference stems from the fact that, in the parametrization of Fig. 1.1 we plot the energy density of cosmic rays per logarithmic interval of E while, in the standard parametrization the plot is in terms of $d\rho_{\text{crays}}/dE$. In the forthcoming years the spectrum above the ankle will be scrutinized by the AUGER experiment [7, 8]. While this book was in its final stages the preliminary results of the AUGER experiment appeared in a pair of papers, i.e. Refs. [9] and [10]. In [9] the collaboration achieved one decisive result for the spectrum of cosmic rays at energies in the range of the EeV. The hypothesis of the pure power law spectrum is rejected with a significance *better than 6 sigma and 4 sigma for minimum energies of $10^{18.6}$ eV and 10^{19} eV respectively* (verbatim from Ref. [9]). In [10] the data were analyzed to search for anisotropies near the direction of the galactic plane at EeV energies. The reported results suggest a highly isotropic distribution and *do not support previous findings of localized excesses in the AGASA and SUGAR data*[d] (verbatim from the abstract of Ref. [10]).

The latest analyses of the AUGER experiment demonstrated a correlation between the arrival directions of cosmic rays with energy above 6×10^{19} eV and the positions of active galactic nuclei within 75 Mpc [11]. At smaller energies it has been convincingly demonstrated, as previously mentioned, that overdensities on windows of 5 deg radius (and for energies $10^{17.9}$ eV $< E < 10^{18.5}$ eV) are compatible with an isotropic distribution. The rejection of the hypothesis of a continuation of the spectrum in the form of a power-law is statistically significant [9]. The position of the ankle (i.e. the spectral break) occurs for $E_a \sim 10^{18.5}$ eV. The combined Auger spectrum can then be parametrized as

$$E^2 \Omega_{\text{crays}}(E) \propto E^{3+\gamma_1}, \qquad E < E_a,$$

$$E^2 \Omega_{\text{crays}}(E) \propto E^{3+\gamma_2} \frac{1}{1 + \exp\left(\frac{\log E - \log E_c}{W_c}\right)}, \qquad E > E_a, \quad (1.3)$$

---

[d]AGASA and SUGAR data are former experiments analyzing cosmic rays in the EeV region.

where log are base-10 logarithms and the energies are expressed in units of eV. In Eq. (1.3) γ_1 and γ_2 are the spectral index before and after the break respectively, E_a is the position of the break, and the second term in the second equation is a flux suppression term where E_c is the energy at which the flux is suppressed 50% compared to a pure power-law, and W_c determines the sharpness of the cutoff. Using Eq. (1.3) to fit the experimental data the collaboration obtains $\gamma_1 = -3.30 \pm 0.06$, $\gamma_2 = -2.56 \pm 0.06$, $\log E_a = 18.65 \pm 0.04$, $\log E_c = 19.74 \pm 0.06$ and $W_c = 0.16 \pm 0.04$. We point out that Eq. (1.3) is a parametrization, not a theoretical prediction. In spite of the fact that cosmic rays are not central to this presentation, we suggest the reader an interesting critical review on the theory of cosmic rays [12]. It is important to stress that while the CMB represents the 93 % of the extragalactic emission, the diffuse x-ray and γ-ray backgrounds are also of upmost importance for cosmology. Various experiments have been dedicated to the study of the x-ray background such as ARIEL, EINSTEIN, GINGA, ROSAT and, last but not least, BEPPO-SAX, an x-ray satellite named after Giuseppe (Beppo) Occhialini.[e] Among γ-ray satellites we shall just mention COMPTON, EGRET and the forthcoming GLAST. In [13, 14] it was actually argued that the typical slope of the γ-ray background as measured by EGRET can be related to the slope of the cosmic ray spectrum.

1.1 The blackbody spectrum and its physical implications

According to Fig. 1.1, in the mm range the electromagnetic spectrum of the Universe is very well fitted by a blackbody spectrum: if we would plot the error bars magnified 400 times they would still be hardly distinguishable from the thickness of the curve. Starting from the discovery of Penzias and Wilson [1] various groups confirmed, independently, the blackbody nature of this emission (see below, in this section, for an oversimplified account of the intriguing history of CMB observations). As it is well known the blackbody has the property of depending only upon one single parameter which is the temperature T_γ of the photon gas. Such a temperature is given by

$$T_\gamma = 2.725 \pm 0.001 \text{ K}. \tag{1.4}$$

---

[e]Part of the present book was prepared in connection with a PhD course at the University of Milan-Bicocca whose physics department is named after Giuseppe Occhialini.

According to Wien's law $\lambda T_\gamma = 2.897 \times 10^{-3}$ m K. Thus, as already remarked the wavelength of the maximum will be $\lambda \simeq$ mm. For a photon gas in thermodynamic equilibrium the energy density of the emitted radiation is given by

$$d\rho_\gamma = g \times \omega \times \frac{d^3\omega}{(2\pi)^3} \times \overline{n}_\omega, \qquad (1.5)$$

where g is the number of intrinsic degrees of freedom ($g = 2$ in the case of photons) and n_ω is the Bose-Einstein occupation number:

$$\overline{n}_\omega = \frac{1}{e^{\omega/T_\gamma} - 1}. \qquad (1.6)$$

Since in natural units, $E_k = k = \omega$, the energy density of the emitted radiation per logarithmic interval of frequency is given by:

$$\frac{d\rho_\gamma}{d\ln k} = \frac{1}{\pi^2} \frac{k^4}{e^{k/T_\gamma} - 1}. \qquad (1.7)$$

Equation (1.7) allows also to compute the total (i.e. integrated) energy density ρ_γ. The differential spectrum of Eq. (1.7) can then be referred to the integrated energy density expressed, in turn, in units of the critical energy density. From Eq. (1.7) the integrated energy density of photons is simply given by

$$\rho_\gamma(t_0) = \frac{T_\gamma^4}{\pi^2} \int_0^\infty \frac{x^3}{e^x - 1} = \frac{\pi^2}{15} T_\gamma^4, \qquad (1.8)$$

where the ratio $x = k/T_\gamma$ has been defined and where the integral in the second equality is given by $\pi^4/15$.

A useful way of measuring energy densities is to refer them to the *critical energy density* of the Universe (see chapter 2 for a more detailed discussion of this important quantity). In short we will also talk about critical density. According to the present data it seems that the critical energy density indeed coincides with the *total* energy density of the Universe. This is just because experimental data seem to favour a spatially flat Universe. The critical energy density today is given by[f]:

$$\rho_{\text{crit}} = \frac{3H_0^2}{8\pi G} = 1.88 \times 10^{-29} h_0^2 \,\text{g cm}^{-3} = 1.05 \times 10^{-5} h_0^2 \,\text{GeV cm}^{-3}, \qquad (1.9)$$

---

[f]To understand more physically the present value of the critical energy density we can say that the vacuum of a particle accelerator is of the order of 10^{-19} g cm^{-3}. Furthermore, we can say that, prior to gravitational collapse, the mean density is of the order of the critical density. After gravitational collapse the mean matter density, for instance within the Milky way, is three or even four orders of magnitude larger than the critical density.

where h_0 (often assumed to be ~ 0.7 for the purpose of numerical estimates along this book) measures the indetermination on the present value of the Hubble parameter

$$H_0 = 100 \frac{\text{km}}{\text{sec Mpc}} h_0. \tag{1.10}$$

From the second equality appearing in Eq. (1.9), recalling that the proton mass is $m_p = 0.938\,\text{GeV}$, it is also possible to deduce

$$\rho_{\text{crit}} = 5.48 \left(\frac{h_0}{0.7}\right)^2 \frac{m_p}{\text{m}^3}, \tag{1.11}$$

showing that, the critical density is, grossly speaking, the equivalent of 6 proton masses per cubic meter. From Eqs. (1.7) and (1.9) we can obtain the energy density of photons per logarithmic interval of energy and in critical units, i.e.

$$\Omega_\gamma(k) = \frac{1}{\rho_{\text{crit}}} \frac{d\rho_\gamma}{d\ln k}. \tag{1.12}$$

Recalling that $E_k = k$ (and neglecting the subscript) we have that

$$\Omega_\gamma(E) = \frac{15}{\pi^4} \Omega_{\gamma 0} \frac{x^4}{e^x - 1}, \tag{1.13}$$

where

$$x = \frac{E}{T_\gamma} = 4.26 \times 10^3 \left(\frac{E}{\text{eV}}\right),$$

$$\Omega_{\gamma 0} = \frac{\rho_\gamma(t_0)}{\rho_{\text{crit}}} = 2.471 \times 10^{-5} h_0^{-2}. \tag{1.14}$$

The quantities $\Omega_\gamma(E)$ and $\Omega_{\gamma 0}$ are physically different:

- $\Omega_{\gamma 0}$ is the ratio between the total (present) energy density of CMB photons and the critical energy density; $\Omega_{\gamma 0}$ is independent on the frequency;
- $\Omega_\gamma(E)$ is the energy spectrum of CMB photons per logarithmic interval of frequency and expressed in critical units; $\Omega_\gamma(E)$, unlike $\Omega_{\gamma 0}$, does depend on the frequency.

Crudely speaking (and up to numerical factors) Eq. (1.13) simply suggests that $\Omega_{\gamma 0}$ sets the overall normalization of $\Omega_\gamma(E)$. It can be explicitly verified that, inserting the numerical value of T_γ and ρ_c (i.e. Eqs. (1.4) and (1.9)), the figure of Eq. (1.14) is swiftly reproduced. The spectrum of Eq. (1.13) can be also plotted in terms of the frequency. Recalling that, in natural units, $k = 2\pi\nu$ the parameter $x = k/T_\gamma$ can be directly expressed in terms of ν. Thus x can be easily written as:

$$x = \frac{k}{T_\gamma} = 0.01765 \left(\frac{\nu}{\text{GHz}}\right). \qquad (1.15)$$

Consider now Eq. (1.13) and multiply both sides by h_0^2. In this way the combination $h_0^2 \Omega_{\gamma 0}$ will appear at the right hand side. This combination, as already remarked, does not depend on the indetermination of the Hubble constant. The logarithm (to base 10) of the obtained expression gives

$$\log h_0^2 \Omega_\gamma(x) = \log\left(\frac{15}{\pi^4}\right) + \log h_0^2 \Omega_{\gamma 0} + 4\log x - \log[e^x - 1]. \qquad (1.16)$$

If we now insert Eq. (1.15) inside Eq. (1.16) and plot the obtained expression as a function of ν/GHz, the curve reported in Fig. 1.2 will be swiftly obtained. It should be borne in mind that the CMB spectrum could be

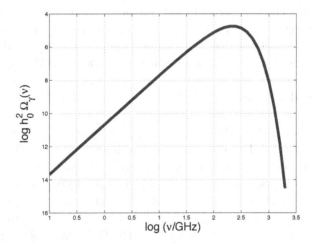

Fig. 1.2 The CMB logarithmic energy spectrum here illustrated in terms of the frequency.

distorted by several energy-releasing processes. These distortions have not been observed so far. In particular we could wonder if a sizable chemical potential is allowed. The presence of a chemical potential will affect the Bose-Einstein occupation number which will become, in our rescaled notations $\overline{n}_k^B = (e^{x+\mu_0} - 1)^{-1}$. The experimental data imply that $|\mu_0| < 9 \times 10^{-5}$ (95% C.L.). It is useful to mention, at this point, the energy density of the CMB in different units and to compare it directly with the cosmic ray spectrum as well as with the energy density of the galactic magnetic field. In particular we will have

$$\rho_\gamma = \frac{\pi^2}{15}T_\gamma^4 = 2 \times 10^{-51}\left(\frac{T_\gamma}{2.725}\right)^4 \text{GeV}^4, \tag{1.17}$$

$$\rho_B = \frac{B^2}{8\pi} = 1.36 \times 10^{-52}\left(\frac{B}{3\mu\text{G}}\right)^2 \text{GeV}^4. \tag{1.18}$$

From Eqs. (1.17) and (1.18) it follows that the CMB energy density is roughly comparable with the magnetic energy density of the galaxy. Furthermore $\rho_{\text{crays}} \simeq \rho_B$.

1.2 A bit of history of CMB observations

The blackbody nature of CMB emission is one of the cornerstones of the Standard Cosmological Model[g] whose essential features will be introduced in chapter 2. The first measurement of the CMB spectrum goes back to the work of Penzias and Wilson [1]. The Penzias and Wilson measurement referred to a wavelength of 7.35 cm (corresponding to 4.08 GHz). They estimated a temperature of 3.5 K. Since the Penzias and Wilson measurement the blackbody nature of the CMB spectrum has been investigated and confirmed for a wide range of frequencies extending from 0.6 GHz [16] (see also [17]) up to 300 GHz. The history of the measurements of the CMB temperature is a subject by itself which has been reviewed in the excellent book of B. Patridge [18]. Before 1990 the measurements of CMB properties have always been conducted through terrestrial antennas or even by means of balloon borne experiments. In the nineties the COBE satellite [19–26] allowed us to measure the properties of the CMB spectrum in a wide range of frequencies including the maximum (see Fig. 1.2). The COBE satellite had two instruments: FIRAS and DMR. The DMR was able to probe the angular power spectrum[h] up to $\ell \simeq 26$. As the name implies, DMR was a differential instrument measuring temperature differences in the microwave sky. The angular resolution of a given instrument, i.e. ϑ, is related to the maximal multipole probed in the sky according to the approximate relation $\vartheta \simeq \pi/\ell$. Consequently:

---

[g]In this book the acronym SCM will often be used instead of standard cosmological model.

[h]While the precise definition of angular power spectrum will be given later on, here it suffices to recall that $\ell(\ell+1)C_\ell/(2\pi)$ measures the degree of inhomogeneity in the temperature distribution per logarithmic interval of ℓ. Consequently, a given multipole ℓ can be related to a given spatial structure in the microwave sky: small ℓ will correspond to low wavenumbers, high ℓ will correspond to larger wavenumbers.

- since the angular resolution of COBE was $7°$, the maximal ℓ accessible to that experiment was $\ell \simeq 180°/7° \sim 26$;
- since the angular resolution of WMAP[i] is $0.23°$, the corresponding maximal harmonic probed by WMAP will be $\ell \simeq 180°/0.23° \sim 783$;
- finally, the Planck Explorer experiment,[j] to be soon launched will achieve an angular resolution of $5'$, implying $\ell \simeq 180°/5' \sim 2160$.

After the COBE mission, various experiments attempted the exploration of smaller angular separation, i.e. larger multipoles. A definite convincing evidence of the existence and location of the first peak in the C_ℓ spectrum came from the Boomerang [27, 28], Dasi [29] and Maxima [30] experiments. Both Boomerang and Maxima were balloon borne (bolometric) experiments. Dasi was a ground based interferometer. The data points of these last three experiments explored multipoles up to 1000, determining the first acoustic oscillation (in the jargon the first Doppler peak) for $\ell \simeq 220$. Another important balloon borne experiments was Archeops [31] providing interesting data for the region characterizing the first rise of the C_ℓ spectrum. Some other useful references on earlier CMB experiments can be found in [32]. The C_ℓ spectrum, as measured by different recent experiments is reported in Fig. 1.3 (adapted from Ref. [33]). At the moment the most accurate determinations of CMB observables are derived from the data of WMAP (Wilkinson Microwave Anisotropy Probe). The first release of WMAP data are the subject of Refs. [35–38]. The three-year release of WMAP data is discussed in Refs. [39, 40]. The WMAP data (filled circles in Fig. 1.3) provided, among other important pieces of information, the precise determination of the position of the first peak (i.e. $\ell = 220.1 \pm 0.8$ [36]) and evidence of the second peak. The WMAP experiment also measured temperature-polarization correlations providing a distinctive signature (the so-called anticorrelation peak in the temperature-polarization power spectrum for $\ell \sim 150$) of primordial adiabatic fluctuations (see chapters 8 and 9 and, in particular, Fig. 9.2). To have a more detailed picture of the

---

[i]WMAP is the acronym for Wilkinson Microwave Anisotropy Probe and it is a satellite mission proposed to NASA in 1996 and launched in June 2001. Various experiments will be mentioned throughout this book and their essential features as well as the meaning of the corresponding acronym can be usefully obtained from the original references which are carefully quoted in the bibliographical section at the end of this book.

[j]The updated science case for the Planck experiment can be found at the following address http://www.rssd.esa.int/index.php?project=PLANCK. The long script that can be downloaded is often referred to, in the jargon, as the *Planck Bluebook*.

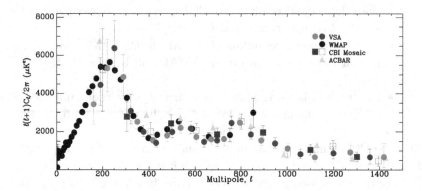

Fig. 1.3 Some CMB anisotropy data are reported (figure adapted from [33]): WMAP data (filled circles); VSA data (shaded circles) [34]; CBI data (squares) [41, 42]; ACBAR data (triangles) [43].

evolution and relevance of CMB experiments we refer the reader to Ref. [44] (for review of the pre-1994 status of the art) and Ref. [45] for a review of the pre-2002 situation). The rather broad set of lectures by Bond [46] may also be usefully consulted.

1.3 The entropy of the CMB and its implications

The pressure of blackbody photons is simply $p_\gamma = \rho_\gamma/3$. Since the chemical potential exactly vanishes in the case of a photon gas at the thermodynamic equilibrium, the entropy density of the blackbody is given, through the fundamental identity of thermodynamics (see Appendix B), by

$$s_\gamma = \frac{S_\gamma}{V} = \frac{\rho_\gamma + p_\gamma}{T_\gamma} = \frac{4}{45}\pi^2 T_\gamma^3, \qquad (1.19)$$

where S_γ is the entropy and V is a fiducial volume. Equation (1.19) implies that the entropic content of the present Universe is dominated by the species that are relativistic today (i.e. photons) and that the total entropy contained in the Hubble volume, i.e. S_γ is *huge*. The Hubble volume can be thought of as the present size of our observable Universe and it is roughly given by $V_H = 4\pi H_0^{-3}/3$. Thus, we will have

$$S_\gamma = \frac{4}{3}\pi s_\gamma H_0^{-3} \simeq 1.43 \times 10^{88} \left(\frac{h_0}{0.7}\right)^{-3}. \qquad (1.20)$$

The figure provided by Eq. (1.20) is still one of the major problems of the standard cosmological model. Why is the entropy of the observable

Universe so large? For the estimate of Eq. (1.20) it is practical to express both T_γ and H_0 in Planck units, namely[k]:

$$T_\gamma = 1.923 \times 10^{-32}\, M_{\rm P}, \qquad H_0 = 1.22 \times 10^{-61}\left(\frac{h_0}{0.7}\right) M_{\rm P}. \qquad (1.21)$$

It is clear that the huge value of the present entropy is a direct consequence of the smallness of H_0 in Planck units. Equation (1.21) implies that $T_\gamma/H_0 \simeq 1.57 \times 10^{29}$. Let us just remark that the present estimate only concerns the thermodynamic entropy. Considerations related to the validity (also in the early Universe) of the second law of thermodynamics seem to suggest also that the entropy of the gravitational field itself may play a decisive role. While some motivations seem compelling there is no consensus, at the moment, on what should be the precise mathematical definition of the entropy of the gravitational field. This remark is necessary since we should keep our minds open. It may well be that the true entropy of the Universe (i.e. the entropy of the sources and of the gravitational field) is larger than the one computed in Eq. (1.20). Along this direction it is possible to think that the maximal entropy that can be stored inside the Hubble radius $r_{\rm H}$ is of the order of a black-hole with Schwartzchild radius $r_{\rm H} \simeq H_0^{-1}$ which would give

$$r_{\rm H}^2 M_{\rm P}^2 \simeq 10^{122}. \qquad (1.22)$$

In connection with Eq. (1.21), it is also useful to point out that the critical density can be expressed directly in terms of the fourth power of the Planck mass, i.e.:

$$\rho_{\rm crit} = \frac{3}{8\pi} H_0^2 M_{\rm P}^2 = 1.785 \times 10^{-123}\left(\frac{h_0}{0.7}\right)^2 M_{\rm P}^4. \qquad (1.23)$$

The huge hierarchy between the critical energy density of the present Universe and the Planckian energy density is, again, a direct reflection of the hierarchy between the Hubble parameter and the Planck mass. Such a hierarchy would not be, by itself, problematic. The rationale for such a statement is connected to the fact that in the SCM the energy densities as well as the related pressures decrease as the Universe expand. However, today, the largest portion of the energy density of the Universe is determined by a component called *dark energy*. The term *dark* is a coded word of astronomy. It means that a given form of matter or energy neither absorbs nor emit radiation. Furthermore the dark energy is homogeneously distributed

---

[k]To derive the second relation from the definition of H_0 (i.e. $H_0 = 100 h_0$ km sec^{-1} Mpc^{-1}) it is practical to recall that Mpc = 3.08×10^{24} cm.

and, unlike *dark matter*, is not concentrated in the galactic halos and in the clusters of galaxies. One of the chief properties of dark energy is that it is not affected by the Universe expansion and this is the reason why it is usually parametrized in terms of a cosmological constant. Measurements tell us that $\rho_\Lambda \simeq 0.7 \rho_{\rm crit}$ which implies, from Eq. (1.23), that

$$\rho_\Lambda \simeq 1.24 \times 10^{-123} \, M_{\rm P}^4. \tag{1.24}$$

Since ρ_Λ *does not* decreases with the expansion of the Universe, we have also to admit that Eq. (1.24) was enforced at any moment in the life of the Universe and, in particular, at the moment when the initial conditions of the SCM were set. A related way of phrasing this impasse relies on the field theoretical interpretation of the cosmological constant. In field theory we do know that the zero-point (vacuum) fluctuations have an energy density (per logarithmic interval of frequency) that goes as k^4. Now, adopting the Planck mass as the ultraviolet cut-off we would be led to conclude that the total energy density of the zero-point vacuum fluctuations would be of the order of $M_{\rm P}^4$. On the contrary, the result of the measurements simply gives us a figure which is 122 orders of magnitude smaller.

The expression of the blackbody spectrum also allows the calculation of the photon concentration. Recalling that, in the case of photons, $dn = (k^3 n_k/\pi^2) d\log k$ we have, after integration over k that the concentration of photons is given by

$$n_{\gamma 0} = \frac{2\zeta(3)}{\pi^2} T_\gamma^3 \simeq 411 \text{ cm}^{-3} \tag{1.25}$$

where $\zeta(r)$ is the Riemann zeta function with argument r.

1.4 The time evolution of the CMB temperature

In summary we can therefore answer, in the first approximation, to the question giving the title of this chapter:

- in the electromagnetic spectrum the contribution of the CMB is by far larger than the other branches and constitutes, roughly speaking, 93 % of the whole emission;
- the CMB energy density is comparable with (but larger than) the energy density of cosmic rays;
- the CMB energy density is a tiny fraction of the total energy density of the Universe (more precisely 24 millionth of the critical energy density);

- the CMB dominates the total entropy of the present Hubble patch: $S_\gamma \simeq 10^{88}$.

The fact that we observe a CMB seems to imply that CMB photons are in thermal equilibrium at the temperature T_γ. This occurrence strongly suggests that the evolution of the whole Universe must somehow be adiabatic. In a preliminary perspective, the following naive observation is rather important. Suppose that the spatial coordinates expand thanks to a time-dependent rescaling. Consequently the wave-numbers will also be rescaled accordingly, i.e.

$$\vec{x}_0 \to \vec{x} = a(t)\vec{x}_0, \qquad \vec{k}_0 \to \vec{k} = \frac{\vec{k}_0}{a(t)}. \qquad (1.26)$$

In the jargon $\vec{k}_0$ is commonly referred to as the comoving wave-number (which is insensitive to the expansion), while $\vec{k}$ is the physical wave-number. Consider then the number of photons contained in an infinitesimal element of the phase-space and suppose that the whole Universe expands according to Eq. (1.26). At a generic time t_1 we will then have

$$dn_k(t_1) = \overline{n}_k(t_1) d^3k_1 d^3x_1. \qquad (1.27)$$

At a generic time $t_2 > t_1$ we will have, similarly,

$$dn_k(t_2) = \overline{n}_k(t_2) d^3k_2 d^3x_2. \qquad (1.28)$$

By looking at Eqs. (1.27) and (1.28) it is rather easy to argue that $dn_k(t_1) = dn_k(t_2)$ *provided* $\overline{n}_k(t_1) = \overline{n}_k(t_2)$. By looking at the specific form of the Bose-Einstein occupation number it is clear that the latter occurrence is verified provided $k(t_1)/T_\gamma(t_1) = k(t_2)/T_\gamma(t_2)$. From this simple argument we can already argue an important fact: the blackbody distribution is preserved under the rescaling (1.26) provided the blackbody temperature evolves as the inverse of the scale factor $a(t)$, i.e.

$$T_{\gamma 0} \to T_\gamma = \frac{T_{\gamma 0}}{a(t)}. \qquad (1.29)$$

The property summarized in Eq. (1.29) holds also in the context of the SCM where $a(t)$ will be correctly defined as the time-dependent scale factor of a Friedmann-Robertson-Walker (FRW) Universe. The physical consequence of Eq. (1.29) is that the temperature of CMB photons is higher at higher redshifts (see Appendix A for a definition of redshift). More precisely:

$$T_\gamma = (1+z) T_{\gamma 0}. \qquad (1.30)$$

This consequence of the theory can be tested experimentally [47]. In short, the argument goes as follows. The CMB will populate excited levels of atomic and molecular species when the energy separations involved are not too different from the peak of the CMB emission. The first measurement of the local CMB temperature was actually made with this method by using the fine structure lines of CN (cyanogen) [48]. Using the same philosophy it is reasonable to expect that clouds of other chemical elements (like Carbon, in Ref. [47]) may be sensitive to CMB photons also at higher redshifts. For instance in [47] measurements were performed at $z = 1.776$ and the estimated temperature was found to be of the order of $T_\gamma(z) \simeq 7.5\ ^0K$. These measurements are potentially very instructive but have been a bit neglected, in the recent past, since the attention of the community focused more on the properties of CMB anisotropies.

1.5 A quick glance to the Sunyaev-Zeldovich effect

For the limitations imposed by the introductory nature of the present script it is not possible to treat in detail the very interesting physics of another important effect that gives us valuable informations concerning the CMB and its primeval origin. It is in fact very important to establish empirically that CMB is not a merely local phenomenon, i.e. a phenomenon that arises in the local Universe. One definite answer, along this direction, is provided by the observation that, at higher redshifts, the putative CMB temperature indeed increases. The other definite answer comes exactly from the Sunyaev-Zeldovich (SZ) effect [49–51]. The physics of the SZ effect is, in a sense, rather simple. Clusters of galaxies have a deep potential well and on the average, by the virial theorem, their kinetic motion is of the order of few keV. So some fraction of the hot gas can get ionized and ionized plasma will be around. This plasma emits x-rays that, for instance, the ROSAT satellite has scutinized.[1] Now the CMB will sweep the whole space. By looking at a direction where there is nothing between the observer and the last scattering surface the radiation arrives basically unchanged except for the effect due to the expansion of the Universe. But if the observation

---

[1] It is actually interesting, incidentally, that from the ROSAT full sky survey (allowing us to determine the surface brightness of various clusters in the x-rays), the average electron density has been determined [52] and this allowed interesting measurements of magnetic fields inside a sample of Abell clusters. The ROSAT satellite was an x-ray satellite flying from June 1990 to February 1999 and exploring the x-ray sky for energies between 0.1 keV and 2.4 keV.

is now made along a direction passing through a cluster of galaxies, some small fraction of the CMB photons (roughly one over 1000 CMB photons) will be scattered by the hot gas. Because the gas is actually hot, there is more probability that photons will be scattered at high energy rather than at low energy. They will also be scattered almost at isotropic angle. The bottom line is that the CMB spectrum along a line of sight that crosses a cluster of galaxies will have a slight excess of high energy photons and a slight deficiency of low energy photons. So if you see this effect (as we do) it means that the CMB photons come from behind the clusters. Some of these clusters are at redshift $0.07 < z < 1.03$. The measurements of the Sunyaev-Zeldovich effect have been attempted for roughly two decades but in the last decade a remarkable progress has been made. As already mentioned, the SZ effect tells that the CMB is really an extra-galactic radiation.

There are excellent long and short reviews on the SZ effect. Here we would like to quote just the classic review of Rephaeli [53] (see also [54]) as well as the non-relativistic treatments of the kinetic equation (which will be used in a moment to derive the non-relativistic expression of the modified spectral intensity) due to Kompaneets [55] (see also the comprehensive book of Peebles [56]). In what follows the simplest non-relativistic set-up will be described. The present treatment closely follows the one of Ref. [53] but within our set of conventions, units and definitions.

To give a simplified and self-contained introduction to the SZ effect, consider the following simplifications:

- assume first the Thompson limit where the frequency of the photons is much smaller than the electron mass; in this limit the cross section will simply be the Thompson cross section;
- derive the evolution of the Bose-Einstein occupation number in the non-relativistic limit where the kinetic equations reduce to the well known form of Fokker-Planck equation.

Using these two approximations we will have that the kinetic equation can be written as [55, 56]:

$$\frac{\partial \overline{n}_k}{\partial t} = \left(\frac{T_\gamma}{m_e}\right) \frac{\sigma_{\text{Th}} n_e}{x^2} \frac{\partial}{\partial x}\left[x^4\left(\frac{T_e}{T_\gamma}\frac{\partial \overline{n}_k}{\partial x} + \overline{n}_k + \overline{n}_k^2\right)\right], \qquad (1.31)$$

where σ_{Th} is the Thompson cross section which will be discussed often in this book starting from chapter 2. In Eq. (1.31), T_γ is the photon temperature and T_e is the electron temperature. Since the temperature of the electrons of the cluster is large in comparison with the photon temperature

the following hierarchies hold:
$$\frac{T_e}{T_\gamma}\frac{\partial \bar{n}_k}{\partial x} \gg \bar{n}_k, \qquad \frac{T_e}{T_\gamma}\frac{\partial \bar{n}_k}{\partial x} \gg \bar{n}_k^2. \tag{1.32}$$
Consequently Eq. (1.31) can be written as
$$\frac{\partial \bar{n}_k}{\partial t} = \frac{T_e}{m_e}\frac{\sigma_{Th} n_e}{x^2}\frac{\partial}{\partial x}\left(x^4 \frac{\partial \bar{n}_k}{\partial x}\right). \tag{1.33}$$
Equations (1.31) and (1.33) also assume that the distribution of the electrons is isotropic in the cluster frame. Now, if the incident radiation (i.e. the CMB) is only weakly scattered by the electrons, the approximate solution of Eq. (1.33) can be obtained by substituting at the right hand side of Eq. (1.33) the Bose-Einstein occupation number, i.e. $\bar{n}_k(x) = (e^x - 1)^{-1}$, and by then integrating along the line of sight to get the modified occupation number. The first step of this procedure leads to the following equation:
$$\frac{\partial \bar{n}_k}{\partial t} = \frac{T_e}{m_e}\sigma_{Th} n_e \mathcal{L}(x), \qquad \mathcal{L}(x) = \frac{x[e^x(x-4)+(x+4)]}{(e^x-1)^3} e^x. \tag{1.34}$$
Notice that the function $\mathcal{L}(x)$ goes to zero. In particular the zero of this function, i.e. $\mathcal{L}(x_0) = 0$ corresponds to $x_0 = 3.83$ which means, using Eq. (1.15) that connects x to ν in GHz units, $\nu_0 = 216.99$ GHz (often called crossover frequency). It is common practice, at this point, to work not with $h_0^2 \Omega_\gamma(\nu)$ but rather with the spectral intensity of the radiation field. There is no deep reason for doing that, however, for the sake of comparison with the notations customarily adopted in the literature we will stick to this convention and define the spectral intensity, in natural units as
$$I = \frac{T_\gamma^3}{2\pi^2} x^3 \bar{n}_k(x). \tag{1.35}$$
The scattered spectral intensity ΔI will then be obtained by integrating Eq. (1.33) along the line of sight with the result that
$$\Delta I = \frac{T_\gamma^3}{2\pi^2} g(x) y, \qquad g(x) = x^3 \mathcal{L}(x) \tag{1.36}$$
where y is the Comptonization[m] parameter, i.e.
$$y = \int \frac{T_e}{m_e} n_e \sigma_{Th} dL = \frac{T_e}{m_e} \sigma_{Th} \text{DM}. \tag{1.37}$$
In the second equality of Eq. (1.37), the so-called dispersion measurement $\text{DM} = \int n_e dL$ has been introduced.[n] The interesting exercise is now to take

---
[m] The COBE-FIRAS data set a bound to the Comptonization parameter which reads, to 95% confidence level, $|y| < 1.2 \times 10^{-5}$.
[n] Sometimes this quantity is also called column density of electrons.

the scattered radiation field given by Eq. (1.37), divide it by the spectral intensity in the absence of scattering (i.e. the spectral intensity of CMB photons that did not cross during their trajectory the hot electrons of a cluster) and evaluate the obtained expression in the limit $x \ll 1$ (which does correspond to the Rayleigh-Jeans region of the spectrum). The result of these simple manipulations is, therefore, the following:

$$\frac{\Delta I}{I} = C(x)y \simeq -2\left(1 + \frac{x}{2}\right)y,$$

$$C(x) = xe^x \frac{[(x-4)e^x + (x+4)]}{(e^x - 1)^2}$$
(1.38)

where the second equality defining $\Delta I/I$ follows in the limit $x \ll 1$. Equation (1.38) shows also the anticipated physics of the SZ effect: while the total photon concentration remains unchanged, photons are transferred from the Rayleigh-Jeans region to the Wien region, i.e. the final spectrum will have a slight deficiency of low-energy photons and a slight excess of high-energy photons.

The spectral change discussed so far had to do with the scattering of photons by hot electrons. It can also happen that the cluster has a peculiar velocity. In this case the modification of the spectral intensity will purely be kinematical and it will be given by [53]:

$$\Delta I = -\frac{T_\gamma^3}{2\pi^2} h(x) v_{\rm r} \, {\rm DM}, \qquad (1.39)$$

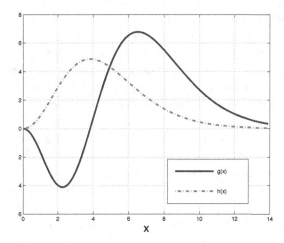

Fig. 1.4 The behaviour of $g(x)$ and $h(x)$ is illustrated.

where v_r is the peculiar velocity of the cluster and where $h(x) = x^4 e^x/(e^x - 1)^2$. In Fig. 1.4 the x-dependence of $g(x)$ and $h(x)$ is illustrated. Notice the $g(x)$ gives a fair approximation for the spectral distortion only if the temperature of the electrons is sufficiently small. In practice this happens for $T_e < 5$ KeV. When $T_e > 5$ keV the inclusion of relativistic corrections is mandatory. Such an inclusion will also correct the cross-over frequency since x_0 inherits relativistic corrections going as T_e/m_e.

1.6 Cosmological parameters

In recent years, thanks to combined observations of CMB anisotropies [35, 36], large scale structure [57, 58], supernovae of type Ia [59], big-bang nucleosyntheis [60], some kind of paradigm for the evolution of the late time (or even present) Universe emerged. It is normally called by practitioners ΛCDM model or even, sometimes, "concordance model". The terminology of ΛCDM refers to the fact that, in this model, the dominant (present) component of the energy density of the Universe is given by a cosmological constant Λ and a fluid of cold dark matter particles. The cosmological constant (or more generally the dark energy) interacts only gravitationally with the other (known) particle species such as baryons, leptons, photons. In the present section we are going to anticipate a portion of the notions which will be discussed later in this book. The main purpose of this strategy is to give some empirical evidence of some of the assumptions (or even conjectures) that will be later on spelled out more clearly. In this perspective, for the moment, the ΛCDM model is simply the Standard Cosmological Model (which will be thoroughly introduced in chapter 2) completed by a (conventional) inflationary extension. By convention, here we mean a model where the only source of inhomogeneity is represented by scalar fluctuations of the geometry and vanishing contribution of the tensors.

According to this paradigm, our understanding of the Universe can be summarized in two sets of cosmological parameters: the first set of parameters refers to the homogeneous background, the second set of parameters to the inhomogeneities. So, on top of the indetermination on the (present) Hubble expansion rate, i.e. h_0, there are various other parameters such as:

- the (present) dark energy density in critical units,° i.e. $h_0^2 \Omega_{\Lambda 0}$;

°Instead of giving the critical fraction of the total energy density alone, it is common practice to multiply this figure by h_0^2 so that the final number will be independent of h_0.

- the (present) cold dark matter (CDM in what follows) energy density, i.e. $h_0^2 \Omega_{c0}$;
- the (present) baryon energy density, i.e. $h_0^2 \Omega_{b0}$;
- the (present) photon energy density (already introduced) $h_0^2 \Omega_{\gamma 0}$;
- the (present) neutrino energy density, i.e. $h_0^2 \Omega_{\nu 0}$;
- the optical depth at reionization (denoted by ϵ but commonly named τ which denotes instead, in the present book, the conformal time coordinate, see section 2);
- the spectral index of the primordial (adiabatic) mode for the scalar fluctuations n_s;
- the amplitude of the curvature perturbations A_s;
- the bias parameter (related to large scale structure).

To this more or less standard set of parameters one can also add other parameters reflecting a finer description of pre-decoupling physics:

- the neutrinos are, strictly speaking, massive and their masses can then constitute an additional set of parameters;
- the dark energy may not be exactly a cosmological constant and, therefore, the barotropic index of dark energy may be introduced as the ratio between the pressure of dark energy and its energy density (similar argument can entail also the introduction of the sound speed of dark energy);
- the spectral index may not be constant as a function of the wavenumber and this consideration implies a further parameter;
- in the commonly considered inflationary scenarios there are not only scalar (adiabatic) modes but also tensor modes and this evidence suggests the addition of the relative amplitude and spectral index of tensor perturbations,[p] i.e., respectively, r and n_T.

Different parameters can be introduced in order to account for even more daring departures from the standard cosmological lore. These parameters include

- the amplitude and spectral index of primordial non-adiabatic perturbations;
- the amplitude and spectral index of the correlation between adiabatic and non-adiabatic modes;

---

[p]These two quantities will be specifically defined and computed in the case of conventional inflationary models. The interested reader may already consult chapter 10 (see, in particular, Eqs. (10.82) and (10.84) and derivations therein).

- a primordial magnetic field which is fully inhomogeneous and characterized, again, by a given spectrum and an amplitude.

This list can be easily completed by other possible (and physically reasonable) parameters. We just want to remark that the non-adiabatic modes represent a whole set of physical parameters since, as it will be swiftly discussed, there are 4 non-adiabatic modes. Consequently, already a thorough parametrization of the non-adiabatic sector will entail, in its most general incarnation, 4 spectral indices, 4 spectral amplitudes and the mutual correlations of each non-adiabatic mode with the adiabatic one. Having said this, it is important to stress that this book will not deal with the problem of data analysis (or parameter extraction from the CMB data). The purpose of the present script will be to use CMB as a guiding theme for the formulation of a consistent cosmological framework which might be in sight but which is certainly not fully understood.

It is useful to collect here some remarks on the values of the cosmological parameters that have been determined by analyzing different sets of data. This discussion has been partially approached in this chapter when we discussed, very briefly, the historical development of CMB physics. Since the theme reported in the present section might be a bit too specific, it could be skipped in the first reading of this script.

The WMAP 3-year [39] data have been combined, so far, with various sets of data. These data sets include the 2dF Galaxy Redshift Survey [62], the combination of Boomerang and ACBAR data [63, 64], the combination of CBI and VSA data [65, 66]. Furthermore the WMAP 3-year data can be also combined with the Hubble Space Telescope Key Project (HSTKP) data [67] as well as with the Sloan Digital Sky Survey (SDSS) [68, 69] data. Finally, the WMAP 3-year data can be also usefully combined with the weak lensing data [70, 71] and with the observations of type Ia supernovae[q](SNIa).

Each of the data sets mentioned in the previous paragraph can be analyzed within different frameworks. The minimal ΛCDM model with no cut-off in the primordial spectrum of the adiabatic mode and with vanishing contribution of tensor modes is the simplest concordance framework.[r] Diverse completions of this minimal model are possible: they include the addition of the tensor modes, a sharp cut-off in the spectrum and so on and so forth.

---

[q]In particular the data of the Supernova Legacy Survey (SNLS) [72] and the so-called Supernova "Gold Sample" (SNGS) [73, 74].

[r]The terminology used in the present section assumes the knowledge of the physics which is actually the main theme of the present book.

All these sets of data (combined with different theoretical models) lead necessarily to slightly different determinations of the relevant cosmological parameters. To have an idea of the range of variations of the parameters the following examples are useful:

- the WMAP 3-year data alone [39] (in a ΛCDM framework) seem to favour a slightly smaller value $h_0^2 \Omega_{M0} = 0.127$;
- if the WMAP 3-year data are combined with the "gold" sample of SNIa [73] (see also [74]) the favoured value is $h_0^2 \Omega_{M0}$ is of the order of 0.134; if the WMAP 3-year data are combined with *all* the data sets $h_0^2 \Omega_{M0} = 0.1324$.
- similarly, if the WMAP data alone are considered, the preferred value of $h_0^2 \Omega_{b0}$ is 0.02229 while this value decreases to 0.02186 if the WMAP data are combined with all the other data sets.

The aforementioned list of statements refers to the case of a pure ΛCDM model. If, for instance, tensors are included, then the WMAP 3-year data combined with CBI and VSA increase a bit the value of $h_0^2 \Omega_{b0}$ which becomes, in this case closer to 0.023.

In what follows a list of values will be presented. In all these tables the slight differences in the various cosmological parameters will be self-evident. Consider, to begin with, the case of the standard ΛCDM model with no tensor. Let us see what this model gives if analyzed in the light of the WMAP data *alone*. The results are:

$$h_0 = 0.732^{+0.031}_{-0.032}, \quad n_s = 0.958 \pm 0.016, \quad A_s = (23.5 \pm 1.3) \times 10^{-10},$$
$$h_0^2 \Omega_{b0} = 0.02229 \pm 0.00073, \quad h_0^2 \Omega_{c0} = 0.1054^{+0.0078}_{-0.0077},$$
$$\Omega_\Lambda = 0.759 \pm 0.034, \quad h_0^2 \Omega_{M0} = 0.1277^{+0.0080}_{-0.0079},$$
$$\sigma_8 = 0.761^{0.049}_{-0.048}, \quad \tau = 0.089 \pm 0.030.$$
(1.40)

In Eq. (1.40) the parameters τ and σ_8 determine, respectively, the optical depth of reionization[s] and the amplitude of matter fluctuations (determined within linear theory) on a scale of $8 h_0^{-1}$ Mpc. The quantities related to the only fluctuations of the geometry (which are the scalars, in the minimal ΛCDM model) are n_s (the scalar spectral index) and A_s (the scalar amplitude). Both quantities must be defined at a given scale which is conventionally chosen to be $k_p = 0.002 \, \text{Mpc}^{-1}$ and is called, in the jargon, *pivot*

---

[s]At the level of notations we stress that, in the following part of the book, the letter τ will denote the conformal time coordinate. Here, however, we prefer to use the standard notations for the benefit of the reader.

wave-number or simply *pivot scale*. Different choices for the pivot scale can be adopted, especially when not only the adiabatic mode is present in the game but also non-adiabatic modes (see chapter 8).

Suppose now to combine the WMAP data with *all* the available data stemming, respectively, from other CMB experiments, from LSS observations, from supernovae and from lensing. The result, always in the framework of the ΛCDM model, will be

$$h_0 = 0.704^{+0.015}_{-0.016}, \quad n_s = 0.947 \pm 0.015, \quad A_s = (23.5 \pm 1.3) \times 10^{-10},$$
$$h_0^2 \Omega_{b0} = 0.02186 \pm 0.00068, \quad h_0^2 \Omega_{c0} = 0.1105^{+0.0039}_{-0.0038},$$
$$\Omega_\Lambda = 0.732 \pm 0.018, \quad h_0^2 \Omega_{M0} = 0.1324^{+0.0042}_{-0.0041}, \quad (1.41)$$
$$\sigma_8 = 0.776^{0.031}_{-0.032}, \quad \tau = 0.073^{0.027}_{-0.028}.$$

Clearly there are slight differences in the determinations of some relevant cosmological parameters. For instance, the indetermination in the Hubble parameter, i.e. h_0 decreases (from Eq. (1.40) to Eq. (1.41)) by almost 3%. Furthermore $h_0^2 \Omega_{M0}$ increases by 1%. The spectral index gets more red.[t]

If the WMAP data are combined with the data stemming from type Ia supernovae (and, in particular, with the gold sample [73, 74]) the cosmological parameters are determined to be, always in the framework of a ΛCDM model:

$$h_0 = 0.701 \pm 0.021, \quad n_s = 0.946 \pm 0.016, \quad A_s = (24.4^{1.3}_{-1.4}) \times 10^{-10},$$
$$h_0^2 \Omega_{b0} = 0.02230^{0.00069}_{-0.00072}, \quad h_0^2 \Omega_{M0} = 0.1349^{+0.0061}_{-0.0060} \quad (1.42)$$
$$\Omega_\Lambda = 0.724 \pm 0.026, \quad \sigma_8 = 0.784^{0.042}_{-0.041}, \quad \tau = 0.079^{0.030}_{-0.029},$$

which means that the combination with the data of type Ia supernovae implies a smaller value for h_0 and a slightly higher value for the rescaled critical fraction of matter. This trend is confirmed, in a non-trivial fashion, by combining the WMAP data with the supernova data stemming from the Supernova Legacy Survey [72]. In this case the obtained cosmological parameters are

$$h_0 = 0.724 \pm 0.023, \quad n_s = 0.950^{+0.016}_{-0.017}, \quad A_s = (23.8 \pm 1.3) \times 10^{-10},$$
$$h_0^2 \Omega_{b0} = 0.02234^{0.00075}_{-0.00074}, \quad h_0^2 \Omega_{M0} = 0.1293 \pm 0.0059, \quad (1.43)$$
$$\Omega_\Lambda = 0.752^{+0.025}_{-0.024}, \quad \sigma_8 = 0.758 \pm 0.041, \quad \tau = 0.085 \pm 0.030.$$

---

[t]In the jargon we talk about red spectra (if the spectral index is smaller than 1), blue spectra (if the spectral index is larger than 1) and white (or scale-invariant spectra) if $n_s = 1$.

Up to now the model used to compare the data with the observations has been the simplest ΛCDM model. However, there are various other possibilities. Take, for instance, the case where a non-vanishing tensor component is allowed. This quantity will be denoted by r (see, in particular, chapter 10 for a derivation). The quantity named r denotes the ratio between the tensor and the scalar power spectrum produced in the framework of the same cosmological model. If only the WMAP data are used for the determination of the cosmological parameters we will have

$$h_0 = 0.787 \pm 0.052, \qquad n_{\rm s} = 0.984^{+0.029}_{-0.028}, \qquad A_{\rm s} = (21.0^{+2.2}_{-2.3}) \times 10^{-10},$$
$$h_0^2 \Omega_{\rm b0} = 0.02233 \pm 0.0010, \qquad h_0^2 \Omega_{\rm M0} = 0.1195^{0.0094}_{-0.0093},$$
$$r = 0.65, \qquad \Omega_\Lambda = 0.803 \pm 0.040,$$
$$\sigma_8 = 0.702 \pm 0.062, \qquad \tau = 0.090 \pm 0.031.$$
(1.44)

It is clear from the reported figures that, when we allow for a tensor component, the values of the rescaled critical fraction of matter decreases, the cosmological constant increases and the Hubble parameter increases sharply. Compare, indeed, Eqs. (1.40) and (1.44) which both refer to the WMAP data alone but analyzed in the light of the two mentioned models, namely the ΛCDM with no tensors and the ΛCDM with a tensor component parametrized in terms of the tensor to scalar ratio.[u]

This trend also depends upon which data sets are combined. For instance, suppose now we take the ΛCDM model with tensors and analyze it in the light of the WMAP data combined with the data of the Sloan Digital Sky Survey [68, 69]. The result will be:

$$h_0 = 0.716 \pm 0.026, \qquad n_{\rm s} = 0.964^{+0.020}_{-0.021}, \qquad A_{\rm s} = (23.0^{+1.6}_{-1.5}) \times 10^{-10},$$
$$h_0^2 \Omega_{\rm b0} = 0.02282^{+0.00080}_{-0.00081}, \qquad h_0^2 \Omega_{\rm M0} = 0.1339^{0.0052}_{-0.0053},$$
$$r = 0.30, \qquad \Omega_\Lambda = 0.803 \pm 0.040,$$
$$\sigma_8 = 0.781 \pm 0.034, \qquad \tau = 0.077^{+0.029}_{-0.030},$$
(1.45)

showing values of h_0 and $h_0^2 \Omega_{\rm M0}$ and r which are substantially smaller than those reported in Eq. (1.44).

This discussion could be extended for various pages and this might even stimulate the interest of some readers for the so-called parameter estimation strategies. This topic is beyond the scope of the present book. Furthermore, the opinion of the author is that at the present stage, it seems really difficult to decide which is the best and most predictive model just on the basis of

---
[u]As with the scalar spectral amplitude, the tensor to scalar ratio is also normally assigned at the pivot scale $k_{\rm p}$.

parameter estimation strategies. Indeed, in the present discussion, just two benchmark models have been quoted but there could be many others. Here is an approximate list of possibilities:

- ΛCDM alone with no tensors;
- ΛCDM with tensors;
- ΛCDM with a cut-off in the primordial scalar spectrum;
- ΛCDM assuming a quadratic inflaton potential with minimal inflationary duration (i.e. 60 e-folds);
- open CDM model.
- wCDM (i.e. CDM plus dark energy parametrized by a constant barotropic index);
- wCDM without perturbations induced by the dark energy;
- w CDM with perturbations induced by dark energy.

Notice that, in the above list, further possibilities can be obtained by combining the different items of the list. For instance it is possible to consider the case where we have wCDM in an open Universe, and so on and so forth. Furthermore, the aforementioned list may become even longer: the choices reported here are by no means exhaustive or fully comprehensive. While we shall get back more precisely on the assumptions of the ΛCDM model, it should be stressed that the only purpose of this discussion is to emphasize that observations, especially in cosmology, are never independent of the underlying model. Therefore, in this context, there are two possible approaches. The first one is to focus on the *minimal* model. By *minimal model* we mean the model with the fewer number of parameters which is consistent with all the sets of data mentioned at the beginning of this section. There is, however, a second approach. Namely the one of selecting a pivot model which is not minimal but which is guided by physical considerations. These two approaches are complementary but none of them is decisive unless data of different quality (i.e. laboratory data) will soon enlighten the observational situation.

Chapter 2

From CMB to the Standard Cosmological Model

Various excellent publications treat the essential elements of the Standard Cosmological Model (SCM in what follows) within different perspectives (see, for instance, [75–78]). The purpose here will not be to present the conceptual foundations SCM but to introduce its main assumptions and its most relevant consequences with particular attention to those aspects and technicalities that are germane to our theme, i.e. CMB physics.

There are a number of relatively ancient papers that can be usefully consulted to dig out both the historical and conceptual foundations of the SCM. In issue number 81 of the "Uspekhi Fizicheskikh Nauk", on the occasion of the seventy-fifth anniversary of the birth of A. A. Friedmann, a number of rather interesting papers were published. Among them there is a review article on the development of Friedmannian cosmology by Ya. B. Zeldovich [79] and the inspiring paper of Lifshitz and Khalatnikov [80] on the relativistic treatment of cosmological perturbations.

Reference [79] describes mainly Friedmann's contributions [81]. Due attention should also be paid to the work of G. Lemaître [82–84] that was also partially motivated by the debate with A. Eddington [85]. According to the idea of Eddington the world evolved from an Einstein static Universe and so developed "infinitely slowly from a primitive uniform distribution in unstable equilibrium" [85]. The point of view of Lemaître was, in a sense, more radical since he suggested, in 1931, that the expansion really did start with the beginning of the entire Universe. Unlike the Universe of some modern big-bang cosmologies, the description of Lemaître did not evolve from a true singularity but from a material pre-Universe, which Lemaître liked to call "primeval atom" [84]. The primeval atom was a unique atom whose atomic weight was the total mass of the Universe. This highly unstable atom would have experienced some type of fission and would have divided into smaller and smaller atoms by some kind of super-radioactive processes. The perspective of Lemaître was that the early expansion of the Universe could be a well defined object of study for natural sciences even in the absence of a proper understanding of the initial singularity. The perspective spelled out in the previous sentence is essentially the one also accepted in the modern approach to cosmology and gravitation. In different words we can say that cosmology can be studied even if the understanding of cosmological singularities is far from being complete. This approach will be the one followed throughout the present script. At the same time it is appropriate to warn the reader that this might not be true. It could happen, for instance, that the solutions to some of the present cosmological puzzles may require some deeper understanding of the dynamics of cosmological singularities.

The discussion of the present chapter follows four main lines:

- firstly the SCM will be formulated in its essential elements;
- then the matter content of the present Universe will be introduced as it emerges in the concordance model;
- the (probably cold) future of our own Universe will be swiftly discussed;
- finally the (hot) past of the Universe will be scrutinized in connection with the properties of the CMB.

Complementary discussions on the concept of distance in cosmology and on the kinetic description of hot plasmas are collected, respectively, in Appendix A and in Appendix B. Indeed, various concepts (like the redshift, the angular diameter distance, the luminosity distance) will be often quoted throughout the chapter but are carefully introduced, in a unified perspective, in Appendix A. Similarly, when dealing with the concepts of thermal, kinetic and chemical equilibrium the treatment presented in Appendix B will be assumed.

2.1 The Standard Cosmological Model (SCM)

The Standard Cosmological Model (SCM) rests on the following three important assumptions :

- for typical length-scales larger than 50 Mpc the Universe is homogeneous and isotropic;
- the matter content of the Universe can be parametrized in terms of perfect barotropic fluids;

- the dynamical law connecting the evolution of the sources to the evolution of the geometry is provided by General Relativity (GR).

In the present section the three assumptions listed above will be specifically discussed one by one.

2.1.1 *Homogeneity and isotropy*

The assumption of homogeneity and isotropy implies that the geometry of the Universe is invariant for spatial roto-translations. In four space-time dimensions the metric tensor will have 10 independent components. Using homogeneity and isotropy the independent components can be reduced from 10 to 4 (having taken into account the 3 spatial rotation and the 3 spatial translations). The most general form of a line element which is invariant under spatial rotations and spatial translations can then be written as:

$$ds^2 = e^\nu dt^2 - e^\lambda dr^2 - e^\mu(r^2 d\vartheta^2 + r^2 \sin^2\vartheta d\varphi^2) + 2e^\sigma dr dt. \quad (2.1)$$

The freedom of choosing a gauge can the be exploited and the metric can be reduced to its canonical Friedmann-Robertson-Walker (FRW) form[a]:

$$ds^2 = g_{\mu\nu} dx^\mu dx^\nu = dt^2 - a^2(t)\left[\frac{dr^2}{1-kr^2} + r^2(d\vartheta^2 + \sin^2\vartheta d\varphi^2)\right], \quad (2.2)$$

where $g_{\mu\nu}$ is the metric tensor of the FRW geometry and $a(t)$ is the scale factor. In the parametrization of Eq. (2.2), $k = 0$ corresponds to a spatially flat Universe; if $k > 0$ the Universe is spatially closed and, finally, $k < 0$ corresponds to a spatially open Universe. The line element (2.2) is invariant under the following transformation:

$$r \to \tilde{r} = \frac{r}{r_0},$$
$$a(t) \to \tilde{a}(t) = a(t) r_0, \quad (2.3)$$
$$k \to \tilde{k} = k r_0^2,$$

where r_0 is a dimensionfull constant. In the parametrization (2.3) the scale factor is dimensionfull and $\tilde{k}$ is 0, +1 or −1 depending on the spatial curvature of the internal space. Throughout this book, the parametrization where the scale factor is dimensionless will be consistently employed. In Eq. (2.2) the time t is the *cosmic* time coordinate. Depending upon the physical problem at hand, different time parametrizations can be also adopted. A particularly useful one (especially in the study of cosmological

---

[a]The transition from Eq. (2.1) to Eq. (2.2) by successive gauge choices can be followed in the book of Tolman [86].

inhomogeneities) is the so-called conformal time parametrization. In the conformal time coordinate τ the line element of Eq. (2.2) can be written as

$$ds^2 = g_{\mu\nu}dx^\mu dx^\nu = a^2(\tau)\left\{d\tau^2 - \left[\frac{dr^2}{1-kr^2} + r^2(d\vartheta^2 + \sin^2\vartheta d\varphi^2)\right]\right\}. \quad (2.4)$$

The line element (2.2) describes a situation where the space-time is homogeneous and isotropic. It is possible to construct geometries that are homogeneous but *not* isotropic. The Bianchi geometries are, indeed, homogeneous but not isotropic. For instance, the Bianchi type-I metric can be written, in Cartesian coordinates, as

$$ds^2 = dt^2 - a^2(t)dx^2 - b^2(t)dy^2 - c^2(t)dz^2. \quad (2.5)$$

Equation (2.5) leads to a Ricci tensor that depends only on time and not on the spatial coordinates. Another, less obvious, example is given by the following line element:

$$ds^2 = dt^2 - a^2(t)dx^2 - e^{2\alpha x}b^2(t)dy^2 - c^2(t)dz^2. \quad (2.6)$$

For $\alpha = -1$ we get the Bianchi III line element while, for $\alpha = -2$ we obtain the Bianchi VI$_{-1}$ line element [87]. In both cases the geometry is homogeneous but not isotropic. This example shows that it is a bit dangerous to infer the homogeneity properties of a given background only by looking at the form of the line element. A more efficient strategy is to scrutinize the properties of the curvature invariants.

2.1.2 Perfect barotropic fluids

The material content of the Universe is often described in terms of perfect fluids (i.e. fluids that are not viscous) which are also barotropic (i.e. with a definite relation between pressure and energy-density). An example of (perfect) barotropic fluid has been already provided: the gas of photons in thermal equilibrium introduced in chapter 1. The energy-momentum tensor of a gas of photons can be indeed written as:

$$T^\nu_\mu = (p_\gamma + \rho_\gamma)u_\mu u^\nu - p_\gamma \delta^\nu_\mu, \quad (2.7)$$

where

$$u^\mu = \frac{dx^\mu}{ds}, \qquad p_\gamma = \frac{\rho_\gamma}{3}, \qquad g_{\mu\nu}u^\mu u^\nu = 1. \quad (2.8)$$

Equations (2.7) and (2.8) are the first example of a radiation fluid. The covariant conservation of the energy-momentum tensor can be written as

$$\nabla_\mu T^\mu_\nu = 0, \quad (2.9)$$

where ∇_μ denotes the covariant derivative with respect to the metric $g_{\mu\nu}$ of Eq. (2.2). Using the definition of covariant derivative, Eq. (2.9) can be also written, in more explicit terms, as

$$\partial_\mu T_\nu^\mu + \Gamma_{\alpha\mu}^\mu T_\nu^\alpha - \Gamma_{\nu\alpha}^\beta T_\beta^\alpha = 0, \tag{2.10}$$

where

$$\Gamma_{\mu\nu}^\alpha = \frac{1}{2} g^{\alpha\beta}(-\partial_\beta g_{\mu\nu} + \partial_\nu g_{\beta\mu} + \partial_\mu g_{\nu\beta}), \tag{2.11}$$

are the Christoffel connections computed from the metric tensor[b] $g_{\mu\nu}$. In the FRW metric, Eq. (2.10) implies

$$\dot{\rho}_\gamma + 3H(\rho_\gamma + p_\gamma) = 0, \tag{2.12}$$

where the overdot denotes a derivation with respect to the cosmic time coordinate t and

$$H = \frac{\dot{a}}{a}, \tag{2.13}$$

is the Hubble rate. The covariant conservation of the energy-momentum tensor implies that the evolution is adiabatic, i.e. $\dot{S} = 0$, where S is the entropy which, today, is effectively dominated by photons (i.e., today, $S = S_\gamma$). Recall, to begin with, that the fundamental thermodynamic identity and the first law of thermodynamics, stipulate, respectively,[c]

$$\mathcal{E} = TS - pV + \mu N, \tag{2.14}$$

$$d\mathcal{E} = TdS - pdV + \mu dN, \tag{2.15}$$

where

- $\mathcal{E}$ is the internal energy of the system ;
- T is the temperature;
- S is the entropy;
- N is the number of particles of the system;
- μ is the chemical potential .

In the case of the photons the chemical potential is zero. The volume V will be then given by a fiducial volume (for instance the Hubble volume)

[b] In the first part of the present discussion we the cosmic time parametrization will be consistently used. However, the conformal time parametrization can also be used and, in this case, the form of the equations will be mathematically different but completely equivalent from the physical point of view. This point will be addressed later in this chapter.

[c] See Appendix B for more details on the kinetic and chemical description of hot plasmas.

rescaled through the third power of the scale factor. In analog terms one can write the energy. In formulas:

$$V(t) = V_0 \left(\frac{a}{a_0}\right)^3, \qquad \mathcal{E} = V(t)\rho_\gamma. \tag{2.16}$$

Thus, using Eq. (2.16) into Eq. (2.15), we do get

$$T\frac{dS}{dt} = V_0 \left(\frac{a}{a_0}\right)^3 [\dot\rho_\gamma + 3H(\rho_\gamma + p_\gamma)]. \tag{2.17}$$

Equation (2.17) shows that $\dot S = 0$ provided the covariant conservation (i.e. Eq. (2.12)) is enforced. Different physical fluids will also imply different equations of state. Still, as long as the total fluid is not viscous, it is possible to write, in general terms, the total energy-momentum tensor as

$$T^\nu_\mu = (p_t + \rho_t)u_\mu u^\nu - p_t \delta^\nu_\mu, \tag{2.18}$$

where u_μ is the peculiar velocity field of the (total) fluid still satisfying $g_{\mu\nu}u^\mu u^\nu = 1$. For instance (see Appendix B) non-relativistic matter (i.e. bosons or fermions in equilibrium at a temperature that is far below the threshold of pair production) leads naturally to an equation of state $p = 0$ (often called dusty equation of state). Another example could be a homogeneous scalar field whose potential vanishes exactly (see section 5). In this case the equation of state is $p = \rho$ (also called stiff equation of state, since, in this case, the sound speed coincides with the speed of light[d]).

Viscous effects, when included, may spoil the homogeneity of the background. This is the case, for instance, of shear viscosity [75]. It is however possible to include viscous effects that do not spoil the homogeneity of the background. Two examples along this direction are the bulk viscosity effects (see [88] for the notion of bulk and shear viscosity), and the possible transfer of energy (and momentum) between different fluids of the mixture. For a single fluid, the total energy-momentum T^ν_μ tensor can then be split into a perfect contribution, denoted in the following by T^ν_μ, and into an imperfect contribution, denoted by ΔT^ν_μ, i.e.

$$\mathcal{T}^\nu_\mu = T^\nu_\mu + \Delta T^\nu_\mu, \tag{2.19}$$

In general coordinates, and within our set of conventions, the contribution of bulk viscous stresses can be written, in turn, as [75]

$$\Delta T^\nu_\mu = \xi \left(\delta^\nu_\mu - u_\mu u^\nu\right)\nabla_\alpha u^\alpha, \tag{2.20}$$

---

[d] The idea that, at early times, the Universe could be dominated by a stiff fluid is originally due to Zeldovich [89] (see also [90]).

where ξ represents the bulk viscosity coefficient [75]. The presence of bulk viscosity can also be interpreted, at the level of the background, as an effective redefinition of the pressure (or, more correctly, of the enthalpy). This discussion follows the spirit of the Eckart approach [91–93]. It must be mentioned that this approach is phenomenological in the sense that the bulk viscosity is not modelled on the basis of a suitable microscopic theory. For caveats concerning the Eckart approach see [94, 95] (see also [96] and references therein). The Eckart approach, however, fits with the phenomenological inclusion of a fluid decay rate that has been also considered recently for related applications to cosmological perturbation theory [97, 98] (see also chapter 12). It is interesting to mention that bulk viscous effects have been used in the past in order to provide an early completion of the SCM [99–101]. According to Eq. (2.20), bulk viscosity modifies the pressure so that the spatial components of the energy-momentum tensor can be written as

$$T_i^j = -\mathcal{P}\delta_i^j, \qquad \mathcal{P} = p - 3H\xi. \tag{2.21}$$

The bulk viscosity coefficient ξ may depend on the energy density and it can be parametrized as $\xi(\rho) = (\rho/\rho_1)^\nu$ where different values of ν will give rise to different cosmological solutions [101]. This parametrization allows for various kinds of unstable quasi-de Sitter solutions [100, 99, 101]. Notice that, according to the parametrization of Eq. (2.21), the covariant conservation of the energy-momentum tensor implies that

$$\dot{\rho} + 3H(\rho + p) = 9H^2 \xi(\rho). \tag{2.22}$$

So far it has been assumed that the energy and momentum exchanges between the different fluids of the plasma are negligible. However, there are situations (rather relevant for CMB physics) where the coupling between different fluids cannot be neglected. An example is the tight-coupling between photons and the lepton-baryon fluid which fits well before hydrogen recombination (see chapters 8 and 9). There are also situations in the early Universe, where it is mandatory to consider the decay of a given species into another species. For instance, massive particles decaying into massless particles. Consider, for this purpose, the situation where the plasma is a mixture of two species whose associated energy-momentum tensors can be written as

$$\begin{aligned} T_{\mathrm{a}}^{\mu\nu} &= (p_{\mathrm{a}} + \rho_{\mathrm{a}}) u_{\mathrm{a}}^\mu u_{\mathrm{a}}^\nu - p_{\mathrm{a}} g^{\mu\nu}, \\ T_{\mathrm{b}}^{\mu\nu} &= (p_{\mathrm{b}} + \rho_{\mathrm{b}}) u_{\mathrm{b}}^\mu u_{\mathrm{b}}^\nu - p_{\mathrm{b}} g^{\mu\nu}. \end{aligned} \tag{2.23}$$

If the fluids are decaying one into the other (for instance the a-fluid decays into the b-fluid), the covariant conservation equation only applies to the global relativistic plasma, while the energy-momentum tensors of the single species are not covariantly conserved and their specific form accounts for the transfer of energy between the a-fluid and the b-fluid:

$$\nabla_\mu T_a^{\mu\nu} = -\Gamma g^{\nu\alpha} u_\alpha (p_a + \rho_a),$$
$$\nabla_\mu T_b^{\mu\nu} = \Gamma g^{\nu\alpha} u_\alpha (p_a + \rho_a), \quad (2.24)$$

where the term Γ is the decay rate that can be both space- and time-dependent; in Eqs. (2.24) u_α represents the (total) peculiar velocity field. Owing to the form of Eqs.(2.24), it is clear that the total energy-momentum tensor of the two fluids, i.e. $T_{\rm tot}^{\mu\nu} = T_a^{\mu\nu} + T_b^{\mu\nu}$ is indeed covariantly conserved. Equations (2.24) can be easily generalized to the description of more complicated dynamical frameworks, where the relativistic mixture is characterized by more than two fluids. Consider the situation where the a-fluid decays as a $\to$ b + c. Then, if a fraction f of the a-fluid decays into the b-fluid and a fraction $(1-f)$ into the c-fluid, Eqs. (2.24) can be generalized as

$$\nabla_\mu T_a^{\mu\nu} = -\Gamma\, g^{\nu\alpha}\, u_\alpha (p_a + \rho_a),$$
$$\nabla_\mu T_b^{\mu\nu} = f\Gamma\, g^{\nu\alpha}\, u_\alpha (p_a + \rho_a), \quad (2.25)$$
$$\nabla_\mu T_c^{\mu\nu} = (1-f)\Gamma\, g^{\nu\alpha}\, u_\alpha (p_a + \rho_a),$$

and so on. In the case of a FRW metric, Eq. (2.24) can be written in explicit terms as:

$$\dot{\rho}_a + 3H(\rho_a + p_a) + \overline{\Gamma}(\rho_a + p_a) = 0 \quad (2.26)$$
$$\dot{\rho}_b + 3H(\rho_b + p_b) - \overline{\Gamma}(\rho_a + p_a) = 0. \quad (2.27)$$

If the a-fluid is identified with dusty matter and the b-fluid with radiation we will have

$$\dot{\rho}_m + (3H + \overline{\Gamma})\rho_m = 0,$$
$$\dot{\rho}_\gamma + 4H\rho_\gamma - \overline{\Gamma}\rho_m = 0. \quad (2.28)$$

Note that $\overline{\Gamma}$ is the homogeneous part of the decay rate. To first-order, the decay rate may be spatially inhomogeneous and this entails various interesting consequences which will be only marginally discussed in this book (see, however, [97, 98] and the discussion of chapter 12). It is relevant to stress that, owing to the form of the FRW metric, the homogeneous decay rate entails only exchange of energy between the fluids of the mixture. To first-order, the peculiar velocity fields will also be affected and the exchange of momentum is explicit.

We shall get back on the possibility of having interacting fluids in the early stages of the life of the Universe. Indeed, in chapter 12 interacting fluids will be studied in connection with the evolution of entropy perturbations for typical wavelengths larger than the Hubble radius.

2.1.3 General Relativity

The moment has come to discuss General Relativity as the guiding dynamical principle connecting the evolution of the geometry with the evolution of the matter sources. It would be rather difficult to introduce General Relativity in one single section and specialized books can be consulted (see, for instance, [102, 103] for two classic introductions). Here just a few important concepts will be recapitulated using, as guiding line of the discussion, the derivation of the Einstein equations from the Einstein-Hilbert action. This discussion will allow to review various technical aspects which will be used later on and which should already be rather familiar from more dedicated treatments of gravitation in its relativistic regime.

Let us therefore start from the total action of gravity supplemented by the action of the matter sources which will be written as

$$S = S_{\text{EH}} + S_{\text{m}}, \tag{2.29}$$

where

$$S_{\text{EH}} = -\frac{1}{16\pi G} \int d^4x \sqrt{-g} R \tag{2.30}$$

is the Einstein-Hilbert action; S_{m} is the generic action for the matter sources; G is the Newton constant. In Eq. (2.30) $R = g^{\mu\nu} R_{\mu\nu}$ is the Ricci scalar and g denotes the determinant of the metric. The functional variation of S_{EH} implies:

$$\delta S_{\text{EH}} = -\frac{1}{16\pi G} \int d^4x \left[\delta(\sqrt{-g}) R + \sqrt{-g} R_{\mu\nu} \delta g^{\mu\nu} + \sqrt{-g} g^{\mu\nu} \delta R_{\mu\nu} \right]. \tag{2.31}$$

To evaluate the variation of $\sqrt{-g}$, we observe that g can be written as $g = \exp\left[\text{Tr} \ln g_{\mu\nu}\right]$ (where Tr denotes the trace). Consequently

$$\delta g = g \delta(\text{Tr} \ln g_{\mu\nu}) = g g^{\mu\nu} \delta g_{\mu\nu}. \tag{2.32}$$

Using Eq. (2.32) it is easy to show that

$$\delta \sqrt{-g} = -\frac{1}{2\sqrt{-g}} \delta g = \frac{1}{2} \sqrt{-g} g^{\mu\nu} \delta g_{\mu\nu} = -\frac{1}{2} \sqrt{-g} g_{\mu\nu} \delta g^{\mu\nu}. \tag{2.33}$$

The last equality in Eq. (2.33) arises from the simple observation that $g_{\alpha\beta}g^{\alpha\beta} = 4$ which also implies that $g^{\alpha\beta}\delta g_{\alpha\beta} = -g_{\alpha\beta}\delta g^{\alpha\beta}$. Using the result of Eq. (2.33) into Eq. (2.31) the functional variation of the Einstein-Hilbert action can be recast in the following form:

$$\delta S_{\rm EH} = -\frac{1}{16\pi G}\int d^4x \left[\sqrt{-g}\left(R_{\mu\nu} - \frac{1}{2}g_{\mu\nu}R\right)\delta g^{\mu\nu} + \sqrt{-g}g^{\mu\nu}\delta R_{\mu\nu}\right]. \quad (2.34)$$

The last term appearing at the right hand side of Eq. (2.34) can be written in a different way. Recall, indeed, that, from the definition of Riemann tensor

$$R^{\alpha}{}_{\mu\gamma\nu} = \partial_{\gamma}\Gamma^{\alpha}_{\mu\nu} - \partial_{\nu}\Gamma^{\alpha}_{\mu\gamma} + \Gamma^{\beta}_{\mu\nu}\Gamma^{\alpha}_{\beta\gamma} - \Gamma^{\beta}_{\mu\gamma}\Gamma^{\alpha}_{\beta\nu} \quad (2.35)$$

the Ricci tensor can be obtained by contraction of the first and third indices, i.e.

$$R_{\mu\nu} = R^{\alpha}{}_{\mu\alpha\nu} \equiv \partial_{\alpha}\Gamma^{\alpha}_{\mu\nu} - \partial_{\nu}\Gamma^{\alpha}_{\mu\alpha} + \Gamma^{\alpha}_{\mu\nu}\Gamma^{\beta}_{\alpha\beta} - \Gamma^{\beta}_{\nu\alpha}\Gamma^{\alpha}_{\beta\mu}. \quad (2.36)$$

From Eq. (2.36), the variation of $R_{\mu\nu}$ appearing in the last term of Eq. (2.34) can be written in terms of the variations of the Christoffel conections:

$$\delta R_{\mu\nu} = \partial_{\alpha}\delta\Gamma^{\alpha}_{\mu\nu} - \partial_{\nu}\delta\Gamma^{\alpha}_{\alpha\mu}$$
$$+ \delta\Gamma^{\alpha}_{\mu\nu}\Gamma^{\beta}_{\alpha\beta} + \Gamma^{\alpha}_{\mu\nu}\delta\Gamma^{\beta}_{\alpha\beta}$$
$$- \delta\Gamma^{\beta}_{\nu\alpha}\Gamma^{\alpha}_{\beta\mu} - \Gamma^{\beta}_{\nu\alpha}\delta\Gamma^{\alpha}_{\beta\mu}. \quad (2.37)$$

Notice now that, by definition of covariant derivative, the following two equalities hold

$$\nabla_{\alpha}\delta\Gamma^{\alpha}_{\mu\nu} = \partial_{\alpha}\delta\Gamma^{\alpha}_{\mu\nu} + \Gamma^{\alpha}_{\alpha\lambda}\delta\Gamma^{\lambda}_{\nu\mu} - \delta\Gamma^{\alpha}_{\lambda\mu}\Gamma^{\lambda}_{\alpha\nu} - \Gamma^{\alpha}_{\mu\lambda}\delta\Gamma^{\lambda}_{\alpha\nu}, \quad (2.38)$$

$$\nabla_{\nu}\delta\Gamma^{\alpha}_{\alpha\mu} = \partial_{\nu}\delta\Gamma^{\alpha}_{\alpha\mu} + \Gamma^{\alpha}_{\nu\lambda}\delta\Gamma^{\lambda}_{\alpha\mu} - \delta^{\alpha}_{\lambda\mu}\Gamma^{\lambda}_{\nu\alpha} - \delta\Gamma^{\alpha}_{\nu\mu}\Gamma^{\lambda}_{\alpha\lambda}. \quad (2.39)$$

By now taking the difference of Eqs. (2.38) and (2.39) and by comparing the obtained result with Eq. (2.37) we get to the following important identity

$$\delta R_{\mu\nu} = \nabla_{\alpha}\delta\Gamma^{\alpha}_{\mu\nu} - \nabla_{\nu}\delta\Gamma^{\alpha}_{\alpha\mu} \quad (2.40)$$

which is often called *Palatini identity*. Since general relativity is a metric theory of gravity we will have that the covariant derivative of the metric tensor vanishes. Therefore, using Eq. (2.40) the last term appearing at the right hand side of Eq. (2.34) can be expressed as

$$\sqrt{-g}g^{\mu\nu}\delta R_{\mu\nu} = \sqrt{-g}\nabla_{\alpha}(g^{\mu\nu}\delta\Gamma^{\alpha}_{\mu\nu}) - \sqrt{-g}\nabla_{\nu}(g^{\mu\nu}\delta\Gamma^{\alpha}_{\alpha\mu}). \quad (2.41)$$

From the definition of covariant derivative of a generic four-dimensional vector V^{α} it can be easily obtained

$$\nabla_{\alpha}V^{\alpha} = \partial_{\alpha}V^{\alpha} + \Gamma^{\alpha}_{\alpha\beta}V^{\beta} \equiv \frac{1}{\sqrt{-g}}\partial_{\alpha}(\sqrt{-g}V^{\alpha}). \quad (2.42)$$

The second equality follows from the definition of the Christoffel connections introduced in Eq. (2.11) by noticing that:

$$\Gamma^\alpha_{\alpha\beta} = \frac{1}{2}g^{\alpha\gamma}\partial_\beta g_{\gamma\alpha} = \partial_\beta(\ln\sqrt{-g}). \tag{2.43}$$

Using Eq. (2.42), Eq. (2.41) can be further modified with the result that

$$\int d^4x \sqrt{-g} g^{\mu\nu} \delta R_{\mu\nu} = \int d^4x \partial_\alpha(\sqrt{-g} V^\alpha), \tag{2.44}$$

where the vector V^α is now defined as:

$$V^\alpha = [g^{\mu\nu} \delta\Gamma^\alpha_{\mu\nu} - g^{\nu\alpha}\delta\Gamma^\beta_{\nu\beta}]. \tag{2.45}$$

In Eqs. (2.44) and (2.45) the freedom of renaming the summation indices has been used. Now the integral of Eq. (2.44) is performed over a four-dimensional volume Ω. Thus, by Gauss theorem,

$$\int_\Omega d^4x \partial_\alpha(\sqrt{-g}V^\alpha) = \int_{\Sigma(\Omega)} d\Sigma_\alpha \sqrt{-g} V^\alpha \tag{2.46}$$

where $d\Sigma_\alpha$ is the infinitesimal element of a three-dimensional hypersurface. But this last integral is zero since, to obtain the field equations, the variation of the action is performed in such a way that $\delta\Gamma$ vanishes on $\Sigma(\Omega)$.

We can then go back to Eq. (2.29) and write the variation of the full action as

$$\delta S = \delta S_{\text{EH}} + \delta S_{\text{m}}. \tag{2.47}$$

Using Eq. (2.34) and dropping the surface term we therefore obtain:

$$\delta S = -\frac{1}{16\pi G} \int d^4x \sqrt{-g} \left[\left(R_{\mu\nu} - \frac{1}{2}g_{\mu\nu}R\right) - 8\pi G T_{\mu\nu}\right] \delta g^{\mu\nu}, \tag{2.48}$$

where it has been used that, by definition of canonical energy-momentum tensor,

$$\delta S_{\text{m}} = \frac{1}{2} \int d^4x \sqrt{-g} T_{\mu\nu} \delta g^{\mu\nu}. \tag{2.49}$$

Finally, from Eq. (2.48) the Einstein equations are obtained to be

$$R^\nu_\mu - \frac{1}{2}\delta^\nu_\mu R = 8\pi G T^\nu_\mu. \tag{2.50}$$

The covariant derivative of the left-hand side gives zero and this is consistent with the covariant conservation of the total energy-momentum tensor of the sources. At the level of notations it is appropriate to remark that the standard introduction of a cosmological term Λ in the action (2.30) implies

$$S_{\text{EH}} = -\frac{1}{16\pi G} \int d^4x \sqrt{-g}(R + 2\Lambda). \tag{2.51}$$

By repeating the same steps outlined above the modified Einstein equations will read

$$R_\mu^\nu - \frac{1}{2}\delta_\mu^\nu R = 8\pi G T_\mu^\nu + \Lambda \delta_\mu^\nu. \qquad (2.52)$$

In this script the cosmological constant will not be viewed as a further term in the gravity action but rather as a further term in the matter part of the action. A specific comment on this perspective will be discussed in the following section. It is often practical to use a unified convention for the right hand side of Einstein equations. Therefore, the so-called Einstein tensor will be defined as

$$\mathcal{G}_\mu^\nu = R_\mu^\nu - \frac{1}{2}\delta_\mu^\nu R. \qquad (2.53)$$

According to this notation Eq. (2.52) can be simply written as

$$\mathcal{G}_\mu^\nu = 8\pi G T_\mu^\nu. \qquad (2.54)$$

Since $\nabla_\mu T_\nu^\mu = 0$, it must also happen that $\nabla_\nu \mathcal{G}_\nu^\mu = 0$. The validity of the latter relation follows by contraction of the so-called Bianchi identity [102].

2.2 Friedmann-Lemaître equations

The three main assumptions of the SCM enter directly the derivation of the so-called Friedmann-Lemaître equations which are nothing but the Einstein equations (2.50) supplemented by the covariant conservation of the total energy-momentum tensor of Eq. (2.18) and written in a FRW metric (see Eq. (2.2)). The steps of the derivation of the Friedmann-Lemaître equations are very simple:

- take the FRW metric of Eq. (2.2) and compute, according to Eq. (2.11) the Christoffel connections;
- from the Christoffel connections the components of the Ricci tensor can be obtained from Eq. (2.36); the contraction of the Ricci tensor leads to the Ricci scalar;
- the explicit expressions for the components of $R_{\mu\nu}$ and R can be inserted into Eq. (2.50) and the explicit form of Einstein equations is obtained once the total energy-momentum tensor is taken, for instance, in the form of Eq. (2.18).

By following this procedure the components of the Ricci tensor and the Ricci scalar are:

$$R_0^0 = -3(H^2 + \dot{H}),$$
$$R_i^j = -\left(\dot{H} + 3H^2 + \frac{2k}{a^2}\right)\delta_i^j, \qquad (2.55)$$
$$R = -6\left(\dot{H} + 2H^2 + \frac{k}{a^2}\right),$$

where H has been defined in Eq. (2.13) and the over-dot denotes, as usual, a derivation with respect to the cosmic time coordinate t.

Equation (2.50) can then be written in explicit terms by specifying the total energy-momentum tensor which will be taken to be the one of a perfect fluid (see Eq. (2.18)). Using Eq. (2.55) into Eq. (2.50), the Friedmann-Lemaître equations (often quoted as FL equations in what follows) are:

$$H^2 = \frac{8\pi G}{3}\rho_t - \frac{k}{a^2}, \qquad (2.56)$$
$$\dot{H} = -4\pi G(\rho_t + p_t) + \frac{k}{a^2}, \qquad (2.57)$$
$$\dot{\rho}_t + 3H(\rho_t + p_t) = 0. \qquad (2.58)$$

While Eq. (2.56) follows from the (00) component of Eq. (2.50), Eq. (2.57) is a linear combination of the the (ij) and (00) components of Eq. (2.50). Eq. (2.58) follows, as already discussed, from the covariant conservation of the total energy-momentum tensor of the sources. Equations (2.56), (2.57) and (2.58) are not all independent once the equation of state is specified. Sometimes a cosmological term is directly introduced in Eq. (2.50). The addition of a cosmological term entails the presence of a term $\Lambda\delta_\mu^\nu$ at the right hand side of Eq. (2.50) (see Eq. (2.52)). In the light of forthcoming applications, it is preferable to think about the Λ term as to a component of the total energy-momentum tensor of the Universe. Such a component will contribute to ρ_t and to p_t with

$$\rho_\Lambda = \frac{\Lambda}{8\pi G}, \qquad p_\Lambda = -\rho_\Lambda. \qquad (2.59)$$

If the evolution of the SCM takes place for positive cosmic times (i.e. $t > 0$):

- the Universe expands when $\dot{a} > 0$;
- the Universe contracts when $\dot{a} < 0$.
- the Universe is said to be accelerating if $\ddot{a} > 0$;
- the Universe is said to be decelerating if $\ddot{a} < 0$.

In the SCM the evolution of the Universe can be parametrized as $a(t) \simeq t^\alpha$ where $0 < \alpha < 1$ and $t > 0$. The power α changes depending upon the different stages of the evolution. As a complementary remark it is useful to mention that, recently, cosmological models inspired by string theory try also to give a meaning to the evolution of the Universe when the cosmic time coordinate is negative, i.e. $t < 0$. What sets the origin of the time coordinate is, in this context, the presence of a curvature singularity that can be eventually resolved into a stage of maximal curvature [104–106] (see also [107–109] and references therein for some recent progress on the evolution of the fluctuations in a regularized pre-big bang background with T-duality invariant dilaton potential). In pre-big bang models it is important to extend the cosmic time coordinate also for negative values so, for instance parametrizations as $a(t) \sim (-t)^{-\gamma}$ are meaningful. Classically there are a number of reasonable conditions to be required on the components of the energy-momentum tensor of a perfect relativistic fluid. These conditions go under the name of *energy conditions* and play an important role in the context of the singularity theorems that can be proved in General Relativity [110, 111]. Some of these energy conditions may be violated once the components of the energy-momentum tensor are regarded as the expectation value of the energy density and of the pressure of a quantum field [112].

The main energy conditions are here listed:

- the weak energy condition (WEC) stipulates that the energy density is positive semi-definite, i.e. $\rho_t \geq 0$;
- the dominant energy condition (DOC) implies, instead, that the enthalpy of the fluid is positive semi-definite, i.e. $\rho_t + p_t \geq 0$;
- the strong energy condition (SEC) demands that $\rho_t + 3p_t \geq 0$.

According to the Hawking-Penrose theorems [110, 111], if the energy conditions are enforced the geometry will develop, in the far past, a singularity where the curvature invariants (i.e. R^2, $R_{\mu\nu}R^{\mu\nu}$, $R_{\mu\nu\alpha\beta}R^{\mu\nu\alpha\beta}$) will all diverge. If some of the energy conditions are not enforced, the geometry may still be singular if the causal geodesics (i.e. null or time-like) are past-incomplete, i.e. if they diverge at a finite value of the affine parameter. A typical example of this phenomenon is the expanding branch of de Sitter space which will be later scrutinized in the context of inflationary cosmology.

The enforcement of the energy conditions (or their consistent violation) implies interesting consequences at the level of the FL equations. For instance, if the DOC is enforced, Eq. (2.57) demands that $\dot{H} < 0$ when the

Universe is spatially flat (i.e. $k = 0$). This means that, in such a case the Hubble parameter is always decreasing for $t > 0$. If the SEC is enforced the Universe always decelerates *in spite of the value of the spatial curvature*. Consider, indeed, the sum of Eqs. (2.56) and (2.57) and recall that $\ddot{a} = (H^2 + \dot{H})a$. The result of this manipulation will be[e]

$$\frac{\ddot{a}}{a} = -\frac{4\pi G}{3}(\rho_t + 3p_t), \quad (2.60)$$

showing that, as long as the SEC is enforced $\ddot{a} < 0$. This conclusion can be intuitively understood since, under normal conditions, gravity is an attractive force: two bodies slow down as they move apart. The second interesting aspect of Eq. (2.60) is that the spatial curvature drops out completely. Again this aspect suggest that inhomogeneities cannot make gravity repulsive. Such a conclusion can be generalized to the situation where the FL equations are written without assuming the homogeneity of the background geometry (see chapter 5 for the first rudiments on this approach). A radiation-dominated fluid (or a matter-dominated fluid) respect both the SEC and the DOC. Hence, in spite of the presence of inhomogeneities, the Universe will always expand (for $t > 0$), it will always decelerate (i.e. $\ddot{a} < 0$) and the Hubble parameter will always decrease (i.e. $\dot{H} < 0$). To have an accelerated Universe the SEC must be violated. By parametrizing the equation of state of the fluid as $p_t = w_t \rho_t$ the SEC will be violated provided $w_t < -1/3$.

Introducing the critical density and the critical parameter at a given cosmic time t

$$\rho_{\mathrm{crit}} = \frac{3H^2}{8\pi G}, \qquad \Omega_t = \frac{\rho_t}{\rho_{\mathrm{crit}}} \quad (2.61)$$

already mentioned in section 1, Eq. (2.56) can be written as

$$\Omega_t = 1 + \frac{k}{a^2 H^2}. \quad (2.62)$$

The critical parameter is is nothing but the total energy density of the Universe expressed in critical units. Equation (2.62) has the following three direct consequences:

- if $k = 0$ (spatially flat Universe), $\Omega_t = 1$ (i.e. $\rho_t = \rho_{\mathrm{crit}}$);
- if $k < 0$ (spatially open Universe), $\Omega_t < 1$ (i.e. $\rho_t < \rho_{\mathrm{crit}}$);
- if $k > 0$ (spatially closed Universe), $\Omega_t > 1$ (i.e. $\rho_t > \rho_{\mathrm{crit}}$).

---

[e]In section 5 we will see how to generalize this result to the case when the spatial gradients are consistently included in the treatment.

Always at the level of the terminology, the *deceleration parameter* is customarily introduced:
$$q(t) = -\frac{\ddot{a}}{a H^2}. \qquad (2.63)$$
Notice the minus sign in the convention of Eq. (2.63):

- if $q < 0$ the Universe accelerates;
- if $q > 0$ the Universe decelerates.

In different applications, it is important to write, solve and discuss the analog of Eqs. (2.56), (2.57) and (2.58) in the conformal time parametrization already introduced in Eq. (2.4). From Eq. (2.55)

$$\begin{aligned}
R_0^0 &= -\frac{3}{a^2}\mathcal{H}', \\
R_i^j &= -\frac{1}{a^2}\left(\mathcal{H}' + 2\mathcal{H}^2 + 2k\right)\delta_i^j, \\
R &= -\frac{6}{a^2}\left(\mathcal{H}' + \mathcal{H}^2 + k\right).
\end{aligned} \qquad (2.64)$$

Using Eq. (2.64) in Eq. (2.50) the conformal time counterpart of Eqs. (2.56), (2.57) and (2.58) become, respectively,

$$\mathcal{H}^2 = \frac{8\pi G}{3}a^2 \rho_t - k, \qquad (2.65)$$

$$\mathcal{H}^2 - \mathcal{H}' = 4\pi G a^2 (\rho_t + p_t) - k, \qquad (2.66)$$

$$\rho_t' + 3\mathcal{H}(\rho_t + p_t) = 0, \qquad (2.67)$$

where the prime denotes a derivation with respect to the conformal time coordinate τ and $\mathcal{H} = a'/a$. Note that Eqs. (2.65), (2.66) and (2.67) can be swiftly obtained from Eqs. (2.56), (2.57) and (2.58) by bearing in mind the following (simple) dictionnary:

$$H = \frac{\mathcal{H}}{a}, \qquad \dot{H} = \frac{1}{a^2}(\mathcal{H}' - \mathcal{H}^2). \qquad (2.68)$$

2.3 Matter content of the SCM

According to the present observational understanding, our own Universe is, to a good approximation, spatially flat. Furthermore, the total energy density receives contribution from three (physically different) components:

$$\rho_t = \rho_M + \rho_R + \rho_\Lambda. \qquad (2.69)$$

Using the definition of the critical density parameter, a critical fraction is customarily introduced for every fluid of the mixture, i.e.

$$\Omega_t = \Omega_M + \Omega_R + \Omega_\Lambda. \qquad (2.70)$$

In Eq. (2.69), ρ_M parametrizes the contribution of non-relativistic species which are today stable and, in particular,

$$\rho_{M0} = \rho_{c0} + \rho_{b0}, \qquad (2.71)$$

i.e. ρ_{M0} (the present matter density) receives contribution from a cold dark matter component (CDM) and from a baryonic component.[f] Both components have the equation of state of non-relativistic matter, i.e. $p_c = 0$ and $p_b = 0$ and the covariant conservation of each species (see Eqs. (2.9) and (2.10)) implies

$$\dot{\rho}_b + 3H\rho_b = 0, \qquad \dot{\rho}_c + 3H\rho_c = 0. \qquad (2.72)$$

The term *cold dark matter* simply means that this component is non-relativistic and it is dark, i.e. it does not emit light and it does not absorb light. CDM particles are *inhomogeneously* distributed. Dark matter may also be hot. However, in this case it is more difficult to form structures because of the higher velocities of the particles. The present abundance of non-relativistic matter can be appreciated by the following illustrative values:

$$h_0^2 \Omega_{M0} = 0.134, \qquad h_0^2 \Omega_{c0} = 0.111, \qquad h_0^2 \Omega_{b0} = 0.023. \qquad (2.73)$$

Notice that, in Eq. (2.73) the abundances are independent on the indetermination of the Hubble parameter h_0. For numerical estimates h_0 will be taken between 0.7 and 0.73. A more complete (but still not comprehensive) discussion on the values of the cosmological parameters can be found at the end of chapter 1 (see section 1.6).

For CDM the main observational evidence come from the rotation curves of spiral galaxies, from the mass to light ratio in clusters and from CMB physics. For baryonic matter an indirect evidence stems from big-bang nucleosynthesis (BBN) (see [113] for a self-contained introduction to BBN). Indeed for temperatures smaller than 1 MeV weak interactions fall out of thermal equilibrium and the neutron to proton ratio decreases via free neutron decay. A bit later the abundances of the light nuclear elements (i.e. ^{4}He, ^{3}He, ^{7}Li and D) start being formed. While the discussion of BBN

---

[f]The subscript 0, when not otherwise stated, denotes the present value of the corresponding quantity.

is rather interesting in its own right (see Appendix B for some further details), it suffices to note here that, within the SCM, the homogeneous BBN only depends in principle upon two parameters: one is the temperature of the plasma (or, equivalently, the expansion rate) and the other is the ratio between the concentration of baryons and the concentration of photons. Recalling the expression for the concentration of CMB photons (see Eq. (1.25)), and recalling that $\rho_b = m_b n_b$ we have

$$\eta_{b0} = \frac{n_{b0}}{n_{\gamma 0}} = 6.27 \times 10^{-10} \left(\frac{h_0^2 \Omega_{b0}}{0.023} \right), \qquad (2.74)$$

having taken the typical baryon mass of the order of the proton mass.

On top of the dark matter component, the present Universe seems to contain also another dark component which is, however, much more homogeneously distributed than dark matter. It is therefore named dark energy and satisfies an equation of state with barotropic index smaller than $-1/3$. In particular, a viable and current model of dark energy is the one of a simple cosmological term[g] with

$$h_0^2 \Omega_{\Lambda 0} = 0.357. \qquad (2.75)$$

Notice that for a fiducial value of $h_0 \simeq 0.7$

$$\Omega_{M0} \simeq 0.27, \qquad \Omega_{c0} \simeq 0.22, \qquad \Omega_{b0} \simeq 0.046, \qquad \Omega_{\Lambda 0} \simeq 0.73. \qquad (2.76)$$

Finally, in the present Universe, as discussed in chapter 1 there is also radiation. Using the same notation employed in Eq. (2.73) we have:

$$h_0^2 \Omega_{R0} = h_0^2 \Omega_{\gamma 0} + h_0^2 \Omega_{\nu 0} + h_0^2 \Omega_{gw0}, \qquad (2.77)$$

where $\Omega_{\nu 0}$ denotes the contribution of neutrinos and Ω_{gw0} the contribution of relic gravitons. The numerical values of the quantities introduced in Eq. (2.77) are given by:

$$h_0^2 \Omega_{\gamma 0} = 2.47 \times 10^{-5}, \qquad h_0^2 \Omega_{\nu 0} = 1.68 \times 10^{-5}, \qquad h_0^2 \Omega_{gw0} < 10^{-11}. \qquad (2.78)$$

The contribution of relic gravitons is, today, smaller than 10^{-11}. This bound stems from the analysis of the integrated Sachs-Wolfe contribution which will be discussed later in this book (see chapters 6 and 7). If neutrinos have masses smaller than the meV they are today non-relativistic and, in principle, should not be counted as radiation. However, since the temperature of CMB was, in the past, much larger (as it will be discussed below), they will be effectively relativistic at the moment when matter decouples

---
[g]A particularly comprehensive review on the role of dark energy can be found in [114].

from radiation. To be more precise, we can say that current oscillation data require at least one neutrino eigenstate to have a mass exceeding 0.05 eV. In this minimal case $h_0^2 \Omega_{\nu 0} \simeq 5 \times 10^{-4}$ so that the neutrino contribution to the matter budget will be negligibly small. However, a nearly degenerate pattern of mass eigenstates could allow larger densities, since oscillation experiments only measure differences in the values of the squared masses.

2.4 The future of the Universe

From the analysis of the luminosity distance (versus the redshift) it appears that type-Ia supernovae are dimmer than expected and this suggests that at high redshifts (i.e. $z \geq 1$) the Universe is effectively accelerating [59]. The redshift z is defined (see Appendix A for further details) as

$$1 + z = \frac{a_0}{a}, \qquad (2.79)$$

where a_0 is the present value of the scale factor and a denotes a generic stage of expansion preceding the present epoch (i.e. $a < a_0$). The concept of redshift (see Appendix A) is related to the observation that, in an expanding Universe, the spectral lines of emitted radiation become more red (i.e. they redshift, they become longer) than in the case when the Universe does not expand. Given the matter content of the present Universe, its destiny can be guessed by using the FL equations and by integrating them forward in time. From Eq. (2.56), with simple algebra, it is possible to obtain the following equation:

$$\frac{d\alpha}{dx} = \sqrt{\frac{\Omega_{M0}}{\alpha} + \Omega_{\Lambda 0}\alpha^2 + \Omega_k + \frac{\Omega_{R0}}{\alpha^2}}, \qquad (2.80)$$

where the rescaled variables have been defined:

$$\alpha = \frac{a}{a_0}, \qquad x = H_0 t, \qquad \Omega_k = -\frac{k}{a_0^2 H_0^2}, \qquad (2.81)$$

and the quantity with subscripts 0 always refer to the present time.[h] To derive Eq. (2.80) it must also be borne in mind that a first integration of the covariant conservation equations leads to the following relations:

$$\rho_R = \rho_{R0}\left(\frac{a_0}{a}\right)^4, \qquad \rho_M = \rho_{M0}\left(\frac{a_0}{a}\right)^3, \qquad \rho_\Lambda = \rho_{\Lambda 0}. \qquad (2.82)$$

---

[h]Notice that k and Ω_k have opposite sign. While it is useful to define Ω_k as a critical fraction, it may also engender unwanted confusion which is related to the fact that, physically, the spatial curvature is not a further form of matter. With these caveats the use of Ω_k is rather practical.

From Eq. (2.80), different possibilities exist for the future dynamics of the Universe. These possibilities depend on the relative weight of the various physical components of the present Universe. In the case $\Omega_{\Lambda 0}$ Eq. (2.80) reduces to

$$\int \frac{\sqrt{\alpha}d\alpha}{\sqrt{\Omega_{M0} + \Omega_k \alpha}} = H_0(t - t_0). \qquad (2.83)$$

From Eq. (2.83) the following conclusions can be easily drawn:

- if $\Omega_k = 0$, $a(t)$ expands forever with $a(t) \sim t^{2/3}$ (decelerated expansion);
- if $\Omega_k < 0$ (closed Universe) the Universe will collapse in the future for a critical value $\alpha_{\rm coll} \simeq \Omega_{M0}/|\Omega_k|$;
- if $\Omega_k > 0$ (open Universe) the geometry will expand forever in a decelerated way.

Notice that, in Eq. (2.83) the role of radiation has been neglected since radiation is subleading today and it will be even more subleading in the future since it decreases faster than matter and faster than the dark energy.

If $\Omega_{\Lambda 0} \neq 0$ and $\Omega_k = 0$ Eq. (2.80) can be solved in explicit terms with the result that

$$\frac{a}{a_0} = \left(\frac{\Omega_{M0}}{\Omega_{\Lambda 0}}\right)^{1/3} \left\{ \sinh\left[\frac{3}{2}\sqrt{\Omega_{\Lambda 0}} H_0(t - t_0)\right] \right\}^{2/3}. \qquad (2.84)$$

This solution interpolates between a matter-dominated Universe expanding in a decelerated way as $t^{2/3}$ and an exponentially expanding Universe which is also accelerating. To get to Eq. (2.84), Eq. (2.80) can be written as

$$\int \frac{\sqrt{\alpha}d\alpha}{\sqrt{1 + \frac{\Omega_{\Lambda 0}}{\Omega_{M0}}\alpha^3}} = \sqrt{\Omega_{M0}}\, dx. \qquad (2.85)$$

By introducing the auxiliary variable

$$\alpha^{3/2} \sqrt{\frac{\Omega_{\Lambda 0}}{\Omega_{M0}}} = y, \qquad (2.86)$$

we obtain

$$\int \frac{dy}{\sqrt{1 + y^2}} = \frac{3}{2}\sqrt{\Omega_{\Lambda 0}}\, H_0(t - t_0). \qquad (2.87)$$

Finally, by introducing a second auxiliary variable $y = \sinh\beta$ the integral can be readily solved and Eq. (2.84) reproduced. While the discussion for $\Omega_k \neq 0$, $\Omega_{\Lambda 0} \neq 0$ and $\Omega_{M0} \neq 0$ is more complicated and will not be treated here, it is also clear that given the matter content of the present

Universe, it is reasonable to expect in the future, the Universe will accelerate while the role of non relativistic matter (and of radiation) will be progressively negligible.

There are a number of ways in which the kinematical features of the present Universe can be observationally accessible. The main tool is represented by the various distance concepts used by astronomers. The three useful distance measures that could be mentioned are (see Appendix A for further details on the derivation of the explicit expressions):

- the distance measure (denoted with $r_e(z)$ in Appendix A and often denoted with $D_M(z)$ in the literature);
- the angular diameter distance $D_A(z)$;
- the luminosity distance.

These three distances are all functions of the redshift z and of the (present) critical fractions of matter, dark energy, radiation and curvature, i.e. respectively, Ω_{M0}, $\Omega_{\Lambda 0}$, Ω_{R0} and Ω_k. In practice, the dependence upon Ω_{R0} can be dropped and it becomes relevant for *very* large redshift, i.e. $z \simeq 10$.

The three distances introduced in the aforementioned list of items are integrated quantities in the sense that they depend upon the integral of the inverse of the Hubble parameter from 0 to the generic redshift z (see Appendix A for a derivation). The angular diameter distance and the luminosity distance are related to $r_e(z)$ as:

$$D_A(z) = \frac{a_0 r_e(z)}{1+z}, \qquad D_L(z) = a_0 r_e(z)(1+z), \tag{2.88}$$

where a_0 is the present value of the scale factor that could be conventionally normalized to 1. The distance measure has been denoted by r_e since it represents the coordinate distance (defined in the FRW line element) once the origin of the coordinate system is placed in the Milky way. The angular diameter distance gives us the possibility of determining the distance of an object by measuring its angular size in the sky. Of course to conduct a measurement successfully we must have a set of standard rulers, i.e. a set of objects that have, at different redshifts, the same size.

The luminosity distance gives us the possibility of determining the distance of an object from its apparent luminosity. Of course, as in the case of the angular diameter distance, to complete the measurement successfully we would need a set of standard candles, i.e. a set of object with the same absolute luminosity.

In Figs. 2.1, 2.2 and 2.3 the three concepts of distance introduced above are illustrated. In Fig. 2.1 the distance measure is illustrated in the case of three models. The lowest (dashed) curve holds in the case of a flat Universe with $\Omega_{M0} = 1$. The intermediate (dot-dashed) curve holds in the case of a flat Universe with $\Omega_{M0} = 1/3$ and $\Omega_{\Lambda 0} = 2/3$. Finally the upper curve (full line) holds in the case of an open Universe dominated by the spatial curvature (i.e. $\Omega_{M0} = \Omega_{\Lambda 0} = 0$ and $\Omega_k = 1$). The angular diameter

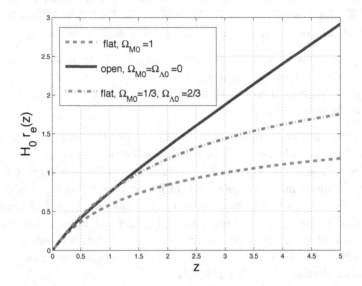

Fig. 2.1 The distance measure as a function of the redshift for three different models of the Universe.

distance is reported in Fig. 2.2 for the same sample of models described by Fig. 2.1. For large redshift, the angular diameter distance may well be decreasing, for some models. This means that the object that is further away may appear larger in the sky. Finally, in Fig. 2.3 the luminosity distance is illustrated.

Let us now briefly address the issue of the comparison of the theoretical curves illustrated in Figs. 2.1, 2.2 and 2.3 with the observational data. In Fig. 2.4 the base 10 logarithm of the luminosity distance is reported for a collection of 194 supernovae [115, 116]. This is a subset of 253 supernovae obtained by imposing the constraint that $z > 0.01$ (which reduces the effect of peculiar velocities) and that extinction effects are not present. So, the data reported in Fig. 2.4 should be compared with the logarithm to base 10 of the curves reported in Fig. 2.3. In doing so there is a caveat to mention. Astronomers measure, as we mentioned, luminosity distances.

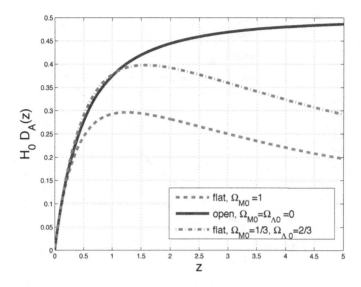

Fig. 2.2 The angular diameter distance as a function of the redshift for the same sample of models discussed in Fig. 2.1.

Therefore there is the need fixing the absolute normalization of the absolute magnitude. In practice this means that if we take blindly the logarithm of the curves appearing in Fig. 2.3 we would get a curve that can certainly be compared to observational data but up to an overall constant. To fix this constant there are, in principle, two different strategies. The first one is to treat the constant as an additional parameter in the fit. The second strategy (which is the one suggested, for instance, in Ref. [117]) is to choose the overall constant by fixing it, once forever, to the best fit value for the zero point magnitude offset. Sometimes the luminosity distance appearing in Figs. 2.3 and 2.4 (i.e. $H_0 D_L(z)$) is also called, in the jargon, Hubble-free luminosity distance to distinguish it from $D_L(z)$.

2.5 The past of the Universe

Even if today $\Omega_{R0} \ll \Omega_{M0}$, in the past history of the Universe radiation was presumably the dominant component. By going back in time, the dark-energy does not increase (or it increases very slowly) while radiation increases faster than the non-relativistic matter. In Fig. 2.5 the evolutions of the critical fractions of matter, radiation and dark energy are reported as-

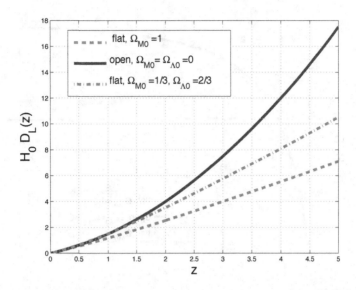

Fig. 2.3 The luminosity distance as a function of the redshift for the same sample of models discussed in Figs. 2.1 and 2.2.

suming, as present values of the illustrated quantities, the numerical values introduced in the present section (see, for instance, Eq. (2.73)). Recalling the evolution of the radiation and matter energy densities, radiation and matter were equally abundant at a redshift

$$1 + z_{\rm eq} = \frac{a_0}{a_{\rm eq}} = \frac{h_0^2 \Omega_{\rm M0}}{h_0^2 \Omega_{\rm R0}} = 3228.91 \left(\frac{h_0^2 \Omega_{\rm M0}}{0.134} \right). \qquad (2.89)$$

In other words:

- for $z > z_{\rm eq}$ (i.e. $a < a_{\rm eq}$) the Universe is effectively dominated by radiation;
- for $z < z_{\rm eq}$ (i.e. $a > a_{\rm eq}$) the Universe is effectively dominated by non-relativistic matter until the moment dark-energy starts being dominant.

Around the equality time, various important phenomena take place in the life of the Universe and they are directly related to CMB physics. For this reason it is practical to solve the FL equations across the transition between radiation and matter. Assuming that the only matter content is given by dust and radiation, and supposing that the Universe is spatially flat,

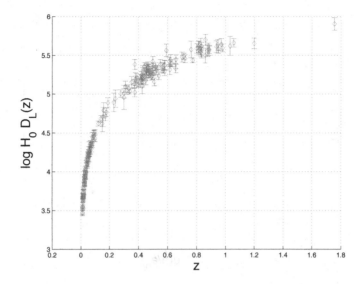

Fig. 2.4 The luminosity distance for 194 supernovae selected among a sample of 253 according to the data published in [115, 116].

Eq. (2.65) implies the following differential equation[i]

$$\left(\frac{da}{d\tau}\right)^2 = \frac{8\pi G}{3} a^4 \left[\rho_{R0}\left(\frac{a_0}{a}\right)^4 + \rho_{M0}\left(\frac{a_0}{a}\right)^3\right], \quad (2.90)$$

where the total energy density has been specified as[j]

$$\rho_t = \rho_R + \rho_M = \rho_{R0}\left(\frac{a_0}{a}\right)^4 + \rho_{M0}\left(\frac{a_0}{a}\right)^3. \quad (2.91)$$

Using the fact that, according to Eq. (2.89), $a_0/a_{eq} = \rho_{R0}/\rho_{M0}$, a_0 can be eliminated from Eq. (2.90). Taking the square root of the resulting expression, Eq. (2.90) implies

$$\frac{1}{H_0}\frac{d}{d\tau}\left(\frac{a}{a_{eq}}\right) = \frac{\Omega_{M0}}{\sqrt{\Omega_{R0}}}\left[\left(\frac{a}{a_{eq}}\right) + 1\right]^{1/2}, \quad (2.92)$$

whose solution is simply:

$$a(\tau) = a_{eq}\left[\left(\frac{\tau}{\tau_1}\right)^2 + 2\left(\frac{\tau}{\tau_1}\right)\right], \quad (2.93)$$

---

[i]Both assumptions are rather well supported by current observational data.
[j]The contribution of dark energy is not essential in the description of the radiation-mattter transition.

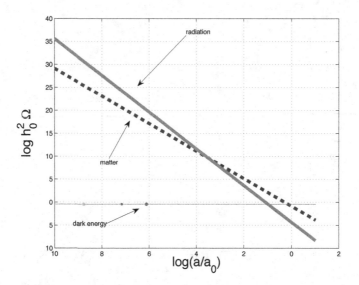

Fig. 2.5 The evolution of the critical fractions of matter, radiation and dark energy as a function of the logarithm (to base 10) of (a/a_0) where a_0 denotes, as usual, the present value of the scale factor.

with
$$\tau_1 = \frac{2}{H_0}\sqrt{\frac{a_{\rm eq}}{\Omega_{\rm M0}}} \simeq 288.25 \left(\frac{h_0^2 \Omega_{\rm M0}}{0.134}\right)^{-1} {\rm Mpc}. \tag{2.94}$$

From Eq. (2.93) $\tau_{\rm eq} = (\sqrt{2} - 1)\tau_1$ and, thus,
$$\tau_{\rm eq} = 119.39 \left(\frac{h_0^2 \Omega_{\rm M0}}{0.134}\right)^{-1} {\rm Mpc}, \qquad \tau_{\rm dec} = 283.47\,{\rm Mpc}, \tag{2.95}$$

where the second relation holds for $h_0^2 \Omega_{\rm M0} = 0.134$. Notice that, for $\tau \ll \tau_1$, $a(\tau) \sim \tau$ (which implies $a(t) \sim t^{1/2}$ in cosmic time). For $\tau \gg \tau_1$, $a(\tau) \sim \tau^2$ (which implies $a(t) \sim t^{2/3}$ in cosmic time). After equality, two important phenomena take place:

- Hydrogen recombination
- the decoupling of radiation from matter.

These will be the last two main topics treated in the present chapter.

2.5.1 *Hydrogen recombination*

After electron positron annihilation, the concentration of electrons can be written as $n_e = x_e n_{\rm B}$ where $n_{\rm B}$ is the concentration of baryons and x_e is the ionization fraction. Before equality, i.e. deep in the radiation-dominated epoch, $x_e = 1$: the concentration of free electrons exactly equals the concentration of protons and the Universe is globally neutral.

After matter-radiation equality, when the temperature drops below the eV, protons start combining with free electrons and the ionization fraction drops from 1 to 10^{-4}–10^{-5}. The drop in the ionization fraction occurs because free electrons are captured by protons to form Hydrogen atoms according to the reaction $e + p \to H + \gamma$. For sake of simplicity we can think that the Hydrogen is formed in its lowest energy level. It would be wrong to guess, however, that this process takes place around 13.2 eV. It takes place, on the contrary, for typical temperatures that are of the order of 0.3 eV. The rationale for this statement is that the pre-factor in the equilibrium concentration of free electrons is actually small and, therefore, the Hydrogen formation cannot be simply estimated from the Boltzmann factor.

The redshift of recombination is defined as the moment at which the ionization fraction drops from his equilibrium value (i.e. $x_e = 1$) to $x_e \simeq 0.1$. The redshift of decoupling is the determined by the requirement that the ionization fraction decreases even more. At $x_e \simeq 10^{-4}$ the decoupling is considered achieved. Let us go through a more quantitative discussion of these figures. When the temperature of the plasma is high enough the reactions of recombination and photodissociation of Hydrogen are in thermal equilibrium, i.e. $e + p \to H + \gamma$ is balanced by $H + \gamma \to e + p$. In this situation the concentrations of Hydrogen, of the protons and of the electrons follow, respectively, from the equlibrium distribution (see Appendix B for further details):

$$n_H = g_H \left(\frac{m_H T}{2\pi}\right)^{3/2} e^{(\mu_H - m_H)/T}, \qquad (2.96)$$

$$n_p = g_p \left(\frac{m_p T}{2\pi}\right)^{3/2} e^{(\mu_p - m_p)/T}, \qquad (2.97)$$

$$n_e = g_e \left(\frac{m_e T}{2\pi}\right)^{3/2} e^{(\mu_e - m_e)/T}, \qquad (2.98)$$

where g_H, g_p and g_e are, respectively, 4, 2 and 2. The exponentials appearing in Eqs. (2.96), (2.97) and (2.98) are often called Boltzmann suppression factors. Since we are in a situation of chemical equilibrium (see Appendix B) we can relate the various chemical potentials according to the order of the reaction, i.e. $\mu_H = \mu_p + \mu_e$. Eliminating μ_H from Eq. (2.96) and using the product of Eqs. (2.97) and (2.98) to express $\exp[(\mu_p + \mu_e)/T]$ in terms of the electron and proton concentrations, the following expression can be obtained:

$$n_{\rm H} = n_e n_p \left(\frac{m_e T}{2\pi}\right)^{-3/2} e^{E_0/T}, \qquad E_0 = m_e + m_p - m_{\rm H} = 13.26 \text{ eV}, \quad (2.99)$$

where E_0 is the absolute value of the binding energy of the hydrogen atoms that corresponds to the energy of the lowest energy level since it has been assumed that hydrogen recombines in the fundamental state. We now observe that:

- the Universe is electrically neutral, hence $n_p = n_e$;
- the total baryonic concentration of the system is $n_{\rm B} = n_{\rm H} + n_p$;
- the concentration of free electrons (or free protons) can be related to the baryonic concentration as $n_e = x_e n_{\rm B}$ where x_e is the ionization fraction.

Concerning the second observation, it should be incidentally remarked that the total baryonic concentration is given, in general terms, by $n_{\rm B} = n_{\rm N} - n_{\overline{\rm N}}$ (where $n_{\rm N}$ and $n_{\overline{\rm N}}$ are, respectively, the concentrations of nucleons and antinucleons). However, for $T < 10$ MeV, $n_{\overline{\rm N}} \ll 1$ and, therefore, $n_{\rm B} = n_n + n_p$. The success of big-bang nucleosynthesis implies, furthermore, that approximately one quarter of all nucleons form nuclei with atomic mass number $A > 1$ (and mostly ^{4}He), while the remaining three quarters are free protons. In similar terms we can also say that for temperatures $T < 10$ keV the concentration of positrons is negligible in comparison with the concentration of electrons. Using all the aforementioned observations, both sides of Eq. (2.99) can be divided by the baryonic concentration $n_{\rm B}$. Then, using of the global charge neutrality of the plasma together with Eq. (2.74), Eq. (2.99) can be written as

$$\frac{1 - x_e}{x_e^2} = \eta_{b0} \frac{4\zeta(3)\sqrt{2}}{\sqrt{\pi}} \left(\frac{T}{m_e}\right)^{3/2} e^{E_0/T}, \qquad (2.100)$$

which is called the Saha equation for the equlibrium ionization fraction. In Eq. (2.100) the baryonic concentration has been expressed through η_{b0}, i.e. the ratio between the concentrations of baryons and photons. Introducing now the dimensionless variable $y = T/\text{eV}$ we have that, using the explicit expression of η_b (i.e. Eq. (2.74)), Eq. (2.100) can be written as

$$\frac{1 - x_e}{x_e^2} = \mathcal{P} y^{3/2} e^{y_0/y}, \qquad \mathcal{P} = 6.530 \times 10^{-18} \left(\frac{h_0^2 \Omega_{b0}}{0.023}\right), \qquad (2.101)$$

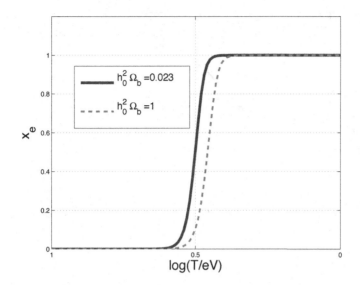

Fig. 2.6 The ionization fraction is illustrated as a function of the rescaled temperature $y = T/\text{eV}$.

where

$$\left(\frac{T}{m_e}\right) = 1.96 \times 10^{-6} y, \qquad y_0 = 13.26. \qquad (2.102)$$

Equation (2.101) stipulates that when $y \simeq 1$ (corresponding to $T \simeq \text{eV}$), $\exp(13.26) \simeq 10^5$ and thus we still have $x_e \simeq 1$. In fact, the smallness of $\mathcal{P}$ appearing in Eq. (2.101) necessarily implies that $1 - x_e \simeq 10^{-13} x_e^2$. This observation shows that atoms do not form for $T \sim 10$ eV nor for $T \sim$ eV but only when the temperature drops well below the eV. Equation (2.101) can be made more explicit by solving with respect to x_e

$$x_e = \left(\frac{-1 + \sqrt{1 + 4\mathcal{P} y^{3/2} e^{y_0/y}}}{2\mathcal{P} y^{3/2}}\right) e^{y_0/y}. \qquad (2.103)$$

From Figs. 2.6 and 2.7 it appears that in order to reduce the ionization fraction to an appreciable value (i.e. $x_e \simeq 10^{-1}$), T must be as low as 0.3 eV. In Fig. 2.6 the ionization fraction is illustrated as a function of the rescaled temperature, while in Fig. 2.7 the logarithm of the same quantity is illustrated. Recalling that $T = T_{\gamma 0}(1 + z)$ we can see that[k]:

---

[k] From now on, without any confusion, we will often drop the subscript γ in the temperature.

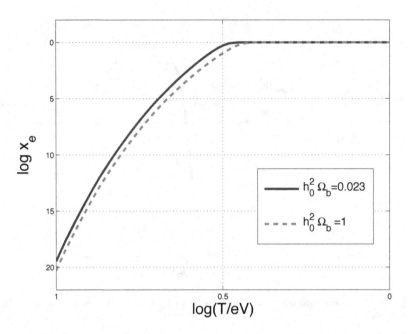

Fig. 2.7 The logarithm to base 10 of the ionization fraction is illustrated as a function of the rescaled temperature $y = T/\text{eV}$.

- $x_e \simeq 10^{-1}$ implies $T_{\text{rec}} \simeq 0.3\,\text{eV}$ and $z_{\text{rec}} \simeq 1300$: this is the moment of hydrogen recombination when photoionization reactions are unable to balance hydrogen formation;
- $x_e \simeq 10^{-4}$ implies $T_{\text{dec}} \simeq 0.2\,\text{eV}$ and $z_{\text{dec}} \simeq 1100$: this is the moment of decoupling when the photon mean free path gets as large as 10^4Mpc (see below in this section and, in particular, Eq. (2.117)).

Often z_{rec} and z_{dec} are used interchangeably when discussing semi-analytical estimates of temperature anisotropies (see chapter 9). Since the most efficient process that can transfer energy and momentum is Thompson scattering, the drop in the ionization fraction entails a dramatic increase of the proton mean free path. Before decoupling the photon mean free path is of the order of the Mpc. After decoupling, the photon mean free path becomes of the order of 10^4 Mpc and the CMB photons may reach our detectors and satellites without being scattered. This is the moment when the Universe becomes transparent to radiation.

2.5.2 Coulomb scattering: the baryon-electron fluid

Before equality electrons and protons are coupled through Coulomb scattering while photons scatter protons and electrons with Thompson cross section. Now, the Coulomb rate of interactions is much smaller than the Hubble rate at the corresponding epoch. Thus, the protons and electrons form a single (globally neutral) component where the velocities of the electrons and of the protons are approximately equal. This is the reason why baryons and leptons will be described in the analysis of CMB anisotropies by a single set of equations, somehow confusingly called baryon fluid.

Photons scatter electrons with Thompson cross section and, in principle, photons scatter also protons with Thompson cross section. However, since the rest mass of the proton is roughly 2000 times larger than the rest mass of the electron, the corresponding cross-section for photon-proton scattering will be much smaller than the cross-section for photon-electron scattering. This observation implies that the mean free path of photons is primarily determined by the photon-electron cross section. Consider then, for $t < t_{\rm eq}$, the Coulomb rate of interactions given by:

$$\Gamma_{\rm Coul} = v_{\rm th} \sigma_{\rm Coul} n_e, \qquad (2.104)$$

where

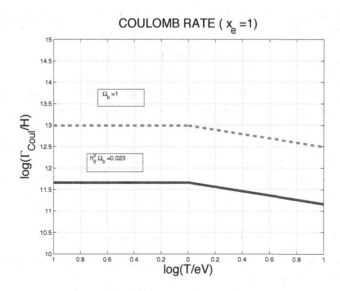

Fig. 2.8 The Coulomb rate is illustrated around equality in the case when $x_e = 1$.

- $v_{\rm th} \simeq \sqrt{T/m_e}$ is the thermal velocity of electrons;
- $\sigma_{\rm Coul} = (\alpha_{\rm em}^2/T^2) \ln \Lambda$ is the Coulomb cross section including the Coulomb logarithm;
- $n_e = x_e n_B$ which may also be written as

$$n_e = \frac{2\zeta(3)}{\pi^2} T^3 x_e \eta_{b0}. \qquad (2.105)$$

Plugging everything into Eq. (2.104) we obtain:

$$\Gamma_{\rm Coul} = 1.15 \times 10^{-17} \, x_e \left(\frac{T}{\rm eV}\right)^{3/2} \left(\frac{h_0^2 \Omega_{b0}}{0.023}\right) \, {\rm eV}. \qquad (2.106)$$

The Coulomb rate may now be compared with the Hubble rate. Since the number of relativistic degrees of feedom is given by $g_\rho \simeq 3.36$, according to the general formula (valid for $t < t_{\rm eq}$ and derived in Eq. (B.44))

$$H = 1.66 \sqrt{g_\rho} \frac{T^2}{M_{\rm P}} = 2.49 \times 10^{-28} \left(\frac{T}{\rm eV}\right)^2 \, {\rm eV}. \qquad (2.107)$$

For $t > t_{\rm eq}$ we will have, instead

$$H = H_{\rm eq} \left(\frac{T}{\rm eV}\right)^{3/2} \, {\rm eV}. \qquad (2.108)$$

Therefore,

$$\frac{\Gamma_{\rm Coul}}{H} = 4.61 \times 10^{11} \left(\frac{T}{\rm eV}\right)^{-1/2} x_e \left(\frac{h_0^2 \Omega_{b0}}{0.023}\right), \quad T > T_{\rm eq}, \qquad (2.109)$$

$$\frac{\Gamma_{\rm Coul}}{H} = 4.61 \times 10^{11} \, x_e \left(\frac{h_0^2 \Omega_{b0}}{0.023}\right), \quad T < T_{\rm eq}. \qquad (2.110)$$

Equations (2.109) and (2.110) are illustrated in Figs. 2.8 and 2.9 where the logarithm (to base 10) of the Coulomb rate is plotted in units of the expansion rate. Figure 2.8 refers to the case when $x_e = 1$. Figure 2.9 refers to the case when $x_e \neq 1$ and it is computed from the Saha equation. We can clearly see that $\Gamma_{\rm Coul} > H$ in the physically interesting range of temperatures. This means, as anticipated, that charged particles are strongly coupled.

2.5.3 Thompson scattering: the baryon-photon fluid

Consider now, always before equality, the Thompson rate of reaction. In this case we will have that

$$\Gamma_{\rm Th} \simeq n_e \sigma_{\rm Th}, \qquad (2.111)$$

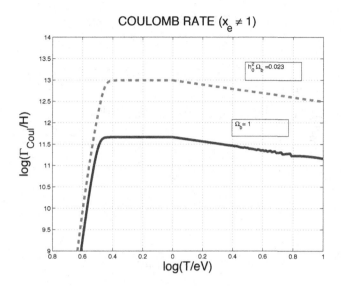

Fig. 2.9 The Coulomb rate is illustrated around equality in the case when $x_e \neq 1$. The dependence of the ionization fraction upon the temperature is determined, as in Fig. 2.8 from the Saha equation.

where

$$\sigma_{\text{Th}} = 0.665 \text{ barn}, \qquad 1\,\text{barn} = 10^{-24}\text{cm}^2. \tag{2.112}$$

Using Eq. (2.112) into Eq. (2.111) we will have

$$\Gamma_{\text{Th}} = 2.6 \times 10^{-25} \, x_e \left(\frac{T}{\text{eV}}\right)^3 \left(\frac{h_0^2 \Omega_{\text{b0}}}{0.023}\right) \text{ eV}, \tag{2.113}$$

which shows that $\Gamma_{\text{Coul}} \gg \Gamma_{\text{Th}}$ and also that

$$\frac{\Gamma_{\text{Th}}}{H} = 1.04 \times 10^3 \left(\frac{T}{\text{eV}}\right) x_e \left(\frac{h_0^2 \Omega_{\text{b0}}}{0.023}\right), \qquad T > T_{\text{eq}}, \tag{2.114}$$

$$\frac{\Gamma_{\text{Th}}}{H} = 1.04 \times 10^3 \left(\frac{T}{\text{eV}}\right)^{3/2} x_e \left(\frac{h_0^2 \Omega_{\text{b0}}}{0.023}\right), \qquad T < T_{\text{eq}}. \tag{2.115}$$

The previous equations also substantiate the statement that the photon mean free path is much larger than the electron mean free path for temperatures $T > \text{eV}$. Thus, Thompson scattering is the most efficient way of transferring energy and momentum. Equations (2.114) and (2.115) are illustrated in Figs. 2.10 and 2.11. It is clear that as soon as the ionization fraction drops, the Thompson rate becomes suddenly smaller than the

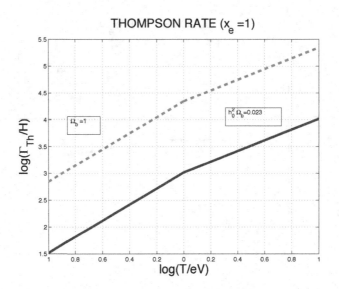

Fig. 2.10 The base-10 logarithm of the Thompson rate of interaction is illustrated around equality in the case $x_e = 1$.

expansion rate. After equality the photon mean free path can be written as

$$\lambda_{\rm Th} \simeq \frac{1}{an_e \sigma_{\rm Th}}, \qquad (2.116)$$

which can also be written, in more explicit terms, as

$$\lambda_{\rm Th} \simeq 1.8\, x_e^{-1} \left(\frac{0.023}{h_0^2 \Omega_{b0}}\right) \left(\frac{1100}{1+z_{\rm dec}}\right)^2 \left(\frac{0.88}{1-Y_p/2}\right)\ {\rm Mpc}. \qquad (2.117)$$

Equation (2.117) shows clearly that as soon as the ionization fraction drops (at recombinantion) the photon mean free path becomes of the order of 10^4–10^5 Mpc. In Eq. (2.117) the mass fraction of ^{4}He appears explicitly and it is denoted by Y_p (typically $Y_p \simeq 0.24$). This is not a surprise since the Helium nucleus contains four nucleons and the ratio of Helium to the total number of nuclei is $Y_p/4$. Each of these absorbs two electrons (one for each proton). Thus when we count the number of free electrons before recombination the estimate of the Thompson reaction rate must be multiplied by $(1 - Y_p/2)$. Note, finally, that in the last estimate the recoil energy of the electron has been neglected. This is justified since the electron rest mass is much larger than the incident photon energy which is, at recombination, of the order of the temperature, i.e. 0.3 eV.

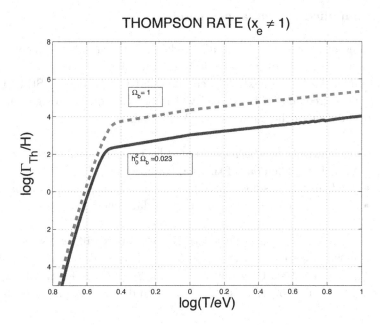

Fig. 2.11 The same quantity reported in Fig. 2.10 is here reported in the case when $x_e(T)$.

In summary, it is important to stress that Coulomb scattering is rather efficient in keeping tight the coupling between baryons and electrons, at least in the standard treatment. This occurrence justifies, at an effective level, for us to consider a single baryon-lepton fluid which is globally neutral but intrinsically charged. The tight-coupling between photons and charged particles (either leptons or baryons) is realized before recombination and it is therefore a very useful analytical tool for the approximate estimate of acoustic oscillations arising in the temperature autocorrelations which will be discussed, under different approximations, in chapter 8 and 9. The (approximate) tight-coupling between photons and charged species allows then, in combination with the largeness of the Coulomb rate, the treatment of a single baryon-lepton-photon fluid or baryon-photon fluid for short. This chain of observations will be turn out to be very useful when writing the evolution equations for the inhomogeneities prior to decoupling. This topic will be discussed in chapters 8 and 9 (see, in particular, before and after Eqs. (8.68), (8.69) and (8.70) when talking about the tight-coupling appoximation).

2.6 Simplified numerical estimates

In the present chapter the essential ingredients of the standard cosmological model have been introduced. This formulation is just preliminary, since, as it will be shown in the next chapter, the SCM has problems. Still, three important ingredients of the model will remain almost unchanged in the final formulation of the minimal ΛCDM paradigm:

- the Universe has been dominated by radiation after weak interactions have fallen out of thermal equlibrium;
- the Universe was dominated by matter at Hydrogen recombination;
- the Universe is today dominated by dark energy.

In the present section some illustrative examples will be discussed. It will be shown how to exploit the exact forms of the scale factors derived in Eqs. (2.84) and (2.93). Consider, first of all, Eq. (2.93). From Eq. (2.93) and from the definition of redshift it is easy to deduce, for instance, that

$$\frac{\tau_{\rm rec}}{\tau_1} = \sqrt{1 + \frac{z_{\rm eq}}{z_{\rm rec}}} - 1. \qquad (2.118)$$

This expression automatically gives the value of the conformal time coordinate at recombination in terms of the corresponding redshifts. Suppose now we wish to evaluate $\mathcal{H}_{\rm rec}\tau_{\rm rec}$. From Eq. (2.93), using the definition of $\mathcal{H} = a'/a$ the following expression can be swiftly obtained

$$\mathcal{H}_{\rm rec}\tau_{\rm rec} = 2\frac{(\tau_{\rm rec}/\tau_1) + 1}{(\tau_{\rm rec}/\tau_1) + 2}. \qquad (2.119)$$

Finally, using Eq. (2.118) into Eq. (2.119) we can obtain

$$\frac{1}{\mathcal{H}_{\rm rec}\tau_{\rm rec}} = \frac{1}{2}\left(1 + \frac{1}{\sqrt{1 + \frac{z_{\rm eq}}{z_{\rm rec}}}}\right). \qquad (2.120)$$

This simple estimate helps to evaluate the semi-analytic form of the so-called visibility function that appears ubiquitously in the Boltzmann treatment of CMB anisotropies (see chapter 9 for a proper definition). In the present notations, indeed, this function will be written as $\exp[-2k^2\sigma^2\tau_{\rm rec}^2]$

where k is the comoving wave-number and where

$$\sigma = \frac{1}{\sqrt{6}\kappa\mathcal{H}_{\rm rec}\tau_{\rm rec}} = 1.48\times 10^{-2}\left(1+\frac{1}{\sqrt{1+\frac{z_{\rm eq}}{z_{\rm rec}}}}\right),$$

$$\kappa = \frac{14400}{1050} = 13.714,$$

$$\frac{z_{\rm eq}}{z_{\rm rec}} = \frac{3228.91}{1050}\left(\frac{h_0^2\Omega_{\rm M0}}{0.134}\right) \quad (2.121)$$

$$= 3.074\times\left(\frac{h_0^2\Omega_{\rm M0}}{0.134}\right) = 12.9\, h_{75}^2\Omega_{\rm M0},$$

where $h_{75} = h/0.75$. Consider now the case of Eq. (2.84). The problem now could be to evaluate the ratio $\tau_0/\tau_{\rm rec}$. It is clear that, to perform this calculation, we not only need Eq. (2.93) but also Eq. (2.84) since, between recombination and the present time, the dark-energy component became dominant in comparison with the matter component. First of all we can write

$$\frac{\tau_0}{\tau_{\rm rec}} = \frac{\tau_0}{\tau_{\rm p}}\frac{\tau_{\rm p}}{\tau_{\rm rec}}, \quad (2.122)$$

where $\tau_{\rm p}$ is a putative time at which the dusty matter is still dominant and the dark energy is on the verge of taking over. Using Eqs. (2.84) and (2.93) it is easy to obtain

$$\frac{\tau_{\rm p}}{\tau_0} = \frac{1}{\sqrt{z_{\rm p}}}\mathcal{T}_\Lambda,$$

$$\frac{\tau_{\rm rec}}{\tau_{\rm p}} = \sqrt{\frac{z_{\rm p}}{z_{\rm eq}}}\left[\sqrt{1+\frac{z_{\rm eq}}{z_{\rm rec}}}-1\right], \quad (2.123)$$

where

$$\mathcal{T}_\Lambda = 3\left(\frac{\Omega_{\Lambda 0}}{\Omega_{\rm M0}}\right)^{1/6}I_\Lambda^{-1}, \quad I_\Lambda = \int_0^{{\rm arcsinh}\sqrt{(\Omega_{\Lambda 0}/\Omega_{\rm M0})}}\frac{dy}{\sinh^{2/3}y}. \quad (2.124)$$

Since $\Omega_{\rm M0} = 1 - \Omega_{\Lambda 0}$ the integral appearing in Eq. (2.124) effectively depends only upon one parameter. The integral parametrized by I_Λ can be performed numerically. It depends only upon one parameter, i.e. $\Omega_{\rm M0}$. As a consequence, the whole expression denoted by $\mathcal{T}_\Lambda$ only depends upon $\Omega_{\rm M0}$. The result of the numerical integration for $\mathcal{T}_\Lambda$ is reported in Fig. 2.12 with the full line. After performing numerically the integral, therefore, a fit to the final result can be found in the form $\mathcal{T}_\Lambda \simeq \Omega_{\rm M0}^{-\gamma_1}$ where $\gamma_1 \simeq 0.0858$ (and for γ_1 as large as 0.09 the fit is still reasonable). In Fig. 2.12 the result obtained in the case $\gamma_1 = 0.0858$ is reported with the dashed line.

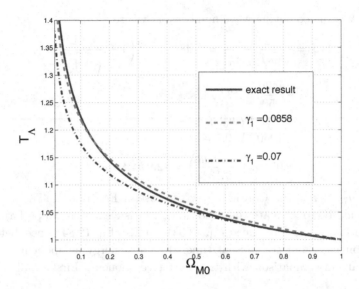

Fig. 2.12 The analytical interpolation and the numerical value of T_Λ.

Always in Fig. 2.12 the dot-dashed line denotes the case $\gamma_1 = 0.7$ which is reported for comparison. Consequently, Eq. (2.122) becomes:

$$\frac{\tau_{\rm rec}}{\tau_0} \simeq \frac{T_\Lambda}{\sqrt{z_{\rm rec}}} \left[\sqrt{1 + \frac{z_{\rm rec}}{z_{\rm eq}}} - \sqrt{\frac{z_{\rm rec}}{z_{\rm eq}}} \right]. \qquad (2.125)$$

These considerations will be relevant when trying to obtain simplified estimates of the temperature autocorrelations of CMB anisotropies for adiabatic initial conditions (see chapters 8 and 9).

Part II

Beyond the Standard Cosmological Model

Beyond the Standard Cosmological Model

Chapter 3

Problems with the SCM

From the quantitative discussions reported in the two previous chapters of the present book, two possible drawbacks of the SCM are already understandable:

- the anisotropies of CMB are not accounted by the SCM (see Fig. 1.3);
- the huge thermodynamic entropy stored in the CMB (see Eq. (1.20) and the related discussion) is not explained within the SCM where the evolution of the Universe is all the time adiabatic (see, for instance, Eqs. (2.12) and (2.17)).

The present hierarchy between the matter and radiation energy density suggests, furthermore, that the Universe was rather hot in the past. This conclusion is indirectly tested through the success of big-bang nucleosynthesis (BBN). As already pointed out, in BBN there are essentially only two free parameters[a]: the temperature and the the baryon to photon ratio η_b (see Eq. (2.74)). After weak interactions fall out of thermal equilibrium the light nuclei start being formed. Since the ^{4}He has the largest binding energy per nucleon for nuclei with atomic number $A < 12$, roughly one quarter of all the protons will end up in ^{4}He while the rest will remain in free protons. Smaller abundances of other light nuclei (i.e. D, ^{3}He and ^{7}Li) can be also successfully computed in the framework of BBN [113]. The

---

[a]This statement holds, strictly speaking, in the simplest (and also most predictive) BBN scenario where the synthesis of light nuclei occurs homogeneously in space and in the absence of matter–antimatter fluctuations. In this scenario the antinucleons have almost completely disappeared by the time weak interactions fall out of thermal equilibrium. There are, however, models where both assumptions have been relaxed (see, for instance [118–120] and references therein). In this case the prediction of BBN will also depend upon the typical inhomogeneity scale of the baryon to photon ratio.

synthesis of light elements is very important since light elements have to turn on the thermonuclear reactions taking place in the stars. However, even if the Universe must be sufficiently hot (and probably as hot as several hundreds GeV to produce a sizable baryon asymmetry) it cannot be dominated by radiation all the way up to the Planck energy scale: this occurrence would lead to logical puzzles in the formulation of the SCM. In what follows some of the problems of the SCM will be discussed in a unified perspective and, in particular, we shall discuss:

- the horizon (or causality) problem;
- the spatial curvature (or flatness) problem;
- the entropy problem;
- the structure formation problem;
- the singularity problem.

The first two problems in the above list of items are often named kinematical problems. It is interesting to notice that both the horizon problem as well as the entropy and structure formation problems are directly related to CMB physics as it will be stressed below in this section.

3.1 The horizon problem

Two important concepts appear in the analysis of the causal structure of cosmological models [111], i.e. the proper distance of the event horizon:

$$d_e(t) = a(t) \int_t^{t_{max}} \frac{dt'}{a(t')}, \quad (3.1)$$

and the proper distance of the particle horizon

$$d_p(t) = a(t) \int_{t_{min}}^t \frac{dt'}{a(t')}, \quad (3.2)$$

(see also Appendix A for further details). The event horizon measures the size over which we can admit *even in the future* a causal connection. The particle horizon measures instead the size of causally connected regions at the time t. In the SCM the particle horizon exists while the event horizon does not exist and this occurrence is the direct cause of a kinematical problem of the standard model. According to the SCM, the Universe, in its past expand in a decelerated way as

$$a(t) \sim t^\alpha, \quad 0 < \alpha < 1, \quad t > 0, \quad (3.3)$$

which implies that $\dot{a} > 0$ and $\ddot{a} < 0$. Inserting Eq. (3.3) into Eqs. (3.1) and (3.2) the following two expressions are swiftly obtained after direct integration:

$$d_e(t) = \frac{t_{\max}}{1-\alpha}\left[\left(\frac{t}{t_{\max}}\right)^\alpha - \left(\frac{t}{t_{\max}}\right)\right], \qquad (3.4)$$

$$d_p(t) = \frac{1}{1-\alpha}\left[t - t_{\min}\left(\frac{t}{t_{\min}}\right)^\alpha\right]. \qquad (3.5)$$

Since $0 < \alpha < 1$, Eqs. (3.4) and (3.5) lead to the following pair of limits

$$\lim_{t_{\min}\to 0} d_p(t) \to \frac{\alpha}{1-\alpha} H^{-1}(t), \qquad (3.6)$$

$$\lim_{t_{\max}\to\infty} d_e(t) \to \infty, \qquad (3.7)$$

where both limits are taken while t is kept fixed. Equations (3.6) and (3.7) show that, in the SCM,

- the event horizon does not exist;
- the particle horizon exists and it is finite.

Because of the existence of the particle horizon, for each time in the past history of the Universe the typical causal patch will be of the order of the Hubble radius, i.e. restoring for a moment the speed of light, $d_p(t) \sim ct$. This simple occurrence represents, indeed, a problem. The present extension of the Hubble radius evolves as the scale factor (i.e. faster than the particle horizon). Let us then see how large was the present Hubble radius at a given reference time at which the evolution of the SCM is supposed to start. Such a reference time can be taken to be, for instance, the Planck time. The Hubble radius at the Planck time will be of the order of the μm, i.e. more precisely:

$$r_H(t_P) = 4.08 \times 10^{-4}\left(\frac{0.7}{h_0}\right)\left(\frac{T_{eq}}{eV}\right) \qquad (3.8)$$

The obtained figure can then be measured in units of the particle horizon at the Planck time, which is the relevant scale set by causality at any given time in the life of the SCM:

$$d_p(t_P) \simeq c t_P \simeq 10^{-33} \text{ cm}. \qquad (3.9)$$

Taking the ratio between (3.8) and (3.9)

$$\frac{r_H(t_P)}{d_p(t_P)} \simeq 4.08 \times 10^{29}\left(\frac{0.7}{h_0}\right)\left(\frac{T_{eq}}{eV}\right). \qquad (3.10)$$

The third power of Eq. (3.10) measures the number of causally disconnected volumes at $t_{\rm P}$. This estimate tells that there are, roughly, to 10^{87} causally disconnected regions at the Planck time. In Fig. 3.1 the physics described by Eq. (3.10) is illustrated in pictorial terms. The Hubble radius at the Planck time has approximate size of the order of the μm and it contains 10^{87} causally disconnected volumes each with approximate size of the order of the particle horizon at the Planck time. A drastic change in the reference time at which initial conditions for the evolution are set does not alter the essence of the problem. Suppose that, indeed, the thermal history of the Universe does not extend up to the Planck temperature. Let us take our reference temperature to be of the order of 200 GeV. For such a temperature all the species of the Glashow-Weinberg-Salam (GWS) model are in thermal equilibrium and the particle horizon is given by

$$d_{\rm p}(t_{\rm ew}) \simeq 35 \sqrt{\frac{106.75}{g_\rho}} \left(\frac{T_{\rm ew}}{200}\right)^{-2} {\rm cm} \qquad (3.11)$$

where $g_\rho(T)$ is the number of relativistic degrees of freedom at the temperature T here taken to be of the order of $T_{\rm ew} \simeq 200$ GeV (see Eqs. (B.35),

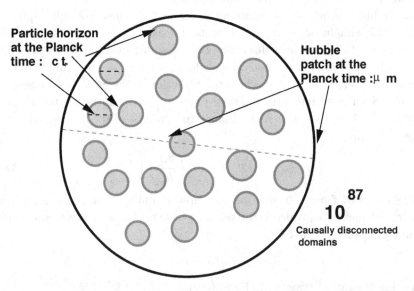

Fig. 3.1 A schematic snapshot of the Hubble patch blueshifted at the Planck time (see Eq. (3.10)). The filled circles represent the typical size of the particle horizon at the corresponding epoch.

(B.43)) and (B.44) of Appendix B). The Hubble radius blueshifted at the temperature $T_{\rm ew} \simeq 200$ GeV will be instead

$$r_H(t_{\rm ew}) \simeq 1.98 \times 10^{13} \left(\frac{0.7}{h_0}\right)\left(\frac{T_{\rm eq}}{\rm eV}\right) \; {\rm cm}. \qquad (3.12)$$

Thus, since $r_H(t_{\rm ew})/d_{\rm p}(t_{\rm ew}) \simeq 10^{12}$, the present Hubble patch will consist, at the temperature $T_{\rm ew}$, of 10^{36} causally disconnected regions. Since the temperature fluctuations in the microwave sky are of the order of $\delta T/T \simeq 10^{-5}$, the density contrast in radiation will be of the order of $\delta\rho_\gamma/\rho_\gamma \sim 10^{-4}$.

How come the CMB is so homogeneous, if, in the past history of the Universe there were so many causally disconnected regions? Is there something else other than causality that can make our Hubble patch homogenous? The answer to this question seems of course to be negative. The final observation to be borne in mind is that the root of the horizon problem resides in the occurrence that, in the SCM, the particle horizon evolves faster than the scale factor. This point is summarized in Fig. 3.2 where the evolution of the particle horizon is compared with the evolution of the scale factor.

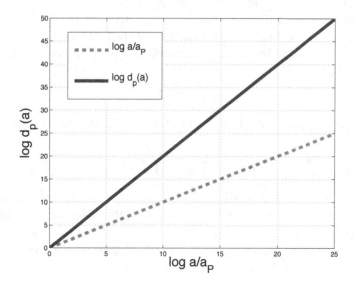

Fig. 3.2 The evolution of the particle horizon in the SCM is compared with the evolution of the scale factor. In the SCM, the particle horizon evolves faster than the scale factor, since, approximately, $d_{\rm p}(a) \sim a^{1/\alpha}$ with $0 < \alpha < 1$. Here, for illustration, the case $\alpha = 1/2$ (radiation dominance) has been assumed.

3.2 The spatial curvature problem

The problem of the spatial curvature can be summarized by the following question: why is the present Universe so close to being flat? From Eq. (2.56), the total energy density in critical units can be written as

$$\Omega_t(a) = 1 + \frac{k}{a^2 H^2}. \qquad (3.13)$$

Equation (3.13) holds at any time where the SCM applies. In particular, at the present time, we will have

$$\Omega_t(t_0) = 1 + \frac{k}{a_0^2 H_0^2}. \qquad (3.14)$$

According to the experimental data[b] [35, 39]

$$\Omega_t(t_0) = 1.02 \pm 0.02. \qquad (3.15)$$

Equation (3.15) implies that the contribution of $|k|(a_0 H_0)^{-2}$ is smaller than 1 (but of order 1). The denominator of the second term at the right hand side of Eq. (3.13) goes as $H^2 a^2 \simeq \dot{a}^2$. So if $a(t) \simeq t^\alpha$ (with $0 < \alpha < 1$), $\dot{a}^2$ will be a decreasing function of the cosmic time coordinate t. But this implies that, overall, the second term at the right hand side of Eq. (3.13) will increase dramatically as time goes by.

As in the case of the horizon problem, a particular reference time may be selected. At this time initial conditions of the SCM are ideally set. Let us take this time to be again, the Planck time and suppose that, at the Planck time

$$\frac{|k|}{a_P^2 H_P^2} \simeq \mathcal{O}(1). \qquad (3.16)$$

If Eq. (3.16) holds around the Planck time, today the same quantity will be:

$$\frac{|k|}{a_0^2 H_0^2} \simeq \frac{|k|}{a_P^2 H_P^2} \left(\frac{a_{eq}}{a_0}\right)^2 \left(\frac{H_{eq}}{H_0}\right)^2 \left(\frac{a_P}{a_{eq}}\right)^2 \left(\frac{H_P}{H_{eq}}\right)^2 \simeq 10^{60} \frac{|k|}{a_P^2 H_P^2}. \qquad (3.17)$$

Equation (3.17) demands that if $|k|a_P^{-2} H_P^{-2}$ is of order 1, today its contribution will be 60 orders of magnitude larger. By reversing the argument it can be argued that if the spatial curvature is (today) smaller than the extrinsic curvature (i.e. $k/a_0^2 \leq H_0^2$), at the Planck time we must require an enormous fine-tuning:

$$\frac{|k|}{a_P^2 H_P^2} \simeq 10^{-60}. \qquad (3.18)$$

---

[b]Notice that this experimental determination is achieved directly from the position of the first Doppler peak of the CMB temperature autocorrelations.

In other words: if the Universe is flat today it must have been even flatter in the past history of the Universe.

Therefore, in summary, since $|k|/\dot{a}^2$ increases during the radiation and matter-dominated epochs, Ω_t must be fine tuned to 1 with a precision of 10^{-60}. It is possible to write an evolution equation directly in terms of Ω_t. The strategy will be to use both the definition of Eq. (2.62) as well as Eq. (2.80). We leave this as an exercise and the result is:

$$\Omega_t(a) - 1 = \frac{\Omega_0 - 1}{1 - \Omega_0 + \Omega_{\Lambda 0}\left(\frac{a}{a_0}\right)^2 + \Omega_{M0}\left(\frac{a_0}{a}\right) + \Omega_{R0}\left(\frac{a_0}{a}\right)^2}, \quad (3.19)$$

where $\Omega_0 = \Omega_t(t_0)$. In the limit $a \to 0$, i.e. $a_0/a \to \infty$, Eq. (3.19) leads to

$$\Omega_t(a) - 1 \simeq \frac{\Omega_0 - 1}{\Omega_{R0}\left(\frac{a_0}{a}\right)^2}. \quad (3.20)$$

According to Eq. (3.20), we need $\Omega(a) \to 1$ with arbitrary precision (for the standards of physics) when $a \to 0$ if we want $\Omega_0 \simeq 1$ today. This is a different way of seeing the fine tuning mentioned before in this section.

3.3 The entropy problem

As discussed in introducing the essential features of the black body emission, the total entropy of the present Hubble patch is enormous and it is of the order of 10^{88} (see, for instance, Eq. (1.20)). This huge number arises since the ratio of $T_{\gamma 0}/H_0 \simeq \mathcal{O}(10^{30})$ (see Eq. (1.21)). The covariant conservation of the energy-momentum tensor (see Eq. (2.12)) implies that the whole Universe is, from a thermodynamic point of view (see Eq. (2.17)), an isolated system where the total entropy is conserved. If the evolution was adiabatic throughout the whole evolution of the SCM, why does the present Hubble patch have such a huge entropy? Really and truly the entropy problem contains, in itself, various other sub-problems that are rarely mentioned. They can be phrased in the following way:

- is the CMB entropy the only entropy that should be included in the formulation of the second law of thermodynamics?
- is the second law of thermodynamics valid throughout the history of the Universe?
- is the gravitational field itself a source of entropy?

- how can we associate an entropy to the gravitational field?

Two mutually exclusive choices then appear: either our own observable Universe originated from a Hubble patch with enormous entropy or the initial entropy was very small so that the initial state of the Universe was highly ordered. The first option applies, of course, if the evolution of the Universe was to a good approximation adiabatic while the second option contemplates a violation of the adiabaticity condition. Even more radically one could wonder if the second law of thermodynamics was indeed enforced in the early stages of the life of the Universe.

3.4 The structure formation problem

The SCM posits that the geometry is isotropic and fairly homogeneous over very large scales. Small deviations from homogeneity are, however, observed. For instance, inhomogeneities arise as spatial fluctuations of the CMB temperature (see, for instance, Fig. 1.3). These fluctuations will grow during the matter-dominated epoch and eventually collapse to form gravitationally bound systems such as galaxies and clusters of galaxies.

In what follows a simplistic description of CMB observables will be introduced.[c] By looking at the plots that are customarily shown in the context of CMB anisotropies (like the one reported in Figs. 1.3 and 3.3) we will try to understand in more detail what is actually plotted on the vertical as well as on the horizontal axis. By looking at Fig. 1.3 (or also at Fig. 3.3) we see that:

- on the horizontal axis the multipole moment ℓ is reported;
- on the vertical axis the corresponding power per logarithmic interval of ℓ is illustrated.

We will now see, in a moment, how to connect the two-point function of temperature fluctuations in real space with the quantity illustrated in the mentioned figures. However, already at this stage, it is important to appreciate that dependence of the angular power spectrum just upon ℓ signals that, indeed, the microwave sky is, to a good approximation, isotropic.

The logic will be to use various successive expansions with the aim of obtaining a reasonably simple parametrization of the CMB temperature

---

[c]The treatment of CMB anisotropies presented here mirrors the approach adopted in a recent review [33] where the main theoretical tools needed for the analysis of CMB anisotropies have been discussed within a consistent set of conventions.

fluctuations. Consider therefore the spatial fluctuations in the temperature of the CMB and expand them in Fourier integral[d]:

$$\Delta_{\rm I}(\vec{x},\hat{n},\tau) = \frac{\delta T}{T}(\vec{x},\hat{n},\tau) = \frac{1}{(2\pi)^{3/2}} \int d^3k \, e^{i\vec{k}\cdot\vec{x}} \Delta_{\rm I}(\vec{k},\hat{n},\tau), \qquad (3.21)$$

where $\hat{k}$ is the direction of the Fourier wave-number, $\hat{n}$ is the direction of the photon momentum. Assuming that the observer is located at a conformal[e] time τ (eventually coinciding with the present time τ_0) and at $\vec{x} = 0$, Eq. (3.21) can be also expanded in spherical harmonics, i.e.

$$\Delta_{\rm I}(\hat{n}) = \sum_{\ell m} a_{\ell m} Y_{\ell m}(\hat{n}) = \frac{1}{(2\pi)^{3/2}} \int d^3k \, \Delta_{\rm I}(\vec{k},\hat{n},\tau). \qquad (3.22)$$

Then, the Fourier amplitude appearing in Eq. (3.21) can be expanded in series of Legendre polynomials according to the well known relation

$$\Delta_{\rm I}(\vec{k},\hat{n},\tau) = \sum_{\ell=0}^{\infty} (-i)^{\ell} (2\ell+1) \Delta_{{\rm I}\ell}(\vec{k},\tau) P_{\ell}(\hat{k}\cdot\hat{n}). \qquad (3.23)$$

Now the Legendre polynomials appearing in Eq. (3.23) can be expressed via the addition theorem of spherical harmonics stipulating that

$$P_{\ell}(\hat{k}\cdot\hat{n}) = \frac{4\pi}{2\ell+1} \sum_{m=-\ell}^{\ell} Y_{\ell m}^*(\hat{k}) Y_{\ell m}(\hat{n}). \qquad (3.24)$$

Inserting now Eq. (3.24) into Eq. (3.23) and recalling the second equality of Eq. (3.22) the coefficients $a_{\ell m}$ are determined to be

$$a_{\ell m} = \frac{4\pi}{(2\pi)^{3/2}} (-i)^{\ell} \int d^3k \, Y_{\ell m}^*(\hat{k}) \Delta_{{\rm I}\ell}(\vec{k},\tau). \qquad (3.25)$$

The two-point temperature correlation function on the sky between two directions conventionally denoted by $\hat{n}_1$ and $\hat{n}_2$, can be written as

$$C(\vartheta) = \langle \Delta_{\rm I}(\hat{n}_1,\tau_0) \Delta_{\rm I}(\hat{n}_2,\tau_0) \rangle, \qquad (3.26)$$

where $C(\vartheta)$ does not depend on the azimuthal angle because of isotropy of the background space-time and where the angle brackets denote a theoretical ensemble average. Since the background space-time is isotropic, the ensemble average of the $a_{\ell m}$ will only depend upon ℓ, not upon m, i.e.

$$\langle a_{\ell m} a_{\ell' m'}^* \rangle = C_{\ell} \delta_{\ell \ell'} \delta_{mm'}, \qquad (3.27)$$

---

[d]In what follows the subscript γ will be dropped from the temperature to match with the conventions that customarily employed.

[e]The dependence of the temperature fluctuations upon the conformal or cosmic time is immaterial for the present calculation. However, it is better to think in terms of the conformal time coordinate since, as we shall see from chapter 6, the treatment of cosmological inhomogeneities simplifies if the conformal time parametrization is consistently used.

where C_ℓ is the angular power spectrum. Thus, the relation (3.27) implies

$$C(\vartheta) = \langle \Delta_I(\hat{n}_1, \tau_0) \Delta_I(\hat{n}_2, \tau_0) \rangle \equiv \frac{1}{4\pi} \sum_\ell (2\ell+1) C_\ell P_\ell(\hat{n}_1 \cdot \hat{n}_2). \quad (3.28)$$

Notice that in Fig. 1.3 the quantity $C_\ell \ell(\ell+1)/(2\pi)$ is directly plotted: as it follows from the approximate equality

$$\sum_\ell \frac{2\ell+1}{4\pi} C_\ell \simeq \int \frac{\ell(\ell+1)}{2\pi} C_\ell d\ln \ell, \quad (3.29)$$

$C_\ell \ell(\ell+1)/(2\pi)$ is roughly the power per logarithmic interval of ℓ. In Fig. 1.3 the angular power spectrum is measured in $(\mu K)^2$. This is simply because instead of discussing Δ_I (which measures the relative temperature fluctuation) one can equally reason in terms of $\Delta T = T_{\gamma 0} \Delta_I$, i.e. the absolute temperature fluctuation. In what follows (and, in particular, in chapters 7, 8 and 9) Δ_I will denote the temperature fluctuations which can be either expressed in absolute terms (as in Figs. 1.3 and 3.3) or in relative terms (i.e. in dimensionless units). Similar quantities can be defined for other observables such as, for instance, the degree of polarization. In Fig. 3.3 the

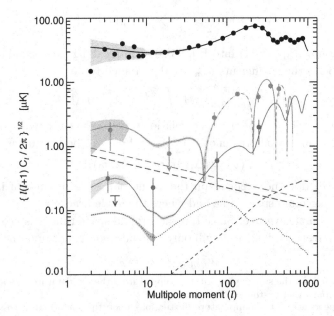

Fig. 3.3 From top to bottom the spectra for the TT, TE and EE correlations. The dashed lines indicate foregrounds of various nature. This figure is adapted from [40]. Reproduced by permission of the AAS.

angular power spectra are reported for (from top to bottom): the temperature autocorrelations (i.e. the quantity we just discussed), the temperature polarization cross-correlation (often indicated as TE spectrum), the polarization autocorrelation (often indicated as EE spectrum).

3.5 The singularity problem

The singularity problem is probably the most serious fundamental drawback of the SCM. While the other problems are certainly very important and manifest diverse logical inconsistencies of the SCM, the singularity problem is fundamental since it is related to the structure of the underlying theory of gravitation, i.e. General Relativity. In the SCM as $t \to 0$

$$\rho_{\text{tot}} \simeq \frac{1}{t^2}, \qquad H^2 \simeq \frac{1}{t^2} \qquad (3.30)$$

and

$$R^2 \simeq R_{\mu\nu} R^{\mu\nu} \simeq R_{\mu\nu\alpha\beta} R^{\mu\nu\alpha\beta} \simeq \frac{1}{t^4}. \qquad (3.31)$$

Thus, in the limit $t \to 0$ the energy density diverges and also the relevant curvature invariants diverge. The Weyl invariant is automatically vanishing since the geometry is isotropic. The singularity problem does not only involve the regularity of the curvature invariants but also the possible completeness (or incompleteness) of causal (i.e. either time-like or null) geodesics. By the Hawking-Penrose theorems [111] the past-incompleteness of causal geodesics is just another diagnostic of a singular space-time.

As already mentioned the singularity problem is deeply rooted in the adoption of General Relativity as underlying gravitational theory. In recent years, in the context of string-inspired cosmological scenarios (see [105] and references therein) a lot of work has been done to see if cosmological singularities can be avoided (or, even more modestly, addressed) in gravity theories that, at early time are different from General Relativity. While the conclusion of these endevours is still far, it is certainly plausible that the ability of string theory in dealing with gravitational interactions can shed some light on the cosmological singularities (and on their possible avoidance). Two key features emerge when string theory is studied in a cosmological framework [105]. The first feature is that string theory demands the existence of a fundamental length-scale (the string scale which 10 or 100 times larger than the Planck length). This occurrence seems to point towards the existence of a maximal curvature scale (and of a maximal energy density) which is the remnant of the general relativistic singularity. While the resolution of cosmological singularities in string theory is still an open problem, there certainly exist amusing (toy) models where the singularity are indeed resolved. The second key feature of string cosmological scenarios is represented by the novelty that gauge couplings are effectively dynamical. This phenomenon has no counterpart in the standard general relativistic treatment but will not be discussed here.

Chapter 4

SCM and Beyond

It is interesting to see how the conceptual problems treated in chapter 3 can be reduced or, at least, partially relaxed in some conventional scenarios which can complement the SCM. In spite of the fact that some of these scenarios (like the inflationary scenario) can cope with the technical problems of the SCM (such as the flatness or the horizon problems) none of these models are able to cope with the deepest of all the problems of the SCM, i.e. the singularity problem. To this statement it should be added that the inflationary solution of the entropy problem relies on the possible decay of inflaton into massless particles with the hope that such a process may produce a sufficiently high reheating temperature. For an introduction to the inflationary paradigm Refs. [121, 122] can be usefully consulted (see also [123–126] for some specific inflationary scenarios). For reasons of space, it will not be possible to treat some of the unconventional approaches to inflation that are rather interesting especially in the light of their connections with string theory. Among them, the pre-big bang scenario (developed in the last fifteen years) represents a rather intriguing option. We refer the reader to the original papers and to some very comprehensive review articles [106] (see also [107, 108]).

4.1 The horizon and the flatness problems

The horizon problem in the SCM has to do with the fact that there exist a particle horizon $d_p(t) \simeq H^{-1}(t)$. Thus, as we go forward in time (and for $t > 0$) the particle horizon evolves faster than the scale factor which evolves, in the SCM, as t^α with $0 < \alpha < 1$. This occurrence also implies that at the moment when the initial conditions are ideally set, our observable Hubble volume consisted of a huge amount of causally disconnected domains (see, for instance, Eqs. (3.10) and (3.12)). A possible way out of this problem is to consider the completion of the SCM by means of a phase where not a particle but an *event* horizon exist. Consider, for instance, a scale factor with power-law behaviour going as

$$a(t) \simeq t^\beta, \qquad \beta > 1, \qquad t > 0, \qquad (4.1)$$

and describing a phase of accelerated expansion (i.e. $\dot{a} > 0$, $\ddot{a} > 0$). The particle and event horizons are given, respectively, by

$$d_\mathrm{p}(t) = \frac{1}{1-\beta}\left[t - t_\mathrm{min}\left(\frac{t}{t_\mathrm{min}}\right)^\beta\right], \qquad (4.2)$$

$$d_\mathrm{e}(t) = \frac{1}{1-\beta}\left[t_\mathrm{max}\left(\frac{t}{t_\mathrm{max}}\right)^\beta - t\right]. \qquad (4.3)$$

From Eqs. (4.2) and (4.3) it immediately follows that the particle horizon does not exist while the event horizon is finite:

$$d_\mathrm{e}(t) \simeq \frac{\beta}{\beta-1} H^{-1}(t). \qquad (4.4)$$

Equation (4.4) follows from Eq. (4.3) in the limit $t_\mathrm{max} \to +\infty$ while in the limit $t_\mathrm{min} \to 0$, $d_\mathrm{p}(t)$ diverges. Similar conclusions follow in the case when the phase of accelerated expansion is parametrized in terms of the (expanding) branch of four-dimensional de Sitter space-time, namely

$$a(t) \simeq e^{H_\mathrm{i} t}, \qquad H_\mathrm{i} > 0. \qquad (4.5)$$

In this case, the particle and event horizons are, respectively,

$$d_\mathrm{p}(t) = H_\mathrm{i}^{-1}\left[e^{H_\mathrm{i}(t-t_\mathrm{min})} - 1\right], \qquad (4.6)$$

$$d_\mathrm{e}(t) = H_\mathrm{i}^{-1}\left[1 - e^{H_\mathrm{i}(t-t_\mathrm{max})}\right]. \qquad (4.7)$$

According to Eq. (4.5) the cosmic time coordinate is allowed to run from $t_\mathrm{min} \to -\infty$ up to $t_\mathrm{max} \to +\infty$. Consequently, for $t_\mathrm{min} \to -\infty$ (at fixed t) the particle horizon will diverge and the typical size of causally connected regions at time t will scale as

$$L_\mathrm{i}(t) \simeq H_\mathrm{i}^{-1} \frac{a(t)}{a(t_\mathrm{min})}. \qquad (4.8)$$

So while in the SCM the particle horizon increases faster than the scale factor, the typical size of causally connected regions scales exactly *as the scale factor*. In the limit $t_{\max} \to \infty$ the event horizon exist and it is given, from Eq. (4.7), by

$$d_e(t) \simeq H_i^{-1}, \qquad (4.9)$$

implying that in the case of de Sitter dynamics the event horizon is constant. Of course, as it will be later pointed out, de Sitter dynamics cannot be exact (see chapter 5). In this case, customarily, we talk about a quasi-de Sitter stage of expansion where H_i is just approximately constant and, more precisely, slightly decreasing. In the last part of chapter 5 it will be explicitly shown how to deal with the case of quasi-de Sitter evolution.

To summarize, the logic to address the horizon problem is then to suppose (or presume) that at some typical time t_i an event horizon is formed with typical size H_i^{-1}. See also Fig. 4.1 for a pictorial illustration. Furthermore, since we are working in General Relativity, we shall also demand that $H_i < M_P$. Now *if* the Universe is sufficiently homogeneous inside the created event horizon, it will remain (approximately) homogeneous also later on, by definition of event horizon. In other words, if, inside the event horizon, $\delta\rho/\rho$ is sufficiently small, we can think of fitting inside a single event horizon at t_i the whole observable Universe. In practice, this condition translates into a typical size of H_i which should be such that $H_i \lesssim 10^{-5} M_P$ or, in equivalent terms, an event horizon that is sufficiently large with respect to the Planck length, i.e. $H_i^{-1} \gg \ell_P$.

To fit the whole observable Universe inside the newly formed event horizon at the onset of inflation, the de Sitter (or quasi-de Sitter) phase must last for a sufficiently large amount of time. In equivalent terms it is mandatory that the scale factor grows of a sufficient amount. Since the growth of the scale factor is exponential (or quasi-exponential) it is common practice to quantify the growth of the scale factor in terms of the number of e-folds, denoted by N and defined as

$$e^N = \frac{a(t_f)}{a(t_i)} \equiv \frac{a_f}{a_i}, \qquad N = \ln\left(\frac{a_f}{a_i}\right). \qquad (4.10)$$

To estimate the condition required on the number of e-folds N we can demand that the whole (present) Hubble volume (blueshifted at the epoch t_i when the event horizon is formed) is smaller than H_i^{-1}. In fully equivalent terms we can demand that H_i^{-1} redshifted at the present epoch is larger than (or comparable with) the present Hubble radius. By following this second path we are led to require that

$$H_i^{-1}\left(\frac{a_i}{a_f}\right)_{dS}\left(\frac{a_f}{a_r}\right)_{reh}\left(\frac{a_r}{a_{eq}}\right)_{rad}\left(\frac{a_{eq}}{a_0}\right)_{mat} \geq H_0^{-1}. \qquad (4.11)$$

In Eq. (4.11) the subscripts appearing in each round bracket indicate the specific phase during which the given amount of redshift is computed. Between the end of the de Sitter stage and the beginning of the radiation-dominated phase there should be an intermediate phase usually called reheating (or pre-heating) where the Universe makes a transition from accelerated to decelerated expansion. The rationale for the existence of this phase stems from the observation that, during the de Sitter phase, any radiation present at t_i is rapidly diluted and becomes soon negligible since, as we saw, ρ_R scales as a^{-4}. In equivalent terms we can easily appreciate that the temperature, as well as the entropy density (possibly present at t_i) decay exponentially (or quasi-exponentially) in cosmic time. Consequently, as soon as the accelerated expansion proceeds, the Universe approaches a configuration where the temperature and the entropy density are exponentially vanishing. There is therefore the need of reheating the Universe at the end of inflation. We can introduce the typical curvature scale of reheating, i.e. $H_{\rm reh}$. Consequently Eq. (4.11) implies that

$$N \geq 67.95 + \frac{1}{3}\ln\xi_1 + \frac{1}{6}\ln\xi_2 - \ln\left(\frac{h_0}{0.7}\right) + \frac{1}{4}\ln\left(\frac{h_0^2 \Omega_{R0}}{4.15 \times 10^{-5}}\right), \quad (4.12)$$

$$\xi_1 = \frac{H_i}{M_P}, \qquad \xi_2 = \frac{H_{\rm reh}}{M_P}. \quad (4.13)$$

To estimate the minimal number of e-folds N we can rely on the sudden reheating approximation where, basically, $a_f \simeq a_r$. Consequently, under this approximation we can write Eq. (4.11) as

$$e^N \geq \left(\frac{H_i}{H_0}\right)\left(\frac{H_{\rm eq}}{H_i}\right)^{1/2}(z_{\rm eq} + 1)^{-1}. \quad (4.14)$$

which can also be expressed, by taking the natural logarithm, as

$$N \geq 62.2 + \frac{1}{2}\ln\left(\frac{\xi}{10^{-5}}\right) - \ln\left(\frac{h_0}{0.7}\right) + \frac{1}{4}\ln\left(\frac{h_0^2 \Omega_{R0}}{4.15\ 10^{-5}}\right). \quad (4.15)$$

Equation (4.15) can also be obtained directly from Eq. (4.12) by setting[a] $\xi_1 = \xi_2 = \xi$. In Eqs. (4.12) and (4.15) the following relations have been used:

$$H_{\rm eq} = \sqrt{2\,\Omega_{M0}}\,H_0\left(\frac{a_0}{a_{\rm eq}}\right)^{3/2} \equiv 1.65 \times 10^{-56}\left(\frac{h_0^2 \Omega_{M0}}{0.134}\right)^2 M_P, \quad (4.16)$$

$$H_0 = 100\,h_0\frac{\rm km}{\rm sec}{\rm Mpc}^{-1} \equiv 1.22 \times 10^{-61}\,M_P \quad (4.17)$$

---

[a]In chapter 2 (as well as in chapter 12) the variable ξ denotes the bulk viscosity coefficient. However, since the two variables never appear in contiguous discussions, a possible clash of notations is avoided.

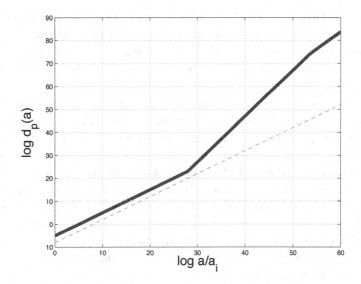

Fig. 4.1 The evolution of the particle horizon in the inflationary case (full line) is reported for a model Universe which passes from a de Sitter stage of expansion to a radiation-dominated stage which evolves, in turn, into a matter-dominated epoch. The evolution of a typical scale smaller than the event horizon is also reported. The first branch (where $d_p(a)$ evolves as the scale factor) illustrates the evolution of the typical size of causally connected regions during inflation. This quantity is formally divergent for $t \to -\infty$.

Equation (4.16) comes directly from Eq. (2.56) by requiring $\rho_M(t_{eq}) = \rho_R(t_{eq})$. Equation (4.17) is just the same relation already derived in Eq. (1.21).

During a quasi-de Sitter stage of expansion, quantum-mechanical fluctuations of the inflaton will be amplified to observable values and their amplitude is exactly controlled by ξ. To ensure that the amplified quantum-mechanical inhomogeneities will match the observed values of the angular power spectrum of temperature inhomogeneities we have to require $\xi \simeq 10^{-5}$ which demands[b] that $N \geq 63$.

The same hierarchy of scales required to address the horizon problem, also relaxes the flatness problem. The flatness problem arises, in the SCM, from the observation that the contribution of the spatial curvature increases sharply, during the radiation and matter-dominated epochs. This observa-

---

[b] We shall be more specific on the more quantitative aspects of this estimate at the end of the present section.

tion entails that if $\Omega_t \simeq \mathcal{O}(1)$ today, Ω_t had also to be fine-tuned to 1 at the onset of the radiation-dominated evolution but with much greater precision. So, if today $\Omega_t \simeq 1$ with an experimental error of, for instance, 0.1, at the Planck scale Ω_t had to be fine-tuned to 1 with accuracy of, roughly, $\mathcal{O}(10^{-60})$.

If the ordinary radiation-dominated evolution is preceded by a de Sitter (or quasi-de Sitter) phase of expansion the spatial curvature will be exponentially (or quasi-exponentially) suppressed with respect to the Hubble curvature H_i^2 which is constant (or slightly decreasing). Thus, if the exponential growth of the scale factor will last for a sufficient number of e-folds, the spatial curvature at the onset of the radiation dominated phase will be sufficiently suppressed to allow for a subsequent growth of $k/(aH)^2$ during the radiation and matter-dominated epochs. The same number of e-folds required to address the horizon problem also guarantees that the spatial curvature will be sufficiently suppressed during the phase of exponential expansion. In fact, while today

$$\Omega_t(t_0) - 1 = \frac{k}{a_0^2 H_0^2}, \tag{4.18}$$

at the onset of the de Sitter phase

$$\Omega_t(t_i) - 1 = \frac{k}{a_i^2 H_i^2}. \tag{4.19}$$

Dividing Eq. (4.18) by Eq. (4.19) we can also obtain rather easily

$$\sqrt{|\Omega_{\rm tot}(t_0) - 1|} = \frac{a_i H_i}{a_0 H_0} \sqrt{|\Omega_{\rm tot}(t_i) - 1|}. \tag{4.20}$$

Now, from Eq. (4.20) it is clear that if $\Omega_{\rm tot}(t_0)$ is tuned to 1 with the precision of, say, 10 %, the pre-factor appearing at the right hand side of Eq. (4.20) must be of the order 0.1 if $|\Omega_{\rm tot}(t_i) - 1|$ was of order 1 at the onset of the de Sitter phase. Thus, more generally, we are led to require that

$$\frac{a_i H_i}{a_0 H_0} < 1, \tag{4.21}$$

which becomes, after making explicit the redshift contribution, exactly Eq. (4.11). We then discover that if $N \geq 63$ the spatial curvature at the end of inflation will be small enough to guarantee that the successive growth (during radiation and matter) will not cause (today) $|\Omega_{\rm tot}(t_0) - 1|$ to be of order 1 (or even larger). In Fig. 4.2 the evolution of $|\Omega_t - 1|$ is reported as a function of the logarithm (to base 10) of the scale factor for the situation where the Universe inflates during 69 e-folds.

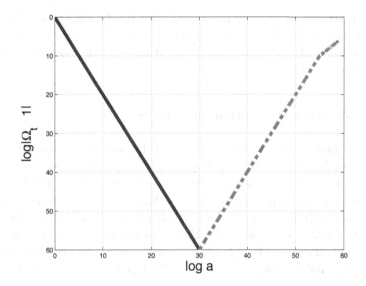

Fig. 4.2 The evolution of the logarithm (to base 10) of $|\Omega_t - 1|$ as a function of the $\log a$ for an inflationary phase where $N \simeq 69$ and in the sudden reheating approximation. The full line denotes the evolution during inflation while the dashed and the dot-dashed lines denote the approximate evolution during radiation and matter.

As it will be discussed in chapter 5 the inflationary dynamics can be modeled in terms of one (or more) minimally (or non-minimally) coupled light scalar degrees of freedom. Here the word light refers to the typical scale of the problem, i.e. H_i so that the mass of the scalar field should be small in units of H_i. So suppose that, at t_i, there is a scalar field which has some typical inhomogeneities over different wavelengths. It is clear that the most generic evolution of such a system represents a tough numerical task under general circumstances. By this we mean that it is not said that the most generic thing a scalar field does will be to inflate. However one can also guess that if the scalar field φ is sufficiently homogeneous over a region H_i^{-1} one of the possibilities will be inflation provided the kinetic energy of the scalar field is sufficiently small in comparison with its potential energy.[c] These observations lead to the following requirements:

$$\frac{|\nabla^2 \varphi|}{a_i^2} \ll \frac{\partial V}{\partial \varphi}, \quad \dot{\varphi}^2 \ll V, \quad \frac{(\nabla \varphi)^2}{a_i^2} \ll \dot{\varphi}^2, \quad (4.22)$$

---

[c]This condition can be, indeed, relaxed by noticing that, in the absence of potential, the (homogeneous) evolution of the inflaton is given by $\ddot{\varphi} + 3H\dot{\varphi} \simeq 0$. This relation (see section 5, Eq. (5.58)) implies that $\dot{\varphi}^2$ scales as a^{-6} and may become, eventually, subleading in comparison with the potential energy.

at the time t_i and over a typical region H_i^{-1}. If the duration of inflation lasts just for 63 (or 65) e-folds it can happen that some initial spatial gradients (i.e. some initial spatial curvature) will still cause inhomogeneities inside the present Hubble volume.

4.2 Classical and quantum fluctuations

Classical and quantum fluctuations, in inflationary cosmology, have similarities but also crucial differences. While classical fluctuations are given, once and forever, on a given space-like hypersurface, quantum fluctuations keep on reappearing all the time thanks to the zero-point energy of various quantum fields that are potentially present in de Sitter space-time. If the accelerated phase lasts just the minimal amount of e-folds required to solve the problem of the Standard Cosmological Model classical fluctuations can definitely have an observational and physical relevance. Suppose, indeed, that classical fluctuations are present prior to the onset of inflation and suppose that their typical wavelength was of the order of H_i^{-1}. Then we can say that their wavelength today is

$$\lambda(t_0) = H_i^{-1} \frac{a_0}{a_i}. \tag{4.23}$$

But a_0/a_i is just the redshift factor required to fit the present Hubble patch inside the event horizon of our de Sitter phase. From Eq. (4.23) it is clear that if $H_i^{-1} = 10^5 \ell_P \simeq 10^{-28}$ cm, then $\lambda(t_0)$ will be comparable with H_0^{-1} if the number of e-folds is just close to minimal.

If the inflationary phase lasts much more than the minimal amount of e-folds the classical fluctuations (possibly present at the onset of inflation) will be, in the future, redshifted to larger and larger length-scales (even much larger than the present Hubble pacth). In the future these wavelengths will be, in some sense, accessible since the Hubble patch increases as time goes by. Therefore, if the inflationary phase lasts much more than the minimal amount of e-folds, the only fluctuations potentially accessible through satellite and terrestrial observations will be quantum-mechanically generated fluctuations which can be parametrically amplified under specific conditions to be discussed in chapter 6. In Fig. 4.3 the evolution of the Hubble radius (in Planck units) is reported as a function of the logarithm of the scale factor. In Fig. 4.3 inflation lasts for the minimal amount (i.e. $N = 63$) while in Fig. 4.4 the duration of inflation is non-minimal (i.e. $N = 85 \gg 63$). From Figs. 4.3 and 4.4 the difference in the behaviour

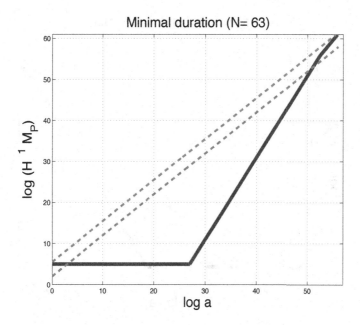

Fig. 4.3 The evolution of the Hubble radius in the case when the duration of inflation is minimal. With the dashed lines we also illustrate the evolution of different typical wavelengths.

of classical and quantum fluctuations is evident. The dashed lines represent the wavelength of a given perturbation (either classical or quantum mechanical). If the duration of inflation is minimal, it is plausible (see Fig. 4.3) that a classical fluctuation crosses the Hubble radius the second time around the epoch of matter-radiation equality. This means that the classical fluctuation may have an observational impact. If, on the contrary, inflation lasts for much more than 63 e-folds (85, in Fig. 4.4) there is no chance that a classical fluctuation present at the onset of inflation will cross the Hubble radius around the epoch of matter-radiation equality. In this second case, the only fluctuations that will be eventually relevant will be the quantum mechanical ones. Summarizing the discussion conducted so far, we can say that there are two physically different situations:

- if the duration of the inflationary phase lasts for the minimal amount of e-folds (i.e. $N \simeq 63$) then it is plausible that some (observable?) relics of a pre-inflationary dynamics can be eventually detected in CMB observations;

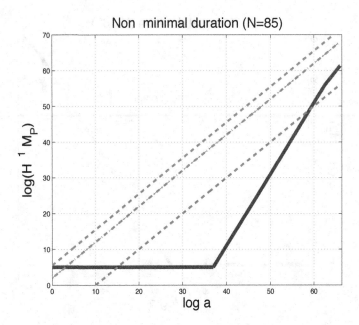

Fig. 4.4 The same quantity illustrated in Fig. 4.3 but for a case when the duration of inflation is non-minimal.

- if, on the contrary, the duration of inflation greatly exceeds the minimal duration we can expect that any memory of the pre-inflationary phase will be lost thanks to the efficiency of exponential expansion.

We conclude this section by an important concept which will be relevant both on a physical and and on a technical ground. It is the concept of *horizon crossing*. This concept can be introduced by looking at Figs. 4.3 and 4.4. In Figs. 4.3 and 4.4 the dashed lines represent, as already mentioned, the evolution of the physical wavelengths while the full line sketches the approximate evolution of the Hubble radius in the concordance model. It is important to stress that

- in the concordance model (i.e. the ΛCDM model where inflation is driven by a single scalar field) the reheating is assumed to be almost sudden;
- between the end of inflation and the onset of radiation there are no exotic phases as, for instance, implied by some models of

quintessence where, after inflation, a relatively long stiff phase[d] can take place.

The moment where the physical wavelength crosses the Hubble radius is called, in the jargon, horizon crossing. We say that this is the jargon (as opposed to a more accurate terminology) because the particle horizon does not exist, strictly speaking, during inflation. Therefore, by definition the horizon crossing is the moment at which

$$\lambda(t) \simeq 2\pi H^{-1}(t), \qquad \frac{k}{\mathcal{H}} \equiv \frac{k}{aH} \simeq 1. \qquad (4.24)$$

The second relation of Eq. (4.24) is written in terms of the comoving wavenumber k. Since $\mathcal{H} \simeq 1/\tau$ we will also have that

- when $k\tau < 1$ the corresponding physical wavelength will be larger than the Hubble radius at the time τ;
- when $k\tau > 1$ the corresponding physical wavelength will be smaller than the Hubble radius at the time τ.

The previous statements can be also dubbed, according to the current terminology, by saying that the physical wavelength is either larger or smaller than the Hubble radius. According to Figs. 4.3 and 4.4 the physical wavelength crosses the Hubble radius twice. It crosses the Hubble radius the first time at early epoch, during inflation, i.e. when the Hubble radius is roughly constant. The second crossing can take place at various epochs. In the minimal concordance model the second crossing (sometimes also called horizon reentry) can take place either during the radiation-dominated phase or during the matter-dominated phase. The wavelengths that become larger than the Hubble radius towards the end of inflation will reenter first and they will make the second crossing during radiation. The wavelengths that left the Hubble radius around 60 e-folds before the end of inflation will reenter around recombination.

In practical problems it is often required to estimate the so-called maximal number of e-folds, i.e. the number of e-folds corresponding to the first horizon crossing of the physical wavelength which is, today, comparable with the present extension of the Hubble radius. This number of e-folds, which will be denoted as $N_{\max}$, defines the maximal number of e-folds which are today accessible by our observations. This does not mean that the effective duration of inflation must coincide with $N_{\max}$. On the

---

[d]A stiff phase can be driven, for instance, by the kinetic energy of the quintessence field which acts effectively as a barotropic fluid whose sound speed equals the speed of light.

contrary there are reasons to believe that inflation should last much more than $N_{\max}$. However, in this case, the physical wavelengths that crossed the horizon at a given $N_1 > N_{\max}$ will reenter in the future.

In the framework of a specific thermodynamic history of the Universe it is possible to compute $N_{\max}$ with reasonable accuracy. In what follows we will compute $N_{\max}$ in the context of the minimal ΛCDM model and under the assumption of sudden reheating. The details of the estimate imply the knowledge of the calculation of the primordial spectrum of curvature perturbations which will be explicitly derived in chapter 10. By using the technique already mentioned in this chapter it turns out that the maximal number of e-folds accessible to our observations coincides with the minimal number of e-folds required to relax the problems of the Standard Cosmological Model, so, we can write

$$N_{\max} = 67.95 + \frac{1}{2}\ln\xi - \ln\left(\frac{h_0}{0.7}\right) + \frac{1}{4}\ln\left(\frac{h_0^2 \Omega_{R0}}{4.15 \times 10^{-5}}\right) \quad (4.25)$$

which is essentially Eq. (4.12) with $\xi_1 = \xi_2 = \xi$ as implied by the sudden reheating approximation. By now assuming that inflation is driven by a single scalar degree of freedom (see chapter 5) we can also write

$$\xi^2 = \frac{H^2}{M_P^2} = \frac{8\pi}{3}\frac{V}{M_P^4}, \quad (4.26)$$

where V denotes the potential of the inflaton, i.e. the scalar degree of freedom which drives inflation. It turns out also that the amplitude of the spectrum of curvature perturbations generated by the (quantum) inhomogeneities of the inflaton and of the geometry is related to V through a small parameter conventionally denoted by ϵ (see Eq. (10.81) of chapter 10):

$$\mathcal{P}_\mathcal{R} \simeq \frac{8}{3M_P^4}\left(\frac{V}{\epsilon}\right)_{k\simeq Ha} \quad (4.27)$$

where $\mathcal{P}_\mathcal{R}$ denotes the power spectrum of curvature perturbations and it is understood that the expression at the right hand side of Eq. (4.27) is evaluated at horizon crossing, i.e. for $k \simeq Ha$. The parameter ϵ is an example of slow-roll parameter. In particular ϵ controls the rate of variation of H, i.e. $\epsilon = -\dot{H}/H^2$. During inflation, as discussed in this chapter, $\epsilon \ll 1$. When $\epsilon \simeq 1$ the inflationary evolution comes to an end. Using Eq. (4.27) to eliminate V from Eq. (4.26), Eq. (4.25) can be written as

$$N_{\max} = 67.951 + \frac{1}{4}\ln(\pi\epsilon\mathcal{P}_\mathcal{R}), \quad (4.28)$$

where the other parameters appearing in Eq. (4.25) have been fixed at their typical values. Now, as we will learn in the following chapters the curvature perturbations generated during inflation will affect the temperature fluctuations of the CMB. Therefore, within the model under consideration, it is possible to estimate $\mathcal{P}_\mathcal{R}$ by assuming that all the anisotropies observed in the CMB are due to the scalar modes of the geometry. This assumption implies that $\mathcal{P}_\mathcal{R} \simeq 2.6 \times 10^{-9}$. Therefore, from Eq. (4.28), $N_{\max}$ can be estimated as:

$$N_{\max} \simeq 63.29 + \frac{1}{4}\ln \epsilon. \tag{4.29}$$

Again the value of the slow-roll parameter can be experimentally bound and we may take $\epsilon \leq 0.05$. For $\epsilon \simeq 0.01$ we will have, for instance, $N_{\max} \simeq 62.13$.

To summarize, by pretending to know which is the thermal history of the Universe the detail of a specific inflationary model can be used to compute $N_{\max}$ with reasonable accuracy. We would be tempted to say that it is highly desirable that $N \simeq N_{\max}$. In other words it might appear simpler, on the theoretical ground, to select the situation where the total number of e-folds coincides with maximal number of e-folds accessible to our extended experience. This is certainly possible. However, within this choice, it remains to be explained why the initial conditions are taken to be of quantum mechanical origin. In fact, if inflation lasts around 62 e-folds, as explained before, classical fluctuations do not have time to be diluted by the accelerated expansion. So in this respect, if $N \simeq N_{\max}$, there is no way of affirming that the initial conditions of the fluctuation observed in the CMB come from quantum mechanics rather than from a classical inhomogeneity present on a given space-like hypersurface at the onset of inflation. For further details on the discussion of $N_{\max}$ the interested reader may consult [127, 128].

4.3 The entropy problem

The entropy of the CMB refers to the entropy of the *matter sources* of the geometry. There could be, in principle, also a truly gravitational entropy associated with the gravitational field itself. The way one can attribute an entropy to the gravitational field is the subject of debate. This entropy, for instance, could be connected to the possibility of activating new degrees of freedom and can be measured, for instance according to Penrose, by the Weyl tensor [134] (see also [135]). So, for the moment, let us focus our attention on the entropy of the sources and let us recall that, today, this entropy seats in photons and it is given, in natural units, by $S_\gamma \simeq 10^{88}$.

Since the evolution of the sources is characterized, in the SCM, by the covariant conservation of the total energy-momentum, the total entropy of the sources will also be conserved. We are therefore in the situation where the entropy at the end of the inflationary phase must be of the order of 10^{88}. During inflation, at the same time, there is no reason why the evolution of the sources must not follow from the covariant conservation of the energy-momentum tensor. If this is the case, the entropy of our observable Universe must be produced, somehow, at the end of inflation. This is, indeed, what we can call the standard lore for the problem of entropy generation within the inflationary proposal. In the standard lore entropy (as well as radiation) is generated at the end of inflation during a phase called reheating.

During reheating the degree of freedom that drives inflation (the inflaton, in single field inflationary models) decays and this process is non-adiabatic. What was the entropy at the beginning of inflation? The answer to this question clearly depends upon the specific inflationary dynamics and, in particular, upon the way inflation starts. Let us try, however, to get a rather general (intuitive) picture of the problem. Suppose that, at some time t_i, an event horizon forms with typical size H_i^{-1}. The source of this dynamics could be, in principle, a cosmological constant or, more realistically, the (almost) constant potential energy of a scalar degree of freedom. In spite of the nature of the source, it can be always argued that its energy density is safely estimated by $H_i^2 M_P^2$. When the event horizon forms, massless particles can be around. Suppose, for a moment, that all the massless species are in thermal equilibrium at a common temperature T_i. Thus, their energy and entropy densities will be estimated, respectively, by T_i^4 and by T_i^3. The total entropy at t_i contained inside the newly formed event horizon can then be quantified as

$$S_i \simeq \left(\frac{T_i}{H_i}\right)^3. \tag{4.30}$$

Yet to be discussed (but already mentioned) the quantum fluctuations amplified in the course of the inflationary evolution force us to a value of $H_i/M_P \simeq 10^{-5}$. On the other hand, we have for inflation to start,

$$H_i^2 M_P^2 \gg g_\rho T_i^4, \tag{4.31}$$

where g_ρ is the effective number of relativistic degrees of freedom at t_i (see Appendix B, Eq. (B.35)). From Eq. (4.31) it is easy to deduce that

$$S_i \simeq \left(\frac{T_i}{H_i}\right)^3, \qquad \frac{T_i}{H_i} \ll g_\rho^{1/4} \left(\frac{H_i}{M_P}\right)^{-1/2}. \tag{4.32}$$

Recalling that $H_i \simeq 10^{-5} M_P$ it is then plausible, under the assumptions mentioned above, that the entropy at the onset of inflation is of order one and, anyway, much smaller than the present entropy sitting in the CMB photons.

During the development of inflation, if there is no significant energy and momentum exchange between the inflaton field and the photons, the temperature, the concentration of photons and the entropy density will all redshift exponentially so that the background will be driven towards a very flat and cold stage[e] where, however, the total entropy is still of order one thanks to the adiabaticity of the evolution. At some point, however, the inflaton will start decaying and massless particles will be produced. Let us now try to estimate the entropy produced in this process. It will be, in general terms, of the order of

$$S_{\rm rh} = \frac{4}{3}\pi \left(\frac{T_{\rm rh}}{H_{\rm rh}}\right)^3, \qquad (4.33)$$

where $T_{\rm rh}$ is the reheating temperature and where $H_{\rm rh}^{-1}$ is the Hubble radius at the reheating. Let us assume, to begin with, that the reheating is instantaneous and perfectly efficient. This amounts to suppose that *all* the energy density of the inflaton is efficiently transformed into radiation at $t_{\rm rh}$. Recalling now that $H_{\rm rh}^{-1}$ can be usefully connected with H_i^{-1} as

$$H_{\rm rh}^{-1} \simeq H_i^{-1}\left(\frac{a_f}{a_i}\right), \qquad (4.34)$$

we will have that the effective number of e-folds should be

$$N \geq 65.9 + \frac{1}{3}\ln\left(\frac{\xi}{10^{-5}}\right) - \ln\left(\frac{T_{\rm rh}}{10^{15}\,{\rm GeV}}\right), \qquad (4.35)$$

where we assumed that $H_{\rm rh} \simeq T_{\rm rh}^2/M_P$. So, if the inflationary phase is sufficiently long, the Hubble radius at reheating will be large enough to match the observed value of the entropy of the sources. There are at least three puzzling features, among others, with the argument we just presented:

- the amount of entropy crucially depends upon the temperature of the reheating which depends, in turn, upon the coupling of the inflaton to the degrees of freedom of the particle physics model;

---

[e]In connection with this point it is interesting to mention the possibility of warm inflation. Since we do not discuss this topic we refer the reader to some of the original papers [129–132]. The possibility of having a successful warm inflation has been questioned in [133].

- the entropy should not exceed the observed one and, consequently, the solution of the entropy problem seems to imply a lower bound on the number of inflationary e-folds;
- the reheating may not be instantaneous and this will entail the possibility that the number of inflationary e-folds may be smaller since, during reheating, the Hubble radius may grow.

In some models of reheating the decay of the inflaton occurs through a phase where the inflaton field oscillates around the minimum of its potential. In this phase of coherent oscillations the Universe becomes, effectively, dominated by radiation and $a(t) \simeq t^{2/3}$. Consequently, the radiation of $H_{\rm rh}^{-1}$ to $H_{\rm i}^{-1}$ will be given by a different equation and, more specifically, by

$$H_{\rm rh}^{-1} = e^N \left(\frac{H_{\rm i}}{H_{\rm rh}} \right)^{2/3} H_{\rm i}^{-1} \qquad (4.36)$$

where the power 2/3 in the last bracket accounts for the evolution during the reheating phase. In this case, requiring that the total entropy exceeds a bit the observed entropy we will obtain the following condition, on N:

$$N \geq 60.1 + \frac{1}{3} \ln \left(\frac{\xi}{10^{-5}} \right) + \frac{4}{3} \ln \left(\frac{T_{\rm rh}}{10^{15}\,{\rm GeV}} \right). \qquad (4.37)$$

Equation (4.35) gives a lower estimate for the number of inflationary e-folds simply because the Universe was redshifting also in the intermediated (reheating) phase by, roughly, 5 effective e-folds.

According to the presented solution of the entropy problem the initial state of the Universe prior to inflation must have been rather ordered. Let us assume, indeed, the validity of the second law of thermodynamics

$$\dot{S}_{\rm m} + \dot{S}_{\rm Gr} \geq 0, \qquad (4.38)$$

where $S_{\rm Gr}$ denotes, quite generically, the entropy of the gravitational field itself. Equation (4.38) is telling us that as we go back in time the Universe had to be always less and less entropic. The conclusion that the pre-inflationary Universe was rather special seems to clash with the idea that the initial conditions of inflation were somehow chaotic [122]. The idea here is that inflation is realized by means of a scalar degree of freedom (probably a condensate, see chapter 5) initially displaced from the minimum of its own potential. In some regions of space the inflaton will be sufficiently displaced from its minimum and its spatial gradients will be large in comparison with the potential. In some other regions the spatial gradients will be sufficiently small. This picture, here only swiftly described, is really chaotic and it is conceptually difficult to imagine that this chaos could also avoid a large entropy of the pre-inflationary stage. A possible way out of this apparent impasse may be to include consistently the entropy of the gravitational field.

There are proposals on the possible measures of the entropy of the gravitational field. On top of the proposal of Penrose already quoted in this section [134, 135], Davies [136, 137] proposed to associate an entropy to the cosmological backgrounds endowed with an event horizon. In this case $S_{\text{Gr}} \simeq d_{\text{e}}^2 M_{\text{P}}^{-2}$. One can easily imagine models where the entropy of the sources decreases but the total entropy (i.e. sources and gravitational field) does not decrease [138–141]. There is also the possibility of associating an entropy to the process of production of relic gravitons [142, 143] but we shall swiftly get back on this point later on in chapter 6.

We conclude this section by mentioning a formulation of the information problem of the present Universe which is particularly lucid and fits well with the discussion we just presented. It is due to G. Smoot [144]. If we look at the Universe today in a naive perspective we would expect that, since the entropy is an extensive quantity, the present information content of the Universe will be of the order of the Hubble volume in Planck units, i.e. 10^{180}. According to the introduced definition of gravitational entropy this figure is drastically reduced to something of the order of 10^{120}. But if we apply the same notion at the event horizon during inflation we will have $M_{\text{P}}^2/H_{\text{i}}^2 \simeq 10^{11}$. The cosmological parameters are roughly 20 (eventually supplemented by the equations describing the evolution of the geometry and of the radiation field). The information content of our Universe then seems to be compressed. This is another possible facet of what goes under the name of entropy problem.

4.4 The problem of geodesic incompleteness

Inflation does not solve the problem of the initial singularity. This statement can be appreciated by noticing that the expanding de Sitter space-time is not past geodesically complete [145–148]. Such an occurrence is equivalent (both technically and physically) to a singularity. The geodesic incompleteness of a given space-time simply means that causal geodesics cannot be extended indefinitely in the past as a function of the affine parameter that shall be denoted with λ. The causal geodesics are either the time-like or null geodesics. Let us therefore consider, for simplicity, the case

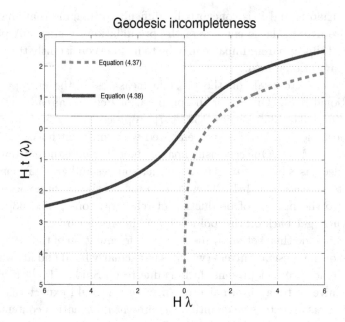

Fig. 4.5 The evolution of null geodesics in the case of Eqs. (4.46) and (4.47)

of null geodesics i.e.

$$dt^2 - a^2(t)d\vec{x}^2 = 0. \qquad (4.39)$$

From the geodesic equation [102]:

$$\frac{d^2 x^\mu}{d\lambda^2} + \Gamma^\mu_{\alpha\beta}\frac{dx^\alpha}{d\lambda}\frac{dx^\beta}{d\lambda} = 0, \qquad (4.40)$$

we immediately obtain the following pair of conditions:

$$\frac{d^2 t}{d\lambda^2} + 2\dot{a}a\left(\frac{d\vec{x}}{d\lambda}\right)^2 = 0, \qquad (4.41)$$

$$\frac{d^2 x^i}{d\lambda^2} + 2\frac{\dot{a}}{a}\frac{dt}{d\lambda}\frac{dx^i}{d\lambda} = 0. \qquad (4.42)$$

Inserting Eq. (4.39) into Eq. (4.41) to eliminate $(d\vec{x}/d\lambda)^2$ we obtain

$$\frac{df}{d\lambda} + \frac{\dot{a}}{a}f^2 = 0, \qquad f(\lambda) = \frac{dt}{d\lambda}. \qquad (4.43)$$

But if we now recall that

$$\frac{df}{d\lambda} = \frac{df}{dt}\frac{dt}{d\lambda} = \frac{df}{dt}f, \qquad (4.44)$$

Eq. (4.43) allows us to express $f(t)$ in terms of the scale factor:

$$f(t) = \frac{dt}{d\lambda} = \frac{1}{a(t)}, \qquad d\lambda = a(t)dt. \tag{4.45}$$

In the case of expanding de Sitter space-time we will have

$$a(t) \simeq e^{Ht}, \qquad t = \frac{1}{H}\ln(H\lambda) \tag{4.46}$$

implying that $t(\lambda) \to -\infty$ for $\lambda \to 0^+$. This means that null geodesics are past-incomplete.

To appreciate what would be a geodesically complete space-time let us consider the following example:

$$a(t) \simeq \cosh Ht, \qquad t = \frac{1}{H}\ln[H\lambda + \sqrt{H^2\lambda^2 + 1}]. \tag{4.47}$$

In the case $\lambda \to 0$, $t(\lambda) \to 0$ and the geodesics are complete. The background discussed in Eq. (4.47) is a solution of Einstein equations but in the presence of positive spatial curvature while, in the present example, we considered, implicitly, a spatially flat background manifold. In the case of pre-big bang models the geometry is geodesically complete in the past and the potentially dangerous (curvature) singularities may arise but not in the far past [105, 106] (see also [107, 108] for some possible mechanism for the regularization of the background). In Fig. 4.5 the null geodesics of Eqs. (4.46) and (4.47) are illustrated, respectively, with the dashed line and with the full line. It is clear from the plot that, in the case of Eq. (4.47) the cosmic time parameter is a regular function of the affine parameter while in the case of Eq. (4.46) there is a singularity for $\lambda \to 0^+$.

Chapter 5

Essentials of Inflationary Dynamics

Some possible dynamical realizations of the inflationary paradigm will now be swiftly discussed. Diverse models have been proposed so far and the purpose of the present chapter is to outline some general aspects that will be useful in the discussion of the evolution of the inhomogeneities.

5.1 Fully inhomogeneous Friedmann-Lemaître equations

The usual presentation of inflationary dynamics deals predominantly with *homogeneous* equations for scalar degrees of freedom in the early Universe. It is then argued that, when the scalar potential dominates over the spatial gradients and over the kinetic energy of the scalar degree of freedom the geometry is led to inflate. In a slightly more quantitative perspective we shall demand that the aforementioned conditions should be verified over a spatial region of typical size $H_i^{-1} > 10^5 \ell_P$ where H_i^{-1}, as explained in section 4, is a newly formed event horizon at the cosmic time t_i. Why should we neglect spatial gradients during a phase of inflationary expansion? The answer to this question can be neatly formulated in terms of the inhomogenous form of Friedmann-Lemaître equations. The *homogeneous* Friedmann-Lemaître equations (see Eqs. (2.56), (2.57) and (2.58)) have been written neglecting all the spatial gradients. A very useful strategy will now be to write the Friedmann-Lemaître equations in a fully inhomogeneous form, i.e. in a form where the spatial gradients are not neglected. From this set of equations it will be possible to expand the metric to a given order in the spatial gradients, i.e. we will have that the zeroth-order solution will not contain any gradient, the first-order iteration will contain two spatial gradients, the second-order solution will contain four spatial gradients and so on. This kind of perturbative expansion has been pioneered, in the late sixties and in the seventies, by Lifshitz, Khalatnikov [149–151], and by Belinskii and Khalatnikov [152–154]. Often this approach is dubbed BKL formalism

There are various applications of this formalism to inflationary cosmology [155–158] as well as to dark energy models (see, for instance, [159–161] and references therein). In the present framework, the fully inhomogeneous approach will be simply employed in order to justify the following statements:

- if the (total) barotropic index of the sources is such that $w > -1/3$, then the spatial gradients will be relevant for large values of the cosmic time coordinate (i.e., formally, $t \to \infty$) but they will be negligible in the opposite limit (i.e. $t \to 0^+$);
- if the total barotropic index is smaller than $-1/3$ the situation is reversed: the spatial gradients will become more and more subleading as time goes by but they will be of upmost importance in the limit of small cosmic times;
- if $w = -1/3$ the contribution of the spatial gradients remains constant.

The second point of the above list of items will justify why spatial gradients can be neglected as inflation proceeds. At the same time, it should be stressed that the announced analysis does not imply that the inflationary dynamics is *generic*. On the contrary it implies that, once inflation takes place, the spatial gradients will be progressively subleading. Similarly, the present analysis will also show that prior to the onset of inflation the spatial gradients cannot be neglected. For the present purposes, a very convenient form of the line element is represented by

$$ds^2 = g_{\mu\nu}dx^\mu dx^\nu = dt^2 - \gamma_{ij}(t,\vec{x})dx^i dx^j, \qquad (5.1)$$

where

$$g_{00} = 1, \qquad g_{ij} = -\gamma_{ij}(t,\vec{x}), \qquad g_{0i} = 0. \qquad (5.2)$$

Since the four-dimensional metric $g_{\mu\nu}$ has ten independent degrees of freedom and since there are four available conditions to fix completely the coordinate system,[a] Eqs. (5.1) and (5.2) encode all the relevant functions allowing a faithful description of the dynamics: the tensor $\gamma_{ij}(t,\vec{x})$ being symmetric, contains 6 independent degrees of freedom. The idea is now, in

---

[a]The freedom of choosing a gauge when the geometry is not fully homogeneous will be addressed, repeatedly, starting from chapter 6 and, more formally, in chapter 11. In the terminology of chapter 11 the choice defined by Eq. (5.2) is an example of synchronous coordinate system.

short, the following. The Einstein equations can be written in a form where the spatial gradients and the temporal gradients are formally separated. In particular, using Eqs. (5.1) and (5.2) it can be easily shown that the Christoffel connections can be written as

$$\Gamma^0_{ij} = \frac{1}{2}\frac{\partial}{\partial t}\gamma_{ij} = -K_{ij}, \qquad \Gamma^j_{0i} = \frac{1}{2}\gamma^{jk}\frac{\partial}{\partial t}\gamma_{ki} = -K^j_i, \qquad (5.3)$$

$$\Gamma^k_{ij} = \frac{1}{2}\gamma^{k\ell}\Big[-\partial_\ell\gamma_{ij} + \partial_j\gamma_{\ell i} + \partial_i\gamma_{j\ell}\Big], \qquad (5.4)$$

where K_{ij} is the so-called extrinsic curvature which is the inhomogeneous generalization of the Hubble parameter. Notice, in fact, that when $\gamma_{ij} = a^2(t)\delta_{ij}$ (as it happens in the homogeneous case) $K^j_i = -H\delta^j_i$ where H is the well known Hubble parameter. Using Eqs. (5.4) the components of the Ricci tensor can be written as

$$R^0_0 = \dot{K} - \text{Tr}K^2, \qquad (5.5)$$

$$R^0_i = \nabla_i K - \nabla_k K^k_i, \qquad (5.6)$$

$$R^j_i = \frac{\partial}{\partial t}K^i_j - KK^j_i - r^j_i. \qquad (5.7)$$

Throughout this discussion the following notations will be employed:

- the overdot denotes a partial derivation with respect to t;
- ∇_i denotes the covariant derivative defined in terms of γ_{ij} and of Eq. (5.4).

While the first convention is rather natural in the light of the previous treatments, the meaning of ∇_i should not be confused with the spatial part of the covariant derivative computed with respect to the full metric tensor.

In Eqs. (5.5), (5.6) and (5.7) the explicit meaning of the various quantities are given by

$$\text{Tr}K^2 = K^j_i K^i_j, \qquad K = K^i_i, \qquad r^j_i = \gamma^{jk}r_{ki}. \qquad (5.8)$$

The three-dimensional Ricci tensor is simply given in terms of the Christoffel connections with spatial indices:

$$r_{ij} = \partial_m \Gamma^m_{ij} - \partial_j \Gamma^m_{mi} + \Gamma^m_{ij}\Gamma^\ell_{m\ell} - \Gamma^\ell_{jm}\Gamma^m_{i\ell}, \qquad (5.9)$$

Equations (5.5), (5.6) and (5.7) allow us to write the Einstein equations in a fully inhomogeneous form. More specifically, assuming that the energy-momentum tensor is a perfect relativistic fluid

$$T^\nu_\mu = (p+\rho)u_\mu u^\nu - p\delta^\nu_\mu, \qquad (5.10)$$

we will have to write, in explicit terms, both Eq. (2.50) and the covariant conservation equation discussed in Eq. (2.9). From the (00) and from the (0i) components of Eq. (2.50), the Hamiltonian and momentum constraints can be derived by using the explicit expressions for the Ricci tensor. The result will be, respectively:

$$K^2 - \text{Tr}K^2 + r = 16\pi G[(p+\rho)u_0 u^0 - p], \tag{5.11}$$

$$\nabla_i K - \nabla_k K_i^k = 8\pi G u_i u^0 (p+\rho). \tag{5.12}$$

It is clear that there two relations are constraints (as opposed to dynamical equations) since they do not contain second derivatives of the metric. The (ij) components of Eq. (2.50) lead instead to

$$(\dot{K}_i^j - K K_i^j - \dot{K}\delta_i^j) + \frac{1}{2}\delta_i^j(K^2 + \text{Tr}K^2) - (r_i^j - \frac{1}{2}r\delta_i^j)$$
$$= -8\pi G[(p+\rho)u_i u^j + p\delta_i^j], \tag{5.13}$$

A trivial remark is that, in Eqs. (5.11), (5.12) and (5.13), the indices are raised and lowered using directly $\gamma_{ij}(t,\vec{x})$. By combining the previous set of equations the following relation can be easily deduced

$$q\,\text{Tr}K^2 = 8\pi G\left[(p+\rho)u_0 u^0 + \frac{p-\rho}{2}\right] \tag{5.14}$$

where

$$q(\vec{x},t) = -1 + \frac{\dot{K}}{\text{Tr}K^2} \tag{5.15}$$

is the inhomogeneous generalization of the deceleration parameter. In fact, in the homogeneous and isotropic limit,

$$\gamma_{ij} = a^2(t)\delta_{ij}, \qquad K_i^j = -H\delta_i^j \qquad q(t) \to -\frac{\ddot{a}a}{\dot{a}^2}. \tag{5.16}$$

Recalling the definition of $\text{Tr}K^2$, it is rather easy to show that

$$\text{Tr}K^2 \geq \frac{K^2}{3} \geq 0, \tag{5.17}$$

where the sign of equality (in the first relation) is reached, again, in the isotropic limit. Since γ^{ij} is always positive semi-definite, it is also clear that

$$u_0 u^0 = 1 + \gamma^{ij} u_i u_j \geq 1, \tag{5.18}$$

which follows from the condition $g^{\mu\nu} u_\mu u_\nu = 1$. From Eq. (5.14) it follows that $q(t,\vec{x})$ is always positive semi-definite if $(\rho + 3p) \geq 0$. This result is physically very important and it shows that spatial gradients cannot make gravity repulsive. One way of making gravity repulsive is instead to change the sources of the geometry and to violate the strong condition. Eqs. (5.14) and (5.15) generalize the relations already obtained in chapter 2 and, in particular, Eqs. (2.60) and (2.63). Always in chapter 2 it has been observed that the acceleration is independent on the contribution of the spatial curvature. Furthermore, it is easy to show that when the (negative) spatial curvature dominates over all the other sources the scale factor expands, at most, linearly in cosmic time (i.e. $a(t) \sim t$) and the deceleration parameter vanishes.

Equations (5.11), (5.13) and (5.12) must be supplemented by the explicit form of the covariant conservation equations written down in chapter 2 (see, in particular, Eq. (2.9)). Thus, from the (i) and (0) components of Eq. (2.9) the resulting explicit expressions will be:

$$\frac{1}{\sqrt{\gamma}}\frac{\partial}{\partial t}[\sqrt{\gamma}(p+\rho)u^0 u^i] - \frac{1}{\sqrt{\gamma}}\partial_k\{\sqrt{\gamma}[(p+\rho)u^k u^i + p\gamma^{ki}]\}$$
$$- 2K^i_\ell u^0 u^\ell(p+\rho) - \Gamma^i_{k\ell}[(p+\rho)u^k u^\ell + p\gamma^{k\ell}] = 0, \quad (5.19)$$

$$\frac{1}{\sqrt{\gamma}}\frac{\partial}{\partial t}\{\sqrt{\gamma}[(p+\rho)u_0 u^0 - p]\} - \frac{1}{\sqrt{\gamma}}\partial_i\{\sqrt{\gamma}(p+\rho)u_0 u^i\}$$
$$- K^\ell_k[(p+\rho)u^k u_\ell + p\delta^k_\ell] = 0, \quad (5.20)$$

where $\gamma = \det(\gamma_{ij})$. It is useful to recall, from the Bianchi identity, that the intrinsic curvature tensor and its trace satisfy the following identity

$$\nabla_j r^j_i = \frac{1}{2}\nabla_i r. \quad (5.21)$$

Note, finally, that combining Eq. (5.11) with the trace of Eq. (5.13) the following equation is obtained:

$$\mathrm{Tr}K^2 + K^2 + r - 2\dot{K} = 8\pi G(\rho - 3p). \quad (5.22)$$

Equation (5.22) allows to re-write Eqs. (5.11), (5.13) and (5.12) as

$$\dot{K} - \mathrm{Tr}K^2 = 8\pi G\left[(p+\rho)u_0 u^0 + \frac{p-\rho}{2}\right], \quad (5.23)$$

$$\frac{1}{\sqrt{\gamma}}\frac{\partial}{\partial t}\left(\sqrt{\gamma}K^j_i\right) - r^j_i = 8\pi G\left[-(p+\rho)u_i u^j + \frac{p-\rho}{2}\delta^j_i\right], \quad (5.24)$$

$$\nabla_i K - \nabla_k K^k_i = 8\pi G(p+\rho)u_i u^0, \quad (5.25)$$

where we used the relation $2K = -\dot{\gamma}/\gamma$. Let us now look for solutions of the previous system of equations in the form of a gradient expansion. We shall be discussing γ_{ij} written in the form

$$\gamma_{ik} = a^2(t)[\alpha_{ik}(\vec{x}) + \beta_{ik}(t, \vec{x})], \quad \gamma^{kj} = \frac{1}{a^2(t)}[\alpha^{kj} - \beta^{kj}(t, \vec{x})], \quad (5.26)$$

where $\beta(\vec{x},t)$ is considered to be the first-order correction in the spatial gradient expansion. Note that from Eq. (5.26) $\gamma_{ik}\gamma^{kj} = \delta_i^j + \mathcal{O}(\beta^2)$. The logic is now very simple: Einstein equations will determine the specific form of $\beta_{ij}(\vec{x},t)$ once the form of $\alpha_{ij}(\vec{x},t)$ is known. Putting Eq. (5.26) into Eqs. (5.3) we obtain

$$K_i^j = -\left(H\delta_i^j + \frac{\dot\beta_i^j}{2}\right) + \mathcal{O}(\beta^2),$$

$$K = -\left(3H + \frac{1}{2}\dot\beta\right) + \mathcal{O}(\beta^2), \tag{5.27}$$

$$\mathrm{Tr}K^2 = 3H^2 + H\dot\beta + \mathcal{O}(\beta^2),$$

where, as usual, $H = \dot a/a$ represents the homogenous part of the Hubble parameter. From Eq. (5.12) it also follows that

$$\nabla_k \dot\beta_i^k - \nabla_i \dot\beta = 16\pi G u_i\, u^0(p+\rho). \tag{5.28}$$

The explicit form of the momentum constraint suggests that we look for the solution in a separable form, namely:

$$\beta_i^j(t,\vec{x}) = g(t)\mathcal{B}_i^j(\vec{x}). \tag{5.29}$$

Thus Eq. (5.28) becomes

$$\dot g(\nabla_k \mathcal{B}_i^k - \nabla_i \mathcal{B}) = 16\pi G u_i u^0(p+\rho). \tag{5.30}$$

Using this parametrization and solving the constraint for u_i, Einstein equations to second order in the gradient expansion reduce then to the following equation:

$$(\ddot g + 3H\dot g)\mathcal{B}_i^j + H\dot g \mathcal{B}\delta_i^j + \frac{2\mathcal{P}_i^j}{a^2} = \frac{w-1}{3w+1}(\ddot g + 2H\dot g)B\delta_i^j. \tag{5.31}$$

In Eq. (5.31) the spatial curvature tensor has been parametrized as

$$r_i^j(\vec{x},t) = \frac{\mathcal{P}_i^j(\vec{x})}{a^2}. \tag{5.32}$$

Recalling that

$$H = H_0 a^{-\frac{3(w+1)}{2}}, \qquad \dot H = -\frac{3(w+1)}{2}H^2, \tag{5.33}$$

the solution for Eq. (5.31) can be written as

$$\mathcal{B}_i^j(\vec{x}) = -\frac{4}{H_0^2(3w+1)(3w+5)}\left(\mathcal{P}_i^j(\vec{x}) - \frac{5+6w-3w^2}{4(9w+5)}\mathcal{P}(\vec{x})\delta_i^j\right), \tag{5.34}$$

$$\mathcal{B}(\vec{x}) = -\frac{\mathcal{P}(\vec{x})}{H_0^2(9w+5)},$$

with $g(t)$ simply given by

$$g(t) = a^{3w+1}. \tag{5.35}$$

Note that, in Eq. (5.34), $H_0 = 2/[3(w+1)t_0]$. Equation (5.34) can be also inverted, i.e. $\mathcal{P}_i^j$ can be easily expressed in terms of $\mathcal{B}_i^j(\vec{x})$ and $\mathcal{B}(\vec{x})$:

$$\mathcal{P}_i^j(\vec{x}) = -\frac{H_0^2}{4}[\mathcal{B}(\vec{x})\delta_i^j(6w+5-3w^2) + \mathcal{B}_i^j(\vec{x})(3w+5)(3w+1)]. \tag{5.36}$$

Using Eq. (5.21) the peculiar velocity field and the energy density can also be written as

$$u^0 u_i = -\frac{3}{8\pi G\rho}\left(\frac{w}{3w+5}\right)a^{3w+1}H\partial_i\mathcal{B}(\vec{x}),$$

$$\rho = \frac{3H_0^2}{8\pi G}\left[a^{-3(w+1)} - \frac{w+1}{2}\mathcal{B}(\vec{x})a^{-2}\right]. \tag{5.37}$$

Let us therefore rewrite the solution in terms of γ_{ij}, i.e.

$$\gamma_{ij} = a^2(t)[\alpha_{ij}(\vec{x}) + \beta_{ij}(\vec{x},t)] = a^2(t)\left[\alpha_{ij}(\vec{x}) + a^{3w+1}\mathcal{B}_{ij}(\vec{x})\right]. \tag{5.38}$$

Concerning this solution a few comments are in order:

- if $w > -1/3$, β_{ij} becomes large as $a \to \infty$ (note that if $w = -1/3$, a^{3w+1} is constant);
- if $w < -1/3$, β_{ij} vanishes as $a \to \infty$;
- if $w < -1$, β_{ij} not only the gradients become sub-leading but the energy density also increases as $a \to \infty$.
- to the following order in the perturbative expansion the time-dependence is easy to show: $\gamma_{ij} \simeq a^2(t)[\alpha_{ij} + a^{3w+1}\mathcal{B}_{ij} + a^{2(3w+1)}\mathcal{E}_{ij}]$ and so on for even higher order terms; clearly the calculation of the curvature tensors will now be just a bit more cumbersome.

Equation (5.38) then proves the statements illustrated at the beginning of the present section and justifies the use of homogeneous equations for the analysis of the inflationary dynamics. Again, the debatable issue on how inflation starts should however be discussed within the inhomogeneous approach. It is finally relevant to mention that the present formalism also answer an important question on the nature of the singularity in the standard cosmological model. Suppose that the evolution of the Universe is always decelerated (i.e. $\ddot{a} < 0$) but expanding (i.e. $\dot{a} > 0$). What should we expect in the limit $a \to 0$? As emphasized in the past by Belinskii,

Lifshitz and Khalatnikov (see for instance [154]) close to the singularity the spatial gradients become progressively less important as also implied by Eq. (5.38). The latter observation means that the standard big-bang may be highly anisotropic but rather homogeneous. In particular, close to the singularity, the solution may fall in one of the metrics of the Bianchi classification [87] (see also Eqs. (2.5) and (2.6)). In more general terms it can also happen that the geometry undergoes anisotropic oscillations that are customarily named BKL (for Belinskii, Khalatnikov and Lifshitz) oscillations.

5.2 Homogeneous evolution of a scalar field

The Friedmann-Lemaître equations imply that the scale factor can accelerate provided $w < -1/3$, where w is the barotropic index of the generic fluid driving the expansion. This condition can be met, for instance, if one (or more) scalar degrees of freedom have the property that their potential dominates over their kinetic energy. Consider, therefore, the simplest case where a single scalar degree of freedom is present in the game. The action can be written as

$$S = \int d^4x \sqrt{-g} \left[-\frac{R}{16\pi G} + \frac{1}{2} g^{\alpha\beta} \partial_\alpha \varphi \partial_\beta \varphi - V(\varphi) \right], \quad (5.39)$$

where φ is the scalar degree of freedom and $V(\varphi)$ its related potential. The scalar field appearing in the action (5.39) is said to be *minimally coupled*. Of course, there are other possibilities. For instance the scalar field φ can be *conformally coupled* or even *non-minimally coupled*. These couplings arise when the scalar field action is written in the form

$$S = \int d^4x \sqrt{-g} \left[-\frac{R}{16\pi G} + \frac{1}{2} g^{\alpha\beta} \partial_\alpha \varphi \partial_\beta \varphi - V(\varphi) - \frac{\alpha}{2} R \varphi^2 \right]. \quad (5.40)$$

Clearly the difference between Eq. (5.39) and Eq. (5.40) is the presence of an extra term, i.e. $-(\alpha/2)R\varphi^2$. If $\alpha = 0$ we recover the case of minimal coupling. If $\alpha = -1/6$ the field is conformally coupled and its evolution equations are invariant under the Weyl rescaling of the metric.[b] In all other cases the field is said to be simply non-minimally coupled. In what follows, for pedagogical reasons, we will stick to the case of minimal coupling. Before going to the case of minimal coupling let us write the field equations derived from the action (5.40) in their full generality, i.e. without assuming a

---
[b]The sign of α is conventional. If the opposite sign would have been chosen the condition of conformal coupling would arise when $\alpha = 1/6$.

specific value of α. In order to do so the derivation leading to Eq. (2.50) must be repeated. The surface term (derived by means of the Palatini identity of Eq. (2.40)) and reported in Eq. (2.44) is now more cumbersome since the term $\alpha\varphi^2 R$ couples directly the Ricci scalar to the scalar field φ. As a result the integration by parts will produce further terms (proportional to α) which will contain further derivatives of φ. The functional variation of Eq. (5.40) with respect to $g^{\mu\nu}$ leads to the modified Einstein equation which can be written as

$$\left(\frac{1}{8\pi G}+\alpha\varphi^2\right)\mathcal{G}_\mu^\nu = (1+2\alpha)\partial_\mu\varphi\partial^\nu\varphi - \left(2\alpha+\frac{1}{2}\right)g^{\alpha\beta}\partial_\alpha\varphi\partial_\beta\varphi\delta_\mu^\nu$$
$$- 2\alpha\varphi(\delta_\mu^\nu g^{\alpha\beta}\nabla_\alpha\nabla_\beta\varphi - \nabla_\mu\nabla^\nu\varphi) + \delta_\mu^\nu V. \quad (5.41)$$

The functional variation of the action (5.40) with respect to φ leads to a generalized form of the Klein-Gordon equation in curved backgrounds which can be written as

$$g^{\alpha\beta}\nabla_\alpha\nabla_\beta\varphi + \alpha R\varphi + \frac{\partial V}{\partial\varphi} = 0. \quad (5.42)$$

If we set $\alpha = 0$ in Eqs. (5.41) and (5.42) we recover, of course, the result obtainable by direct functional variation of Eq. (5.39) with respect to $g^{\mu\nu}$ and φ:

$$R_\mu^\nu - \frac{1}{2}\delta_\mu^\nu R = 8\pi G T_\mu^\nu, \quad (5.43)$$

$$g^{\alpha\beta}\nabla_\alpha\nabla_\beta\varphi + \frac{\partial V}{\partial\varphi} = 0, \quad (5.44)$$

where

$$\nabla_\alpha\nabla_\beta\varphi = \partial_\alpha\partial_\beta\varphi - \Gamma_{\alpha\beta}^\sigma\partial_\sigma\varphi, \quad (5.45)$$

$$T_\mu^\nu = \partial_\mu\varphi\partial^\nu\varphi - \delta_\mu^\nu\left[\frac{1}{2}g^{\alpha\beta}\partial_\alpha\varphi\partial_\beta\varphi - V(\varphi)\right]. \quad (5.46)$$

The components of Eq. (5.46) can be written, in a spatially flat FRW metric, as

$$T_0^0 \equiv \rho_\varphi = \left(\frac{\dot\varphi^2}{2}+V\right) + \frac{1}{2a^2}(\partial_k\varphi)^2, \quad (5.47)$$

$$T_i^j = -\frac{1}{a^2}\partial_i\varphi\partial^j\varphi - \left(\frac{\dot\varphi^2}{2}-V\right)\delta_i^j + \frac{1}{2a^2}(\partial_k\varphi)^2\delta_i^j \quad (5.48)$$

$$T_i^0 = \dot\varphi\partial_i\varphi \quad (5.49)$$

where, for the moment, the spatial gradients have been kept. To correctly identify the pressure and energy density of the scalar field the components of T^ν_μ can be written as

$$T^0_0 = \rho_\varphi, \qquad T^j_i = -p_\varphi \delta^j_i + \Pi^j_i(\varphi) \tag{5.50}$$

where Π^j_i is a traceless quantity and it is called anisotropic stress.[c] By comparing Eqs. (5.47), (5.48) and (5.49) with Eq. (5.50) we will have

$$\rho_\varphi = \left(\frac{\dot\varphi^2}{2} + V\right) + \frac{1}{2a^2}(\partial_k\varphi)^2, \tag{5.51}$$

$$p_\varphi = \left(\frac{\dot\varphi^2}{2} - V\right) - \frac{1}{6a^2}(\partial_k\varphi)^2, \tag{5.52}$$

$$\Pi^j_i(\varphi) = -\frac{1}{a^2}\left[\partial_i\varphi\partial^j\varphi - \frac{1}{3}(\partial_k\varphi)^2\delta^j_i\right]. \tag{5.53}$$

Equations (5.51) and (5.52) imply that the effective barotropic index for the scalar system under discussion is simply given by

$$w_\varphi = \frac{p_\varphi}{\rho_\varphi} = \frac{\left(\frac{\dot\varphi^2}{2} - V\right) - \frac{1}{6a^2}(\partial_k\varphi)^2}{\left(\frac{\dot\varphi^2}{2} + V\right) + \frac{1}{2a^2}(\partial_k\varphi)^2}. \tag{5.54}$$

Concerning Eq. (5.54) three comments are in order:

- if $\dot\varphi^2 \gg V$ and $\dot\varphi^2 \gg (\partial_k\varphi)^2/a^2$, then $p_\varphi \simeq \rho_\varphi$: in this regime the scalar field behaves as a stiff fluid;
- if $V \gg \dot\varphi^2 \gg (\partial_k\varphi)^2/a^2$, then $w_\varphi \simeq -1$: in this regime the scalar field is an inflaton candidate;
- if $(\partial_k\varphi)^2/a^2 \gg \dot\varphi^2$ and $(\partial_k\varphi)^2/a^2 \gg V$, then $w_\varphi \simeq -1/3$: in this regime the system is gradient-dominated and, according to the previous results the inhomogeneous deceleration parameter $q(t,\vec{x}) \simeq 0$.

Of course also intermediate situations are possible (or plausible). If the scalar potential dominates both over the gradients and over the kinetic energy for a sufficiently large event horizon at a given time the subsequent evolution is therefore likely to be rather homogeneous and the relevant equations will simply be the fully homogenous ones. The various components of Eq. (5.43) can the be written in the case of the metric of Eq. (2.2).

---

[c]The anisotropic stress is rather relevant for the correct discussion of the pre-decoupling physics and, as we shall see, is mainly due, after weak interactions have fallen out of thermal equilibrium, to the quadrupole moment of the neutrino phase space distribution.

This derivation mirrors exactly what has been already discussed in chapter 2 when we derived the Friedmann-Lemaître equations from Eq. (2.50). The only new element of the derivation is represented by Eq. (5.44) which should now be written in the metric (2.2). For this purpose the trick is to recall that the covariant derivative of a scalar (φ in our case) is just an ordinary derivative. With this observation in mind it is immediate to see that

$$g^{\alpha\beta}\nabla_\alpha\nabla_\beta\varphi = g^{\alpha\beta}(\partial_\alpha\partial_\beta\varphi - \Gamma^\sigma_{\alpha\beta}\partial_\sigma\varphi) \equiv \ddot\varphi + 3H\dot\varphi, \qquad (5.55)$$

where the second equality follows from the explicit form of the metric (2.2) and of the Christoffel connections. Therefore the explicit form of Einsten equations which is directly comparable with Eqs. (2.56), (2.57) and (2.58) is given by:

$$\overline{M}_{\rm P}^2 H^2 = \frac{1}{3}\left[\frac{\dot\varphi^2}{2} + V\right] - \frac{k\,\overline{M}_{\rm P}^2}{a^2}, \qquad (5.56)$$

$$\overline{M}_{\rm P}^2 \dot H = -\frac{\dot\varphi^2}{2} + \frac{k\,\overline{M}_{\rm P}^2}{a^2}, \qquad (5.57)$$

$$\ddot\varphi + 3H\dot\varphi + \frac{\partial V}{\partial \varphi} = 0, \qquad (5.58)$$

where the reduced Planck mass has been defined according to the following chain of equalities:

$$\overline{M}_{\rm P}^2 = \frac{1}{8\pi G} = \frac{M_{\rm P}^2}{8\pi}. \qquad (5.59)$$

Even if it is not desirable to introduce different definitions of the Planck mass, the conventions adopted in Eq. (5.59) are widely used in the study of inflationary dynamics so we will stick to them. Because of the factor $\sqrt{8\pi}$ in the denominator, $\overline{M}_{\rm P}$ will be roughly 5 times smaller than $M_{\rm P}$.

Before passing to a classification of the inflationary backgrounds in the minimally coupled case let us try to understand what it implies to have $\alpha \neq 0$. To understand physically what is going on in the case $\alpha \neq 0$ the simplest thing to do is to look at the analog of Eq. (5.58) in the case when $\alpha \neq 0$. Before doing so, it is useful for the present purposes to rephrase Eq. (5.58) in the conformal time parametrization and then compare the resulting expression with its generalization derivable from Eq. (5.42). By recalling that, in the conformal time parametrization $dt = a(\tau)d\tau$ (see Eqs. (2.4) and (2.64)), Eq. (5.58) can be written as

$$\varphi'' + 2\mathcal{H}\varphi' - \nabla^2\varphi = 0, \qquad (5.60)$$

where, for our convenience, we neglected the potential but we included the contribution of the spatial gradient which was, on the contrary, not taken into account in Eq. (5.58). The notations of Eq. (5.60) are the same as the one already discussed at the end of chapter 2, i.e. $\mathcal{H} = a'/a$ and the prime denotes a derivation with respect to τ. Let us now compute the free Klein-Gordon equation as it emerges from Eq. (5.42). The result of this straightforward manipulation implies

$$\varphi'' + 2\mathcal{H}\varphi' + \alpha \overline{R} a^2 \varphi - \nabla^2 \varphi = 0, \qquad (5.61)$$

where $\overline{R}$ denotes the Ricci scalar evaluated on the (spatially flat) FRW metric. Recalling now Eq. (2.64) we have Eq. (5.61) which can be simply written as

$$\varphi'' + 2\mathcal{H}\varphi' - 6\alpha \frac{a''}{a} \varphi - \nabla^2 \varphi = 0. \qquad (5.62)$$

By now defining the rescaled field $\varphi_1 = a\varphi$ the final equation reads

$$\varphi_1'' - (6\alpha + 1)\frac{a''}{a}\varphi_1 - \nabla^2 \varphi_1 = 0. \qquad (5.63)$$

It is then clear, by looking at Eq. (5.63) why the case $\alpha = -1/6$ is special. In this case the Klein-Gordon equation for φ_1 has the same form of the Klein-Gordon equation in a *flat Minkowski space-time*. When $\alpha = -1/6$ it is said in the jargon that the scalar field φ is conformally coupled in the sense that, by an appropriate conformal transformation of the curved metric, the Klein-Gordon equation can be brought in its Minkowskian form. There is a second very important case where Eq. (5.63) coincides with its Minkowskian analog, i.e. the case $a'' = 0$. In the latter case, in spite of the value of α (but always in the situation of vanishing potential term) Eq. (5.63) assumes the same form it would have in Minkowski space-time. But the case $a'' = 0$ is nothing but the case of a radiation-dominated Universe described in terms of the conformal time coordinate.

Viable inflationary models can be constructed not only in the case of minimal coupling but also in the cases when $\alpha \neq 0$. The interested reader may usefully consult, on this particular problem, Refs. [162–164].

5.3 Classification(s) of inflationary backgrounds

Inflationary backgrounds can be classified either in *geometric* or in *dynamical* terms. The geometric classification is based on the evolution of the Hubble parameter (or of the extrinsic curvature). The conditions $\ddot{a} > 0$

and $\dot{a} > 0$ can be realized for different evolutions of the Hubble parameter. Three possible cases arise naturally:

- de Sitter inflation (realized when $\dot{H} = 0$);
- power-law inflation (realized when $\dot{H} < 0$);
- superinflation (realized when $\dot{H} > 0$).

The case of *exact* de Sitter inflation is a useful simplification but it is, in a sense, unrealistic. On one hand it is difficult, for instance by means of a (single) scalar field, to obtain a pure de Sitter dynamics. On the other hand, if $\dot{H} = 0$ only the tensor modes of the geometry are excited by the time evolution of the background geometry.[d] This observation would imply that the scalar modes (so important for the CMB anisotropies) will not be produced.

The closest situations to a pure de Sitter dynamics is realized by means of a *quasi-de Sitter* phase of expansion where $\dot{H} \lesssim 0$. Quasi-de Sitter inflation is closely related to power-law inflation where the scale factor exhibits a power-law behaviour and $\dot{H} < 0$. If the power of the scale factor is much larger than 1, i.e.

$$a(t) \simeq t^\beta, \qquad \beta \gg 1, \tag{5.64}$$

the quasi-de Sitter phase is essentially a limit of the power-law models which may be realized, for instance, in the case of exponential potentials as we shall see in a moment. Finally, an unconventional case is the one of super-inflation. In standard Einstein-Hilbert gravity superinflation can only be achieved (in the absence of spatial curvature) if the dominant energy condition is violated, i.e. if the effective enthalpy of the sources is negative definite. This simple observation (stemming directly form Eq. (2.57)) implies that, in Einstein-Hilbert gravity, scalar field sources with positive kinetic terms cannot give rise to superinflationary dynamics. This impasse can be overcome in two different (but complementary) ways. If internal dimensions are included in the game, the overall solutions differ substantially from the simple four-dimensional case contemplated along this book. This possibility arises naturally in string cosmology and has been investigated [165, 166] in the context of the evolution of fundamental strings in curved backgrounds [167]. If the Einstein-Hilbert theory is generalized to include a fundamental scalar field (the dilaton) different frames arise naturally in the problem. In this context, superinflation arises as a solution

---

[d]This statement follows from the considerations developed in the following chapters of this book and, in particular, from the discussions reported in chapters 6 and 10.

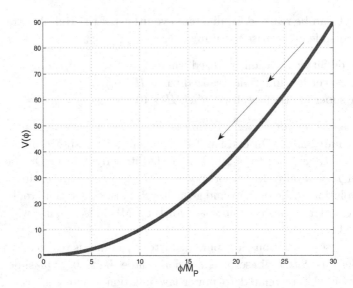

Fig. 5.1 A schematic illustration of the inflationary potential in the case of large field models. In this case the slow roll parameters are small since the value of $\varphi/M_{\rm P}$ is initially large.

in the string frame as a result of the dynamics of the dilaton (which is, in turn, connected with the dynamics of the gauge coupling). This is the path followed, for instance, in the context of pre-big bang models (see [105, 107] and references therein).

If inflation is realized by means of one (or many) scalar degrees of freedom, the classification of inflationary models is usually described in terms of the properties of the scalar potential. This is a more dynamical classification which is, however, narrower than the geometric one introduced above. The rationale for this statement is that while the geometric classification is still valid in the presence of many (scalar) degrees of freedom driving inflation, the dynamical classification may slightly change depending upon the number and nature of the scalar sources introduced in the problem. Depending on the way the slow-roll dynamics is realized, two cases are customarily distinguished

- large field models (see Fig. 5.1);
- small field models (see Fig. 5.2).

In small field models the value of the scalar field in Planck units is usually smaller than one. In large field models the value of the scalar field at the

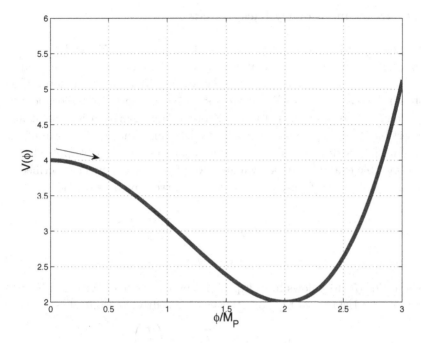

Fig. 5.2 A schematic illustration of the inflationary potential in the case of small field models.

onset of inflation is usually larger than one (ore even much larger than one) in Planck units. In Figs. 5.1 and 5.2 a schematic view of the large and small field models is provided. In what follows examples of large and small field models will be given. Let us finally comment on the relevance of spatial curvature. Inflation is safely described in the absence of spatial curvature since, as we saw, the inflationary dynamics washes out the spatial gradients quite efficiently. In spite of this statement, there can be situations (see second example in the following section) where the presence of spatial curvature leads to *exact* inflationary backgrounds. Now, these solutions, to be phenomenologically relevant, should inflate for the *minimal* amount of e-folds. If, on the contrary, inflation lasts much more than the required 60 or 65 e-folds, the consequences of inflationary models endowed with spatial curvature will be indistinguishable, for practical purposes, from the consequences of those models where the spatial curvature is absent from the very beginning.

5.4 Exact inflationary backgrounds

As Eqs. (2.56), (2.57) and (2.58), also Eqs. (5.56), (5.57) and (5.58) are not all independent. To illustrate inflationary dynamics, the following game can be, in some cases, successfully played: specify a given geometry, then obtain the scalar field profile by integrating (with respect to the cosmic time coordinate) Eq. (5.57). If the result of this manipulation is explicit and invertible, then Eq. (5.56) allows us to determine the specific form of the potential. The drawback of this strategy concerns the range of applicability: few solvable examples are known and two of them will now be described in pedagogical terms. Consider the following power-law background:

$$a(t) = a_1 \left(\frac{t}{t_1}\right)^\beta, \qquad \beta > 1, \qquad t > 0, \tag{5.65}$$

$$H = \frac{\beta}{t}, \qquad \dot{H} = -\frac{\beta}{t^2}, \tag{5.66}$$

where Eq. (5.66) follows from Eq. (5.65) by using the definition of Hubble parameter. Using now Eq. (5.57), $\dot{\varphi}$ can be swiftly determined as

$$\varphi(t) = \varphi_0 + \sqrt{2\beta} \overline{M}_{\rm P} \ln\left(\frac{t}{t_1}\right). \tag{5.67}$$

Inverting Eq. (5.67) we can easily obtain:

$$\left(\frac{t}{t_1}\right) = \exp\left[\frac{\varphi - \varphi_0}{\sqrt{2\beta}\,\overline{M}_{\rm P}}\right]. \tag{5.68}$$

With the help of Eq. (5.68), Eq. (5.56) can be used to determine the specific form of the potential, i.e.

$$V(\varphi) = V_0 e^{-\sqrt{\frac{2}{\beta}}\frac{\varphi}{\overline{M}_{\rm P}}} \tag{5.69}$$

$$V_0 = \beta(3\beta - 1)\overline{M}_{\rm P}^2 \, t_1^{-2} e^{\sqrt{\frac{2}{\beta}}\frac{\varphi_0}{\overline{M}_{\rm P}}}. \tag{5.70}$$

With the geometry (5.66), Eq. (5.44) is automatically satisfied provided φ is given by (5.67) and the potential is the one determined in Eqs. (5.69)–(5.70). The example developed in this paragraph goes often under the name of power law inflation [168–171].

Consider now a different example where the scale factor is given by

$$a(t) = a_1 \cosh(H_1 t), \qquad H_1 > 0, \tag{5.71}$$

$$H(t) = H_1 \tanh(H_1 t), \qquad \dot{H} = \frac{H_1^2}{\cosh^2(H_1 t)}. \tag{5.72}$$

This type of solution is not compatible with Eq. (5.57) if the spatial curvature vanish (or if it is negative). In these cases, in fact, $\dot{H}$ would be positive semi-definite and it should equal, by Eq. (5.57), $-\dot{\varphi}^2/2$ which is, overall negative definite. Taking thus into account the necessary contribution of the spatial curvature, Eq. (5.57) gives us $\dot{\varphi}$ and, after explicit integration, also $\varphi(t)$. The result of this procedure is that

$$\dot{\varphi} = \frac{A_1}{\cosh H_1 t}, \tag{5.73}$$

$$\varphi(t) = \varphi_0 + \frac{A_1}{H_1} \arctan[\sinh(H_1 t)], \tag{5.74}$$

$$A_1 = \sqrt{2}\left(\frac{k}{a_1^2} - H_1^2\right)^{1/2} \overline{M}_\mathrm{P}. \tag{5.75}$$

Recalling now that, from Eq. (5.74),

$$\sinh(H_1 t) = \tan\tilde{\varphi}, \qquad \tilde{\varphi} = \frac{H_1}{A_1}(\varphi - \varphi_0), \tag{5.76}$$

Eq. (5.56) can be used to determine the potential which is

$$V(\tilde{\varphi}) = \overline{M}_\mathrm{P}^2 H_1^2[3 - 2\cos^2\tilde{\varphi}] + 2\frac{k\,\overline{M}_\mathrm{P}^2}{a_1^2}\cos^2\tilde{\varphi}. \tag{5.77}$$

The last example developed in this section goes also under the name of de Sitter bounce and has been studied in different contexts [172, 173] (see also [174, 175] for more general considerations on inflationary models with non-vanishing spatial curvature).

Before ending this section it is amusing to mention another exact solution which can be obtained by thinking that the inflationary dynamics is smoothly connected with a radiation-dominated Universe. This is, in some sense, what should really happen. The toy example we ought to discuss involves, however, only the dynamics of a single scalar degree of freedom. So, in this example, the inflaton does not decay at the end of inflation. To derive the result is useful to integrate directly the evolution equations in the conformal time parametrization. Recalling the results derived in chapter 2 (see, in particular, Eqs. (2.4) and (2.64)), Eqs. (5.56), (5.57) and (5.58) can be expressed as:

$$\overline{M}_\mathrm{P}^2 \mathcal{H}^2 = \frac{1}{3}\left[\frac{\varphi'^2}{2} + Va^2\right], \tag{5.78}$$

$$\overline{M}_\mathrm{P}^2(\mathcal{H}^2 - \mathcal{H}') = \frac{\varphi'^2}{2}, \tag{5.79}$$

$$\varphi'' + 2\mathcal{H}\varphi' + \frac{\partial V}{\partial \varphi}a^2 = 0, \tag{5.80}$$

where the prime denotes, as usual, a derivation with respect to the conformal time coordinate τ. The linear combination of Eqs. (5.78) and (5.79) allow to write the following pair of equations:

$$V = \frac{\overline{M}_P^2}{a^2}(2\mathcal{H}^2 + \mathcal{H}'), \qquad (5.81)$$

$$\varphi'^2 = 2\overline{M}_P^2(\mathcal{H}^2 - \mathcal{H}'). \qquad (5.82)$$

Suppose then to have the (dimensionless) scale factor in the form

$$a(\tau) = a_i(x + \sqrt{x^2 + 1}), \qquad x = \frac{\tau}{\tau_i}. \qquad (5.83)$$

It can be easily checked that in the limit $\tau \to -\infty$ and $\tau \to +\infty$ the scale factor of Eq. (5.83) implies

$$\lim_{\tau \to -\infty} a(\tau) \simeq \frac{a_i}{2}\left(-\frac{\tau_i}{\tau}\right), \qquad \lim_{\tau \to \infty} a(\tau) \simeq 2a_i\left(\frac{\tau}{\tau_i}\right). \qquad (5.84)$$

But this means that the scale factor (5.83) smoothly interpolates between an inflating de Sitter phase and a radiation-dominated phase. Recall, in fact (see, for instance, Eq. (2.93)) that, during a radiation dominated phase, the scale factor goes linearly in the conformal time parametrization. To appreciate that, indeed, the behavior $a(\tau) \sim (-\tau)^{-1}$ is just the conformal time version of $a(t) \simeq \exp(H_i t)$ let us recall the relation between the cosmic time t and the conformal time τ. Since $a(\tau)d\tau = dt$ we can also write that

$$\tau = \int \frac{dt}{a(t)} \simeq -\frac{1}{H_i}e^{-H_i t}, \qquad a(\tau) = -\frac{1}{H_i \tau}. \qquad (5.85)$$

The two limiting forms of the scale factor reported in Eq. (5.84) approximate very well the interpolating form given in Eq. (5.83) for conformal time intervals $|\tau| > \tau_i$.

It is easy to obtain, from Eq. (5.83), that

$$\mathcal{H} = \frac{1}{\tau_i}\frac{1}{\sqrt{x^2 + 1}}, \qquad \mathcal{H}' = -\frac{1}{\tau_i^2}\frac{x}{(x^2 + 1)^{3/2}}. \qquad (5.86)$$

Using Eq. (5.86) into Eqs. (5.81) and (5.82) the following pair of relations emerges

$$V(x) = \frac{\overline{M}_P^2}{a_i^2 \tau_i^2}\frac{2\sqrt{x^2 + 1} - x}{(x^2 + 1)^{3/2}[x + \sqrt{x^2 + 1}]^2}, \qquad (5.87)$$

$$\varphi'^2 = \frac{2\overline{M}_P^2}{\tau_i^2}\frac{(x + \sqrt{x^2 + 1})}{(x^2 + 1)^{3/2}}. \qquad (5.88)$$

Now the constructive method spelled out in the present section will be, once more, applied to Eqs. (5.87) and (5.88). More precisely, it will be shown that Eq. (5.88) can be exactly integrated. By inverting the obtained result we will be able to obtain the explicit form of the potential in terms of φ. From Eq. (5.88), by taking the square root and by choosing the positive sign it is easy to obtain that

$$\varphi(\tau) = \sqrt{2}\,\overline{M}_{\rm P} I_1(x) + \varphi_0, \tag{5.89}$$

$$I_1(x) = \int \frac{\sqrt{x + \sqrt{x^2 + 1}}}{(x^2 + 1)^{3/4}} dx. \tag{5.90}$$

The integral appearing in Eq. (5.90) can be performed with elementary methods. In particular, a first change of variables $x = \sinh y$ allows us to express $I_1(x)$ as

$$I_1[y(x)] = \int \frac{e^y dy}{\sqrt{e^{2y} + 1}}. \tag{5.91}$$

Then a further change of variable (i.e. $z = e^y$) implies a usual integral that can be solved if we posit $z = \sinh w$ where w is a further auxiliary variable. The final result will be that

$$I_1(x) = \sqrt{2}\operatorname{arcsinh}[e^{\operatorname{arcsinh}(x)}] = \sqrt{2}\operatorname{arcsinh}[x + \sqrt{x^2 + 1}]. \tag{5.92}$$

Using Eq. (5.92) into Eq. (5.89) the wanted relation can be easily obtained, i.e.

$$\varphi(\tau) = \varphi_0 + 2\overline{M}_{\rm P}\operatorname{arcsinh}[x + \sqrt{x^2 + 1}], \tag{5.93}$$

which can be easily inverted as

$$x(\overline{\varphi}) = \frac{1}{2}\left(\sinh\overline{\varphi} - \frac{1}{\sinh\overline{\varphi}}\right), \qquad \overline{\varphi} = \frac{\varphi - \varphi_0}{2\overline{M}_{\rm P}}. \tag{5.94}$$

Noticing now that

$$x^2 + 1 = \frac{\cosh^4\overline{\varphi}}{4\sinh^2\overline{\varphi}}, \qquad x + \sqrt{x^2 + 1} = \sinh\overline{\varphi} \tag{5.95}$$

Eq. (5.87) can be used to obtain $V(\varphi)$ and the final result will be

$$V(\varphi) = V_0 \frac{1 + \cosh^2\overline{\varphi}}{\cosh^6\overline{\varphi}}, \qquad V_0 = \frac{4\overline{M}_{\rm P}^2}{a_i^2 \tau_i^2}. \tag{5.96}$$

Since Eq. (5.80) is just a consequence of the two equations that have been explicitly solved (i.e. Eqs. (5.78) and (5.79)), it will be automatically solved.

As already stressed, solvable examples are rather uncommon. It is therefore mandatory to devise general procedure allowing the discussion of the scalar field dynamics even in the situation when the exact solution is lacking. This will be the theme of the following section.

5.5 Slow-roll dynamics

In the previous section it has been pointed out that the expanding de Sitter phase used for the first description of inflationary dynamics may not be exact and, therefore, we talked about quasi-de Sitter dynamics. In inflationary dynamics a number of slow-roll parameters are customarily defined. They have the property of being small during the (quasi)-de Sitter stage of expansion. Thus they can be employed as plausible expansion parameters. As an example consider the following choice[e]:

$$\epsilon = -\frac{\dot{H}}{H^2}, \qquad \eta = \frac{\ddot{\varphi}}{H\dot{\varphi}}. \tag{5.97}$$

As we shall see, in the literature these parameters are often linearly combined. The smallness of these two (dimensionless) parameters define the range of validity of a given inflationary solution characterized by the dominance of the potential term in the field equations. In other words during the (slow-roll) inflationary phase $|\epsilon| \ll 1$ and $|\eta| \ll 1$. As soon as $\epsilon \simeq \eta \simeq 1$ inflation ends.

During a slow-roll phase the (effective) evolution equations for the homogeneous part of the inflaton background can be written as

$$H^2 \overline{M}_{\rm P}^2 \simeq \frac{V}{3}, \tag{5.98}$$

$$3H\dot{\varphi} + \frac{\partial V}{\partial \varphi} = 0. \tag{5.99}$$

A naive example of slow-roll dynamics characterized by the following single-field potential:

$$V(\varphi) = V_1 - \frac{m^2}{2}\varphi^2 + \frac{\lambda}{4}\varphi^4 + ..., \tag{5.100}$$

where V_1 is a constant. In the jargon, this is a rather simplistic example of what is called a small field model. The solution of Eqs. (5.98) and (5.99) implies, respectively, that

$$a(t) \simeq e^{H_1 t}, \qquad H_1 \simeq \frac{\sqrt{V_1}}{\overline{M}_{\rm P}\sqrt{3}}, \tag{5.101}$$

$$\varphi \simeq \varphi_1 e^{\frac{m^2}{3H_1}t}, \qquad \frac{m^2}{3H_1}t < 1. \tag{5.102}$$

---

[e] Concerning the notation employed for the second slow-roll parameter η we remark that the same Greek letter has been also used to denote the ratio between the concentration of baryons and photons (i.e. $\eta_{\rm b}$) introduced in Eq. (2.74). No confusion is possible both because of the subscript and because the two variables never appear together in this discussion. We warn the reader that, however, very often $\eta_{\rm b}$ is simply denoted by η in the existing literature and, therefore, it will only be the context to dictate the correct signification of the symbol.

The slow-roll phase lasts until the scalar field is approximately constant, i.e. until the cosmic time t_f that can be read-off from Eq. (5.102):

$$t_f \simeq \frac{3H_1}{m^2}, \qquad H_1 t_f \simeq \frac{3H_1^2}{m^2}. \tag{5.103}$$

From Eq. (5.103) the number of e-folds of this toy model can be computed and it is given by

$$N \simeq \frac{3H_1^2}{m^2} > 65, \qquad m^2 \leq \frac{3H_1^2}{65} \tag{5.104}$$

which shows that m should be sufficiently small in units of H_1 to get a long enough inflationary phase. In the case of the exact inflationary background discussed in Eq. (5.66) the definitions of the slow-roll parameters given in Eq. (5.97) lead quite simply to the following expressions:

$$\epsilon = \frac{1}{\beta}, \qquad \eta = -\frac{1}{\beta}, \tag{5.105}$$

which can be smaller than 1 provided $\beta > 1$ as already required in the process of deriving the solution.

We conclude this section by mentioning a rather naive observation. In chapter 4 it was observed that, during the stage of reheating the Universe might be effectively dominated by dusty matter. This is due to the fact that prior to its decay the inflaton may experience a phase of so-called coherent oscillations. It is not our purpose to enter the details of reheating (which are, anyway, model-dependent). We want just to justify the statement that the coherent oscillations of the inflaton might give rise to a scale factor that expands, in average, like the one of dusty matter. Consider therefore the evolution of φ in a potential that can be approximated around its minimum φ_0 as $V(\varphi) = (m^2/2)(\varphi - \varphi_0)^2$:

$$\ddot{\varphi} + 3H\dot{\varphi} + m^2(\varphi - \varphi_0) = 0. \tag{5.106}$$

Defining the auxiliary variable $\overline{\varphi} = a^{3/2}\varphi$, Eq. (5.106) can also be written as

$$\ddot{\overline{\varphi}} + \left[m^2 - a^{-3/2}\frac{d^2 a^{3/2}}{dt^2}\right]\overline{\varphi} = 0. \tag{5.107}$$

If $H < m$ the second term in the square bracket of Eq. (5.107) can be approximately neglected and the solution of Eq. (5.107) becomes

$$\varphi(t) = \varphi_0 + A_1 \frac{\cos mt}{a^{3/2}} + A_2 \frac{\sin mt}{a^{3/2}}. \tag{5.108}$$

If we choose to set initial conditions in such a way that, for instance, $A_2 = 0$, the solution will lead, in the limit $mt > 1$, to an energy density and pressure given by:

$$\rho_\varphi = \frac{m^2 A_1^2}{2a^3}, \qquad p_\varphi = -\frac{m^2 A_1^2}{2a^3} \cos 2mt. \tag{5.109}$$

If we now average over one period of oscillations $\langle p_\varphi \rangle = 0$ while $\langle \rho_\varphi \rangle \propto a^{-3}$ which shows that, effectively, the scalar field behaves, in this regime, like dusty matter. Of course this naive example depends crucially upon the form of the potential near the minimum. If the potential would be quartic (rather than quadratic, as assumed above) the effective equation of state would change and it would be, in particular, the one of radiation.

It has been realized that the reheating phase is actually preceded by a phase customarily called preheating. The observation is here the following. Suppose that the inflaton field φ is coupled to other (massive) degrees of freedom. Suppose, for simplicity, that these fields are scalars. Then the resulting evolution equations for the massive fields coupled to a coherently oscillating inflaton will be of Mathieu type [215]. Mathieu equation is characterized by resonant bands and this occurrence implies the possibility of the so called parametric resonance[176–178]. During preheating the classical coherently oscillating inflaton field decays into massive particles due to parametric resonance. This stage is also dubbed as explosive decay. To be precise the explosive decay only arises when the resonant band of the corresponding Mathieu equation is sufficiently broad. Then the produced (massive) particles will decay. Because of Pauli blocking, fermions cannot experience explosive decay. Further discussions on preheating can be found in Refs. [179, 180].

Without entering the details of preheating studies we can say that one of the attempts of this approach is to stress that there indeed exist models where reheating is rather fast and efficient so that, after all, the sudden decay approximation is not so insane and it is, indeed, widely used in order to assess, for instance, the maximal number of e-folds accessible through the analysis of CMB anisotropies. There exist, however, radically different possibilities also. For instance it can happen that, after inflation, a relatively long stiff phase takes place [181]. This can occur, for instance, in quintessential inflationary models [182] (see also [183] and [184]). The logic here is that the inflaton field does not decay and after inflation is identified with the quintessence field. Consequently these models have been dubbed quintessential inflationary models [182]. In these models the stage of reheating takes place after a stiff phase that may be rather long. The mechanism invoked for instance in [182] is due to the production of quanta in the transition from the inflationary to the stiff phase. If the coupling of the degrees of freedom that evolve during the quasi-de Sitter stage of expansion is nearly conformal [185], then the typical energy density will scale as H_i^4. The gravitationally produced fluctuations will then scale as radiation. So, when the background is dominated, after inflation, by the kinetic energy of the inflaton field, the background energy density will scale as a^{-6} while the energy density of the produced quantum fluctuations will scale as a^{-4}, i.e. at a slower rate. This implies that, at some point, the energy density of the redshifted quantum fluctuations will become dominant providing and will reheat the Universe.

5.6 Slow-roll parameters

The slow-roll parameters of Eq. (5.97) can be directly expressed in terms of the potential and of its derivatives by using Eqs. (5.98) and (5.99). The result of this calculation is that

$$\epsilon = -\frac{\dot{H}}{H^2} = \frac{\overline{M}_P^2}{2}\left(\frac{V_{,\varphi}}{V}\right)^2, \tag{5.110}$$

$$\eta = \frac{\ddot{\varphi}}{H\dot{\varphi}} = \epsilon - \overline{\eta}, \qquad \overline{\eta} = \overline{M}_P^2\left(\frac{V_{\varphi\varphi}}{V}\right), \tag{5.111}$$

where $V_{,\varphi}$ and $V_{,\varphi\varphi}$ denote, respectively, the first and second derivatives of the potential with respect to φ. Equations (5.110) and (5.111) follow from Eqs. (5.97) by using Eqs. (5.98) and (5.99). From the definition of ϵ (i.e. Eqs. (5.97)) we can write

$$\epsilon = -\frac{1}{H^2}\frac{\partial H}{\partial \varphi}\dot{\varphi} = \frac{1}{3H^3}\left(\frac{\partial H}{\partial \varphi}\right)\left(\frac{\partial V}{\partial \varphi}\right). \tag{5.112}$$

But from Eq. (5.98) it also follows that

$$H\frac{\partial H}{\partial \varphi} = \frac{1}{6\overline{M}_P^2}\frac{\partial V}{\partial \varphi} \tag{5.113}$$

Inserting now Eq. (5.113) into Eq. (5.112) and recalling Eq. (5.98), Eq. (5.110) is swiftly obtained.

Consider next the definition of η as it appears in Eq. (5.97) and let us write it as

$$\eta = -\frac{1}{H\dot{\varphi}}\frac{\partial}{\partial t}\left[\frac{1}{3H}\frac{\partial V}{\partial \varphi}\right] = -\frac{1}{H\dot{\varphi}}\left[-\frac{\dot{H}}{3H^2}\frac{\partial V}{\partial \varphi} + \frac{\dot{\varphi}}{3H}\frac{\partial^2 V}{\partial \varphi^2}\right], \tag{5.114}$$

where the second time derivative has been made explicit. Recalling now the definition of ϵ as well as Eq. (5.99), Eq. (5.114) can be written as

$$\eta = \epsilon - \overline{\eta}, \qquad \overline{\eta} = \overline{M}_{\rm P}^2 \frac{V_{,\varphi\varphi}}{V}. \tag{5.115}$$

It is now possible to illustrate the use of the slow-roll parameters by studying, in rather general terms, the total number of e-folds and by trying to express it directly in terms of the slow-roll parameters. Consider, first of all, the following way of writing the total number of e-folds:

$$N = \int_{a_i}^{a_f} \frac{da}{a} = \int_{t_i}^{t_f} H\, dt = \int_{\varphi_i}^{\varphi_f} \frac{H}{\dot\varphi} d\varphi. \tag{5.116}$$

Using now Eq. (5.98) and, then, Eq. (5.99) inside Eq. (5.116) we do get the following chain of equivalent expressions:

$$N = -\int_{\varphi_i}^{\varphi_f} \frac{3H^2}{V_{,\varphi}} d\varphi = \int_{\varphi_f}^{\varphi_i} \frac{d\varphi}{\overline{M}_{\rm P}\sqrt{2\epsilon}} \tag{5.117}$$

Let us now give a simple and well known example, i.e. the case of a monomial potential.[f] Recently of the form $V(\varphi) \propto \varphi^n$. In this case Eqs. (5.110) and (5.117) imply, respectively,

$$\epsilon = \frac{\overline{M}_{\rm P}^2}{2}\frac{n^2}{\varphi^2}, \qquad N = \frac{\varphi_i^2 - \varphi_f^2}{2n\,\overline{M}_{\rm P}^2}. \tag{5.118}$$

Let us now ask the following pair of questions:

- what was the value of φ say 60 e-folds before the end of inflation?
- what was the value of ϵ 60 e-folds before the end of inflation?

To answer the first question let us recall that inflation ends when $\epsilon(\varphi_f) \simeq 1$. Thus from Eq. (5.118) we will have, quite simply, that

$$\varphi_{60}^2 = \frac{n(n+240)}{2}\overline{M}_{\rm P}^2. \tag{5.119}$$

Consequently, the value of ϵ corresponding to 60 e-folds before the end of inflation is given by

$$\epsilon(\varphi_{60}) = \frac{n}{n+240}, \tag{5.120}$$

which is, as it should be, smaller than one.

---

[f]It is clear that monomial potentials are not so realistic for various reasons. A more general approach to the study of generic polynomial potentials has been recently developed [186–191]. In this framework the inflaton field is not viewed as a fundamental field but rather as a condensate. Such a description bears many analogy with the Landau-Ginzburg description of superconducting phases.

This last example, together with the definition of slow-roll parameters suggests a second class of inflationary models which has been illustrated in Fig. 5.1. The slow-roll dynamics is also realized if, in the case of monomial potential φ is sufficiently large in Planck units. These are the so-called large field models. Notice that to have a field $\varphi > \overline{M}_P$ does not imply that the energy density of the field is larger than the Planck energy density.

The slow-roll algebra introduced in this section allows us to express the spectral indices of the scalar and tensor modes of the geometry in terms of ϵ and $\overline{\eta}$. The technical tools appropriate for such a discussion are collected in chapter 10. The logic is, in short, the following. The slow-roll parameters can be expressed in terms of the derivatives of the potential. Now, the spectra of the scalar and tensor fluctuations of the geometry (allowing the comparison of the model with the data of the CMB anisotropies) can be expressed, again, in terms of ϵ and $\overline{\eta}$. Consider, as an example, the case of single-field inflationary models. In this case the scalar and tensor power spectra (i.e. the Fourier transforms of the two-point functions of the corresponding fluctuations) are computed in chapter 10 (see, in particular, the final formulas reported in Eqs. (10.81), (10.82) and (10.83)). Therefore, according to the results derived in chapter 10 we will have, in summary, that the power spectra of the scalar and tensor modes can be parametrized as

$$\mathcal{P}_T \simeq k^{n_T}, \qquad \mathcal{P}_\mathcal{R} \simeq k^{n_s - 1}, \qquad (5.121)$$

where n_T and n_s are, respectively, the tensor and scalar spectral indices. Using the slow-roll algebra of this section and following the derivation of Appendix D, n_s and n_T can be related to ϵ and $\overline{\eta}$ as

$$n_s = 1 - 6\epsilon + 2\overline{\eta}, \qquad n_T = -2\epsilon. \qquad (5.122)$$

The ratio of the scalar and tensor power spectra is usually called r and it is also a function of ϵ, more precisely (see section 10),

$$r = \frac{\mathcal{P}_T(k_p)}{\mathcal{P}_\mathcal{R}(k_p)} = 16\epsilon = -8n_T. \qquad (5.123)$$

Equation (5.123) is often named consistency condition and the wave-number k_p is the pivot scale at which the scalar and tensor power spectra are normalized. A possible choice in order to parametrize the inflationary predictions is to assign the amplitude of the scalar power spectrum, the scalar and tensor spectral indices and the the r-parameter.

The development of this script can now follow two complementary (but logically very different) approaches. In the first approach we may want to assume that the whole history of the Universe is known and it consists of an inflationary phase almost suddenly followed by a radiation-dominated stage of expansion which is replaced, after equality, by the matter and by the dark energy epochs. In this first approach the initial conditions for the scalar and tensor fluctuations of the geometry will be set during inflation. There is a second approach where the initial conditions for CMB anisotropies are set after weak interactions have fallen out of thermal equilibrium. In this second approach the scalar and tensor power spectra are taken as free parameters and are assigned when the relevant wavelengths of the perturbations are still larger than the Hubble radius after matter-radiation equality but prior to decoupling. In what follows the second approach will be developed. In chapter 10 the inflationary power spectra will instead be computed within the first approach.

Part III

Physics of the Cosmic Microwave Background anisotropies

Chapter 6

Inhomogeneities in FRW Models

Up to now the essentials of the SCM (and of its conventional inflationary completions) have been introduced dealing, predominantly, with homogeneous background geometries. However, as it has been repeatedly mentioned in the previous chapter, the observed Universe might possess, in its infancy, minute inhomogeneities that are not accounted for in the context of a fully homogeneous paradigm. The path that will now be undertaken concerns, therefore, the evolution of the inhomogeneities in the case of FRW backgrounds. This topic is not only interesting per se but it is absolutely mandatory for the calculation of CMB anisotropies both in conventional and unconventional inflationary models.

In the present chapter we will also discuss how to use quantum mechanics to normalize the initial inhomogeneities of the the geometry. Usually this idea is attributed to inflationary models but this is not exactly the case. A. D. Sakharov [192] has been one of the first to invoke quantum mechanics to set initial conditions for the evolution of metric fluctuations. The title of the paper reported in Ref. [192] is *The initial stage of an expanding Universe and the appearance of a non-uniform distribution of matter*. Quoting verbatim from the abstract of Ref. [192] *"A hypothesis for the creation of astronomical bodies as a result of gravitational instability of the expanding Universe is investigated. It is assumed that the initial inhomogeneities arise as a result of quantum fluctuations of cold baryon-lepton matter..."*. The model adopted by Sakharov was, according to the modern perspective, not so reasonable. Still, in spite of questionable premises the quantum nature of initial conditions represents both a very natural and intriguing option. In this paper, among other things, Sakharov also develops (in rather obscure manner) the theory of Doppler oscillations and this is the reason why Doppler oscillations are sometimes dubbed Sakharov oscillations. It is amusing to notice that [192] is rarely quoted in current literature where, on the contrary, the most popular paper of Sakharov is regarded to be the one spelling out the successful conditions for the generation of the baryon asymmetry of the Universe [193].

The content of the present chapter can be viewed, in general terms, as an introduction to the treatment of inhomogeneities in FRW backgrounds. After the first two general sections, the tensor modes of the geometry will be directly analyzed. More specifically, in the present chapter the following topics will be addressed:

- decomposition of inhomogeneities in FRW Universes;
- gauge issues for the scalar modes;
- evolution of the tensor modes;
- quantum mechanical treatment of the tensor modes;
- spectra of relic gravitons;
- the quantum state of cosmological fluctuations.

At the end of the present chapter two complementary topics will also be treated:

- a digression on the possible ambiguities connected with the quantum mechanical normalization of the tensor modes;
- a numerical approach that allows to compute the graviton spectra.

The evolution of the tensor modes and their quantum-mechanical normalization will lead to the calculation of the spectral properties of relic gravitons. For practical reasons it is better to study first the evolution of the tensor modes. They have the property of not being coupled with the matter sources. In the simplest case of FRW models they are only sensitive to the evolution of the curvature. Moreover, the amplification of quantum-mechanical (tensor) fluctuations is technically easier. The analog phenomenon (arising in the case of the scalar modes of the geometry) will be separately discussed for the simplest case of the fluctuations induced by a (single) scalar field (see, in particular, chapter 10 and part of Appendix C). For technical reasons, the conformal time parametrization is more convenient for the treatment of the fluctuations of FRW geometries The background equations in the conformal time parametrization have been derived in Eqs. (2.4) and (2.65)–(2.67).

6.1 Decomposition of inhomogeneities in FRW Universes

Given a conformally flat background metric of FRW type

$$\overline{g}_{\mu\nu}(\tau) = a^2(\tau)\eta_{\mu\nu}, \qquad (6.1)$$

its first-order fluctuations can be written as

$$\delta g_{\mu\nu}(\tau, \vec{x}) = \delta_{\rm s} g_{\mu\nu}(\tau, \vec{x}) + \delta_{\rm v} g_{\mu\nu}(\tau, \vec{x}) + \delta_{\rm t} g_{\mu\nu}(\tau, \vec{x}), \qquad (6.2)$$

where the subscripts define, respectively, the scalar, vector and tensor perturbations classified according to rotations in the three-dimensional Euclidean sub-manifold. Being a symmetric rank-two tensor in four-dimensions, the perturbed metric $\delta g_{\mu\nu}$ has, overall, 10 independent components whose explicit form will be parametrized as[a]

$$\delta g_{00} = 2a^2\phi, \qquad (6.3)$$
$$\delta g_{ij} = 2a^2(\psi\delta_{ij} - \partial_i\partial_j E) - a^2 h_{ij} + a^2(\partial_i W_j + \partial_j W_i), \qquad (6.4)$$
$$\delta g_{0i} = -a^2\partial_i B - a^2 Q_i, \qquad (6.5)$$

together with the conditions

$$\partial_i Q^i = \partial_i W^i = 0, \qquad h_i^i = \partial_i h_j^i = 0. \qquad (6.6)$$

The decomposition expressed by Eqs. (6.3)–(6.5) and (6.6) is the one normally employed in the Bardeen formalism [194–198] (see also [199, 200]) and it is the one adopted in [33] to derive, consistently, the results relevant for the theory of CMB anisotropies. In the Appendix (see, in particular, Appendix C and Appendix D) the fluctuations of the Christoffel connections as well as of the Ricci and Einstein tensors are reported within the decomposition expressed by Eqs. (6.3)–(6.6). If a different decomposition is adopted, it will suffice to express the relevant degrees of freedom in the new decomposition in terms of the ones appearing in Eqs. (6.3)–(6.5). An explicit example of this trivial observation will be given in chapter 11 in connection with the synchronous coordinate system.

Concerning Eqs. (6.3)–(6.5) few comments are in order:

- the scalar fluctuations of the geometry are parametrized by 4 scalar functions, i. e. ϕ, ψ, B and E;

---
[a]The partial derivations with respect to the spatial indices arise as a result of the explicit choice of dealing with a spatially flat manifold. In the case of a spherical or hyperbolic spatial manifold they will be replaced by the appropriate covariant derivative defined on the given spatial section.

- the vector fluctuations are described by the two (divergenceless) vectors in three (spatial) dimensions W_i and Q_i, i.e. by 4 independent degrees of freedom;
- the tensor modes are described by the three-dimensional rank-two tensor h_{ij}, leading, overall, to 2 independent components because of the last two conditions of Eq. (6.6).

The strategy will then be to obtain the evolution equations for the (separate) scalar, vector and tensor contributions. To achieve this goal we can either perturb the most appropriate form of the Einstein equation to first-order in the amplitude of the fluctuations, or we may perturb the action to second-order in the amplitude of the same fluctuations. Schematically, within the first approach we are led to compute

$$\delta^{(1)} R_\mu^\nu - \frac{1}{2} \delta_\mu^\nu \delta^{(1)} R = 8\pi G \delta^{(1)} T_\mu^\nu, \qquad (6.7)$$

where $\delta^{(1)}$ denotes the first-order variation with respect either to the scalar, vector or tensor modes. Of course it will be very convenient to perturb also the covariant conservation of the sources which will give rise to a supplementary set of equations that are not independent from Eqs. (6.7) (see also last part of Appendix D).

The form of the energy-momentum tensor depends on the specific physical application. For instance around recombination, the matter sources are represented by the total energy-momentum tensor of the fluid (i.e. baryons photons, neutrinos and dark matter). During inflation, the matter content will be given by the scalar degrees of freedom whose dynamics produces the inflationary evolution. In the simplest case of a *single* scalar degree of freedom this analysis will be discussed in chapter 10.

Instead of perturbing the equations of motion to first-order in the amplitude of the fluctuations, it is possible to perturb the gravitational and matter parts of the action to second-order in the amplitude of the fluctuations, i.e. formally

$$\delta^{(2)} S = \delta^{(2)} S_\text{g} + \delta^{(2)} S_\text{m}. \qquad (6.8)$$

From the second order action of Eq. (6.8) the canonical normal modes of the system can be found and promoted to the status of quantum-mechanical operators. For the success of such an approach it is essential to perturb the action to second order in the amplitude of the fluctuations. This step will give us the Hamiltonian of the fluctuations leading, ultimately, to the evolution of the field operators in the Heisenberg representation.

6.2 Gauge issues for the scalar modes

The discussion of the perturbations on a given background geometry is complicated by the fact that, for infinitesimal coordinate transformations of the type

$$x^\mu \to \tilde{x}^\mu = x^\mu + \epsilon^\mu, \tag{6.9}$$

the fluctuation of a rank-two (four-dimensional) tensor changes according to the Lie derivative in the direction ϵ^μ. It can be easily shown that the fluctuations of a tensor $T_{\mu\nu}$ change, under the transformation (6.9) as (see, for instance, [102]):

$$\delta T_{\mu\nu} \to \tilde{\delta T}_{\mu\nu} = \delta T_{\mu\nu} - T_\mu^\lambda \nabla_\nu \epsilon_\lambda - T_\nu^\lambda \nabla_\lambda \epsilon_\mu - \epsilon^\lambda \nabla_\lambda T_{\mu\nu}, \tag{6.10}$$

where the covariant derivatives are performed by using the background metric which is given, in our case, by Eq. (6.1). Thus, for instance, we will have

$$\nabla_\mu \epsilon_\nu = \partial_\mu \epsilon_\nu - \overline{\Gamma}^\sigma_{\mu\nu} \epsilon_\sigma, \tag{6.11}$$

where $\overline{\Gamma}^\sigma_{\mu\nu}$ are Christoffel connections computed using the background metric (6.1) and they are

$$\overline{\Gamma}^0_{ij} = \mathcal{H}\delta_{ij}, \quad \overline{\Gamma}^0_{00} = \mathcal{H}, \quad \overline{\Gamma}^j_{i0} = \mathcal{H}\delta^j_i, \quad \mathcal{H} = \frac{a'}{a}. \tag{6.12}$$

If $T_{\mu\nu}$ coincides with the metric tensor, then the metricity condition allows to simplify (6.10) which then becomes:

$$\delta g_{\mu\nu} \to \tilde{\delta g}_{\mu\nu} = \delta g_{\mu\nu} - \nabla_\mu \epsilon_\nu - \nabla_\nu \epsilon_\mu, \tag{6.13}$$

where

$$\epsilon_\mu = a^2(\tau)(\epsilon_0, -\epsilon_i), \tag{6.14}$$

is the shift vector that induces the explicit transformation of the coordinate system, namely:

$$\tau \to \tilde{\tau} = \tau + \epsilon_0, \quad x^i \to \tilde{x}^i = x^i + \epsilon^i. \tag{6.15}$$

Equation (6.13) can be also written as

$$\delta g_{\mu\nu} \to \tilde{\delta g}_{\mu\nu} = \delta g_{\mu\nu} - \Delta_\epsilon, \tag{6.16}$$

where, Δ_ϵ is the Lie derivative in the direction ϵ_μ. The functions ϵ_0 and ϵ_i are often called gauge parameters since the infinitesimal coordinate transformations of the type (6.15) form a group which is in fact the gauge group of gravitation. The gauge-fixing procedure amounts, in four space-time dimensions, to fix the four independent functions ϵ_0 and ϵ_i. As they are, the three gauge parameters ϵ_i (one for each axis) will affect both scalars and three-dimensional vectors. To avoid this possible confusion, the gauge parameters ϵ_i can be separated into their divergenceless and divergencefull parts, i.e.

$$\epsilon_i = \partial_i \epsilon + \zeta_i, \tag{6.17}$$

where $\partial_i \zeta^i = 0$. Two relevant remarks are now in order:

- the gauge transformations involving ϵ_0 and ϵ preserve the scalar nature of the fluctuations;
- the gauge transformations parametrized by ζ_i preserve the vector nature of the fluctuation.

The fluctuations in the tilded coordinate system, defined by the transformation of Eq. (6.15), can then be written as

$$\phi \to \tilde{\phi} = \phi - \mathcal{H}\epsilon_0 - \epsilon'_0, \tag{6.18}$$

$$\psi \to \tilde{\psi} = \psi + \mathcal{H}\epsilon_0, \tag{6.19}$$

$$B \to \tilde{B} = B + \epsilon_0 - \epsilon', \tag{6.20}$$

$$E \to \tilde{E} = E - \epsilon, \tag{6.21}$$

in the case of the scalar modes of the geometry. Equations (6.18), (6.19), (6.20) and (6.21) show, as anticipated, that the scalar gauge transformations only involve ϵ_0 and ϵ. Under a coordinate transformation preserving the vector nature of the fluctuation, i.e.

$$x^i \to \tilde{x}^i = x^i + \zeta^i, \qquad \partial_i \zeta^i = 0, \tag{6.22}$$

the rotational modes of the geometry transform as

$$Q_i \to \tilde{Q}_i = Q_i - \zeta'_i, \tag{6.23}$$

$$W_i \to \tilde{W}_i = W_i + \zeta_i. \tag{6.24}$$

Again, Eqs. (6.23) and (6.24) show that the vector gauge transformations only involve ζ_i and its first derivative so that the gauge transformations generated by ζ_i preserve the vector nature of the fluctuations.

Finally, the tensor fluctuations, in the parametrization of Eq. (6.4), are automatically invariant under infinitesimal diffeomorphisms, i.e. $\tilde{h}_{ij} = h_{ij}$. It is possible to select appropriate combinations of the fluctuations introduced in Eqs. (6.3)–(6.5) that are invariant under infinitesimal coordinate transformations. This possibility is particularly clear in the case of the vector modes. If we define the quantity

$$V_i = Q_i + W'_i, \tag{6.25}$$

we will have, according to Eqs. (6.23) and (6.24), that
$$x^i \to \tilde{x}^i = x^i + \zeta^i, \qquad \tilde{V}_i = V_i, \tag{6.26}$$
i.e. V_i is invariant for infinitesimal coordinate transformations or, for short, gauge-invariant. The same trick can be used in the scalar case. In the scalar case the most appropriate gauge-invariant fluctuations depend upon the specific problem at hand. An example of fully gauge-invariant fluctuations arising, rather frequently, in the treatment of scalar fluctuations is given in section 10 (see in particular Eqs. (10.10), (10.9) and (10.11)). Furthermore, in chapter 11 various gauge-invariant and gauge-dependent treatments will be discussed mainly for the case of scalar fluctuations. The perturbed components of the energy-momentum tensor can be written, for a single species λ, as:
$$\delta T_0^0 = \delta\rho_\lambda, \quad \delta T_i^j = -\delta p_\lambda \delta_i^j, \quad \delta T_0^i = (p_\lambda + \rho_\lambda)\partial^i v^{(\lambda)}, \tag{6.27}$$
where we defined $\delta u_i^{(\lambda)} = \partial_i v^{(\lambda)}$ and where the index λ denotes the specific component of the fluid characterized by a given barotropic index and by a given sound speed. It is also appropriate, for applications, to work directly with the divergence of the peculiar velocity field by defining a variable $\theta_\lambda = \nabla^2 v^{(\lambda)}$. Under the infinitesimal coordinate transformations of Eq. (6.15) the fluctuations given in Eq. (6.27) transform according to Eq. (6.10) and the explicit results are
$$\delta\rho_\lambda \to \delta\tilde{\rho}_\lambda = \delta\rho_\lambda - \rho'_\lambda \epsilon_0, \tag{6.28}$$
$$\delta p_\lambda \to \delta\tilde{p}_\lambda = \delta p_\lambda - w_\lambda \rho'_\lambda \epsilon_0, \tag{6.29}$$
$$\theta_\lambda \to \tilde{\theta}_\lambda = \theta_\lambda + \nabla^2 \epsilon'. \tag{6.30}$$
Using the covariant conservation equation for the background fluid density, the gauge transformation for the density contrast, i.e. $\delta_{(\lambda)} = \delta\rho_{(\lambda)}/\rho_{(\lambda)}$, follows easily from Eq. (6.28):
$$\tilde{\delta}_{(\lambda)} = \delta_{(\lambda)} + 3\mathcal{H}(1 + w_{(\lambda)})\epsilon_0. \tag{6.31}$$
There are now, schematically, three possible strategies:

- a specific gauge can be selected by fixing (completely or partially) the coordinate system; this will fix, in the scalar case, the two independent functions ϵ_0 and ϵ;
- gauge-invariant fluctuations of the sources and of the geometry can be separately defined;
- gauge-invariant fluctuations mixing the perturbations of the sources and of the geometry can be employed.

The vector modes are not so relevant in the conventional scenarios. If the Universe is expanding, the vector modes will always be damped depending upon the barotropic index of the sources of the geometry. This result has been obtained long ago [201]. However, if the geometry contracts or if internal dimensions are present in the game [202, 203], such a statement is no longer true. These topics, however, involve unconventional completions of the standard cosmological model and will not receive here specific attention.

The problems related to the choice of a gauge for the scalar modes will be the guiding theme of chapter 11. So we encourage the interested reader to find there a more complete discussion on these themes.

6.3 Super-adiabatic amplification

The evolution of the tensor modes of the geometry can be obtained, as stressed before, either from the Einstein equations perturbed to first-order (i.e. Eq. (6.7)) or from the action perturbed to second order (i.e. Eq. (6.8)). Consider, to begin with, the case of the tensor modes of the geometry, i.e., according to Eq. (6.6), the two polarizations of the graviton:

$$\delta_t g_{ij} = -a^2 h_{ij}, \qquad \delta_t g^{ij} = \frac{h^{ij}}{a^2}. \tag{6.32}$$

The tensor contribution to the fluctuation of the connections can then be expressed as[b]

$$\delta_t \Gamma^0_{ij} = \frac{1}{2}(h'_{ij} + 2\mathcal{H} h_{ij}), \qquad \delta_t \Gamma^j_{i0} = \frac{1}{2} h_i^{j\prime},$$

$$\delta_t \Gamma^k_{ij} = \frac{1}{2}[\partial_j h_i^k + \partial_i h_j^k - \partial^k h_{ij}]. \tag{6.33}$$

Inserting these results into the perturbed expressions of the Ricci tensors it is easy to obtain:

$$\delta_t R_{ij} = \frac{1}{2}[h''_{ij} + 2\mathcal{H} h'_{ij} + 2(\mathcal{H}' + 2\mathcal{H}^2)h_{ij} - \nabla^2 h_{ij}], \tag{6.34}$$

$$\delta_t R_i^j = -\frac{1}{2a^2}[h_i^{j\prime\prime} + 2\mathcal{H} h_i^{j\prime} - \nabla^2 h_i^j], \tag{6.35}$$

where $\nabla^2 = \partial_i \partial^i$ is the usual four-dimensional Laplacian. In order to pass from Eq. (6.34) to Eq. (6.35) we may recall that

$$\delta_t R_i^j = \delta_t(g^{jk} R_{ki}) = \delta_t g^{jk} \overline{R}_{ki} + \overline{g}^{jk} \delta_t R_{ij}, \tag{6.36}$$

---

[b]See also Appendix D for a more complete collection of formulae involving the tensor fluctuations of the geometry.

where the relevant Ricci tensor, i.e. $\overline{R}_{ij}$ is simply given (see also Eqs. (2.64)):

$$\overline{R}_{ij} = (\mathcal{H}' + 2\mathcal{H}^2)\delta_{ij}. \tag{6.37}$$

Since both the fluid sources and the scalar fields do not contribute to the tensor modes of the geometry the evolution equation for h_i^j becomes, in Fourier space, by

$$h_i^{j''} + 2\mathcal{H} h_i^{j'} + k^2 h_i^j = 0. \tag{6.38}$$

Thanks to the conditions $\partial_i h_j^i = h_k^k = 0$ (see Eq. (6.6)), the direction of propagation can be chosen to lie along the third axis and, in this case, the two physical polarizations of the graviton will be

$$h_1^1 = -h_2^2 = h_\oplus, \qquad h_1^2 = h_2^1 = h_\otimes, \tag{6.39}$$

where $h_\oplus$ and $h_\otimes$ obey the same evolution equation (6.38) and will be denoted, in the remaining part of this section, by h. Equation (6.38) can also be written in one of the following two equivalent forms:

$$h_k'' + 2\mathcal{H} h_k' + k^2 h_k = 0, \tag{6.40}$$

$$\mu_k'' + \left[k^2 - \frac{a''}{a}\right]\mu_k = 0, \tag{6.41}$$

where $\mu_k = a h_k$. Equation (6.41) simply follows from Eq. (6.40) by eliminating the first time derivative. Equations (6.40) and (6.41) are useful to highlight different properties of the solutions. For instance Eq. (6.40) can be solved in the limit when the relevant wavelengths are larger than the Hubble radius[c] (i.e. $|k\tau| \ll 1$). In this approximation the term containing k^2 can be neglected in Eq. (6.41). Therefore h_k will have a constant mode for $|k\tau| \ll 1$. In this sense h is the tensor counterpart of the gauge-invariant variable which will be denoted by $\mathcal{R}$ and which measures the fluctuations of the scalar curvature on comoving orthogonal hypersurfaces (see chapter 11 and discussions therein). Equation (6.40) can be written also in a third form. By defining the new time variable [204–206]

$$dr = \frac{d\tau}{a^2(\tau)} = \frac{dt}{a^3(t)}, \tag{6.42}$$

---

[c]Here it is appropriate to recall the concept of horizon exit and horizon reentry that has been already introduced, in qualitative terms, at the end of chapter 4. Here it will turn out that a more precise description of the amplification process implies an analogy with a potential barrier. According to this analogy the wavelengths that are larger than the Hubble radius will be under the potential barrier while those wavelengths that are inside the Hubble radius will be outside the potential barrier. This will be one of the themes of the present section.

Eq. (6.40) can be rearranged as:

$$\frac{d^2 h_k}{dr^2} + \Omega^2(k,r) h_k = 0, \qquad \Omega(k,r) = k a^2(r), \qquad (6.43)$$

leading to an oscillating behaviour in the short wavelength limit (i.e. $|k\tau| > 1$) and to a solution $h_k \simeq C_1(k) + C_2(k) r$ in the long wavelength limit.

Concerning Eq. (6.41), which will be often employed in the forthcoming sections of this chapter, two comments are in order:

- in the limit where k^2 dominates over $|a''/a|$ the solution of the Eq. (6.41) are simple plane waves;
- in the opposite limit, i.e. $|a''/a| \gg k^2$ the solution may exhibit, under certain conditions, a growing mode.

In more quantitative terms, the solutions of Eq. (6.41) in the two aforementioned limits are

$$k^2 \gg |a''/a|, \qquad \mu_k(\tau) \simeq C_\pm(k) e^{\pm i k \tau} \qquad (6.44)$$

$$k^2 \ll |a''/a|, \qquad \mu_k(\tau) \simeq A_k a(\tau) + B_k a(\tau) \int^\tau \frac{dx}{a^2(x)}. \qquad (6.45)$$

The oscillatory regime is sometimes called *adiabatic* since, in this regime, $h_k \simeq a^{-1}$. If the initial fluctuations are normalized to quantum mechanics (see the following part of the present section) $\mu_k \simeq 1/\sqrt{k}$ initially and, therefore

$$\delta_h \simeq k^{3/2} |h_k(\tau)| \simeq \frac{k}{a} \simeq \omega, \qquad (6.46)$$

where $\omega(\tau) = k/a(\tau)$ denotes, in the present context, the physical wave-number while k denotes the comoving wave-number. Recalling that $\mu_k = a h_k$, Eq. (6.45) implies that

$$h_k(\tau) \simeq A_k + B_k \int^\tau \frac{dx}{a^2(x)}. \qquad (6.47)$$

This solution describes what is often named *super-adiabatic amplification*. In particular cases (some of which of practical interest) Eq. (6.47) implies the presence of a decaying mode and of a constant mode. Since in the adiabatic regime $h_k \simeq 1/a$, the presence of a constant mode would imply a growth with respect to the adiabatic solution hence the name super-adiabatic [207, 208]. We pause here for a moment to say that the adjective *adiabatic* is sometimes used not in direct relation with thermodynamic notions as we shall also see in the case of the so-called adiabatic perturbations. It should be recalled that the first author to notice that the tensor modes of the geometry can be amplified in FRW backgrounds was L. P. Grishchuk [207, 208] (see also [209–212]).

Equation (6.41) suggests an interesting analogy for the evolution of the tensor modes of the geometry since it can be viewed, for practical purposes, as a Schrödinger-like equation where the analog of the wave-function does not depend on a spatial coordinate (like in the case of one-dimensional potential barriers) but on a time coordinate (the conformal time in the case of Eq. (6.41)). The counterpart of the potential barrier is represented by the term a''/a sometimes also called *pump field*. The physics of the process is therefore rather simple: energy is transferred from the background geometry to the corresponding fluctuations. However, the success and effectiveness of such a transfer depends ultimately on the peculiar features of the background geometry. For instance, the pump field a''/a vanishes in the case of a radiation-dominated Universe. In this case the evolution equations of the tensor modes are said to be conformally invariant (or, more correctly, Weyl invariant) since, with an appropriate rescaling the evolution equations have the same form they would have in the Minkowskian space-time. On the contrary, in the case of de Sitter expansion[d]

$$a(\tau) \simeq (-\tau_1/\tau), \qquad \frac{a''}{a} = \frac{2}{\tau^2}. \tag{6.48}$$

It should be appreciated that expanding de Sitter space-time supports the evolution of the tensor modes of the geometry while the scalar modes, in the pure de Sitter case, are not amplified. To get amplification of scalar modes during inflation it will be mandatory to have a phase of quasi-de Sitter expansion.

A more realistic model of the evolution of the background geometry can be achieved by a de Sitter phase that evolves into a radiation-dominated epoch which is replaced, in turn, by a matter-dominated stage of expansion. In the latter case the evolution of the scale factor can be parametrized as:

$$a_{\rm i}(\tau) = \left(-\frac{\tau}{\tau_1}\right)^{-\beta}, \qquad \tau \leq -\tau_1, \tag{6.49}$$

$$a_{\rm r}(\tau) = \frac{\beta\tau + (\beta+1)\tau_1}{\tau_1}, \qquad -\tau_1 \leq \tau \leq \tau_2, \tag{6.50}$$

$$a_{\rm m}(\tau) = \frac{[\beta(\tau+\tau_2) + 2\tau_1(\beta+1)]^2}{4\tau_1[\beta\tau_2 + (\beta+1)\tau_1]}, \qquad \tau > \tau_2, \tag{6.51}$$

where the subscripts in the scale factors refer, respectively, to the inflationary, radiation and matter-dominated stages. As already discussed, a

---
[d]In the cosmic time parametrization the (expanding) de Sitter metric is parametrized as $a(t) = e^{H_1 t}$. Recalling that $\tau = \int dt/a(t)$ it follows, after integration, that $a(\tau) \simeq \tau^{-1}$.

generic power-law inflationary phase is characterized by a power β. In the case $\beta = 1$, from Eq. (6.49), the case of the expanding branch of de Sitter space can be recovered. During the radiation-dominated epoch the scale factor expands linearly in the conformal time parametrization while during matter it expands quadratically (see Eqs. (6.50) and (6.51)). The form of the scale factors given in Eqs. (6.49)–(6.51) is continuous and differentiable at the transition points, i.e.

$$a_i(-\tau_1) = a_r(-\tau_1), \qquad a'_i(-\tau_1) = a'_r(-\tau_1),$$
$$a_r(-\tau_2) = a_m(-\tau_2), \qquad a'_r(-\tau_2) = a'_m(-\tau_2). \tag{6.52}$$

The continuity of the scale factor and its derivative prevents the presence of divergences in the pump field, given by a''/a. In Fig. 6.1 the structure

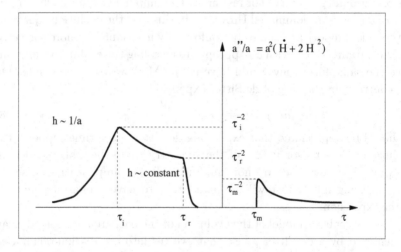

Fig. 6.1 The effective "potential" appearing in Eq. (6.41) is illustrated as a function of the conformal time coordinate τ in the case when the background passes through different stages of expansion. Being conformally invariant in the case of radiation, $a''/a = 0$ in the central part of the plot.

of the potential barrier is reproduced. By using known identities together with the definition of the conformal time τ in terms of the cosmic time t, it is possible to express a''/a in terms of the Hubble parameter and its first (cosmic) time derivative

$$\frac{a''}{a} = \mathcal{H}^2 + \mathcal{H}' = a^2[\dot{H} + 2H^2], \tag{6.53}$$

where, as usual, the prime denotes a derivation with respect to τ and a dot denotes a derivation with respect to t.

6.4 Quantum mechanical description of the tensor modes

The tensor modes of the geometry will now be discussed along a quantum-mechanical perspective. Such a treatment is essential in order to normalize properly the fluctuations, for instance during an initial inflationary phase or during any initial stage of the Universe where the only relevant fluctuations are the ones of quantum mechanical origin. The calculation proceeds, in short, along the following steps:

- obtain the action perturbed to second order in the amplitude of the tensor modes of the geometry;
- define the appropriate normal modes and promote them to the status of (quantum) field operators in the Heisenberg representation;
- solve the evolution of the system and compute the number of produced particles.

To comply with the first step, let us observe that the second-order action can be written, up to (non-covariant) total derivatives, as

$$\delta_t^{(2)} S = \frac{1}{64\pi G} \int d^4x\, a^2(\tau) \eta^{\alpha\beta} \partial_\alpha h_i^j \partial_\beta h_j^i, \qquad (6.54)$$

where $\eta_{\alpha\beta}$ is, as usual, the four-dimensional Minkowski metric with signature mostly minus (i.e. $(+,-,-,-)$). Recalling that the polarization can be chosen as

$$h_1^1 = -h_2^2 = h_\oplus, \qquad h_1^2 = h_2^1 = h_\otimes. \qquad (6.55)$$

and recalling the definition of reduced Planck mass (see Eq. (5.59))

$$\ell_P = \sqrt{8\pi G} \equiv \frac{1}{\overline{M}_P} \equiv \frac{\sqrt{8\pi}}{M_P}, \qquad (6.56)$$

Eq. (6.54) can be rewritten as

$$\delta_t^{(2)} S = \frac{1}{4\ell_P^2} \int d^3x\, d\tau\, a^2(\tau) \big[{h'_\oplus}^2 + {h'_\otimes}^2 - (\partial_i h_\oplus)^2 - (\partial_i h_\otimes)^2\big], \qquad (6.57)$$

becoming, for a single tensor polarization,

$$S_{\rm gw} = \frac{1}{2} \int d^3x\, d\tau\, a^2(\tau) \big[{h'}^2 - (\partial_i h)^2\big], \qquad (6.58)$$

where

$$h = \frac{h_\oplus}{\sqrt{2}\ell_P} = \frac{h_\otimes}{\sqrt{2}\ell_P}, \qquad (6.59)$$

denotes, indifferently, each of the two polarization of the graviton. Defining now the appropriate canonical normal mode of the action (6.58), i.e. $\mu = ah$ we get to the action

$$\tilde{S}_{\text{gw}} = \frac{1}{2} \int d^3x \, d\tau \left[\mu'^2 + \mathcal{H}^2 \mu^2 - 2\mathcal{H}\mu\mu' - (\partial_i \mu)^2 \right]. \qquad (6.60)$$

From Eq. (6.60) non-covariant total derivatives can be dropped. With this method it is clear that the term going as $-2\mathcal{H}\mu\mu'$ can be traded for $(\mathcal{H}\mu^2)'$ by paying the prize of a new term proportional to $\mathcal{H}'\mu^2$. Hence, up to total derivatives, Eq. (6.60) gives:

$$S_{\text{gw}} = \frac{1}{2} \int d^3x \, d\tau \left[\mu'^2 + (\mathcal{H}^2 + \mathcal{H}')\mu^2 - (\partial_i \mu)^2 \right]. \qquad (6.61)$$

From Eq. (6.61) it follows that the Lagrangian and the Hamiltonian of the tensor modes can be expressed in terms of the appropriate Lagrangian and Hamiltonian densities, namely,

$$L_{\text{gw}}(\tau) = \int d^3x \, \mathcal{L}_{\text{gw}}(\tau, \vec{x}), \qquad H_{\text{gw}}(\tau) = \int d^3x \left[\pi \mu' - \mathcal{L}_{\text{gw}} \right]. \qquad (6.62)$$

The quantity π appearing in Eq. (6.62) is the conjugate momentum. From the actions (6.60) and (6.61) the derived conjugate momenta are different and, in particular, they are, respectively:

$$\tilde{\pi} = \mu' - \mathcal{H}\mu, \qquad \pi = \mu'. \qquad (6.63)$$

Equation (6.63) implies that the form of the Hamiltonian changes depending on the specific form of the action. This is a simple reflection of the fact that, in the Lagrangian formalism, the inclusion (or exclusion) of a total derivative does not affect the Euler-Lagrange equations. Correspondingly, in the Hamiltonian formalism, the total Hamiltonian will necessarily change by a time derivative of the generating functional of the canonical transformation. This difference will have, however, no effect on the Hamilton equations. Therefore, the Hamiltonian derived from Eq. (6.61) can be simply expressed as:

$$H_{\text{gw}}(\tau) = \frac{1}{2} \int d^3x \left[\pi^2 - \frac{a''}{a} \mu^2 + (\partial_i \mu)^2 \right]. \qquad (6.64)$$

The Hamiltonian derived from Eq. (6.60) can be instead written as:

$$\tilde{H}_{\text{gw}}(\tau) = \frac{1}{2} \int d^3x \left[\tilde{\pi}^2 + 2\mathcal{H}\mu\tilde{\pi} + (\partial_i \mu)^2 \right]. \qquad (6.65)$$

Suppose than to start with the Hamiltonian of Eq. (6.65) and define the appropriate generating functional of the (time-dependent) canonical transformation, i.e.

$$\mathcal{F}(\mu, \pi, \tau) = \int d^3x \left[\mu\pi - \frac{\mathcal{H}}{2}\mu^2\right], \qquad (6.66)$$

which is, by definition, a functional of the new momenta (i.e. $\tilde{\pi}$). Thus, we will have

$$\tilde{\pi} = \frac{\delta \mathcal{F}}{\delta \mu} = \pi - \mathcal{H}\mu, \qquad (6.67)$$

$$H_{\text{gw}}(\tau) = \tilde{H}_{\text{gw}}(\tau) + \frac{\partial \mathcal{F}}{\partial \tau}. \qquad (6.68)$$

Equation (6.67) gives the new momenta as a function of the old ones so that, if we start with Eq. (6.65) we will need to bear in mind that $\pi = \tilde{\pi} + \mathcal{H}\mu$ and substitute into $\tilde{H}_{\text{gw}}(\tau)$. Equation (6.68) will then allow us to get the $H_{\text{gw}}(\tau)$ reported in Eq. (6.64), as it can be directly verified.

To complete the cases, it should be mentioned that yet another class of Hamiltonians appears naturally in the treatment of the tensor modes of the geometry in FRW backgrounds. Indeed, directly from the Lagrangian density associated with the action reported in Eq. (6.58), the canonical momenta will simply be $\Pi = a^2 h'$ and the Hamiltonian, here denoted with $\overline{H}_{\text{gw}}$ will simply be

$$\overline{H}_{\text{gw}}(\tau) = \frac{1}{2}\int d^3x \left[\frac{\Pi^2}{a^2} + a^2(\partial_i h)^2\right]. \qquad (6.69)$$

This digression on the canonical properties of time-dependent Hamiltonians is useful not so much at the classical level (since, by definition of canonical transformation, the Hamilton equations are invariant) but rather at the quantum level [213]. Indeed, the vacuum will be the state minimizing a given Hamiltonian. It happens that some non-carefully selected Hamiltonians may lead to initial vacua that, indeed, give rise to an energy density of the initial state which is (possibly) larger than the one of the background geometry [214]. The Hamiltonian of Eq. (6.64) is valuable in this respect since the initial vacuum (i.e. the state minimizing (6.64)) possesses an energy density which is usually much smaller than the background, as it should be to have a consistent picture (see also Ref. [33] for the discussion of the so-called transplankian ambiguities). It should be finally remarked that all the imaginable Hamiltonians (connected by time-dependent canonical transformations) lead always to the same quantum evolution either in the Heisenberg or in the Schrödinger description. There are, however, practical differences. For instance, Eq. (6.65) seems more convenient in the Schrödinger description. Indeed at the quantum level the time evolution operator would contain, in the exponential, operator products as $\tilde{\pi}\mu$ which are directly related to the so-called squeezing operator in the theory of optical coherence (see the last part of the present section). In a complementary perspective and always at a practical level, the Hamiltonian defined in Eq. (6.64) is more suitable for the Heisenberg description. In what follows the attention will be focussed on the Hamiltonian (6.64). The interested reader may also consult the last section of the present chapter as well as the last section of chapter 10. In these two sections the minimization of different (but always canonically related) Hamiltonians will be swiftly explored both for the tensor modes and for the scalar modes. The bottom line of the two aforementioned sections will be that the class of Hamiltonians of the type (6.64), when minimized, will lead to an acceptable initial state with negligible back-reaction. The connection of this problem with the so-called trans-Planckian effects (or trans-Planckian ambiguities) will be swiftly explored at the end of chapter 10.

The quantization of the canonical Hamiltonian of Eq. (6.64) is performed by promoting the normal modes of the action to field operators in the Heisenberg description and by imposing (canonical) equal-time commutation relations:

$$[\hat{\mu}(\vec{x},\tau), \hat{\pi}(\vec{y},\tau)] = i\delta^{(3)}(\vec{x}-\vec{y}). \tag{6.70}$$

The operator corresponding to the Hamiltonian (6.64) becomes:

$$\hat{H}(\tau) = \frac{1}{2}\int d^3x \left[\hat{\pi}^2 - \frac{a''}{a}\hat{\mu}^2 + (\partial_i\hat{\mu})^2\right]. \tag{6.71}$$

In Fourier space the quantum fields $\hat{\mu}$ and $\hat{\pi}$ can be expanded as

$$\hat{\mu}(\vec{x},\tau) = \frac{1}{2(2\pi)^{3/2}}\int d^3k \left[\hat{\mu}_{\vec{k}}e^{-i\vec{k}\cdot\vec{x}} + \hat{\mu}_{\vec{k}}^\dagger e^{i\vec{k}\cdot\vec{x}}\right],$$

$$\hat{\pi}(\vec{y},\tau) = \frac{1}{2(2\pi)^{3/2}}\int d^3p \left[\hat{\pi}_{\vec{k}}e^{-i\vec{p}\cdot\vec{y}} + \hat{\pi}_{\vec{k}}^\dagger e^{i\vec{p}\cdot\vec{y}}\right]. \tag{6.72}$$

Demanding the validity of the canonical commutation relations of Eq. (6.70), the Fourier components must obey:

$$[\hat{\mu}_{\vec{k}}(\tau), \hat{\pi}_{\vec{p}}^\dagger(\tau)] = i\delta^{(3)}(\vec{k}-\vec{p}), \qquad [\hat{\mu}_{\vec{k}}^\dagger(\tau), \hat{\pi}_{\vec{p}}(\tau)] = i\delta^{(3)}(\vec{k}-\vec{p}),$$
$$[\hat{\mu}_{\vec{k}}(\tau), \hat{\pi}_{\vec{p}}(\tau)] = i\delta^{(3)}(\vec{k}+\vec{p}), \qquad [\hat{\mu}_{\vec{k}}^\dagger(\tau), \hat{\pi}_{\vec{p}}^\dagger(\tau)] = i\delta^{(3)}(\vec{k}+\vec{p}). \tag{6.73}$$

Inserting now Eq. (6.72) into Eq. (6.64) the Fourier space representation of the quantum Hamiltonian[e] can be obtained:

$$\hat{H}(\tau) = \frac{1}{4}\int d^3k \left[(\hat{\pi}_{\vec{k}}\hat{\pi}^\dagger_{\vec{k}} + \hat{\pi}^\dagger_{\vec{k}}\hat{\pi}_{\vec{k}}) + \left(k^2 - \frac{a''}{a}\right)(\hat{\mu}_{\vec{k}}\hat{\mu}^\dagger_{\vec{k}} + \hat{\mu}^\dagger_{\vec{k}}\hat{\mu}_{\vec{k}})\right]. \quad (6.74)$$

The evolution of $\hat{\mu}$ and $\hat{\pi}$ is therefore dictated, in the Heisenberg representation, by

$$i\hat{\mu}' = [\hat{\mu}, \hat{H}], \qquad i\hat{\pi}' = [\hat{\pi}, \hat{H}], \quad (6.75)$$

where, as usual, units $\hbar = 1$ are assumed. Using now the mode expansion (6.72) and the Hamiltonian in the form (6.74) the evolution for the Fourier components of the operators is

$$\hat{\mu}'_{\vec{k}} = \hat{\pi}_{\vec{k}}, \qquad \hat{\pi}'_{\vec{k}} = -\left(k^2 - \frac{a''}{a}\right)\hat{\mu}_{\vec{k}}, \quad (6.76)$$

implying

$$\hat{\mu}''_{\vec{k}} + \left[k^2 - \frac{a''}{a}\right]\hat{\mu}_{\vec{k}} = 0. \quad (6.77)$$

It is not a surprise that the evolution equations of the field operators, in the Heisenberg description, reproduces, for $\hat{\mu}_{\vec{k}}$ the classical evolution equation derived before in Eq. (6.41). The general solution of the system is then

$$\hat{\mu}_{\vec{k}}(\tau) = \hat{a}_{\vec{k}}(\tau_0) f_i(k,\tau) + \hat{a}^\dagger_{-\vec{k}}(\tau_0) f_i^*(k,\tau), \quad (6.78)$$

$$\hat{\pi}_k(\tau) = \hat{a}_{\vec{k}}(\tau_0) g_i(k,\tau) + \hat{a}^\dagger_{-\vec{k}}(\tau_0) g_i^*(k,\tau), \quad (6.79)$$

where the mode function f_i obeys

$$f_i'' + \left[k^2 - \frac{a''}{a}\right]f_i = 0, \quad (6.80)$$

and[f] $g_i = f_i'$. In the case when the scale factor has a power dependence, in cosmic time, the scale factor will be, in conformal time $a(\tau) = (-\tau/\tau_1)^{-\beta}$ (with $\beta = p/(p-1)$ and $a(t) \simeq t^p$). The solution of Eq. (6.80) is then

$$f_i(k,\tau) = \frac{\mathcal{N}}{\sqrt{2k}}\sqrt{-x}H^{(1)}_\mu(-x), \quad (6.81)$$

$$g_i(k,\tau) = f_k' = -\mathcal{N}\sqrt{\frac{k}{2}}\sqrt{-x}\left[H^{(1)}_{\mu-1}(-x) + \frac{(1-2\mu)}{2(-x)}H^{(1)}_\mu(-x)\right], \quad (6.82)$$

---

[e]In order to derive the following equation, the relations $\hat{\mu}^\dagger_{-\vec{k}} \equiv \hat{\mu}_{\vec{k}}$ and $\hat{\pi}^\dagger_{-\vec{k}} \equiv \hat{\pi}_{\vec{k}}$ should be used.

[f]Of course if the form of the Hamiltonian is different by a time-dependent canonical transformation, also the canonical momenta will differ and, consequently, the relation of g_i to f_i may be different.

where $x = k\tau$ and

$$\mathcal{N} = \sqrt{\frac{\pi}{2}} e^{\frac{i}{2}(\mu + 1/2)\pi}, \qquad \mu = \beta + \frac{1}{2}. \tag{6.83}$$

The function

$$H_\mu^{(1)}(-x) = J_\mu(-x) + iY_\mu(-x) \tag{6.84}$$

(where $J_\mu(-x)$ and $Y_\mu(-x)$ are the ordinary Bessel functions) of index μ and argument $(-x)$ is the Hankel function of first kind [215, 216] and the other linearly independent solution will be $H_\mu^{(2)}(z) = H^{(1)*}_\mu(z)$. The phases appearing in Eqs. (6.81) and (6.82) are carefully selected in such a way that for $\tau \to -\infty$, $f_i \to e^{-ik\tau}/\sqrt{2k}$.

A possible application of the formalism developed so far is the calculation of the energy density of the gravitons produced, for instance, in the transition from a de Sitter stage of inflation and a radiation-dominated stage of (decelerated) expansion. This corresponds to a scale factor that, for $\tau < -\tau_1$ goes as in Eq. (6.49) with $\beta = 1$. For $\tau > -\tau_1$ the scale factor is, instead, exactly the one reported in Eq. (6.50). Consequently, from Eq. (6.81) and (6.82), the mode functions

$$f_i(k, \tau) = \frac{1}{\sqrt{2k}}\left(1 - \frac{i}{k\tau}\right)e^{-ik\tau}, \qquad \tau \leq -\tau_1, \tag{6.85}$$

$$g_i(k, \tau) = \sqrt{\frac{k}{2}}\left(\frac{i}{k^2\tau^2} - \frac{1}{k\tau} - i\right)e^{-ik\tau}, \qquad \tau \leq -\tau_1. \tag{6.86}$$

For $\tau > -\tau_1$ the field operators can be expanded in terms of a new set of creation and annihilation operators, i.e.

$$\hat{\mu}_{\vec{k}}(\tau) = \hat{b}_{\vec{k}}(\tau_1)\tilde{f}_r(k, \tau) + \hat{b}^\dagger_{-\vec{k}}(\tau_1)\tilde{f}^*_r(k, \tau), \qquad \tau > -\tau_1$$

$$\hat{\pi}_{\vec{k}}(\tau) = \hat{b}_{\vec{k}}(\tau_1)\tilde{g}_r(k, \tau) + \hat{b}^\dagger_{-\vec{k}}(\tau_1)\tilde{g}^*_r(k, \tau), \qquad \tau > -\tau_1, \tag{6.87}$$

where, $f_r(k, \tau)$ are now simply appropriately normalized plane waves since in this phase, $a'' = 0$:

$$\tilde{f}_r(k, \tau) = \frac{1}{\sqrt{2k}}e^{-iy}, \qquad \tilde{g}_i(k, \tau) = -i\sqrt{\frac{k}{2}}e^{-iy}, \qquad \tau > -\tau_1, \tag{6.88}$$

where $y = k[\tau + 2\tau_1]$. Since the creation and annihilation operators must always be canonical, $\hat{b}_{\vec{k}}$ and $\hat{b}^\dagger_{\vec{k}}$ can be expressed as a linear combination of $\hat{a}_{\vec{k}}$ and $\hat{a}^\dagger_{\vec{k}}$, i.e.

$$\begin{aligned}\hat{b}_{\vec{k}} &= B_+(k)\hat{a}_{\vec{k}} + B_-(k)^*\hat{a}^\dagger_{-\vec{k}}, \\ \hat{b}^\dagger_{\vec{k}} &= B_+(k)^*\hat{a}^\dagger_{\vec{k}} + B_-(k)\hat{a}_{-\vec{k}}.\end{aligned} \tag{6.89}$$

Equation (6.89) is a special case of a Bogoliubov-Valatin transformation. But because

$$[\hat{a}_{\vec{k}}, \hat{a}_{\vec{p}}^\dagger] = \delta^{(3)}(\vec{k}-\vec{p}), \qquad [\hat{b}_{\vec{k}}, \hat{b}_{\vec{p}}^\dagger] = \delta^{(3)}(\vec{k}-\vec{p}) \qquad (6.90)$$

we must also strictly demand that

$$|B_+(k)|^2 - |B_-(k)|^2 = 1. \qquad (6.91)$$

Equation (6.89) can be inserted into Eq. (6.87) and the following expressions can be easily obtained:

$$\hat{\mu}_{\vec{k}}(\tau) = \hat{a}_{\vec{k}}[B_+(k)f_r + B_-(k)f_r^*] + \hat{a}^\dagger_{-\vec{k}}[B_+(k)^* f_r^* + B_-(k)^* f_r], \qquad (6.92)$$

$$\hat{\pi}_k(\tau) = \hat{a}_{\vec{k}}[B_+(k)g_r + B_-(k)g_r^*] + \hat{a}^\dagger_{-\vec{k}}[B_+(k)^* g_r^* + B_-^*(k)g_r]. \qquad (6.93)$$

Since the evolution of the canonical fields must be continuous, Eqs. (6.78)–(6.79) together with Eqs. (6.92)–(6.93) imply

$$f_i(-\tau_1) = B_+(k)f_r(-\tau_1) + B_-(k)f_r^*(-\tau_1),$$
$$g_i(-\tau_1) = B_+(k)g_r(-\tau_1) + B_-(k)g_r^*(-\tau_1), \qquad (6.94)$$

which allows to determine the coefficients of the Bogoliubov transformation $B_\pm(k)$, i.e.

$$B_+(k) = e^{2ix_1}\left[1 - \frac{i}{x_1} - \frac{1}{2x_1^2}\right],$$
$$B_-(k) = \frac{1}{2x_1^2}, \qquad (6.95)$$

where $x_1 = k\tau_1$ and where Eq. (6.91) is trivially satisfied. Between $B_+(k)$ and $B_+(k)$ the most important quantity is clearly $B_-(k)$ since it defines the amount of "mixing" between positive and negative frequencies. In the case when the gravitational interaction is switched off, the positive/negative frequencies will not mix and $B_-(k)$ would vanish. The presence of a time-dependent gravitational field, however, implies that outgoing waves will mix, in a semiclassical language, with ingoing waves. This mixing simply signals that energy has been transferred from the background geometry to the quantum fluctuations (of the tensor modes, in this specific example). This aspect can be appreciated by computing the mean number of produced pairs of gravitons. Indeed, if a graviton with momentum $\vec{k}$ is produced, also a graviton with momentum $-\vec{k}$ is produced so that the total momentum of the vacuum (which is zero) is conserved:

$$\overline{n}_{\vec{k}}^{\text{gw}} = \frac{1}{2}\langle 0|\hat{N}|0\rangle = \frac{1}{2}\langle 0|[\hat{b}_{\vec{k}}^\dagger \hat{b}_{\vec{k}} + \hat{b}^\dagger_{-\vec{k}}\hat{b}_{-\vec{k}}]|0\rangle = |B_-(k)|^2. \qquad (6.96)$$

Now the total energy of the produced gravitons can be computed recalling that

$$d\rho_{\rm gw} = 2 \times \frac{d^3 k}{(2\pi)^3} \overline{n}_k^{\rm gw}, \qquad (6.97)$$

where the factor 2 counts the two helicities. Using now the result of Eq. (6.95) into Eqs. (6.96) and (6.97) we do obtain the following interesting result, i.e.

$$\frac{d\rho_{\rm gw}}{d\ln k} = \frac{k^4}{\pi^2} \overline{n}_k = \frac{H_1^4}{\pi^2} \qquad (6.98)$$

where $|H_1/a_1| = \tau_1^{-1} = H_1$ (since in our parametrization of the scale factor $a(-\tau_1) = 1$). The result expressed by Eq. (6.98) implies that, in conventional inflationary models, the spectrum of relic gravitons is, in the best case,[g] flat. In more realistic cases, in fact, it is quasi-flat (i.e. slightly decreasing) since the de Sitter phase, most likely, is not exact. As it is evident from the pictorial illustration of the effective potential the modes inheriting flat spectrum are the ones leaving the potential barrier during the de Sitter stage and re-entering during the radiation-dominated phase, i.e. comoving wave-numbers $k_2 < k < k_1$ where $k_1 = \tau_1^{-1}$ and $k_2 = \tau_2^{-1}$. It is also clear that for sufficiently infra-red modes, we must also take into account the second relevant transition of the background from radiation to matter-dominated phase. This second transition will lead, for $k < k_2$ a slope k^{-2} in terms of the quantity defined in Eq. (6.98). In analogy with what done in the case of black-body emission it is practical to parametrize the energy density of the relic gravitons in terms of the differential energy spectrum in critical units

$$\Omega_{\rm GW}(\nu, \tau) = \frac{1}{\rho_{\rm crit}} \frac{d\rho_{\rm gw}}{d\ln\nu}, \qquad (6.99)$$

where $\nu = k/[2\pi a(\tau)]$ is the physical frequency which is conventionally evaluated at the present time since our detectors of gravitational radiation are at the present epoch. In Fig. 6.2 different models are illustrated in terms of their energy spectrum. The calculation performed above estimates the flat plateau labeled by "conventional inflation". To compare directly the plot with the result of the calculation one must, however, also take into

---

[g]This remark has to be understood, of course, in the light of the detection prospects. If the spectrum decreases (as opposed to being constant) as a function of the frequency, the signal at higher frequencies will be smaller. Since the detectors of gravitational waves are typically operating at high enough frequencies, the best possibility achievable in the context of conventional inflationary models is to presume that the spectrum is indeed flat.

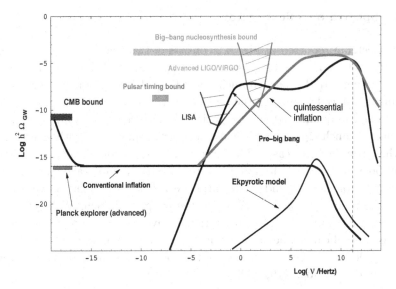

Fig. 6.2 The logarithmic energy spectrum of relic gravitons is illustrated in different models of the early Universe as a function of the present frequency, ν.

account the redshift of the energy density during the matter-dominated phase.

Before passing to the discussion of the spectra of relic gravitons, it is appropriate to comment on the way the mixing coefficients behave in the limit when the wave-number is much larger than the height of the potential barrier. The mixing coefficients determined in the approximation of a sudden change of the background geometry (taking place, for instance, in $-\tau_1$) lead to an ultraviolet divergence in the number of produced gravitons [217]. For modes of comoving wave-number much larger than the height of the barrier the sudden approximation is not adequate. In this regime the mixing coefficients should be computed using a smooth function interpolating between the two regimes. The standard analysis [217, 218] leads to a number of particles which is exponentially suppressed as $\exp[-qk\tau_1]$ where q is of order 1 and depends upon the details of the smooth interpolation. So the wave-numbers that never hit the potential barrier can be, in the first approximation, neglected. The analytical calculations discussed in the present section will be corroborated by direct numerical integration in the last section of the present chapter.

6.5 Spectra of relic gravitons

The spectrum reported in Fig. 6.2 consists of two branches: a soft branch ranging between $\nu_0 \simeq 10^{-18} h_0$ Hz and $\nu_{\rm dec} \simeq 10^{-16}$ Hz. For $\nu > \nu_{\rm dec}$ we have instead the hard branch consisting of high frequency gravitons mainly produced thanks to the transition from the inflationary regime to the radiation-dominated stage. In the soft branch $\Omega_{\rm GW}(\nu, \tau_0) \sim \nu^{-2}$. In the hard branch $\Omega_{\rm GW}(\nu, \tau_0)$ is constant in frequency (or slightly decreasing in the quasi-de Sitter case). The large-scale observation of the first (thirty) multipole moments of the temperature anisotropy imply a bound for the relic graviton background. The rationale for this statement is very simple since relic gravitons contribute to the integrated Sachs-Wolfe effect as discussed in chapter 7. The gravitational wave contribution to the Sachs-Wolfe integral cannot be larger than the (measured) amount of anisotropy directly detected. The soft branch of the spectrum is then constrained and the bound reads [33, 219, 220]

$$h_0^2 \Omega_{\rm GW}(\nu, \tau_0) \lesssim 6.9 \times 10^{-11}, \qquad (6.100)$$

for $\nu \sim \nu_0 \sim 10^{-18}$Hz. The very small size of the fractional timing error in the arrivals of the millisecond plusar's pulses imply that also the hard branch is bounded according to [221]

$$\Omega_{\rm GW}(\nu, \tau_0) \lesssim 10^{-8}, \qquad (6.101)$$

for $\nu \sim 10^{-8}$ Hz corresponding, roughly, to the inverse of the observation time during which the various millisecond pulsars have been monitored.

The two constraints of Eqs. (6.100) and (6.101) are reported in Fig. 6.2, at the two relevant frequencies. The Sachs-Wolfe and millisecond pulsar constraints are differential since they limit, locally, the logarithmic derivative of the gravitons energy density. There exists also an integral bound coming from standard BBN analysis and constraining the integrated graviton energy spectrum [222]:

$$h_0^2 \int_{\nu_{\rm n}}^{\nu_{\rm max}} \Omega_{\rm GW}(\nu, \tau_0) d\ln \nu \lesssim 0.2 \; h_0^2 \Omega_\gamma \simeq 10^{-6}. \qquad (6.102)$$

In Eq. (6.102) $\nu_{\rm max}$ corresponds to the (model dependent) ultra-violet cut-off of the spectrum and $\nu_{\rm n}$ is the frequency corresponding to the horizon scale at nucleosynthesis. Notice that the BBN constraint of Eq. (6.102) has been derived in the context of the simplest BBN model, namely, assuming that no inhomogeneities and/or matter anti–matter domains are present at the onset of nucleosynthesis. In the presence of matter–antimatter domains for scales comparable with the neutron diffusion scale this bound is relaxed [120].

From Fig. 6.2 we see that also the global bound of Eq. (6.102) is satisfied and the typical amplitude of the logarithmic energy spectrum in critical units for frequencies $\nu_I \sim 100$ Hz (and larger) cannot exceed 10^{-14}. This amplitude has to be compared with the LIGO sensitivity to a flat $\Omega_{\mathrm{GW}}(\nu_I, \tau_0)$ which could be *at most* of the order of $h_0^2 \Omega_{\mathrm{GW}}(\nu_I, \tau_0) = 5 \times 10^{-11}$ after four months of observation with 90% confidence. At the moment there is no direct detection of relic gravitons (and more generally of GW) from any detectors. For an introduction to various detectors of gravitational waves see, for instance, [223] (see also [33]).

Even if gravitational waves of high frequency are not central for the present discussion, it should be borne in mind that there exist cosmological scenarios where, for frequencies larger than 10^{-3} Hz, $\Omega_{\mathrm{GW}}(\nu)$ can deviate from the inflationary (nearly scale-invariant) spectrum. In particular, in Fig. 6.2 the expected signals from quintessential inflationary models [224, 225] (see also [226]) and from pre-big bang models [227–229] are reported. In quintessential inflationary models the rise in the spectrum occurs since the inflaton and the quintessence field are unified in a single scalar degree of freedom. Consequently, during a rather long phase (after inflation and before the radiation epoch) the Universe is dominated by the kinetic energy of the inflaton. This dynamics enhances the graviton spectrum at high frequencies. In fact, pre big-bang models are formulated in the framework of the low-energy string effective action where the Einstein-Hilbert term is naturally coupled to the dilaton. The evolution of the tensor modes will then be slightly different from the one derived in the present section and will be directly sensitive to the evolution of the dilaton. Both in quintessential inflation and in pre-big bang models the spectrum of relic gravitons is larger at high frequencies suggesting that superconducting cavities are a promising tool for the experimental investigation in this range of frequencies (see [230–232] and references therein). Another very interesting (complementary) approach along this direction has been reported in [233, 234] where a prototype detector working in the 100 MHz region has been described.

6.6 Quantum state of cosmological perturbations

The evolution of the cosmological inhomogeneities has been described, so far, in the Heisenberg representation. To investigate the correlation properties of the fluctuations and their semiclassical limit it is often useful to work within the Schrödinger representation where the evolution can be pictured as the spreading of a quantum mechanical wave-functional. The initial wave-functional will be constructed as the direct product of states minimizing the indetermination relations for the different harmonic oscillators forming the quantum field. The quantum mechanical states of the fluctuations will then be a generalization of the concept of coherent state firstly introduced in [235, 236]. These states are essentially coherent states associated to Lie algebras of non-compact groups (such as $SU(1,1)$ which is isomorphic to the algebra of $SO(2,1)$ and $SL(2,R)$). Since their discovery, it has been understood that their typical quantum mechanical property was to minimize the indetermination relations [236]. It was then appreciated that these states can be obtained, as the coherent states, by the action of a unitary operator acting on the vacuum. Following the pioneering work of Yuen [237] the squeezed states have been experimentally investigated in quantum optics with the hope of obtaining "squeezed light". This light could be of upmost importance for various devices since it would allow to have one of the conjugate (quantum) variables fluctuating above the quantum limit while the the other variable fluctuates below the quantum limit preserving, overall the minimal uncertainty. To have the flavor of the manifold applications of squeezed states to quantum optics the reader can consult two classical textbooks [238, 239] and also two (not so recent) review articles [240, 241]. While the experimental evidence is that squeezed light is rather hard to produce for large values of the squeezing parameter[h] (which would be the interesting range for applications), the squeezed states formalism has been applied with success to the analysis of the correlation properties of quantum fluctuations produced in the early Universe (see, for instance, [242–246] and references therein). In particular a natural definition of coarse grained entropy arises in the squeezed state formalism [142, 143].

In what follows, instead of giving the full discussion of the problem in the Schrödinger representation, the squeezed states will be analyzed not in the case of a quantum field but in the case of a single (quantum) harmonic oscillator. This will allow to get an interesting physical interpretation of Eq. (6.89). To be more precise, the analog of Eq. (6.89) can be realized with a two-mode squeezed state. However, to be even simpler, only one-mode squeezed states will be discussed. Let us therefore rewrite Eq. (6.89)

---

[h]In the language of this discussion $r \gg 1$, see below.

in its simplest form, namely

$$\hat{b} = c(r)\hat{a} + s(r)\hat{a}^\dagger, \qquad \hat{b}^\dagger = s(r)\hat{a} + c(r)\hat{a}^\dagger, \qquad (6.103)$$

where $c(r) = \cosh r$ and $s(r) = \sinh r$; r is the so-called squeezing parameter.[i] Since $[\hat{a}, \hat{a}^\dagger] = 1$ and $[\hat{b}, \hat{b}^\dagger] = 1$, the transformation that allows to pass from the $\hat{a}$ and $\hat{a}^\dagger$ to the $\hat{b}$ and $\hat{b}^\dagger$ is clearly unitary (extra phases may appear in Eq. (6.103) whose coefficients may be complex; in the present exercise we will stick to the case of real coefficients). In Eq. (6.103) the index referring to the momentum has been suppressed since we are dealing here with a single harmonic oscillator with Hamiltonian

$$\hat{H}_a = \frac{\hat{p}^2}{2} + \frac{\hat{x}^2}{2} = \hat{a}^\dagger \hat{a} + \frac{1}{2}, \qquad (6.104)$$

where

$$\hat{a} = \frac{1}{\sqrt{2}}(\hat{x} + i\hat{p}), \qquad \hat{a}^\dagger = \frac{1}{\sqrt{2}}(\hat{x} - i\hat{p}). \qquad (6.105)$$

Equation (6.103) can also be written as

$$\hat{b} = S^\dagger(z)\, \hat{a}\, S(z), \qquad \hat{b}^\dagger = S^\dagger(z)\, \hat{a}^\dagger\, S(z), \qquad (6.106)$$

where the (unitary) operator $S(z)$ is the so-called squeezing operator defined as

$$S(z) = e^{\frac{1}{2}(z\hat{a}^{\dagger 2} - z^* \hat{a}^2)}. \qquad (6.107)$$

In Eq. (6.107), in general, $z = re^{i\vartheta}$. In the case of Eq. (6.103) $\vartheta = 0$ and $z = r$. A squeezed state is, for instance, the state $|z\rangle = S(z)|0\rangle$. The same kind of operator arises in the field theoretical description of the process of production of gravitons (or phonons, as we shall see, in the case of scalar fluctuations). The state $|0\rangle$ is the state annihilated by $\hat{a}$ and minimizing the Hamiltonian (6.104). In the coordinate representation, therefore, the wave-function of the vacuum will be, in the coordinate representation

$$\psi_0(x) = \langle x|0\rangle = N_0 e^{-\frac{x^2}{2}}, \qquad \hat{a}|0\rangle = 0, \qquad (6.108)$$

where N_0 is a constant fixed by normalizing to 1 the integral over x of $|\psi_0(x)|^2$. Obviously the wave-function in the p-representation will also be Gaussian. Equation (6.108) is simply obtained from the condition $\hat{a}|0\rangle = 0$ by recalling Eq. (6.105) (where $\hat{p} = -i\frac{\partial}{\partial x}$). By applying the same trick,

---

[i]There should be no ambiguity in this notation. It is true that, in the present script, the variable r may also denote the ratio between the tensor and the scale power spectrum arising in the consistency condition (see Eqs. (5.123) and (10.83)). However, since the squeezing parameter and the tensor to scalar ratio are never used in the same context, there is no possible confusion.

we can obtain the wave-function (in the coordinate representation) for the state $|z\rangle$. By requiring $\hat{b}|0\rangle = 0$, using Eq. (6.103) together with Eq. (6.105) we will have the following simple differential equation

$$[c(r) + s(r)]x\psi_z + [c(r) - s(r)]\frac{\partial \psi_z}{\partial x} = 0, \qquad (6.109)$$

whose solution, up to the normalization constant N_z will be

$$\psi_z(x) = \langle x|z\rangle = N_z e^{-\frac{x^2}{2\sigma^2}}, \qquad (6.110)$$

implying that the wave-function will still be Gaussian but with a different variance (since $\sigma = e^{-r}$) and with a different normalization (since $N_z \neq N_0$). In this case the wave-function gets squeezed in the x-representation while it gets broadened in the p-representation in such a way that the indetermination relations $\Delta\hat{x}\Delta\hat{p} = 1/2$. It should be borne in mind that the broadening (or squeezing) of the Gaussian wave-function(al) corresponds to a process of particle production when we pass, by means of a unitary transformation, from one vacuum to the other. In fact, using Eq. (6.103) we also have

$$\langle 0|\hat{a}^\dagger \hat{a}|0\rangle = 0, \qquad \langle 0|\hat{b}^\dagger \hat{b}|0\rangle = \sinh^2 r. \qquad (6.111)$$

So, while the initial vacuum has no particles, the "new" vacuum is, really and truly, a many-particle system. In the case of the amplification of fluctuations driven by the gravitational field in the early Universe the squeezing parameter is always much larger than one and the typical mean number of particles per Fourier mode can be as large as 10^4–10^5.

Squeezed states, unitarily connected to the vacuum, minimize the indetermination relations as the well known coherent states introduced by Glauber (see, for instance, [238]). The usual coherent states can be obtained in many different ways, but the simplest way to introduce them is to define the so-called Glauber displacement operator:

$$\mathcal{D}(\alpha) = e^{\alpha \hat{a}^\dagger - \alpha^* \hat{a}}, \qquad |\alpha\rangle = \mathcal{D}(\alpha)|0\rangle, \qquad (6.112)$$

where α is a complex number and $|\alpha\rangle$ is a coherent state such that $\hat{a}|\alpha\rangle = \alpha|\alpha\rangle$. It is clear that while the squeezing operator of Eq. (6.107) is quadratic in the creation and annihilation operators, the Glauber operator is linear in $\hat{a}$ and $\hat{a}^\dagger$. By using the Baker-Campbell-Hausdorff (BCH) formula it is possible to get, from Eq. (6.112) the usual expression of a coherent state in whatever basis (such as the Fock basis or the coordinate basis). In the coordinate basis, coherent states are also Gaussians but rather than squeezed they are simply not centered around the origin. From the BCH formula it is possible to understand that the Glauber operator is simply given by the product of the generators of the Heisenberg algebra, i.e. $\hat{a}$, $\hat{a}^\dagger$ and the identity. So the Glauber coherent states are related to the Heisenberg algebra. The squeezed states, arise, instead in the context of non-compact groups. To see this intuitively, consider the Hamiltonian

$$\hat{H}_b = \hat{b}^\dagger \hat{b} + \frac{1}{2}, \tag{6.113}$$

and express it in terms of $\hat{a}$ and $\hat{a}^\dagger$ according to Eq. (6.103). The result of this simple manipulation is

$$\hat{H}_b = \cosh 2r \left(\hat{a}^\dagger \hat{a} + \frac{1}{2} \right) + \sinh 2r \left(\frac{\hat{a}^2}{2} + \frac{\hat{a}^{\dagger\,2}}{2} \right). \tag{6.114}$$

But, in Eq. (6.114), the operators

$$\left(\hat{a}^\dagger \hat{a} + \frac{1}{2} \right) = L_0,$$

$$\frac{\hat{a}^2}{2} = L_-, \qquad \frac{\hat{a}^{\dagger\,2}}{2} = L_+, \tag{6.115}$$

form a realization of the $SU(1,1)$ Lie algebra since, as it can be explicitly verified:

$$[L_+, L_-] = -2L_0,$$
$$[L_0, L_\pm] = \pm L_\pm. \tag{6.116}$$

Note that the squeezing operator of Eq. (6.107) can be written as the exponential of $L_\pm$ and L_0, i.e. more precisely

$$S(z) = e^{\frac{1}{2}(z\hat{a}^{\dagger\,2} - z^*\hat{a}^2)} = e^{\tanh r L_+} \, e^{-\ln[\cosh r] L_0} \, e^{-\tanh r L_-}, \tag{6.117}$$

where the second equality follows from the BCH relation [247, 248]. This is the rationale for the statement that the squeezed states are the coherent states associated with SU(1,1). Of course more complicated states can be obtained from the squeezed vacuum states. These have various applications in cosmology. For instance we can have the squeezed coherent states, i.e.

$$|\alpha, z\rangle = S(z)\, \mathcal{D}(\alpha)|0\rangle. \tag{6.118}$$

Similarly one can define the squeezed number states (squeezing operator applied to a state with n particles) or the squeezed thermal states (squeezed states of a thermal state [244]. The squeezed states are also important to assess precisely the semiclassical limit of quantum mechanical fluctuations. On a purely formal ground the semiclassical limit arises in the limit $\hbar \to 0$. However, on a more operational level, the classical limit can be addressed more physically by looking at the number of particles produced via the pumping action of the gravitational field. In this second approach the squeezed states represent an important tool. The squeezed state formalism therefore suggests that what we call *classical* fluctuations are the limit of quantum mechanical states in the same sense a laser beam is formed by coherent photons. Also the laser beam has a definite classical meaning, however, we do know that coherent light is different from thermal (white) light. This kind of distinction follows, in particular, by looking at the effects of second-order interference such as the Hanbury-Brown-Twiss effect [238, 239].

In closing this section it is amusing to get back to the problem of entropy [142, 143]. In fact, in connection with squeezed states of the tensor modes of the geometry it is possible to define a coarse-grained entropy in which the loss of information associated to the reduced density matrix is represented by an increased dispersion in a superfluctuant field operator which is the field-theoretical analog of the quantum mechanical momentum $\hat{p}$. The estimated entropy goes in this case as $r(\nu)$ (now a function of the present frequency) i.e. as $\ln \bar{n}_\nu$ where $\bar{n}_\nu$ is the number of produced gravitons. Consequently the entropy can be estimated by integrating over all the frequencies of the graviton spectrum presented, for instance, in Fig. 6.2. The result will be that

$$S_{\rm gr} = V \int_{\nu_0}^{\nu_1} r(\nu)\, \nu^2 \, d\nu \simeq (10^{29})^3 \left(\frac{H_1}{M_{\rm P}}\right)^{3/2}. \qquad (6.119)$$

The factor 10^{29} arises from the hierarchy between the lower frequency of the spectrum (i.e. $\nu_0 \simeq 10^{-18}$ Hz) and the higher frequency (i.e. $\nu_1 \simeq 10^{11}\, (H_1/M_{\rm P})^{1/2}$ Hz). From Eq. (6.119) it follows that this gravitational entropy is of the same order of the thermodynamic entropy provided the curvature scale at the inflation-radiation transition is sufficiently close to the Planck scale.

6.7 Digression on different vacua

After having discussed the quantum treatment of the tensor modes of the geometry as defined by minimizing the Hamiltonian of Eq. (6.64) it is now appropriate to scrutinize slightly different possibilities. The ideal prosecution of the present section can be found at the end of chapter 10 where the analog discussion will be reported in the case of the scalar modes and where the relevance of the present discussion for the so-called trans-Planckian ambiguities will be made more explicit. The content of the present section closely follows the treatment reported in [213] and in [214].

According to the discussion reported in the previous sections, a possible selection of different Hamiltonians for the tensor modes[j]

$$H^{(1)}(\tau) = \int d^3x \frac{1}{2}\left[\frac{\Pi^2}{a^2} + a^2(\partial_i \Psi)^2\right], \quad (6.120)$$

$$H^{(2)}(\tau) = \int d^3x \frac{1}{2}\left[\pi^2 + 2\mathcal{H}\psi\pi + (\partial_i \psi)^2\right], \quad (6.121)$$

$$H^{(3)}(\tau) = \int d^3x \frac{1}{2}\left[\tilde{\pi}^2 + (\partial_i \psi)^2 - \frac{a''}{a}\psi^2\right], \quad (6.122)$$

where

$$\Pi = a^2 \Psi', \quad \psi = a\Psi, \quad \pi = \frac{\Pi}{a}, \quad \tilde{\pi} = \psi', \quad (6.123)$$

are the relations between the different canonical fields. It is not difficult to check that it is possible to go from one Hamiltonian to the other through a suitable canonical transformation. For instance, the transformation $H^{(2)}(\tau) \to H^{(3)}(\tau)$ is generated, in a standard way, by

$$\mathcal{F}_{2\to 3}(\psi, \tilde{\pi}, \tau) = \int d^3x \left(\psi\tilde{\pi} - \frac{\mathcal{H}}{2}\psi^2\right), \quad (6.124)$$

a functional of the old fields and of the new momenta $\tilde{\pi}$. By differentiating the generating functional, we obtain the relation between the old momenta (i.e. π) and the new ones, as well as a change in the Hamiltonian

$$\pi = \tilde{\pi} - \mathcal{H}\psi, \quad (6.125)$$

$$H^{(2)}(\psi, \pi, \tau) \to H^{(3)}(\psi, \tilde{\pi}, \tau) = H^{(2)}(\psi, \pi, \tau) + \frac{\partial \mathcal{F}_{2\to 3}}{\partial \tau}. \quad (6.126)$$

Bearing in mind Eqs. (6.124) and (6.125), the right-hand side of Eq. (6.126) leads exactly to Eq. (6.122). With similar considerations, all the Hamiltonians (6.120)–(6.122) can be related to one another by suitable canonical transformations.

---

[j]To keep the discussion as general as possible we will slightly change the notations and we will introduce two putative fields Ψ and ψ whose relation with h and μ is evident by comparing Eqs. (6.120), (6.121) and (6.122) with the analog expressions obtained, for instance, in Eqs. (6.69), (6.65) and (6.64). This notation will just be used in this section since, as we shall see in chapters (7) and (10) the variable ψ and Ψ will denote, respectively, one of the longitudinal fluctuations of the metric and its gauge-invariant generalization (see also Eq. (6.19)). Always at the level of notations, the subscript "gw" (referring to the tensor nature of the fluctuations) will be omitted since, with the appropriate modifications of the respective pump fields, the present discussion can be largely applied also to the scalar fluctuations of the geometry and to their quantum mechanical normalization (see last part of chapter 10).

Fully equivalent classical evolutions should be expected by solving the appropriate Hamilton equations with the Hamiltonians (6.120)–(6.122). This statement is also true at the quantum level (see e.g. [256]) because the different classical actions corresponding to Eqs. (6.120)–(6.122) only differ by total derivatives. Hence, thinking for instance in terms of a functional integral approach, the transitions amplitudes differ, at most, by a field-dependent phase since the total derivatives appearing in the classical action, once inserted in the path integral, can be explicitly integrated.

As an example, consider the case of $H^{(2)}(\tau)$ whose associated Lagrangian density is

$$\mathcal{L}^{(2)}(\vec{x}, \tau) = \frac{1}{2}\left[\psi'^2 - 2\mathcal{H}\psi\psi' + \mathcal{H}^2\psi^2 - (\partial_i\psi)^2\right]. \qquad (6.127)$$

Comparing Eq. (6.127) to the Lagrangian density associated with $H^{(3)})(\tau)$, we can notice that they differ by a total derivative

$$\mathcal{L}^{(3)}(\vec{x}, \tau) = \mathcal{L}^{(2)}(\vec{x}, \tau) + \frac{d\mathcal{D}}{d\tau}, \qquad \mathcal{D} = \frac{\mathcal{H}}{2}\psi^2. \qquad (6.128)$$

The wave-functional computed with (6.128) only differs from the one computed with (6.127) by a field-dependent phase, namely

$$\Phi^{(3)}[\psi] = e^{i\mathcal{D}(\psi,\tau)}\Phi^{(2)}[\psi]. \qquad (6.129)$$

The expectation value of a generic observable $\mathcal{O}[\hat{\psi}, \hat{\pi}]$ can be computed using either $\Phi^{(2)}[\psi]$ or $\Phi^{(3)}[\psi]$. However, if we compute such an expectation value using $\Phi^{(3)}[\psi]$, we have to bear in mind that the canonical momentum ($\hat{\pi}$) acts non-trivially on it, its action being fully specified by the transformation (6.125). The conclusion is that the expectation value of any operator $\mathcal{O}[\hat{\psi}, \hat{\pi}]$ is independent on the Hamiltonian one is using (we shall see below an example of this fact in the context of cosmological perturbations). In spite of the above equivalence, the state minimizing one of the Hamiltonians at the initial time τ_0 does depend on which one of (6.120)–(6.122) is chosen. The time τ_0 has been taken before independent on the comoving wave-number. There are, however, situations where τ_0 does depend on k.

Let us suppose, for instance, that the effective description of the tensor modes of the geometry is only valid up to some typical scale M. This assumption amounts to demanding a cut-off on the physical momenta of the gravitons, i.e.

$$k/a(\tau) \leq M, \qquad (6.130)$$

In any kind of inflationary background, Eq. (6.130) is saturated at some time $\tau_0(k)$ defining an unconventional hypersurface on which initial conditions for the fluctuations are given. This hypersurface has been dubbed in

[213] new physics hypersurface (NPH, for short). Working in Fourier space,[k]

$$\hat{\psi}(\vec{x},\tau) = \frac{1}{(2\pi)^{3/2}} \int d^3k \, \hat{\psi}_{\vec{k}}(\tau) e^{-i\vec{k}\cdot\vec{x}},$$

$$\hat{\pi}(\vec{x},\tau) = \frac{1}{(2\pi)^{3/2}} \int d^3k \, \hat{\pi}_{\vec{k}}(\tau) e^{-i\vec{k}\cdot\vec{x}},$$
(6.131)

with $\hat{\psi}_{\vec{k}} = \hat{\psi}^{\dagger}_{-\vec{k}}, \hat{\pi}_{\vec{k}} = \hat{\pi}^{\dagger}_{-\vec{k}}$, the evolution in the Heisenberg representation reads

$$\begin{pmatrix} \hat{\psi}_{\vec{k}}(\tau) \\ \hat{\pi}_{\vec{k}}(\tau) \end{pmatrix} = \begin{pmatrix} A_{\vec{k}}(\tau,\tau_0) & B_{\vec{k}}(\tau,\tau_0) \\ C_{\vec{k}}(\tau,\tau_0) & D_{\vec{k}}(\tau,\tau_0) \end{pmatrix} \begin{pmatrix} \hat{\psi}_{\vec{k}}(\tau_0) \\ \hat{\pi}_{\vec{k}}(\tau_0) \end{pmatrix},$$
(6.132)

where

$$A_{\vec{k}}(\tau,\tau_0) = i\left[g_k(\tau_0)f_k^{\star}(\tau) - g_k^{\star}(\tau_0)f_k(\tau)\right],$$

$$B_{\vec{k}}(\tau,\tau_0) = i\left[f_k(\tau)f_k^{\star}(\tau_0) - f_k^{\star}(\tau)f_k(\tau_0)\right],$$

$$C_{\vec{k}}(\tau,\tau_0) = i\left[g_k(\tau_0)g_k^{\star}(\tau) - g_k^{\star}(\tau_0)g_k(\tau)\right],$$

$$D_{\vec{k}}(\tau,\tau_0) = i\left[g_k(\tau)f_k^{\star}(\tau_0) - g_k^{\star}(\tau)f_k(\tau_0)\right].$$
(6.133)

In Eqs. (6.133) $f_k(\tau)$ and $g_k(\tau)$ denote the mode functions that are, respectively, solutions of the Heisenberg evolution equations for a (generic) pair $(\hat{\psi}, \hat{\pi})$ of canonically conjugated operators. At every time, for consistency with the canonical commutation relations, the phases and amplitudes of the mode functions are subjected to the Wronskian condition

$$f_k(\tau)g_k^{\star}(\tau) - f_k^{\star}(\tau)g_k(\tau) = i.$$
(6.134)

In Eqs. (6.133), with the condition (6.134), for $\tau \to \tau_0$, $C_{\vec{k}}(\tau_0,\tau_0) = B_{\vec{k}}(\tau_0,\tau_0) = 0$ and $A_{\vec{k}}(\tau_0,\tau_0) = D_{\vec{k}}(\tau_0,\tau_0) = 1$. Clearly, each of the Hamiltonians (6.120)–(6.122) leads to different $f_k(\tau)$ and $g_k(\tau)$ all satisfying (6.134).

The time $\tau_0(k)$ will be on the NPH and different Hamiltonians will be minimized at that time. It should already be clear at this point that the two-point function of, say, $\hat{\psi}_{\vec{k}}$ will depend on the choice of the initial state through Eqs. (6.132) and (6.133), since minimizing different Hamiltonians corresponds to imposing different conditions on the field operators at $\tau_0(k)$.

---
[k]The Fourier decomposition of the quantum operators used in Eq. (6.132) is fully consistent with the one used in Eq. (6.72).

In order to deal with simple mode functions, we will confine our attention to the case of an inflationary background with a scale factor parametrized, in conformal time, as

$$a(\tau) = \left(-\frac{\tau}{\tau_1}\right)^{-\beta}, \quad \tau < -\tau_1, \tag{6.135}$$

where τ_1 marks the end of the inflationary epoch. The pure de Sitter case corresponds to $\beta = 1$. The inequality $k/a(\tau) \leq \Lambda$ is saturated, by definition, at the time $\tau_0(k)$. We shall refer to this time as the time of exit from the NPH, not to be confused, of course, with the more standard "horizon-exit" time τ_{ex}. In our case,

$$\tau_0(k) = -\tau_1 \left(\frac{M}{k}\right)^{1/\beta}. \tag{6.136}$$

Let us then discuss the minimization of the different Hamiltonians starting with (6.120) but suppressing the label $^{(1)}$ for simplicity. In Fourier space

$$\hat{H}_k(\tau) = \frac{1}{4}\left[\frac{1}{a^2}(\hat{\Pi}_{\vec{k}}\hat{\Pi}_{\vec{k}}^\dagger + \hat{\Pi}_{\vec{k}}^\dagger\hat{\Pi}_{\vec{k}}) + k^2 a^2(\hat{\Psi}_{\vec{k}}\hat{\Psi}_{\vec{k}}^\dagger + \hat{\Psi}_{\vec{k}}^\dagger\hat{\Psi}_{\vec{k}})\right], \tag{6.137}$$

with $\hat{H}(\tau) = \int d^3k \hat{H}_k(\tau)$. The appropriately normalized mode functions are now[1]

$$f_k(\tau) = \frac{\mathcal{N}}{a(\tau)\sqrt{2k}}\sqrt{-x}H_\mu^{(1)}(-x),$$

$$g_k(\tau) = a^2(\tau) f_k', \tag{6.138}$$

$$\mathcal{N} = \sqrt{\frac{\pi}{2}} e^{\frac{i}{2}(\mu+1/2)\pi}, \quad \beta = \mu - \frac{1}{2}, \tag{6.139}$$

where $x = k\tau$, and $H_\mu^{(1)}(-x)$ are, as usual, the first-order Hankel functions of index μ [215]. In order to minimize the Hamiltonian (6.120) let us consider the auxiliary operator

$$\hat{Q}_{\vec{k}} = \frac{1}{\sqrt{2k}}\left[\frac{\hat{\Pi}_{\vec{k}}}{a} - iak\hat{\Psi}_{\vec{k}}\right]. \tag{6.140}$$

Equation (6.140) allows Eq. (6.137) to be expressed as

$$\hat{H}_k = \frac{k}{2}\left[\hat{Q}_{\vec{k}}^\dagger \hat{Q}_{\vec{k}} + \hat{Q}_{\vec{k}} \hat{Q}_{\vec{k}}^\dagger\right], \tag{6.141}$$

while canonical commutation relations between conjugate field operators,

$$[\hat{\psi}(\vec{x},\tau), \hat{\pi}(\vec{y},\tau)] = i\delta^{(3)}(\vec{x}-\vec{y}), \tag{6.142}$$

---

[1] In Eq. (6.138) the mode functions have been also named f_k and g_k. However, the present mode functions, refer to a different Hamiltonian.

imply $[\hat{Q}_{\vec{k}}, \hat{Q}_{\vec{p}}^\dagger] = \delta^{(3)}(\vec{k}-\vec{p})$. Consequently, the state minimizing (6.120) is the one annihilated by $\hat{Q}_{\vec{k}}$ (provided it is normalizable).

In order to evaluate the corrections induced on the two-point function by this particular initial state, we have to compute

$$\langle 0^{(1)}, \tau_0 | \hat{h}(\vec{x}, \tau) \hat{h}(\vec{y}, \tau) | \tau_0, 0^{(1)} \rangle$$
$$= \frac{\ell_P^2}{4\pi^3} \int d^3k \int d^3p \langle \, \hat{\Psi}_{\vec{k}}(\tau) \, \hat{\Psi}_{\vec{p}}(\tau) \, \rangle e^{-i(\vec{k}\cdot\vec{x}+\vec{p}\cdot\vec{y})}, \qquad (6.143)$$

where $\langle ... \rangle \equiv \langle 0^{(1)}, \tau_0 | ... | \tau_0, 0^{(1)} \rangle$ means that the expectation values should be evaluated over the state minimizing $H^{(1)}$ at the time τ_0. Inserting Eqs. (6.132) into Eq. (6.143), we obtain

$$\langle 0^{(1)}, \tau_0 | \hat{h}(\vec{x}, \tau) \hat{h}(\vec{y}, \tau) | \tau_0, 0^{(1)} \rangle$$
$$= \frac{\ell_P^2}{4\pi^3} \int d^3k \int d^3p \bigg[A_k(\tau,\tau_0) A_p(\tau,\tau_0) \langle \hat{\Psi}_{\vec{k}}(\tau_0) \hat{\Psi}_{\vec{p}}(\tau_0) \rangle$$
$$+ B_k(\tau,\tau_0) B_p(\tau,\tau_0) \langle \hat{\Pi}_{\vec{k}}(\tau_0) \hat{\Pi}_{\vec{p}}(\tau_0) \rangle + B_k(\tau,\tau_0) A_p(\tau,\tau_0) \langle \hat{\Pi}_{\vec{k}}(\tau_0) \hat{\Psi}_{\vec{p}}(\tau_0) \rangle$$
$$+ A_k(\tau,\tau_0) B_p(\tau,\tau_0) \langle \hat{\Psi}_{\vec{k}}(\tau_0) \hat{\Pi}_{\vec{p}}(\tau_0) \rangle \bigg] e^{-i(\vec{k}\cdot\vec{x}+\vec{p}\cdot\vec{y})}. \qquad (6.144)$$

The various expectation values appearing in (6.144) can be computed using the relation between the canonical operators (evaluated at the initial time τ_0) and the operators (6.140) annihilating the initial state. Defining now the power spectrum (i.e. the Fourier transform of the two-point function) by

$$\langle 0^{(1)}, \tau_0 | \hat{h}(\vec{x}, \tau) \hat{h}(\vec{y}, \tau) | \tau_0, 0^{(1)} \rangle = \int \frac{dk}{k} |\delta_h(k,\tau)|^2 \frac{\sin kr}{kr}, \qquad (6.145)$$

we obtain, from Eq. (6.144) and with the help of Eqs. (6.138)–(6.139),

$$|\delta_h(k,\tau)|^2 = |\Delta_h(k,\tau)|^2 \left[1 + \frac{\beta}{x_0} \sin(2x_0 + \beta\pi) \right], \qquad (6.146)$$

where

$$|\Delta_h(k,\tau)|^2 = \frac{2^{4\beta-1}}{\pi^3} (2\beta)^{-2\beta} \Gamma(\beta+1/2)^2 \left(\frac{H_1}{M_P}\right)^2 \left(\frac{k}{k_1}\right)^{2(1-\beta)}. \qquad (6.147)$$

In Eq. (6.146) $\frac{k}{k_1} = \frac{\omega}{\omega_1}$ is the ratio of the generic proper frequency to the one corresponding to the end-point of the spectrum $\omega_1 = H_1/a$.

In order to derive Eq. (6.146) the limit $x = k\tau \ll 1$, corresponding to looking at the correlation function at late times, has been taken. Also, since $|x_0| \gg 1$, only the leading correction in $1/x_0$ has been kept. Furthermore, using $k/a(\tau_0) = \Lambda$, according to Eq. (6.135) the initial time τ_0 can be easily related to the value of Λ by

$$|x_0| = |k\tau_0| = \beta \frac{M}{H_{\text{ex}}^{\text{NPH}}}, \qquad (6.148)$$

where $H_{\text{ex}}^{\text{NPH}} = H(t_0(k))$ denotes the Hubble parameter at the time $t_0(k)$ when a given scale "exits" the NPH. Note that x_0 depends on k except in the case of pure de Sitter. We see that, as a consequence, corrections to the standard results are larger at small k for power law inflation (corresponding to larger values of $H_{\text{ex}}^{\text{NPH}}$), while the opposite is true for superinflation ($0 < \beta < 1$). For $\beta = 1$, Eq. (6.146) gives exactly

$$|\delta_h(k,\tau)|^2 = \frac{1}{2\pi^2}\left(\frac{H}{\overline{M}_{\text{P}}}\right)^2\left[1 - \frac{\sin 2x_0}{x_0}\right]. \qquad (6.149)$$

Let us now repeat the same procedure in the case of $H^{(2)}$. In the case of (6.121) we have

$$\hat{H}_k(\tau) = \frac{1}{4}\bigg[(\hat{\pi}_{\vec{k}}\hat{\pi}_{\vec{k}}^\dagger + \hat{\pi}_{\vec{k}}^\dagger\hat{\pi}_{\vec{k}}) + k^2(\hat{\psi}_{\vec{k}}\hat{\psi}_{\vec{k}}^\dagger + \hat{\psi}_{\vec{k}}^\dagger\hat{\psi}_{\vec{k}})$$

$$+ kF(x)(\hat{\pi}_{\vec{k}}\hat{\psi}_{\vec{k}}^\dagger + \hat{\pi}_{\vec{k}}^\dagger\hat{\psi}_{\vec{k}} + \hat{\psi}_{\vec{k}}\hat{\pi}_{\vec{k}}^\dagger + \hat{\psi}_{\vec{k}}^\dagger\hat{\pi}_{\vec{k}})\bigg], \qquad (6.150)$$

where

$$kF(x) = \mathcal{H}. \qquad (6.151)$$

Solving the evolution in the Heisenberg representation, the mode functions can be written as [215]

$$f_k(\tau) = \frac{\mathcal{N}}{\sqrt{2k}}\sqrt{-x}H_\mu^{(1)}(-x), \qquad (6.152)$$

$$g_k(\tau) = -\mathcal{N}\sqrt{\frac{k}{2}}\sqrt{-x}H_{\mu-1}^{(1)}(-x). \qquad (6.153)$$

In order to minimize the Hamiltonian (6.150) at the initial time τ_0, we introduce

$$\hat{Q}_{\vec{k}} = \frac{1}{\sqrt{2k}}\left[e^{-i\gamma}\hat{\pi}_{\vec{k}} - ie^{i\gamma}k\hat{\psi}_{\vec{k}}\right], \qquad (6.154)$$

where γ is a time-dependent parameter. Using Eq. (6.154), the Hamiltonian (6.150) can be put in the same form as (6.141) provided the following relation is imposed between γ and $F(x)$ of Eq. (6.150):

$$\sin 2\gamma = F(x). \qquad (6.155)$$

The canonical commutation relations Eq. (6.142) now imply $[\hat{Q}_{\vec{k}}, \hat{Q}_{\vec{p}}^\dagger] = \cos 2\gamma \delta^{(3)}(\vec{k} - \vec{p})$, so that the initial state minimizing (6.150) is again the one annihilated by $\hat{Q}_{\vec{k}}$.

The wave-functional of the initial state can be easily derived and, for each mode, it has a Gaussian form:

$$\Phi[\psi_{\vec{k}}] = N \exp\left(-\sum_k \frac{k}{2}(\psi_{\vec{k}}\psi_{-\vec{k}})e^{-2i\gamma}\right). \tag{6.156}$$

This state is normalizable provided $|\gamma| < \pi/4$. Using Eq. (6.155), we see that $|\gamma| = \pi/4$ corresponds to a time τ_0 for which $|F(x_0)| = 1$, which is basically equivalent to the condition of (standard) horizon crossing. Consequently, provided the modes of the field are inside the horizon at the "initial" time τ_0, the state (6.156) is normalizable.

The two-point function to be computed now is

$$\langle 0^{(2)}, \tau_0 | \hat{h}(\vec{x}, \tau) \hat{h}(\vec{y}, \tau) | \tau_0, 0^{(2)} \rangle$$
$$= \frac{\ell_P^2}{4\pi^3 \, a(\tau)^2} \int d^3k \int d^3p \langle \, \hat{\psi}_{\vec{k}}(\tau) \, \hat{\psi}_{\vec{p}}(\tau) \, \rangle e^{-i(\vec{k}\cdot\vec{x} + \vec{p}\cdot\vec{y})}, \tag{6.157}$$

and the related power spectrum evaluated at late times ($x = k\tau \ll 1$) is

$$|\delta_h(k,\tau)|^2 = |\Delta_h(k,\tau)|^2 \left[1 - \beta \frac{\cos(2x_0 + \beta\pi)}{2x_0^2}\right]. \tag{6.158}$$

In the de Sitter case, $\beta = 1$, Eq. (6.158) gives

$$|\delta_h(k,\tau)|^2 = \frac{1}{2\pi^2}\left(\frac{H}{\overline{M}_P}\right)^2 \left[1 + \frac{\cos 2x_0}{2x_0^2}\right]. \tag{6.159}$$

The Hamiltonian (6.122) can be minimized following the same procedure already discussed in the case of Eqs. (6.120) and (6.121). Defining the function

$$\omega^2(x) = \left(1 - \frac{1}{k^2}\frac{a''}{a}\right), \tag{6.160}$$

the Hamiltonian (6.122) can be written in the simple form[m]

$$\hat{H}_k(\tau) = \frac{1}{4}\left[(\hat{\pi}_{\vec{k}}\hat{\pi}_{\vec{k}}^\dagger + \hat{\pi}_{\vec{k}}^\dagger\hat{\pi}_{\vec{k}}) + k^2\omega^2(x)(\hat{\psi}_{\vec{k}}\hat{\psi}_{\vec{k}}^\dagger + \hat{\psi}_{\vec{k}}^\dagger\hat{\psi}_{\vec{k}})\right]. \tag{6.161}$$

Defining now the operator

$$\hat{Q}_{\vec{k}} = \frac{1}{\sqrt{2k}}\left[\hat{\pi}_{\vec{k}} - ik\omega\hat{\psi}_{\vec{k}}\right], \tag{6.162}$$

---
[m]From now on the tilde in the momentum operators will be omitted for the sake of simplicity.

the Hamiltonian can again be expressed in the same form previously discussed, namely, the one given by Eq. (6.141) with the caveat that now the operator (6.162), if compared to the one defined in Eq. (6.154) has a different expression in terms of the canonical fields. The commutation relations are now $[Q_{\vec{k}}, Q_{\vec{p}}^\dagger] = \omega \delta^{(3)}(\vec{k} - \vec{p})$. The mode functions $f_k(\tau)$ are the same as the ones given in Eq.(6.152), while g_k is given by

$$g_k(\tau) = -\mathcal{N}\sqrt{\frac{k}{2}}\sqrt{-x}\left[H^{(1)}_{\mu-1}(-x) + \frac{(1-2\mu)}{2(-x)}H^{(1)}_\mu(-x)\right], \quad (6.163)$$

Repeating the steps used in the two previous cases we arrive at the power spectrum:

$$|\delta_h(k,\tau)|^2 = |\Delta_h(k,\tau)|^2\left[1 - \frac{\beta(\beta+1)}{2x_0^3}\sin(2x_0 + \pi\beta)\right]. \quad (6.164)$$

In the de Sitter case, we have:

$$|\delta_h(k,\tau)|^2 = \frac{1}{2\pi^2}\left(\frac{H}{\overline{M}_P}\right)^2\left[1 + \frac{\sin 2x_0}{x_0^3}\right]. \quad (6.165)$$

Comparing Eq. (6.164) with (6.158) we see that the corrections are even smaller than the ones obtained using the Hamiltonian of Eq. (6.150). Furthermore, both Eqs. (6.164) and (6.158) lead to effects smaller than (6.146).

6.8 Numerical estimates of the mixing coefficients

In various problems it is often difficult to obtain analytical expressions for the mode functions and, a fortiori, for the mixing coefficients. In these situations, to determine the graviton spectrum, one has to resort to numerical methods. In this section we are going to illustrate, with an example, how it is possible to compute numerically the mixing coefficients. This method allows, actually, not only the evaluation of the tensor power spectra but also the evaluation of the scalar power spectra [108, 257, 258]. For reasons of presentation we will stick here to the tensor case. Notice, moreover, that the numerical results will both corroborate and enlighten the results previously obtained (in the present section) within the sudden approximation. Actually it will be shown, for instance, that the estimate of the mixing coefficient obtained in the transition from an expanding de Sitter stage to a radiation dominated phase (i.e. Eq. (6.95)) is correctly reproduced. Consider, therefore, a smooth transition between a de Sitter stage of expansion and a radiation-dominated stage of expansion. As discussed

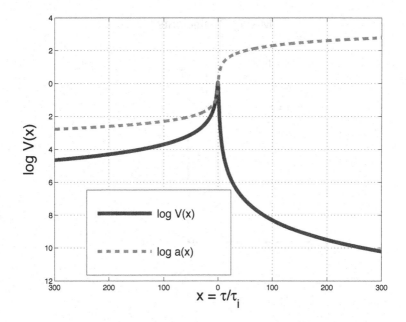

Fig. 6.3 The effective potential $\mathcal{V}(x)$ reported in Eq. (6.170) is illustrated with the full line. With the dashed line we report the scale factor $a(x)$ discussed in chapter 5 (see Eq. (5.83) and derivations therein).

in chapter 4 the smooth transition can be parametrized, in conformal time as in Eq. (5.83). Consequently the evolution equation for the (complex) mode function $f_k(\tau)$ will be given by

$$\frac{d^2 f_k}{dx^2} + \left[\kappa^2 - \frac{\sqrt{x^2+1}-x}{(x^2+1)^{3/2}}\right] f_k = 0 \qquad (6.166)$$

where, according to the notations of Eq. (5.83), $x = \tau/\tau_i$; in Eq. (6.166) it is also practical to define the rescaled wavenumber $\kappa = k\tau_i$. Now the strategy is very simple. We can select a given τ_b at which the numerical integration begins. Equation (6.166) has to be integrated for different wavenumbers and initial conditions will therefore be set in such a way that, for $\tau = \tau_b$,

$$k\tau_b = \kappa x_b \gg 1. \qquad (6.167)$$

Equation (6.167) implies that the given mode k, for $\tau = \tau_b$ is inside the Hubble radius. In terms of Eq. (6.166) this is equivalent to require that κ^2 is larger than the effective potential barrier. For $x = x_b$ it will also be required that the mode functions are appropriately normalized as $1/\sqrt{2k}$. This can

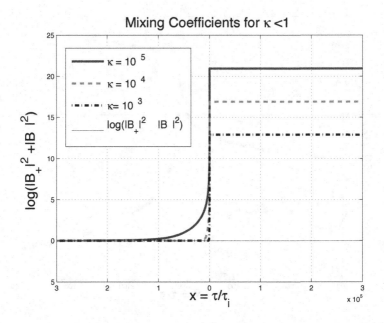

Fig. 6.4 The mixing coefficients are reported for three different values of $\kappa < 1$. With the full (thin) line at the bottom of the plot the base-10 logarithm of the Wronskian is reported.

be actually done in an even more precise fashion. In the limit $x \to -\infty$ it must actually happen that the effective potential vanishes as x^{-2} not only in the present case but also under more general circumstances. Therefore, in this regime, it is always possible to find the correctly normalized mode function and express it in terms of Hankel functions evaluated in x_b. These will be the initial conditions of the numerical integration provided $\kappa x_b \gg 1$. By introducing the real and imaginary parts of f_k and g_k as

$$f_k(x) = f_1(x) + if_2(x), \qquad g_k(x) = g_1(x) + ig_2(x), \qquad (6.168)$$

they will satisfy the following set of equations:

$$f_1' = g_1, \qquad g_1' = [\mathcal{V}(x) - \kappa^2]f_1,$$
$$f_2' = g_2, \qquad g_2' = [\mathcal{V}(x) - \kappa^2]f_2 \qquad (6.169)$$

where, with obvious notation,

$$\mathcal{V}(x) = \tau_i^2 \frac{a''}{a} \equiv \frac{\sqrt{x^2+1} - x}{(x^2+1)^{3/2}}. \qquad (6.170)$$

By now integrating Eqs. (6.169) forward in time the functions $f_k(x)$ and $g_k(x)$ can be computed. Even if the numerical integration will be carried

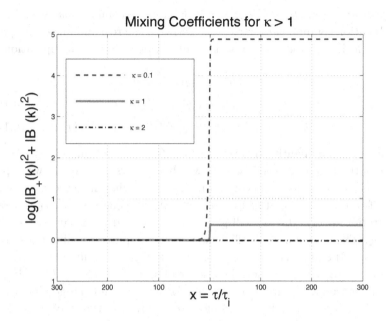

Fig. 6.5 The mixing coefficients are reported for $\kappa = 0.1$ (dashed line), for $\kappa = 1$ (full line) $\kappa = 2$ (dot-dashed line). As in Fig. 6.4, on the vertical axis base-10 logarithms are reported.

out when the potential is given as in Eq. (6.170), the method described in this section can be applied for any smooth potential $\mathcal{V}(x)$. The typical duration of the transition regime is controlled by τ_i. Since $x = \tau/\tau_i$ the transition regime will coincide, in the present parametrization, with the region $|x| < 1$. For $x \gg 1$ (i.e. asymptotically for $x \to +\infty$) it is possible to define the mixing coefficients as

$$f_k(x) = \frac{1}{\sqrt{2\kappa}}[B_+(\kappa)e^{-i\kappa x} + B_-(\kappa)e^{i\kappa x}],$$

$$g_k(x) = -i\sqrt{\frac{k}{2}}[B_+(\kappa)e^{-i\kappa x} - B_-(\kappa)e^{i\kappa x}].$$

(6.171)

From Eqs. (6.171) it is immediate to obtain that

$$B_+(\kappa) = \frac{i}{\sqrt{2\kappa}}e^{i\kappa x}(g_k - ikf_k),$$

$$B_-(\kappa) = -\frac{i}{\sqrt{2k}}e^{-i\kappa x}(g_k + ikf_k).$$

(6.172)

Since f_k and g_k are both complex, it is practical to compute the two relevant combinations of mixing coefficients in terms of the real and imaginary parts defined in Eq. (6.168). The result of this straightforward manipulation is:

$$|B_+(\kappa)|^2 + |B_-(\kappa)|^2 = \frac{1}{k}[g_1(x)^2 + g_2(x)^2]$$
$$+ k[f_1(x)^2 + f_2(x)^2], \qquad (6.173)$$
$$|B_+(k)|^2 - |B_-(k)|^2 = 2[g_1(x)f_2(x) - g_2(x)f_1(x)]. \qquad (6.174)$$

Notice that the quantity reported in Eq. (6.174) is nothing but the Wronskian of the two solutions f_k and f_k^* which is always equal to 1 if we impose quantum mechanical initial conditions. Of course the whole procedure must be self-consistent. So, after imposing the initial conditions at $x_{\rm b}$, forward integration of Eq. (6.171) will give us the real and imaginary parts of f_k and g_k. Then the mixing coefficients can be computed for different k as in Eqs. (6.173) and (6.174). To close the circle it must occur that for $x \gg 1$, $|B_+(\kappa)|^2 + |B_-(\kappa)|^2$ are indeed constant in x and $|B_+(\kappa)|^2 - |B_-(\kappa)|^2 = 1$. The latter condition must be satisfied for every x and the goodness with which is satisfied actually determines the accuracy of the whole algorithm. Before presenting the results of the numerical calculation let us recall that, in the sudden approximation, we expect exponential suppression for $\kappa > 1$ and, for $\kappa < 1$,

$$|B_+(k)|^2 + |B_-|^2 = \frac{1}{2\kappa^4} + 1. \qquad (6.175)$$

Equation (6.174) follows from Eq. (6.95) by identifying τ_1 with τ_i. This identification is very rough. There are more refined ways to compare the results of the sudden approximation with the numerical results. They are described, in some detail, in Ref. [257]. The main strategy is, in short, the following:

- compute the mixing coefficient in the sudden approximation in terms of the typical time-scale of the transition (i.e. what we called τ_1 in Eq. (6.95);
- compute the mixing coefficients numerically;
- the two procedure will always lead, by definition, to the same k dependence but the normalization will be determined with greater accuracy only by the numerical results;
- finally, the putative time scale of the sudden approximation can be used as a parameter for the fit of the numerical results in terms of the sudden results.

With these caveats in mind we can say that it is justified to presume, for the sake of comparison, that $\tau_i \simeq \tau_1$.

In Fig. 6.4 the base-10 logarithm of $|B_+(k)|^2 + |B_-(k)|^2$ is reported as a function of the rescaled (conformal) time coordinate $x = \tau/\tau_i$. The fact that, for $x \gg 1$, $|B_+(k)|^2 + |B_-(k)|^2$ is a constant and that $|B_+(k)|^2 - |B_-(k)|^2 = 1$ shows that the whole procedure is self-consistent. It is also clear that, because of the previous discussion in this section,

$$\log[|B_+(k)|^2 + |B_-(k)|^2] = \log[2\overline{n}_k^{\text{gw}} + 1], \qquad (6.176)$$

where $\overline{n}_k^{\text{gw}}$ denotes the mean number of graviton pairs per Fourier mode. From Fig. 6.4 it can be argued that the estimate of Eq. (6.95) is indeed reproduced by the numerical calculation. In fact, smaller κ are more amplified (since they spend more time under the potential barrier). Moreover, it is clearly visible from Fig. 6.4 that by decreasing κ of one order of magnitude the mean number of produced pairs increases by a factor 10^4, as predicted by Eq. (6.95). By taking different values of κ it is possible to determine the spectrum numerically with the wanted accuracy. In Fig. 6.5 the modes $\kappa > 1$ are considered. In this region the sudden approximation would imply an exponential suppression of the mixing coefficients and this is basically what can be argued from Fig. 6.5.

Chapter 7

The First Lap in CMB Anisotropies

In this chapter (and in chapters 8 and 9) the description of the CMB anisotropies will be introduced by successive approximations. In the present chapter a simplified account of the physics of CMB anisotropies will be derived in a two-fluid treatment. In chapter 8 a more realistic account of pre-decoupling physics will be outlined in the framework of an (improved) fluid description. In chapter 9 the Einstein-Boltzmann hierarchy will be introduced.

The discussion of the present chapter is organized around the Sachs-Wolfe effect and its estimate for different initial conditions of the cosmological perturbations. The following derivations will be specifically analyzed

- the Sachs-Wolfe effect for the tensor modes of the geometry;
- the Sachs-Wolfe effect for the scalar modes of the geometry;
- the evolution equations of the scalar fluctuations in the pre-decoupling phase;
- a simplified solution of the system allowing the estimate of the (scalar) Sachs-Wolfe contribution;
- the concept of adiabatic and non-adiabatic modes.

The logic will be to estimate first the Sachs-Wolfe (SW) effect for the tensor modes which have been already treated in chapter 6. Then the same physical considerations will be applied to the case of the scalar modes of the geometry. Finally, the scalar SW effect will be computed in a simplified description where the neutrinos (as well as the baryons) are supposed to be absent. The only two fluid components of the mixture will be cold dark matter particles and radiation. Still, as it will be shown, this oversimplified description allows to compute the so-called Sachs-Wolfe plateau which is an essential ingredient to set the large-scale normalization of CMB anisotropies.

Before plunging into the discussion a technical warning is in order. As far as the scalar modes of the geometry are concerned, the discussion will be conducted in the longitudinal coordinate system. However this choice is not essential. In chapter 11 the usefulness of different gauge choices will be extensively highlighted.

7.1 Tensor Sachs-Wolfe effect

After decoupling, the photon mean free path becomes comparable with the actual size of the present Hubble patch (see, for instance, Eq. (2.117) and derivations therein). Consequently, the photons will travel to our detectors and satellites without suffering any scattering. In this circumstance the photon geodesics are slightly perturbed by the presence of inhomogeneities. The temperature fluctuation induced by scalar and tensor modes of the geometry can then be estimated. Also vector fluctuations may induce relevant sources of anisotropy but they will be neglected in the first approximation mainly for the reason that, in the conventional scenario, the vector modes are always decaying both during radiation and, a fortiori, during the initial inflationary phase [249, 250] (see also [251, 252] for the case when the pre-decoupling sources support vector modes).

Since the Coulomb rate is much larger than Thompson rate of interactions around equality (see Eqs. (2.109)–(2.114) and (2.110)–(2.115)), baryons and electrons are more tightly coupled than photons and baryons. Still, prior to decoupling, it is rather plausible to treat the whole baryon-lepton-photon fluid as a unique physical entity.

Let us therefore start by studying the null geodesics in a conformally flat metric of FRW type where $g_{\alpha\beta} = a^2(\tau)\tilde{g}_{\alpha\beta}$ where $\tilde{g}_{\alpha\beta}$ coincides, in the absence of metric fluctuations, with the Minkowski metric. If metric fluctuations are present $\tilde{g}_{\alpha\beta}$ will have the form of a (slightly inhomogeneous) Minkowski metric. The latter observation implies that:

$$\tilde{g}_{00} = 1 + \delta_s \tilde{g}_{00}, \qquad \delta_s \tilde{g}_{00} = 2\phi, \tag{7.1}$$

$$\tilde{g}_{ij} = -\delta_{ij} + \delta_s \tilde{g}_{ij} + \delta_t \tilde{g}_{ij}, \qquad \delta_s \tilde{g}_{ij} = 2\psi \delta_{ij}, \qquad \delta_t \tilde{g}_{ij} = -h_{ij}, \tag{7.2}$$

with $\partial_i h^i_j = 0 = h^i_i$ (see Eq. (6.6)) and where the scalar fluctuations of the geometry have been introduced in the longitudinal gauge characterized by the two non-vanishing degrees of freedom ϕ and ψ (see Eqs. (6.3) and (6.4)).

Neglecting the inhomogeneous contribution, the lowest-order geodesics of the photon in the background $g_{\alpha\beta} = a^2(\tau)\eta_{\alpha\beta}$ are

$$\frac{d^2 x^\mu}{d\lambda^2} + \Gamma^\mu_{\alpha\beta} \frac{dx^\alpha}{d\lambda} \frac{dx^\beta}{d\lambda} = 0, \tag{7.3}$$

$$g_{\alpha\beta} \frac{dx^\alpha}{d\lambda} \frac{dx^\beta}{d\lambda} = 0, \tag{7.4}$$

where λ denotes the affine parameter. Recalling Eq. (6.12), the (0) component of Eq. (7.3) and Eq. (7.4) can be written, respectively, as

$$\frac{d^2\tau}{d\lambda^2} + \mathcal{H}\left(\frac{d\vec{x}}{d\lambda}\right)^2 + \mathcal{H}\left(\frac{d\tau}{d\lambda}\right)^2 = 0. \tag{7.5}$$

$$\left(\frac{d\vec{x}}{d\lambda}\right)^2 = \left(\frac{d\tau}{d\lambda}\right)^2. \tag{7.6}$$

Using Eq. (7.6), Eq. (7.5) can be usefully rearranged as

$$\frac{dF}{d\lambda} + 2\mathcal{H}F^2 = 0, \qquad F = \frac{d\tau}{d\lambda}. \tag{7.7}$$

With a simple manipulation Eq. (7.7) can be solved; the result will then be

$$F = \frac{d\tau}{d\lambda} = \frac{1}{a^2(\tau)}. \tag{7.8}$$

Equation (7.7) implies that if the affine parameter and the metric are changed as

$$d\lambda \to a^2(\tau) d\tau, \qquad g_{\alpha\beta} \to a^2(\tau) \tilde{g}_{\alpha\beta} \tag{7.9}$$

the new geodesics will be exactly the same as before. In particular, as a function of τ we will have that the unperturbed geodesics will be

$$x^\mu = n^\mu \tau, \qquad n^\mu = (n^0, n^i), \tag{7.10}$$

where $n^{0^2} = n_i n^i = 1$. Consider now the energy of the photon as measured in the reference frame of the baryonic fluid, i.e.[a]

$$\mathcal{E} = g_{\mu\nu} u^\mu P^\nu, \tag{7.11}$$

where u^μ is the four-velocity of the fluid and P^ν is the photon momentum defined as

$$P^\mu = \frac{dx^\mu}{d\lambda} = \frac{E}{a^2} \frac{dx^\mu}{d\tau}, \tag{7.12}$$

---

[a]The internal energy of a thermodynamic system has been denoted by $\mathcal{E}$ in Appendix B. Now $\mathcal{E}$ will denote the energy of a photon in the reference frame of the baryon fluid. These possible ambiguity is harmless since the two concepts will never interfere in the present treatment.

where E is a parameter (not to be confused with one of the off-diagonal entries of the perturbed metric) defining the photon energy. If the geodesic is perturbed by a tensor fluctuation we will have that Eq. (7.12) becomes

$$P^\mu = \frac{E}{a^2}\left[n^\mu + \frac{d\delta_t x^\mu}{d\tau}\right]. \qquad (7.13)$$

Since the condition $u^\mu u^\nu g_{\mu\nu} = 1$ implies that $u^0 = 1/a$, Eqs. (7.11) and (7.13) lead, respectively, to the following two more explicit expressions:

$$\mathcal{E} = \frac{E}{a}\left[1 + \frac{d\delta_t x^0}{d\tau}\right], \qquad \frac{d^2 \delta_t x^0}{d\tau^2} = -\delta\tilde{\Gamma}^0_{ij} n^i n^j. \qquad (7.14)$$

The quantity $\delta_t \tilde{\Gamma}^0_{ij}$ can be computed from $\delta_t \Gamma^0_{ij}$ (see, in particular, the first expression in Eq. (6.33)) by setting $\mathcal{H} = 0$. The result is:

$$\delta_t \tilde{\Gamma}^0_{ij} = \frac{1}{2} h'_{ij}. \qquad (7.15)$$

Consequently Eq. (7.14) can be be rearranged as

$$\frac{d\delta_t x^0}{d\tau} = -\frac{1}{2}\int_{\tau_i}^{\tau_f} h'_{ij} n^i n^j d\tau, \qquad \mathcal{E} = \frac{E}{a}\left[1 - \frac{1}{2}\int_{\tau_i}^{\tau_f} h'_{ij} n^i n^j d\tau\right], \qquad (7.16)$$

where $\tau_i = \tau_0$ and $\tau_f = \tau_{\rm dec}$. The temperature fluctuation due to the tensor modes of the geometry can then be computed as

$$\left(\frac{\Delta T}{T}\right) = \frac{a_f \mathcal{E}_f - a_i \mathcal{E}_i}{a_i \mathcal{E}_i}. \qquad (7.17)$$

By making use of Eq. (7.16), Eq. (7.17) simply becomes:

$$\left(\frac{\Delta T}{T}\right)_t = -\frac{1}{2}\int_{\tau_i}^{\tau_f} h'_{ij} n^i n^j d\tau. \qquad (7.18)$$

The only contribution to the tensor Sachs-Wolfe effect is given by the Sachs-Wolfe integral. It is clear that since during the matter dominated stage the evolution of the tensor modes is not conformally invariant there will be a tensor contribution to the SW effect. The absence of positive detection places bounds on the possible existence of a stochastic background of gravitational radiation for present frequencies of the order of 10^{-18} Hz (see Fig. 6.2).

7.2 Scalar Sachs-Wolfe effect

According to the observations discussed in Eqs. (7.2), in the case of the scalar modes of the geometry the perturbed geodesics can be written as

$$\frac{d^2 \delta_s x^\mu}{d\tau^2} + \delta_s \tilde{\Gamma}^\mu_{\alpha\beta} \frac{dx^\alpha}{d\tau} \frac{dx^\beta}{d\tau} = 0, \quad (7.19)$$

where now δ_s denotes the scalar fluctuation of the Christoffel connection computed with respect to $\tilde{g}_{\alpha\beta}$ which is the inhomogeneous Minkowski metric. In the conformally Newtonian gauge the fluctuations of the Christoffel connections of a perturbed Minkowski metric are[b]:

$$\delta_s \tilde{\Gamma}^0_{00} = \phi', \qquad \delta_s \tilde{\Gamma}^0_{0i} = \partial_i \phi, \qquad \delta_s \tilde{\Gamma}^0_{ij} = -\psi'. \quad (7.20)$$

Thus, using Eq. (7.20), Eq. (7.19) becomes

$$\frac{d}{d\tau}\left[\frac{d\delta_s x^0}{d\tau}\right] = -\delta_s \tilde{\Gamma}^0_{00} n^0 n^0 - \delta_s \tilde{\Gamma}^0_{ij} n^i n^j - 2\delta_s \tilde{\Gamma}^0_{i0} n^i n^0$$
$$\equiv \psi' - \phi' - 2\partial_i \phi n^i. \quad (7.21)$$

Since

$$\frac{d\phi}{d\tau} = \phi' + n^i \partial_i \phi, \quad (7.22)$$

we will also have

$$\frac{d\delta x^0}{d\tau} = \int_{\tau_i}^{\tau_f} (\psi' + \phi') d\eta - 2\phi. \quad (7.23)$$

The quantity to be computed, as previously anticipated, is the photon energy as measured in the frame of reference of the fluid. Defining u^μ as the four-velocity of the fluid and P^ν as the photon four-momentum, the photon energy will exactly be the one given in Eq. (7.11) but with a different physical content for P^ν and u^μ. The rationale for this statement is that while the tensor modes do not contribute to u^μ, the scalar modes affect the 0-component of u^μ. The logic will now be to determine u^μ, $g_{\mu\nu}$ and P^μ to first order in the scalar fluctuations of the geometry. This analysis allows to compute the right hand side of Eq. (7.11) in terms of the inhomogeneities of the metric. The first-order variation of $g^{\mu\nu} u_\mu u_\nu = 1$ leads to

$$\delta_s g_{00} u^0 = -2\delta_s u^0 g_{00}, \quad (7.24)$$

---

[b]These expressions can be obtained from Eqs. (C.2) and (C.3) of Appendix C by setting $\mathcal{H} = 0$.

so that in the longitudinal coordinate system Eq. (7.24) gives, to first-order in the metric fluctuations,

$$u^0 = \frac{1}{a}(1-\phi). \tag{7.25}$$

The divergencefull peculiar velocity field is given by

$$\delta_s u^i = \frac{v^i}{a} \equiv \frac{1}{a}\partial^i v. \tag{7.26}$$

The relevant peculiar velocity field will be, in this derivation, the baryonic peculiar velocity since this is the component emitting and observing (i.e. absorbing) the radiation. In the following this identification will be understood and, hence, $v^i = v_b^i$.

The energy of the photon in the frame of reference of the fluid becomes, then

$$\mathcal{E} = g_{\mu\nu}u^\mu P^\nu = g_{00}u^0 P^0 + g_{ij}u^i P^j. \tag{7.27}$$

Recalling the explicit forms of g_{00} and g_{ij} to first order in the metric fluctuations we have

$$\mathcal{E} = \frac{E}{a}\left[1 + \phi - n_i v_b^i + \frac{d\delta x^0}{d\tau}\right]. \tag{7.28}$$

Assuming, as previously stated, that the observer located at the end of a photon geodesic, is at $\vec{x} = 0$, Eq. (7.28) can be expressed as

$$\mathcal{E} = \frac{E}{a}\left\{1 - \phi - n_i v_b^i + \int_{\tau_i}^{\tau_f}(\psi' + \phi')d\tau\right\}. \tag{7.29}$$

The temperature fluctuation can be expressed by taking the difference between the final and initial energies, i.e.

$$\frac{\delta T}{T} = \frac{a_f \mathcal{E}(\tau_f) - a_i \mathcal{E}_i}{a_i \mathcal{E}_i}. \tag{7.30}$$

The final and initial photon energy are also affected by an intrinsic contribution, i.e.

$$\frac{a_f E_f}{a_i E_i} = \frac{T_0 - \delta T_f}{T_i - \delta T_i} \tag{7.31}$$

where we wrote that

$$T_0 = T_f + \delta_s T_f, \qquad T_{\text{dec}} = T_i + \delta_s T_i. \tag{7.32}$$

Now the temperature variation at the present epoch can be neglected while the intrinsic temperature variation at the initial time (i.e. the last scattering surface) cannot be neglected and it is given by

$$\frac{\delta_s T_i}{T_i} = \frac{\delta_\gamma}{4}(\tau_i), \tag{7.33}$$

since $\rho_\gamma(\tau_i) \simeq T_i^4$; δ_γ is the fractional variation of photon energy density.

The final expression for the SW effect induced by scalar fluctuations can be written as

$$\left(\frac{\Delta T}{T}\right)_s = \frac{\delta_r(\tau_i)}{4} - [\phi]_{\tau_i}^{\tau_f} - [n_i v_b^i]_{\tau_i}^{\tau_f} + \int_{\tau_i}^{\tau_f} (\psi' + \phi')d\tau. \qquad (7.34)$$

Sometimes, for simplified esitmates, the temperature fluctuation can then be written, in explicit terms, as

$$\left(\frac{\Delta T}{T}\right)_s = \left[\frac{\delta_r}{4} + \phi + n_i v_b^i\right]_{\tau_i} + \int_{\tau_i}^{\tau_f} (\psi' + \phi')d\tau. \qquad (7.35)$$

Equation (7.35) has three contribution

- the ordinary SW effect given by the first two terms at the right hand side of Eq. (7.35) i.e. $\delta_r/4$ and ϕ;
- the Doppler term (third term in Eq. (7.35));
- the integrated SW effect (last term in Eq. (7.35)).

The ordinary SW effect is both due to the intrinsic temperature inhomogeneities on the last scattering surface and to the inhomogeneities of the metric. On large angular scales the ordinary SW contribution dominates. The Doppler term arises thanks to the relative velocity of the emitter and of the receiver. At large angular scales its contribution is subleading but it becomes important at smaller scales, i.e. in multipole space, for $\ell \sim 200$ corresponding to the first peak in Fig. 1.3. The SW integral contributes to the temperature anisotropy if ψ and ϕ depend on time. Recalling the notations of chapter 3 (and, more specifically, Eq. (3.21) and derivations therein) it is practical, for the forthcoming applications, to separate the contributions of the ordinary SW effect from the ones arising, respectively, from the Doppler term and from the integrated SW effect. Therefore we can write,

$$\Delta_I^{(SW)}(\vec{x}_0, \hat{n}, \tau_0) = \Delta_I^{(SW)}(\vec{x}_{dec}, \hat{n}, \tau_{dec})$$
$$= \frac{\delta_\gamma(\vec{x}_{dec}, \tau_{dec})}{4} + \phi(\vec{x}_{dec}, \tau_{dec}), \qquad (7.36)$$

$$\Delta_I^{(ISW)}(\vec{x}_0, \hat{n}, \tau_0) = \Delta_I^{(ISW)}(\vec{x}_{dec}, \hat{n}, \tau_{dec})$$
$$= \int_{\tau_{dec}}^{\tau_0} (\psi' + \phi')d\tau, \qquad (7.37)$$

$$\Delta_I^{(Dop)}(\vec{x}_0, \hat{n}, \tau_0) = \Delta_I^{(Dop)}(\vec{x}_{dec}, \hat{n}, \tau_{dec})$$
$$= n_i v_b^i(\vec{x}_{dec}, \tau_{dec}). \qquad (7.38)$$

Notice that, in Eq. (7.36) the radiation density contrast has been identified with the one attributed to photons. This is both consistent with the employed approximations and generally correct since, as it will be shown in the context of the Boltzmann hierarchy, the neutrino density contrasts does not enter the ordinary SW term. The definition of $\Delta_{\rm I}$ can differ slightly for different authors. Indeed, we may choose to work directly with the (dimensional) temperature fluctuation $\delta_{\rm s} T$. In this case, as already discussed in chapter 3, the C_ℓ spectrum will have dimensions of $(\mu{\rm K})^2$.

It is elementary to obtain, from Eqs. (7.36), (7.37) and (7.38), the corresponding expressions in Fourier space. Even if this somehow straightforward it is wise to do this calculation step by step especially in the light of the derivations reported at the end of the present chapter. The definition of Fourier transform will be the one reported in Eq. (3.21) in chapter 3. So let us take, as an example, the ordinary Sachs-Wolfe contribution of Eq. (7.36). At the present time, from Eq. (3.21) it follows that

$$\Delta_{\rm I}^{\rm (SW)}(\vec{x}_0, \hat{n}, \tau_0) = \frac{1}{(2\pi)^{3/2}} \int d^3 k \Delta_{\rm I}^{\rm (SW)}(\vec{k}, \hat{n}, \tau_0) e^{i\vec{k}\cdot\vec{x}_0}. \tag{7.39}$$

At the decoupling time, the Fourier transform of the ordinary Sachs-Wolfe term reads instead:

$$\Delta_{\rm I}^{\rm (SW)}(\vec{x}_{\rm dec}, \hat{n}, \tau_{\rm dec}) = \frac{1}{(2\pi)^{3/2}} \int d^3 k \Delta_{\rm I}^{\rm (SW)}(\vec{k}, \hat{n}, \tau_{\rm dec}) e^{i\vec{k}\cdot\vec{x}_{\rm dec}}. \tag{7.40}$$

Since, according to Eq. (7.36)

$$\Delta_{\rm I}^{\rm (SW)}(\vec{x}_0, \hat{n}, \tau_0) = \Delta_{\rm I}^{\rm (SW)}(\vec{x}_{\rm dec}, \hat{n}, \tau_{\rm dec}), \tag{7.41}$$

we must also have

$$\Delta_{\rm I}^{\rm (SW)}(\vec{k}, \hat{n}, \tau_0) e^{i\vec{k}\cdot\vec{x}_0} = \Delta_{\rm I}^{\rm (SW)}(\vec{k}, \hat{n}, \tau_{\rm dec}) e^{i\vec{k}\cdot\vec{x}_{\rm dec}}. \tag{7.42}$$

But recall now that the photons follow, in the first approximation, unperturbed (null) geodesics (see Eq. (7.10)). Hence we will have

$$\vec{x}_0 = \vec{x}_{\rm dec} + \hat{n}(\tau_0 - \tau_{\rm dec}). \tag{7.43}$$

Using Eq. (7.43) into Eq. (7.42) it emerges immediately that

$$\Delta_{\rm I}^{\rm (SW)}(\vec{k}, \hat{n}, \tau_0) = \Delta_{\rm I}^{\rm (SW)}(\vec{k}, \hat{n}, \tau_{\rm dec}) e^{ik\mu(\tau_{\rm dec}-\tau_0)}, \tag{7.44}$$

where $\mu = \hat{n} \cdot \hat{k}$ is the projection of the photon momentum on the direction defined by the wave-vector. As we will see in chapter 9 this result can be obtained also by solving the (perturbed) Boltzmann equation in the sudden decoupling limit.

7.3 Scalar modes in the pre-decoupling phase

To estimate the various scalar Sachs-Wolfe contributions, the evolution of the metric fluctuations and of the perturbations of the sources must be discussed when the background is already dominated by matter, i.e. after equality but before decoupling. In this regime, for typical wavelengths larger than the Hubble radius at the corresponding epoch, the primeval fluctuations produced, for instance, during inflation, will serve as initial conditions for the fluctuations of the various plasma variables such as the density contrasts and the peculiar velocities.[c] As already mentioned in different circumstances the wavelengths of the fluctuations are larger than the Hubble radius provided that the corresponding wave-numbers satisfy the condition $k\tau < 1$ where τ is the conformal time coordinate that has been consistently employed throughout the whole discussion of inhomogeneities in FRW models. The first step along this direction is to write down the evolution equations of the metric perturbations which will now be treated in the longitudinal gauge. For the moment the explicit expression for the fluctuations of the matter sources will be left unspecified.

The perturbed components the Christoffel connections are obtained in Appendix C (see Eqs. (C.2) and (C.3)). The perturbed form of the components of the Ricci tensor can be readily obtained (see Eqs. (C.4) and (C.9)). Finally, from the first-order form of the components of the Einstein tensor (see Eqs. (C.10), (C.11) and (C.12)), the perturbed Einstein equations can then be formally written as:

$$\delta_{\rm s}\mathcal{G}_0^0 = 8\pi G \delta_{\rm s} T_0^0, \qquad (7.45)$$

$$\delta_{\rm s}\mathcal{G}_i^j = 8\pi G \delta_{\rm s} T_i^j, \qquad (7.46)$$

$$\delta_{\rm s}\mathcal{G}_0^i = 8\pi G \delta_{\rm s} T_0^i. \qquad (7.47)$$

The fluctuations of total the energy-momentum tensor are written as the sum of the fluctuations over the various species composing the plasma, i.e. according to Eqs. (C.19)–(C.23),

---

[c]The amplification of the fluctuations during an early stage of inflationary expansion will be addressed in chapter 10. The approach pursued here will be more model-independent. In other words, the initial conditions of cosmological fluctuations will not be set during inflation but later on, i.e. when the relevant modes have wavelengths larger than the Hubble radius during the radiation-dominated stage of expansion.

$$\delta_s T^0_0 = \delta\rho = \sum_\lambda \rho_\lambda \delta_\lambda, \qquad (7.48)$$

$$\delta_s T^j_i = -\delta^j_i \delta p + \Pi^j_i = -\delta^j_i \sum_\lambda w_\lambda \rho_\lambda \delta_\lambda + \Pi^j_i, \qquad (7.49)$$

$$\delta_s T^i_0 = (p+\rho)v^i = \sum_\lambda (1+w_\lambda)\rho_\lambda v^i_\lambda. \qquad (7.50)$$

Concerning Eqs. (7.48)–(7.50) few comments are in order:

- the sum over λ runs, in general, over the different species of the plasma (in particular, photon, baryons, neutrinos and CDM particles);
- δ_λ denotes the density contrast for each single species of the plasma (see Eq. (6.31) for the properties of δ_λ under infinitesimal gauge transformations);
- the term Π^j_i denotes the contribution of the anisotropic stress to the spatial components of the (perturbed) energy-momentum tensor of the fluid mixture.

Concerning the last point of this list, it should be borne in mind that the relevant physical situation is the one where Universe evolves for temperatures smaller than the MeV. In this case neutrinos have already decoupled and form a quasi-perfect (collisionless) fluid. For this reason, neutrinos will be the dominant source of anisotropic stress and such a contribution will be directly proportional to the quadrupole moment of the neutrino phase-space distribution.

In what follows the full content of the plasma will be drastically reduced with the purpose of obtaining (simplified) analytical estimates. In chapters 8 and 9 the simplifying assumptions adopted in the present section will be relaxed. Equations (7.45)–(7.47) lead then to the following system:

$$\nabla^2 \psi - 3\mathcal{H}(\mathcal{H}\phi + \psi') = 4\pi G a^2 \delta\rho, \qquad (7.51)$$

$$\nabla^2(\mathcal{H}\phi + \psi') = -4\pi G a^2 (p+\rho)\theta, \qquad (7.52)$$

$$\left[\psi'' + \mathcal{H}(2\psi' + \phi') + (2\mathcal{H}' + \mathcal{H}^2)\phi + \frac{1}{2}\nabla^2(\phi - \psi)\right]\delta^j_i$$
$$-\frac{1}{2}\partial_i\partial^j(\phi - \psi) = 4\pi G a^2 (\delta p \delta^j_i - \Pi^j_i). \qquad (7.53)$$

where the divergence of the total velocity field has been defined as

$$(p+\rho)\theta = \sum_\lambda (p_\lambda + \rho_\lambda)\theta_\lambda, \qquad (7.54)$$

with $\theta = \partial_i v^i$ and $\theta_\lambda = \partial_i v_\lambda^i$. Equations (7.51) and (7.52) are, respectively, the Hamiltonian and the momentum constraint. The enforcement of these two constraints is crucial for the regularity of the initial conditions. Taking the trace of Eq. (7.53) and recalling that the anisotropic stress is, by definition, traceless (i.e. $\Pi_i^i = 0$) it is simple to obtain

$$\psi'' + 2\mathcal{H}\psi' + \mathcal{H}\phi' + (2\mathcal{H}' + \mathcal{H}^2)\phi + \frac{1}{3}\nabla^2(\phi - \psi) = 4\pi G a^2 \delta p. \quad (7.55)$$

The difference of Eqs. (7.53) and (7.55) leads to

$$\frac{1}{6}\nabla^2(\phi - \psi)\delta_i^j - \frac{1}{2}\partial_i \partial^j(\phi - \psi) = -4\pi a^2 G \Pi_i^j. \quad (7.56)$$

By now applying to both sides of Eq. (7.56) the differential operator $\partial_j \partial^i$ we are led to the following expression

$$\nabla^4(\phi - \phi) = 12\pi G a^2 \partial_j \partial^i \Pi_i^j. \quad (7.57)$$

The right hand side of Eq. (7.57) can be usefully parametrized as

$$\partial_j \partial^i \Pi_i^j = \sum_\lambda (p_\lambda + \rho_\lambda)\nabla^2 \sigma_\lambda. \quad (7.58)$$

This parametrization may now appear baroque but it is helpful since σ_λ, in the case of neutrinos, is easily related to the quadrupole moment of the (perturbed) neutrino phase space distribution (see section 9). Equations (7.51)–(7.53) may be supplemented with the perturbation of the covariant conservation of the energy-momentum tensor[d]:

$$\delta_\lambda' = (1 + w_\lambda)(3\psi' - \theta_\lambda) + 3\mathcal{H}\left[w_\lambda - \frac{\delta p_\lambda}{\delta \rho_\lambda}\right]\delta_\lambda, \quad (7.59)$$

$$\theta_\lambda' = (3w_\lambda - 1)\mathcal{H}\theta_\lambda - \frac{w_\lambda'}{w_\lambda + 1}\theta_\lambda - \frac{1}{w_\lambda + 1}\frac{\delta p_\lambda}{\delta \rho_\lambda}\nabla^2 \delta_\lambda$$
$$+ \nabla^2 \sigma_\lambda - \nabla^2 \phi. \quad (7.60)$$

In Eq. (7.60) σ_λ appears and it stems directly from the correct fluctuation of the spatial components of energy-momentum tensor of the fluid mixture. In Eqs. (7.59) and (7.60) the energy and momentum exchange has been assumed to be negligible between the various components of the plasma. This approximation is not so realistic as far as the baryon-photon system is concerned as it will be explained in chapter 8.

---

[d] See Appendix C (and, in particular, Eqs. (C.31), (C.32) and (C.33)) for further details on the derivation.

7.3.1 Scale crossing and CMB initial conditions

In the problem of the analysis of CMB initial conditions there exist three kinds of wavelengths (or, for short, scales). They are illustrated schematically in the cartoon of Fig. 7.1:

- physical wavelengths that are much larger than the Hubble radius at recombination; in Fig. 7.1 they correspond to the full line and they are generically labeled by A;
- physical wavelengths that reentered the Hubble radius after equality (but before decoupling); in Fig. 7.1 they correspond to the dot-dashed line and they are generically labeled by B;
- physical wavelengths that reentered the Hubble radius prior to equality (i.e. during the radiation-dominated epoch); in Fig. 7.1 they correspond to the dashed line and they are generically labeled by C.

The scales illustrated with the full line in Fig. 7.1 are still outside the Hubble radius (i.e. $k\tau \leq 1$) at recombination, i.e. $\tau \simeq \tau_{\text{rec}}$. Since these wavelengths are on the verge of reentering at recombination, they felt the radiation-matter transition when they were still larger than the Hubble radius. Similar observations apply for those wavelengths that reentered the Hubble radius after equality (see the dot-dashed line in Fig. 7.1). These scales are the ones that determine the essential features of the Sachs-Wolfe plateau which will be derived later on in this chapter.

The scales illustrated with the dashed line in Fig. 7.1 reentered the Hubble radius prior to equality and felt the radiation-matter transition when they were already inside the Hubble radius. These simple remarks are rather important for the exercises that will be discussed in the remaining sections of this chapter. When the relevant wavelengths are larger than the Hubble radius the evolution equations of the system greatly simplify and analytical solutions are possible. Furthermore, the cartoon reported in Fig. 7.1 suggests that the most relevant regime is somehow the one when the Universe is already dominated by matter. Crudely speaking this is correct, however, for physical reasons that will be more completely discussed in chapter 8, it will be more appropriate to set the CMB initial conditions deep in the radiation dominated phase also for those modes that reenter during the matter epoch.

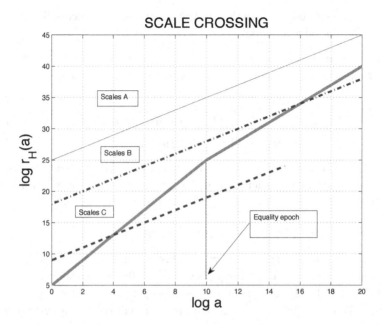

Fig. 7.1 The crossing of different wavelengths is schematically illustrated around the epoch of radiation-matter equality.

7.4 CDM-radiation system

The content of the plasma is formed by four different species, namely dark matter particles, photons, neutrinos and baryons. The following simplifying assumptions will now be proposed:

- the pre-decoupling plasma is only formed by a radiation component (denoted by a subscript r) and by a CDM component (denoted by a subscript c);
- neutrinos will be assumed to be a component of the radiation fluid but their anisotropic stress will be neglected: hence, according to Eqs. (7.56) and (7.57), the two longitudinal fluctuations of the metric will then be equal, i.e. $\phi = \psi$;
- the energy-momentum exchange between photons and baryons will be neglected.

These three assumptions will allow two interesting exercises:

- a simplified estimate of the large-scale Sachs-Wolfe contribution;

- a simplified introduction to the distinction between adiabatic and non-adiabatic modes.

Since the neutrinos are absent, then we can set $\psi = \phi$ in Eqs. (7.51), (7.52) and (7.53) whose explicit form will become, in Fourier space

$$-k^2\psi - 3\mathcal{H}(\mathcal{H}\psi + \psi') = 4\pi G a^2 \delta\rho, \tag{7.61}$$

$$\psi'' + 3\mathcal{H}\psi' + (\mathcal{H}^2 + 2\mathcal{H}')\psi = 4\pi G a^2 \delta p, \tag{7.62}$$

$$k^2(\mathcal{H}\psi + \psi') = 4\pi G a^2 (p + \rho)\theta. \tag{7.63}$$

Since the only two species of the plasma are, in the present discussion, radiation and CDM particles we will have, according to Eqs. (7.48) and (7.54):

$$(p + \rho)\theta = \frac{4}{3}\rho_r \theta_r + \rho_c \theta_c, \qquad \delta\rho = \rho_r \delta_r + \rho_c \delta_c. \tag{7.64}$$

There are two different regimes where this system can be studied, i.e. either before equality or after equality. Typically, as we shall see, initial conditions for CMB anisotropies are set deep in the radiation-dominated regime. However, in the present example we will solve the system separately during the radiation and the matter-dominated epochs. During the radiation-dominated epoch, i.e. prior to $\tau_{\rm eq}$, the evolution equation for ψ can be solved exactly by noticing that, in this regime, $3\delta p = \delta\rho$. Thus, by linearly combining Eqs. (7.61) and (7.62) to eliminate the contribution of the fluctuations of the energy density and of the pressure, the (decoupled) evolution equation for ψ can be written as:

$$\psi'' + 4\mathcal{H}\psi' + \frac{k^2}{3}\psi = 0. \tag{7.65}$$

Since during radiation $a(\tau) \sim \tau$, Eq. (7.65) can be solved as a combination of Bessel functions of order 3/2 (connected with spherical Bessel functions) which can be expressed, in turn, as a combination of trigonometric functions weighted by inverse powers of their argument [215, 216]

$$\psi(k,\tau) = A_1(k)\frac{y\cos y - \sin y}{y^3} + B_1(k)\frac{y\sin y + \cos y}{y^3}, \tag{7.66}$$

where $y = k\tau/\sqrt{3}$. Since Eq. (7.66) plays a relevant role in the subsequent semi-analytical estimates let us derive it with some detail by recalling some result already obtained in a totally different context. Indeed, as we shall see in a moment, Eq. (7.65) can be rearranged in a form which is indeed very close to the situation already encountered in chapter 6 (see, in particular, Eqs. (6.85) and (6.86)) when describing the evolution of the tensor modes of the geometry in a phase of de Sitter expansion. Consider, therefore, Eq. (7.65). It is easy to show, by eliminating the first derivative, that Eq. (7.65) ca be rearranged as

$$f'' + \left[\frac{k^2}{3} - \frac{(a^2)''}{a^2}\right]f = 0, \qquad f = a^2\psi. \tag{7.67}$$

Since we are at the radiation epoch (i.e. $a(\tau) \simeq \tau$), Eq. (7.67) becomes

$$f'' + \left[\frac{k^2}{3} - \frac{2}{\tau^2}\right]f = 0, \qquad f = a^2\psi, \tag{7.68}$$

whose solutions have exactly the same form of Eq. (6.85), i.e. more specifically

$$f(y) = \left(1 - \frac{i}{y}\right)e^{-iy}, \qquad f^*(y) = \left(1 + \frac{i}{y}\right)e^{iy}, \tag{7.69}$$

where, now, unlike Eq. (6.85), $y = k\tau/\sqrt{3}$. The solution of Eq. (7.69) can be written in terms of the real and imaginary part of f by writing that $f(y) = f_1(y) + if_2(y)$ (and, analogously, $f^*(y) = f_1(y) - if_2(y)$). Consequently, $f_1(y)$ and $f_2(y)$ become, after simple algebra:

$$f_1(y) = \cos y - \frac{\sin y}{y}, \qquad f_2(y) = -\left(\sin y + \frac{\cos y}{y}\right). \tag{7.70}$$

The functions $f_1(y)$ and $f_2(y)$ are exactly the functions appearing in Eq. (7.66), by noticing that the minus sign appearing in Eq. $f_2(y)$ can be reabsorbed in the definition of $B_1(k)$. Consider now the solution denoted by the integration constant $A_1(k)$, i.e.

$$\psi(k,\tau) = A_1(k)\frac{y\cos y - \sin y}{y^3}, \tag{7.71}$$

and take the limit $y \ll 1$. This limit corresponds to physical wavelengths that are larger than the Hubble radius *before* equality. In this limit the longitudinal fluctuation of the metric goes to a constant, i.e. $\psi(k,\tau) \to \psi_{\rm r}(k)$ for $y \ll 1$. For future convenience we can therefore express $A_1(k)$ in terms of $\psi_{\rm r}(k)$: by taking the limit $y \ll 1$ in Eq. (7.71) and by requiring that, in this limit, $\psi(k,\tau) \to \psi_{\rm r}(k)$, the constant $A_1(k)$ is determined to be $A_1(k) = -3\psi_{\rm r}(k)$, i.e. Eq. (7.71) becomes

$$\psi(k,\tau) = -3\psi_{\rm r}(k)\frac{y\cos y - \sin y}{y^3}. \tag{7.72}$$

This is the case of purely adiabatic initial conditions. If, on the contrary, $A_1(k)$ is set to zero, then ψ will not go to a constant. This second solution is important in the case of the non-adiabatic modes. At the end of this section the physical distinction between adiabatic and non-adiabatic modes will be specifically discussed.

In the case of adiabatic fluctuations the constant mode ψ_r matches to a constant mode during the subsequent matter dominated epoch. In fact, during the matter dominated epoch and under the same assumptions of absence of anisotropic stresses the equation for ψ is

$$\psi'' + 3\mathcal{H}\psi' = 0. \tag{7.73}$$

Since, after equality, $a(\tau) \sim \tau^2$, the solution of Eq. (7.73) is then

$$\psi(k,\tau) = \psi_\mathrm{m}(k) + D_1(k)\left(\frac{\tau_\mathrm{eq}}{\tau}\right)^5, \tag{7.74}$$

where $\psi_\mathrm{m}(k)$ is a constant in time. The values of $\psi_\mathrm{r}(k)$ and $\psi_\mathrm{m}(k)$ are different but can be easily connected. In fact we are interested in wavenumbers $k\tau < 1$ after equality and, in this regime[e]

$$\psi_\mathrm{m}(k) = \frac{9}{10}\psi_\mathrm{r}(k). \tag{7.75}$$

Disregarding the complication of an anisotropic stress (i.e. from Eqs. (7.56)–(7.57), $\phi = \psi$) from Eqs. (7.59)–(7.60), the covariant conservation equations become

$$\delta_\mathrm{c}' = 3\psi' - \theta_\mathrm{c}, \tag{7.76}$$

$$\theta_\mathrm{c}' = -\mathcal{H}\theta_\mathrm{c} + k^2\psi, \tag{7.77}$$

$$\delta_\mathrm{r}' = 4\psi' - \frac{4}{3}\theta_\mathrm{r}, \tag{7.78}$$

$$\theta_\mathrm{r}' = \frac{k^2}{4}\delta_\mathrm{r} + k^2\psi. \tag{7.79}$$

Combining Eqs. (7.78) and (7.79) in the presence of the constant adiabatic mode ψ_m and during the matter-dominated phase

$$\delta_\mathrm{r}'' + k^2 c_\mathrm{s}^2 \delta_\mathrm{r} = -4c_\mathrm{s}^2 k^2 \psi_\mathrm{m}, \tag{7.80}$$

where $c_\mathrm{s} = 1/\sqrt{3}$. The solution of Eq. (7.80) can be obtained with elementary methods. In particular it will be

$$\delta_\mathrm{r}(k,\tau) = c_1 \cos k c_\mathrm{s}\tau + c_2 \sin k c_\mathrm{s}\tau$$
$$- 4c_\mathrm{s}^2 k^2 \psi_\mathrm{m} \int_0^\tau d\xi [\cos k c_\mathrm{s}\xi \sin k c_\mathrm{s}\tau - \sin k c_\mathrm{s}\xi \cos k c_\mathrm{s}\tau]. \tag{7.81}$$

---

[e]Equation (7.75) follows from the conservation of curvature perturbations in the long wavelength limit, i.e. $k\tau \ll 1$. The explicit derivation of this result will be given in a moment, see, in particular, Eqs. (7.103) and (7.104).

The full solution of Eqs. (7.77)–(7.79) will then be:

$$\delta_c(k,\tau) = -2\psi_m - \frac{\psi_m}{6}k^2\tau^2, \tag{7.82}$$

$$\theta_c(k,\tau) = \frac{k^2\tau}{3}\psi_m, \tag{7.83}$$

$$\delta_r(k,\tau) = \frac{4}{3}\psi_m[\cos(kc_s\tau) - 3], \tag{7.84}$$

$$\theta_r(k,\tau) = \frac{k\psi_m}{\sqrt{3}}\sin(kc_s\tau). \tag{7.85}$$

Notice that:

- for $k\tau \ll 1$, $\theta_c \simeq \theta_r$;
- for $k\tau \ll 1$, $\delta_r = 4\delta_c/3$.

These relations have a rather interesting physical interpretation that will be scrutinized in the following section of this chapter. The ordinary SW effect can now be roughly estimated. Consider Eq. (7.34) in the case of the pure adiabatic mode. Since the longitudinal degrees of freedom of the metric are roughly constant and equal, inserting the solution of Eqs. (7.82)–(7.85) into Eq. (7.34) the following result can be obtained:

$$\left(\frac{\Delta T}{T}\right)^{ad}_{k,s} = \left(\frac{\delta_r}{4} + \psi\right)_{\tau \simeq \tau_{dec}}$$

$$\equiv \frac{\psi_m}{3}\cos(kc_s\tau_{dec}) = \frac{3}{10}\psi_r\cos(kc_s\tau_{dec}), \tag{7.86}$$

where the third equality follows from the relation between the constant modes during radiation and matter, i.e. Eq. (7.75). Concerning Eq. (7.86) few comments are in order:

- for superhorizon modes the baryon peculiar velocity does not contribute to the leading result of the SW effect;
- for $kc_s\tau_{dec} \ll 1$ the temperature fluctuations induced by the adiabatic mode are simply $\psi_m/3$;
- even if more accurate results on the temperature fluctuations on small angular scales can be obtained from a systematic expansion in the inverse of the differential optical depth,[f] Eq. (7.86) suggests that the first true peak in the temperature fluctuations is located at $kc_s\eta_{dec} \simeq \pi$.

---

[f]This expansion, called tight coupling expansion, will be discussed in chapters 8 and 9.

In this discussion, the role of the baryons has been completely neglected. In chapter 8 a more refined picture of the acoustic oscillations will be developed and it will be shown that the inclusion of baryons induces a shift of the first Doppler peak.

7.5 Adiabatic and non-adiabatic modes: an example

The solution obtained in Eqs. (7.82), (7.83), (7.84) and (7.85) obeys, in the limit $k\tau \ll 1$ the following interesting condition

$$\delta_{\rm r}(k,\tau) = \frac{4}{3}\delta_{\rm c}(k,\tau). \qquad (7.87)$$

A solution obeying Eq. (7.87) for the radiation and matter density contrasts is said to be *adiabatic*. A distinction playing a key rôle in the theory of the CMB anisotropies is the one between *adiabatic* and *isocurvature*[g] modes. Consider, again, the idealized case of a plasma where the only fluid variables are the ones associated with CDM particles and radiation. The entropy per dark matter particle will then be given by $\varsigma = T^3/n_{\rm c}$ where $n_{\rm c}$ is the number density of CDM particles and $\rho_{\rm c} = m_{\rm c} n_{\rm c}$ is the associated energy density. Recalling that $\delta_{\rm r} = \delta\rho_{\rm r}/\rho_{\rm r}$ and $\delta_{\rm c} = \delta\rho_{\rm c}/\rho_{\rm c}$ are, respectively the density contrast in radiation and in CDM, the fluctuations of the specific entropy will then be

$$\mathcal{S} = \frac{\delta\varsigma}{\varsigma} = 3\frac{\delta T}{T} - \delta_{\rm c} = \frac{3}{4}\delta_{\rm r} - \delta_{\rm c}, \qquad (7.88)$$

where the second equality follows recalling that $\rho_{\rm r} \propto T^4$. If the fluctuations in the specific entropy vanish, at large-scales, then a chacteristic relation between the density contrasts of the various plasma quantities appears, i.e. for a baryon-photon-lepton fluid with CMD particles,

$$\delta_\gamma(k,\tau) \simeq \delta_\nu(k,\tau) \simeq \frac{4}{3}\delta_{\rm c}(k,\tau) \simeq \frac{4}{3}\delta_{\rm b}(k,\tau). \qquad (7.89)$$

Eq. (7.88) can be generalized to the case of a mixture of different fluids with arbitrary equation of state. For instance, in the case of two fluids a and b with barotropic indices $w_{\rm a}$ and $w_{\rm b}$ the fluctuations in the specific entropy are

$$\mathcal{S}_{\rm a\,b}(k,\tau) = \frac{\delta_{\rm a}(k,\tau)}{1+w_{\rm a}} - \frac{\delta_{\rm b}(k,\tau)}{1+w_{\rm b}}, \qquad (7.90)$$

---

[g]To avoid misunderstanding it would be more appropriate to use the terminology non-adiabatic since the term isocurvature may be interpreted as denoting a fluctuation giving rise to a uniform curvature. In the following the common terminology will however be used.

where δ_a and δ_b are the density contrasts of the two species. It is appropriate to stress that, according to Eq. (6.31), giving the gauge variation of the density contrast of a given species, $\mathcal{S}_{ab}$ is gauge-invariant (see, in fact, Eq. (6.31)). As a consequence of the mentioned distinction the total pressure density can be connected to the total fluctuation of the energy density as

$$\delta p = c_s^2 \delta \rho + \delta p_{\text{nad}}, \tag{7.91}$$

where

$$c_s^2 = \left(\frac{\delta p}{\delta \rho}\right)_\varsigma = \left(\frac{p'}{\rho'}\right)_\varsigma, \tag{7.92}$$

is the speed of sound computed from the variation of the total pressure and energy density at constant specific entropy, i.e. $\delta\varsigma = 0$. The second term appearing in Eq. (7.91) is the pressure density variation produced by the fluctuation in the specific entropy at constant energy density, i.e.

$$\delta p_{\text{nad}} = \left(\frac{\delta p}{\delta \varsigma}\right)_\rho \delta \varsigma, \tag{7.93}$$

accounting for the non-adiabatic contribution to the total pressure perturbation. If only one species is present with equation of state $p = w\rho$, then it follows from the definition that $c_s^2 = w$ and the non-adiabatic contribution vanishes. As previously anticipated around Eq. (7.88), a sufficient condition in order to have $\delta p_{\text{nad}} \neq 0$ is that the fluctuation in the specific entropy $\delta\varsigma$ is not vanishing. Consider, for simplicity, the case of a plasma made of radiation and CDM particles. In this case the speed of sound and the non-adiabatic contribution can be easily computed and they are:

$$c_s^2 = \frac{p'}{\rho'} = \frac{p_r' + p_c'}{\rho_r' + \rho_c'} \equiv \frac{4}{3}\left(\frac{\rho_r}{3\rho_c + 4\rho_r}\right), \tag{7.94}$$

$$\varsigma\left(\frac{\delta p}{\delta \varsigma}\right)_\rho = \frac{4}{3}\left(\frac{\delta\rho_r}{3\frac{\delta\rho_r}{\rho_r} - 4\frac{\delta\rho_c}{\rho_c}}\right) \equiv \frac{4}{3}\left(\frac{\rho_c \rho_r}{3\rho_c + 4\rho_r}\right) \equiv \rho_c c_s^2. \tag{7.95}$$

To obtain the final expression appearing at the right-hand-side of Eq. (7.94) the conservation equations for the two species (i.e. $\rho_r' = -4\mathcal{H}\rho_r$ and $\rho_c' = -3\mathcal{H}\rho_c$) have been used. Concerning Eq. (7.95) the following remarks are in order:

- the first equality follows from the fluctuation of the specific entropy computed in Eq. (7.88);
- the second equality appearing in Eq. (7.95) follows from the observation that the increment of the pressure should be computed for constant (total) energy density, i.e. $\delta\rho = \delta\rho_r + \delta\rho_c = 0$, implying $\delta\rho_c = -\delta\rho_r$;

- the third equality (always in Eq. (7.95)) is a mere consequence of the explicit expression of c_s^2 obtained in Eq. (7.94).

As in the case of Eq. (7.90), the analysis presented up to now can be easily generalized to a mixture of fluids "a" and "b" with barotropic indices w_a and w_b. The generalized speed of sound is then given by

$$c_s^2 = \frac{w_a(w_a+1)\rho_a + w_b(w_b+1)\rho_b}{(w_a+1)\rho_a + (w_b+1)\rho_b}. \tag{7.96}$$

As we shall see in a while, there are several generalizations of the concept of non-adiabatic mode. These generalizations depend upon the other possible components of the fluid mixture. In the CDM-radiation system there are only two fluids. Consequently, only one non-adiabatic mode can be constructed. If three (or even four) fluids are simultaneously present (as it happens in the realistic situation) the structure of the non-adiabatic modes will be richer. With these caveats in mind let us study in more depth the CDM-radiation system with the goal of estimating the value of curvature perturbations after equality. This calculation will allow, at the end of this chapter, the estimate of the ordinary and integrated SW contributions.

In the case of the CDM-radiation system the barotropic index and the sound speed can be written in a very simple way. Indeed, recalling the definition of barotropic index we have[h]

$$w_t = \frac{p_t}{\rho_t} = \frac{\rho_r}{3(\rho_r + \rho_c)} = \frac{1}{3(\alpha+1)}$$
$$c_{st}^2 = \frac{p_t'}{\rho_t'} = w_t - \frac{1}{3}\frac{d\ln(w_t+1)}{d\ln\alpha} = \frac{4}{3(3\alpha+4)}. \tag{7.97}$$

Equation (7.97) expresses the (total) barotropic index and the (total) sound speed solely in terms of α which is defined, recalling Eqs. (2.89) and (2.93), as $\alpha(\tau) = a(\tau)/a_{eq}$. Therefore, $\alpha(\tau)$ is nothing but the rescaled scale factor. The expressions reported in Eq. (7.97) are exact and are independent of the present normalization of the scale factor (i.e. we do not need to assume, for instance, that $a_0 = 1$) since a_0, ρ_{r0} amd ρ_{c0} have been eliminated in favour of a_{eq} (see Eq. (2.89) and discussions therein). At the level of notation it is also useful to remark that, within the present approximations, $\rho_r = \rho_R \equiv \rho_\gamma$ (since neutrinos have been neglected) and $\rho_c = \rho_M$ (since baryons have been neglected). This last remark allows a direct matching between the present notations and the general discussion of chapter 2. Finally, it is useful to

---

[h]In Eq. (7.97) the notations w_t and c_{st}^2 have been used. This convention allows to stress that, in the CDM-radiation syatem, the evolution of curvature perturbations depends indeed upon the total barotropic index and upon the total sound speed.

express the critical parameters for CDM and radiation in terms of $\alpha(\tau)$, i.e.

$$\Omega_{\rm c}(\tau) = \Omega_{\rm M}(\tau) = \frac{\alpha(\tau)}{\alpha(\tau)+1} = \frac{x^2+2x}{(x+1)^2},$$
$$\Omega_{\rm r}(\tau) = \Omega_\gamma(\tau) = \frac{1}{\alpha(\tau)+1} = \frac{1}{(x+1)^2}, \qquad (7.98)$$

where, recalling Eq. (2.93), $x = \tau/\tau_1$ is the rescaled conformal time coordinate. Using Eqs. (7.93)–(7.95) the non-adiabatic contribution to the total pressure fluctuation becomes

$$\delta p_{\rm nad}(k,\tau) = \frac{4}{3}\rho_{\rm c}\frac{\mathcal{S}_*(k)}{3\alpha+4}, \qquad (7.99)$$

where the definition given in Eq. (7.88), i.e. $\mathcal{S}_* = (\delta\varsigma)/\varsigma$, has been used. Using the splitting of the total pressure density fluctuation into a adiabatic and a non-adiabatic parts, Eq. (7.51) can be multiplied by a factor $c_{\rm s}^2$ and subtracted from Eq. (7.55). The result of this operation leads to a formally simple expression for the evolution of curvature fluctuations in the longitudinal gauge, namely:

$$\psi'' + \mathcal{H}[\phi' + (2+3c_{\rm st}^2)\psi'] + [\mathcal{H}^2(1+2c_{\rm st}^2) + 2\mathcal{H}']\phi$$
$$-c_{\rm st}^2\nabla^2\psi + \frac{1}{3}\nabla^2(\phi-\psi) = 4\pi Ga^2\delta p_{\rm nad}, \qquad (7.100)$$

which is independent of the specific form of $\delta p_{\rm nad}$. The left hand side of Eq. (7.100) can be written as the (conformal) time derivative of a single scalar function whose specific form is ,

$$\mathcal{R} = -\left(\psi + \frac{\mathcal{H}(\psi'+\mathcal{H}\phi)}{\mathcal{H}^2-\mathcal{H}'}\right). \qquad (7.101)$$

Taking now the first (conformal) time derivative of $\mathcal{R}$ as expressed by Eq. (7.101) and using the definition of $c_{\rm s}^2$, we arrive at the following expression

$$\mathcal{R}' = -\frac{\mathcal{H}}{4\pi Ga^2(\rho_{\rm t}+p_{\rm t})}\{\psi'' + \mathcal{H}[(2+3c_{\rm st}^2)\psi' + \phi']$$
$$+ [2\mathcal{H}' + (3c_{\rm st}^2+1)\mathcal{H}^2]\phi\}. \qquad (7.102)$$

Comparing now Eqs. (7.102) and (7.100), it is clear that Eq. (7.102) reproduces Eq. (7.100) but only up to the spatial gradients. Hence, using Eq. (7.102) into Eq. (7.100) the following final expression can be obtained:

$$\mathcal{R}' = -\frac{\mathcal{H}}{p_{\rm t}+\rho_{\rm t}}\delta p_{\rm nad} - \frac{k^2\mathcal{H}}{12\pi Ga^2(p_{\rm t}+\rho_{\rm t})}(\phi-\psi)$$
$$+ \frac{c_{\rm st}^2\mathcal{H}}{4\pi Ga^2(p_{\rm t}+\rho_{\rm t})}k^2\psi. \qquad (7.103)$$

Equation (7.103) is very useful in different situations. Suppose, as a simple exercise, we consider the evolution of modes with wavelengths larger than the Hubble radius at the transition between matter and radiation. Suppose also that $\delta p_{\text{nad}} = 0$. In this case Eq. (7.103) implies, quite simply, that across the radiation-matter transition $\mathcal{R}$ is constant up to corrections of order of $k^2 \tau^2$ which are small when the given wavelngths are larger than the Hubble radius. Now it happens so that the relevant modes for the estimate of the ordinary Sachs-Wolfe effect are exactly the ones that are still larger than the Hubble radius at the transition between matter and radiation. This observation allows us to derive Eq. (7.75). In fact, using the definition of $\mathcal{R}$ and recalling that during radiation and matter the longitudinal fluctuations of the geometry are constants we will have

$$\mathcal{R}_{\text{m}}(k, \tau) = -\frac{5}{3} \psi_{\text{m}}(k), \qquad \mathcal{R}_{\text{r}}(k, \tau) = -\frac{3}{2} \psi_{\text{r}}(k). \qquad (7.104)$$

But since $\mathcal{R}_{\text{m}}(k, \tau) = \mathcal{R}_{\text{r}}(k, \tau)$, Eq. (7.75) easily follows. The result expressed by Eqs. (7.75) and (7.104) holds in the case when neutrinos are not taken into account. This result can be however generalized to the case where neutrinos are present in the system, as it will be discussed in chapter 8. Equation (7.103) can also be used in order to obtain the evolution of ψ. Consider again the case of adiabatic initial conditions. In this case, as already mentioned, deep in the radiation-dominated regime $\mathcal{R}_{\text{r}} = -3\psi_{\text{r}}/2$. From the definition of $\mathcal{R}$ we can write the evolution for ψ using, as integration variable, the scale factor. Using the same trick exploited in the derivation of Eq. (7.97), we have, across the equality time

$$\frac{\rho_{\text{t}} + p_{\text{t}}}{\rho_{\text{t}}} = \frac{3\alpha + 4}{3(\alpha + 1)}. \qquad (7.105)$$

Thus, plugging Eqs. (7.97) and (7.105) Eq. (7.101) it is easy to obtain the following (first order) differential equation:

$$\frac{d\psi}{d\alpha} + \frac{5\alpha + 6}{2\alpha(\alpha + 1)} \psi = \frac{3}{4} \left(\frac{3\alpha + 4}{\alpha + 1} \right) \psi_{\text{r}}(k), \qquad (7.106)$$

which can be also written as

$$\frac{\sqrt{\alpha + 1}}{\alpha^3} \frac{d}{d\alpha} \left(\frac{\alpha^3}{\sqrt{\alpha + 1}} \psi \right) = \frac{3}{4} \psi_{\text{r}}(k) \left(\frac{3\alpha + 4}{\alpha(\alpha + 1)} \right). \qquad (7.107)$$

By integrating once the result for $\psi(\alpha)$ is[i]

$$\psi(k,\alpha) = \frac{\psi_{\rm r}(k)}{10\alpha^3}\{16(\sqrt{\alpha+1}-1) + \alpha[\alpha(9\alpha+2)-8]\}; \qquad (7.108)$$

the limit for $\alpha \to \infty$ (matter-dominated phase) of the right hand side of Eq. (7.108) leads to $(9/10)\psi_{\rm r}$. In the simplistic case of CDM-radiation plasma a rather instructive derivation of the gross features of the non-adiabatic mode can also be obtained. If $\delta p_{\rm nad}$ is given by Eq. (7.99), Eq. (7.103) can be simply written as

$$\frac{d\mathcal{R}}{d\alpha} = -\frac{4\mathcal{S}_*(k)}{(3\alpha+4)^2} + \mathcal{O}(k^2\tau^2). \qquad (7.109)$$

Eq. (7.109) can be easily obtained inserting Eq. (7.99) into Eq. (7.103) and recalling that, in the physical system under consideration, $(p_{\rm t}+\rho_{\rm t}) = \rho_{\rm c} + (4/3)\rho_{\rm r}$. In the case of the CDM-radiation isocurvature mode, the non-adiabatic contribution is non-vanishing and proportional to $\mathcal{S}_*(k)$. Furthermore, it can be easily shown that the fluctuations of the entropy density, $\mathcal{S}_*(k)$ are roughly constant (up to logarithmic corrections) for $k\tau \ll 1$, i.e. for the modes which are relevant for the SW effect after equality.[j] This conclusion can be easily derived by subtracting Eq. (7.76) from 3/4 of Eq. (7.78). Recalling the definition of $\mathcal{S}_*$ the result is

$$\mathcal{S}' = -(\theta_{\rm r} - \theta_{\rm c}). \qquad (7.110)$$

Since $\theta_{\rm r}$ and $\theta_{\rm c}$ vanish in the limit $k\tau \ll 1$, $\mathcal{S}_*$ is indeed constant. Eq. (7.109) can then be integrated in explicit terms, across the radiation-matter transition

$$\mathcal{R}(k,\tau) = -4\mathcal{S}_*(k)\int_0^\alpha \frac{d\beta}{(3\beta+4)^2} \equiv -\mathcal{S}_*(k)\frac{\alpha}{3\alpha+4}, \qquad (7.111)$$

implying that $\mathcal{R} \to 0$ for $\alpha \to 0$ (pre-equality limit) and that $\mathcal{R} \to -\mathcal{S}_*/3$ for $\alpha \to \infty$ (matter-dominated limit). Recalling again the explicit form of $\mathcal{R}$ in terms of ψ, i.e. Eq. (7.101), Eq. (7.111) leads to a simple equation giving the evolution of ψ for modes $k\tau \ll 1$, i.e.

$$\frac{d\psi}{d\alpha} + \frac{5\alpha+6}{2\alpha(\alpha+1)}\psi = \frac{\mathcal{S}_*(k)}{2(\alpha+1)}. \qquad (7.112)$$

---

[i]It should be remarked that direct integration of Eq. (7.107) demands for us to fix the boundary condition $\alpha^3\psi(\alpha)$ in the limit $\alpha \to 0$. This boundary term will always give vanishing contribution, within the present approximations. Indeed, there are two possible cases. Either $\psi(0) \neq 0$ or $\psi(0) \to 0$. Incidentally, the first case corresponds to the adiabatic mode and the second case to the CDM-radiation non-adiabatic mode. In both cases $\alpha^3\psi(\alpha) \to 0$ for $\alpha \to 0$ *unless* $\psi(\alpha)$ diverges, at early times.

[j]Owing to the constancy of $\mathcal{S}$ in the limit $k\tau \ll 1$ we denote the constant value of $\mathcal{S}$ by $\mathcal{S}_*(k)$.

The solution of Eq. (7.112) can be simply obtained by imposing the isocurvature boundary condition, i.e. $\psi(0) \to 0$:

$$\psi(k,\alpha) = \frac{S_*(k)}{5\alpha^3}\{16(1 - \sqrt{\alpha+1}) + \alpha[8 + \alpha(\alpha - 2)]\}. \tag{7.113}$$

Eq. (7.113) is similar to Eq. (7.108) but with few crucial differences. According to Eq. (7.113) (and unlike Eq. (7.108)), $\psi(\alpha)$ vanishes, for $\alpha \to 0$, as $S_*\alpha/8$. In the limit $\alpha \to \infty$ $\psi(\alpha) \to S_*/5$. This is the growth of the adiabatic mode triggered during the transition from radiation to matter by the presence of the non-adiabatic pressure density fluctuation.

Having obtained the evolution of ψ, the evolution of the total density contrasts and of the total peculiar velocity field can be immediately obtained by solving the Hamiltonian and momentum constraints of Eqs. (7.51) and (7.52) with respect to $\delta\rho$ and θ

$$\delta = \frac{\delta\rho}{\rho} \equiv \frac{\delta_r}{\alpha+1} + \delta_c\frac{\alpha}{\alpha+1} = -2\left(\psi + \frac{d\psi}{d\ln\alpha}\right), \tag{7.114}$$

$$\theta = \frac{2k^2(\alpha+1)}{(3\alpha+4)}\left(\psi + \frac{d\psi}{d\ln\alpha}\right), \tag{7.115}$$

which also implies that for $k\tau \ll 1$,

$$\theta = -\frac{k^2(\alpha+1)}{(3\alpha+4)}\delta. \tag{7.116}$$

Equation (7.116) is indeed consistent with the result that the total velocity field is negligible for modes outside the horizon. Inserting Eq. (7.113) into Eqs. (6.28)–(7.60) it can be easily argued that the total density contrast goes to zero for $\alpha \to 0$, while for $\alpha \to \infty$ we have the following relations

$$\delta_c \simeq -\frac{2}{5}S_* \simeq -2\psi \simeq -\frac{1}{2}\delta_r \tag{7.117}$$

The first two equalities in Eq. (7.117) follow from the asymptotics of Eq. (7.114), the last equality follows from the conservation law (valid for isocurvature modes) which can be derived from Eq. (7.78), i.e.

$$\delta_r(k,\alpha) \simeq 4\psi(k,\alpha). \tag{7.118}$$

Thanks to the above results, the contribution to the scalar Sachs-Wolfe effect can be obtained in the case of the CDM-radiation non-adiabatic mode. From Eq. (7.34) we have

$$\left(\frac{\Delta T}{T}\right)^{\text{nad}}_{k,s} = \left(\frac{\delta_r}{4} + \psi\right)_{\tau\simeq\tau_{\text{dec}}} \equiv 2\psi_{\text{nad}} \equiv \frac{2}{5}S_*, \tag{7.119}$$

where the second equality follows from Eq. (7.118) and the third equality follows from Eq. (7.113) in the limit $\alpha \to \infty$ (i.e. $a \gg a_{\text{eq}}$).

The following comments are in order:

- as in the case of the adiabatic mode, also in the case of non-adiabatic mode in the CDM-radiation system, the peculiar velocity does not contribute to the SW effect;
- for $k\, c_s\, \tau_{\rm dec} \ll 1$ the temperature fluctuations induced by the adiabatic mode are simply $2\psi_{\rm nad}$ (unlike the adiabatic case);
- Equation (7.119) suggests that the first true peak in the temperature fluctuations is located at $kc_s\eta_{\rm dec} \simeq \pi/2$.

The last conclusion comes from an analysis similar to the one conducted in the case of the adiabatic mode but with the crucial difference that, in the case of the isocurvature mode, ψ vanishes as τ at early times. This occurrence implies the presence of sinusoidal (rather than cosinusoidal) oscillations. This point will be further discussed in section 8. In Fig. 7.2 the evolution of $\psi(\alpha)$ is illustrated for the adiabatic and for the non-adiabatic mode. In the case of the adiabatic mode, deep in the radiation epoch (i.e. $a \to 0$) $\psi \to \psi_{\rm r}$ (conventionally chosen to be 1 in Fig. 7.2). Always in the case of the adiabatic mode, for $a \to \infty$ (i.e. during the matter epoch) $\psi_{\rm m} = 9\psi_{\rm r}/10$. In Fig. 7.3 the evolution of $\mathcal{R}(\alpha)$ is reported. For the adiabatic mode $\mathcal{R}$ is constant. In fact, according to Eq. (7.103) the non-

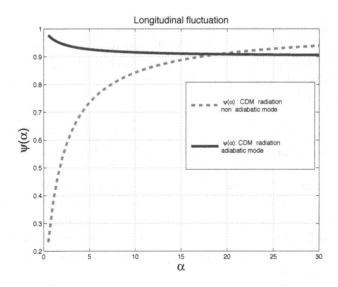

Fig. 7.2 The longitudinal fluctuation of the metric is plotted as a function of the scale factor in the case of the adiabatic mode (see Eq. (7.108)) and in the case of the non-adiabatic mode (see Eq. (7.113)).

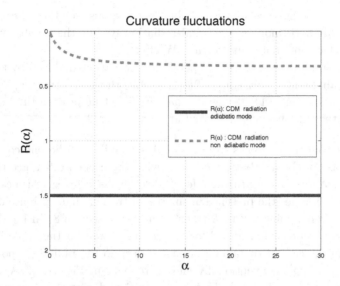

Fig. 7.3 The curvature fluctuation $\mathcal{R}$ is plotted as a function of the scale factor in the case of the adiabatic (full line) and in the case of the non-adiabatic (dashed line) modes. In both cases the curves are reported for the CDM-radiation system.

adiabtic pressure variation vanishes. In the case of the non-adiabatic mode $\delta p_{\text{nad}} \neq 0$ and $\mathcal{R}$ goes to zero deep in the radiation epoch (i.e. $\alpha \to 0$) and it goes to a constant in the matter epoch. Correspondingly, in the case of the non-adiabatic mode $\psi(\alpha)$ goes to zero in the limit $\alpha \to 0$ and it goes to a constant for $\alpha \to \infty$.

7.6 Sachs-Wolfe plateau: mixture of initial conditions

In the present section, building up on the results obtained so far, we will compute the ordinary SW contribution in the case of a mixture of adiabatic and non-adiabatic initial conditions. In the case of the CDM-radiation system this is indeed the most general situation. However, as already mentioned, there could be even richer mixtures of fluids and these situations will be partially discussed in chapter 8.

Some of the derivations performed in the previous section will then be repeated but in a slightly different perspective. Let us therefore introduce a different variable, i.e. the density contrast on uniform curvature hypersurfaces that will be conventionally denoted by ζ:

$$\zeta = -\psi - \frac{\mathcal{H}\delta\rho}{\rho'}. \tag{7.120}$$

In Eq. (7.120) the variable ζ has been defined in terms of quantities defined in the longitudinal gauge. The geometric meaning of $\mathcal{R}$ and ζ will be discussed in full detail in chapter 11. For the moment let us take Eq. (7.120) as a useful variable whose properties will be exploited in the present derivations.

Inserting Eqs. (7.120) and (7.101) into Eq. (7.51) the relation between ζ and $\mathcal{R}$ can be obtained since Eq. (7.51) becomes:

$$\zeta = \mathcal{R} + \frac{\nabla^2 \psi}{12\pi G a^2 (p+\rho)}. \tag{7.121}$$

When the relevant modes are outside the Hubble radius (i.e. $k\tau \ll 1$) Eq. (7.121) implies, in Fourier space, $\zeta(k) \simeq \mathcal{R}(k) + \mathcal{O}(k^2\tau^2)$. So the two variables can be used interchangeably as long as $k\tau < 1$. But this is exactly the limit we are interested in when the evaluation of the SW plateau is concerned.

The evolution equation of ζ can be swiftly obtained from the evolution of the total density fluctuation $\delta\rho$. The equation for $\delta\rho$ follows from the perturbation of the covariant conservation equation. The result of this calculation is reported in Appendix D (see, in particular, Eq. (D.50)). By setting $E = 0$ in Eq. (D.50) we obtain the evolution equation for $\delta\rho$ in the longitudinal gauge:

$$\delta\rho' - 3\psi'(p+\rho) + (p+\rho)\theta + 3\mathcal{H}(\delta\rho + \delta p) = 0, \tag{7.122}$$

where θ is the divergence of the total peculiar velocity of the fluid introduced in the momentum constraint (see Eqs. (7.52), (7.54) and (7.64)). Recalling now that, according to the decomposition of Eq. (7.91), $\delta p = c_{st}^2 \delta\rho + \delta p_{nad}$, Eq. (7.122) rapidly becomes

$$\zeta' = -\frac{\mathcal{H}}{p+\rho}\delta p_{nad} - \frac{\theta}{3}. \tag{7.123}$$

Equation (7.123) is obtained by directly inserting Eq. (7.120) into Eq. (7.122) and by expressing the pressure fluctuation in terms of the adiabatic and non-adiabatic contribution. Since $\theta \simeq \mathcal{O}(k^2\tau^2)$, the dynamical content of the evolution equation of ζ and of $\mathcal{R}$ (see Eq. (7.103)) is the same, as anticipated from the inspection of Eq. (7.121), when $k\tau \ll 1$. Furthermore, it can be directly verified with some algebra that if we insert the Hamiltonian constraint (expressed in the form (7.121) into Eq. (7.123), the dependence upon ζ can be eliminated and the resulting equation for $\mathcal{R}$ coincides with Eq. (7.103) (provided the momentum constraint (7.52) is used to eliminate θ).

With these specifications in mind, we have the solution for the evolution of $\mathcal{R}$ (or of ζ) that can be written as

$$\mathcal{R}(k,\tau) \simeq \zeta(k,\tau) = \mathcal{R}_*(k) - \frac{\alpha}{3\alpha+4}\mathcal{S}_*(k), \qquad (7.124)$$

where $\mathcal{R}_*(k)$ and $\mathcal{S}_*(k)$ are constant and will be taken to be characterized solely in terms of their power spectrum, i.e.

$$\langle \mathcal{R}_*(\vec{k})\mathcal{R}_*(\vec{p})\rangle = \frac{2\pi^2}{k^3}\mathcal{P}_\mathcal{R}(k)\delta^{(3)}(\vec{k}+\vec{p}),$$
$$\langle \mathcal{S}_*(\vec{k})\mathcal{S}_*(\vec{p})\rangle = \frac{2\pi^2}{k^3}\mathcal{P}_\mathcal{S}(k)\delta^{(3)}(\vec{k}+\vec{p}). \qquad (7.125)$$

The form of $\mathcal{P}_\mathcal{R}(k)$ and $\mathcal{P}_\mathcal{S}(k)$ will be taken to be of the type

$$\mathcal{P}_\mathcal{R}(k) = \mathcal{A}_\mathcal{R}\left(\frac{k}{k_\mathrm{p}}\right)^{n_r-1}, \qquad (7.126)$$

$$\mathcal{P}_\mathcal{S}(k) = \mathcal{A}_\mathcal{S}\left(\frac{k}{k_\mathrm{p}}\right)^{n_s-1}, \qquad (7.127)$$

In Eqs. (7.126) and (7.127) the typical wave-number k_p is called pivot scale and it represents the scale at which the spectral amplitudes are defined. By assuming a specific thermal history of the Universe and by selecting a specific inflationary model the spectral amplitudes $\mathcal{A}_\mathcal{R}$ and $\mathcal{A}_\mathcal{S}$ can be estimated. Also the slopes n_r and n_s can be estimated once the early evolution of the geometry is assumed. This discussion is postponed to chapter 10 where the power spectra of the scalar modes of the geometry will be discussed in the context of conventional inflationary models. In what follows we will concentrate more on the CMB side and we will show that it is also productive to study CMB observables in terms of a set of (undetermined) initial conditions. According to the parametrizations of Eqs. (7.126) and (7.127) the scale-invariant limit is realized when $n_r \to 1$ and $n_s \to 1$. This scale invariant limit is often dubbed Harrison-Zeldovich spectrum. Finally, it should be mentioned that, in the most general case (but always within the CDM-radiation system) the adiabatic and non-adiabatic contributions can be correlated in the sense that

$$\langle \mathcal{S}_*(\vec{k})\mathcal{R}_*(\vec{p})\rangle = \frac{2\pi^2}{k^3}\mathcal{P}_{SR}(k)\delta^{(3)}(\vec{k}+\vec{p}),$$
$$\mathcal{P}_{SR}(k) = \mathcal{A}_{SR}\left(\frac{k}{k_\mathrm{p}}\right)^{n_{rs}-1}, \qquad (7.128)$$

where $\mathcal{A}_{\mathcal{SR}}$ is spectral the amplitude and n_{rs} the slope of the cross-correlation. In the simplest situation $n_{rs} = (n_r + n_s)/2$. To estimate the ordinary and integrated SW contributions we do need the values of $\delta_\gamma(k,\tau_{\rm dec}) = \delta_{\rm r}(k,\tau_{\rm dec})$ as well as the values of ψ at the same epoch. Following the same steps discussed in the previous section, the value of $\psi(k,\tau)$ can be obtained by direct integration. The previous equation also implies

$$\frac{\alpha^3}{\sqrt{\alpha+1}}\psi(k,\tau) = -\frac{\mathcal{R}_*(k)}{2}\mathcal{I}_1(\alpha) + \frac{\mathcal{S}_*(k)}{2}\mathcal{I}_2(\alpha), \qquad (7.129)$$

where

$$\begin{aligned}\mathcal{I}_1 &= \int_0^\alpha \frac{\beta^2(3\beta+4)}{(\beta+1)^{3/2}}d\beta,\\ \mathcal{I}_2 &= \int_0^\alpha \frac{\beta^3}{(\beta+1)^{3/2}}d\beta.\end{aligned} \qquad (7.130)$$

By using the obvious change of variables $y = \beta + 1$ both integrals can be evaluated with elementary methods:

$$\begin{aligned}\mathcal{I}_1(\alpha) &= \int_1^{\alpha+1}\left[3y\sqrt{y} - 5\sqrt{y} + \frac{1}{\sqrt{y}} + \frac{1}{y\sqrt{y}}\right]dy,\\ \mathcal{I}_2(\alpha) &= \int_1^{\alpha+1}\left[y\sqrt{y} + 3\sqrt{y} - \frac{1}{y\sqrt{y}} - \frac{3}{\sqrt{y}}\right]dy.\end{aligned} \qquad (7.131)$$

By performing the elementary integrals (7.131) the final result will be:

$$\begin{aligned}\mathcal{I}_1(\alpha) &= \frac{2\{16[\sqrt{\alpha+1}-1] + \alpha[\alpha(9\alpha+2) - 8]\}}{15\sqrt{\alpha+1}},\\ \mathcal{I}_2(\alpha) &= \frac{2\{16[1-\sqrt{\alpha+1}] + \alpha[8+\alpha(\alpha-2)]\}}{5\sqrt{\alpha+1}}.\end{aligned} \qquad (7.132)$$

Thus, the dependence of $\psi(k,\tau)$ will be

$$\psi(k,\tau) = -\frac{\mathcal{R}_*(k)}{15\alpha^3}\{16[\sqrt{\alpha+1}-1] + \alpha[\alpha(9\alpha+2)-8]\}$$

$$+ \frac{\mathcal{S}_*(k)}{5\alpha^3}\{16[1-\sqrt{\alpha+1}] + \alpha[8+\alpha(\alpha-2)]\}. \qquad (7.133)$$

In the two physically interesting limits we will have, therefore,

$$\lim_{\alpha \gg 1} \psi(k,\tau) = -\frac{3}{5}\mathcal{R}_*(k) + \frac{\mathcal{S}_*(k)}{5}. \qquad (7.134)$$

$$\lim_{\alpha \ll 1} \psi(k,\tau) = -\frac{2}{3}\mathcal{R}_*(k) + \frac{\alpha}{8}\left[\frac{\mathcal{R}_*(k)}{3} + \mathcal{S}_*(k)\right]. \qquad (7.135)$$

Notice that $\mathcal{R}_*(k)$ appears also in the correction which goes as α.

The quantity we need to estimate is then according to Eq. (7.36),

$$\Delta_I^{(SW)}(\vec{k},\hat{n},\tau_{dec}) = \frac{\delta_r(k,\tau_{dec})}{4} + \psi(k,\tau_{dec}). \qquad (7.136)$$

In Eq. (7.136) we have already the value of ψ after equality (see Eqs. (7.133) and (7.134)). We need the value of the radiation density contrast after equality. From Eq. (7.78) we do know that, for typical wavelengths larger than the Hubble radius,

$$\delta_r' - 4\psi' \simeq 0 \qquad (7.137)$$

which also means, integrating once upon the conformal time coordinate,

$$\delta_r(k,\tau_{dec}) = 4\psi(k,\tau_{dec}) + \delta_r(k,\tau_i) - 4\psi(k,\tau_i), \qquad (7.138)$$

where $\psi(k,\tau_i)$ and $\delta_r(k,\tau_i)$ denote the longitudinal fluctuation of the metric and the radiation density contrast for $\tau_i \ll \tau_{eq}$, i.e. in the limit $\alpha \to 0$.

Using then Eq. (7.138) into Eq. (7.136) we get the following intermediate expression:

$$\Delta_I^{(SW)}(\vec{k},\hat{n},\tau_{dec}) = \frac{\delta_r(k,\tau_i)}{4} + 2\psi(k,\tau_{dec}) - \psi(k,\tau_i). \qquad (7.139)$$

Two out of the three contributions appearing in Eq. (7.139) have already been estimated. The last quantity to be estimated is actually $\delta_r(k,\tau_i)$. To do that we may simply observe, from Eq. (7.120), that the *total* density contrast can be written as

$$\delta_t(k,\tau_i) = 3(w_t + 1)[\mathcal{R}_*(k) + \psi(k,\tau_i)]. \qquad (7.140)$$

But now it is enough to observe that, well before equality, the total density contrast coincides with the radiation density contrast and $w_t = 1/3$. Consequently,

$$\delta_r(k,\tau_i) \simeq 4[\mathcal{R}_*(k) + \psi(k,\tau_i)] = \frac{4}{3}\mathcal{R}_*(k), \qquad (7.141)$$

since, according to Eq. (7.135) the only contribution to ψ (in the limit $\alpha \to 0$) comes from the adiabatic component and it is $\psi(k,\tau_i) = -(2/3)\mathcal{R}_*(k)$.

Therefore, from Eq. (7.139) the following expression can be easily obtained by using, simultaneously, Eqs. (7.141) together with Eqs. (7.134)–(7.135):

$$\Delta_I^{(SW)}(\vec{k},\hat{n},\tau_{dec}) = \left[-\frac{\mathcal{R}_*(k)}{5} + \frac{2}{5}\mathcal{S}_*(k)\right]. \qquad (7.142)$$

From this expression it is easy to compute the angular power spectrum since

$$C_\ell^{(SW)} = \frac{2}{\pi}\int k^3 d\ln k |\Delta_{I\ell}^{(SW)}(k,\tau_0)|^2 \qquad (7.143)$$

Recalling now that, from Eq. (7.44)

$$\Delta_{\rm I}^{\rm (SW)}(\vec{k},\hat{n},\tau_0) = \Delta_{\rm I}^{\rm (SW)}(\vec{k},\hat{n},\tau_{\rm dec})e^{-ik\mu(\tau_0-\tau_{\rm rec})]}, \qquad (7.144)$$

we can obtain the specific form of $\Delta_{\rm I\ell}^{\rm (SW)}(k,\tau_0)$ by expanding the plane wave as[k]

$$e^{-ik\mu(\tau_0-\tau_{\rm dec})} = \sum_\ell (-i)^\ell (2\ell+1) j_\ell(k\tau_0) P_\ell(\mu) \qquad (7.145)$$

where $j_\ell(k\tau_0)$ are the spherical Bessel functions [215] and they are defined, in terms of the standard solutions of the Bessel equation, as

$$j_\ell(k\tau_0) = \sqrt{\frac{\pi}{2k\tau_0}} J_{\ell+1/2}(k\tau_0). \qquad (7.146)$$

We recall that, in Eq. (7.145) (as in Eq. (7.44)), $\mu = \hat{k}\cdot\hat{n}$ is the projection of the photon momentum on the direction of the Fourier wave-vector. The quantity appearing in the integrand of Eq. (7.143) can then be written, using Eqs. (7.125) and (7.128)

$$|\Delta_{\rm I\ell}^{\rm (SW)}(k,\tau_0)|^2 = \frac{2\pi^2}{k^3} \left\{ \frac{\mathcal{P}_\mathcal{R}(k)}{25} + \frac{4}{25} \mathcal{P}_\mathcal{S}(k) \right.$$
$$\left. - \frac{4}{25} \mathcal{P}_\mathcal{R\mathcal{S}}(k) \cos\gamma_{rs} \right\} j_\ell^2(k\tau_0). \qquad (7.147)$$

Concerning Eq. (7.147) few remarks are in order:

- the spectrum of the curvature perturbations appears explicitly but some authors prefer to write down the Sachs-Wolfe plateau in terms of the spectrum of ψ; it is clear that while $\mathcal{R}$ is constant along the radiation-matter transition (for the modes relevant for the SW plateau), ψ is not constant and Eq. (7.104) will then have to be used to relate the two spectra appropriately;
- in Eq. (7.36) $\cos\gamma_{\rm rs}$ is the correlation cosine which can be either positive or negative or even 0 (if the adiabatic and non-adiabatic modes are uncorrelated);
- notice finally that we named differently the spectral indices of the adiabatic and non-adiabatic modes (i.e. respectively, n_r and n_s); when only *one adiabatic mode* is present we will denote, in chapter 10, the scalar spectral index as $n_{\rm s}$.

---
[k]Notice that $|\tau_{\rm dec}| \ll \tau_0$ so that the dependence upon $\tau_{\rm dec}$ can be neglected in the argument of the spherical Bessel function.

Inserting Eq. (7.147) into Eq. (7.143) we easily obtain:

$$C_\ell^{(\text{SW})} = \frac{\mathcal{A}_\mathcal{R}}{25} \mathcal{Z}_\mathcal{R}(n_r, \ell) + \frac{4}{25} \mathcal{A}_\mathcal{S} \mathcal{Z}_\mathcal{S}(n_s, \ell)$$
$$- \frac{4}{25} \mathcal{A}_{\mathcal{R}\mathcal{S}} \mathcal{Z}_{\mathcal{R}\mathcal{S}}(n_{rs}, \ell) \cos\gamma_{rs} \qquad (7.148)$$

The various functions appearing in Eq. (7.148) are

$$\mathcal{Z}_\mathcal{R}(n_r, \ell) = 4\pi \left(\frac{k_0}{k_p}\right)^{n_r - 1} \int_0^\infty \frac{dy_0}{y_0} y_0^{n_r - 1} j_\ell^2(y_0),$$
$$\mathcal{Z}_\mathcal{S}(n_r, \ell) = 4\pi \left(\frac{k_0}{k_p}\right)^{n_s - 1} \int_0^\infty \frac{dy_0}{y_0} y_0^{n_s - 1} j_\ell^2(y_0), \qquad (7.149)$$
$$\mathcal{Z}_{\mathcal{R}\mathcal{S}}(n_{rs}, \ell) = 4\pi \left(\frac{k_0}{k_p}\right)^{n_{rs} - 1} \int_0^\infty \frac{dy_0}{y_0} y_0^{n_s - 1} j_\ell^2(y_0),$$

where $y_0 = k\tau_0$. The integrals appearing in Eqs. (7.149) lead to the same expression but evaluated for different values of the spectral index. Therefore, calling the spectral index, generically, n we will have

$$\mathcal{Z}(n, \ell) = 2\pi^2 \left(\frac{k_0}{k_p}\right)^{n-1} \int_0^\infty y_0^{n-3} J_{\ell+1/2}^2(y_0)\, dy_0. \qquad (7.150)$$

The integral appearing in Eq. (7.150) can be performed for $-3 < n < 3$ with the result

$$\int_0^\infty dy_0 y_0^{n-3} J_{\ell+1/2}^2(y_0) = \frac{1}{2\sqrt{\pi}} \frac{\Gamma(3 - n/2)\Gamma(\ell + n/2 - 1/2)}{\Gamma(4 - n/2)\Gamma(5/2 + \ell - n/2)}. \qquad (7.151)$$

The duplication formula for the Euler Γ function implies [215, 216]

$$\Gamma\left(\frac{3-n}{2}\right) = \frac{\sqrt{2\pi}\,\Gamma(3-n)}{2^{5/2-n}\Gamma(4-n/2)}. \qquad (7.152)$$

Inserting now Eq. (7.152) into Eq. (7.151) we do get, from Eq. (7.149) the following general expression for $\mathcal{Z}(n, \ell)$

$$\mathcal{Z}(n, \ell) = \frac{\pi^2}{4} \left(\frac{k_0}{k_p}\right)^{n-1} 2^n \frac{\Gamma(3-n)\Gamma(\ell + n/2 - 1/2)}{\Gamma^2(4 - n/2)\Gamma(5/2 + \ell - n/2)}. \qquad (7.153)$$

Concerning Eqs. (7.148) and (7.153) few comments are in order:

- the estimates presented in this section hold for sufficiently large angular scales, i.e. $\ell < 30$;
- the correlation angle introduced in Eq. (7.148) may be equal to $\pi/2$ and in this case the adiabatic and the non-adiabatic mode are said to be uncorrelated;
- if $\cos\gamma_{rs} > 0$ the cross-correlation reduces the value of the SW plateau while, in the opposite case, the cross-correlation increases the SW plateau.

It should be pointed out that we normally talk about the SW plateau because, indeed, when the spectra are all scale invariant, i.e. $n_r = n_s = n_{rs} = 1$, then Eq. (7.148) implies that

$$\frac{\ell(\ell+1)}{2\pi} C_\ell = \frac{\mathcal{A}_\mathcal{R}}{25} + \frac{4}{25} \mathcal{A}_\mathcal{S} - \frac{4}{25} \mathcal{A}_{\mathcal{R}\mathcal{S}} \cos\gamma_{rs}. \tag{7.154}$$

But recalling now Eq. (3.29) we clearly see that, in the case of scale invariant spectrum, the power per logarithmic interval of ℓ also exhibits a plateau. We conclude this section by presenting a quick estimate of the integrated SW effect. From Eq. (7.37) it easily follows, in the absence of neutrinos, that

$$\Delta_I^{(\text{ISW})}(k, \tau_{\text{dec}}) = 2\psi(k, \tau_{\text{dec}}) - 2\psi(k, \tau_i). \tag{7.155}$$

Using now the Eqs. (7.134) and (7.135), Eq. (7.155) implies that

$$\Delta_I^{(\text{ISW})}(k, \tau_{\text{dec}}) = \frac{2}{15}\mathcal{R}_*(k) + \frac{2}{5}\mathcal{S}_*(k). \tag{7.156}$$

According to Eq. (7.156), the ISW mimics the ordinary SW effect. More specifically, the ISW tends to partially cancel the ordinary SW effect. This conclusion can easily be reached by comparing Eqs. (7.142) and (7.156) in the case of vanishing non-adiabatic contribution (i.e. $\mathcal{S}_* = 0$). It should be stressed that here the main focus have been the large angular scales. However, the ISW effect contributes about 20 % on all angular scales up to the Doppler region.

The possibility of having a mixture of modes (adiabatic and isocurvature) as initial condition for the CMB anisotropies led to a number of mode-independent analysis of various sets of data. In [259] it was argued, on the basis of Maxima data, that mixture of isocurvature and adiabatic modes could not be excluded and in [260] Enqvist and Kurki-Suonio pointed out that the Planck explorer mission could detect isocurvature modes thanks to its foreseen sensitivity to polarization. CMB polarization is basically induced by scattering processes (either last Thompson scattering or late reionization scattering). This occurrence allows us to eliminate possible contamination with line-of-sight (integrated) effects and makes polarization a valuable tool in order to constrain isocurvature modes. In [261] CDM isocurvature modes in open and closed FRW backgrounds have also been considered.

In [262, 263] the importance of polarization was also stressed in view of the Planck explorer planned sensitivities. A general (model independent) analysis of the recent WMAP data (combined with other sets of data including 2dF galaxy redhift survey) was recently performed in [264, 265]. The results can be roughly summarized by saying that the isocurvature fraction allowed by the present data ranges from about 10 % (when only one isocurvature mode is present on top of the adiabatic mode) up to 40 % (when two isocurvature modes are allowed together with their cross-correlations). Finally the fraction of isocurvature modes rises to about 60 % when three isocurvature modes are allowed. Notice, as a remark, that in [265] the authors indeed allowed for general correlations between adiabatic and isocurvature modes while the power spectrum was parametrized by means of a power-law. From the analyses reported so far it seems that the constraints on isocurvature modes are rather dependent upon the possible correlations with the adiabatic modes. In [266] (see also [267]), following the analysis of Ref. [259], the possibility of (correlated) CDM and adiabatic modes has been discussed in light of WMAP and large-scale structure data. Different values of the spectral indices for the power of each mode (and of their correlation) have been scrutinized. As a consequence, the bounds reported in [259] became more stringent by, roughly, a factor of 1/6.

One can also address the question on how some correlated mixture of adiabatic and isocurvature modes may arise in the early Universe. This subject will be also partially addressed in chapter 13 where some general ideas on how isocurvature modes may be excited in the early Universe will be introduced. As an example, consider Ref. [268] where a model dependent analysis has been performed. In the context of minimal curvaton models (see chapter 13) the authors found that correlated isocurvature modes seems to be disfavoured with respect to pure adiabatic modes. However, the presence of a (totally correlated) baryon isocurvature mode does not seem to be ruled out (see also [269, 270]).

Chapter 8

Improved Fluid Description of Pre-Decoupling Physics

The results obtained in the previous chapter are only meaningful for sufficiently large scales or, equivalently, for sufficiently small harmonics ℓ. For larger harmonics (i.e. for smaller length-scales) the approach introduced in chapter 7 fails. The angular separation appearing in the expression of the angular power spectrum of Eq. (3.26) is related to the harmonics as

$$\theta \simeq \frac{\pi}{\ell}, \qquad k \simeq \ell\, h_0\, 10^{-4}\ \mathrm{Mpc}^{-1} \qquad (8.1)$$

where the second relation gives the comoving wave-number in terms of ℓ for a Universe with $\Omega_{\mathrm{M}0} \simeq 0.3$ and $\Omega_{\Lambda 0} \simeq 0.7$. In chapter 7 the general system of fluctuations has been (artificially) reduced to the case when only radiation and CDM particles were present. In the present chapter this assumption will be dropped and the main theme will therefore be a consistent improvement of the approximations already exploited in chapter 7. More specifically the list of topics to be addressed shall include:

- the general four-components plasma;
- a more quantitative discussion of the CDM component;
- the tight-coupling between photons and baryons;
- the general solution for the adiabatic mode;
- some useful numerical solutions in the tightly coupled regime.

In the last part of the chapter the physics of non-adiabatic modes (already introduced in chapter 7) will be further discussed. Since here both baryons and neutrinos will be included, qualitatively new solutions of the pre-equality evolution equations will arise. These solutions can then be used to set initial conditions of the numerical integrations leading, ultimately, to the estimate of the temperature autocorrelations. These initial conditions can be viewed either as an alternative or as a completion of the standard adiabatic solution. If they are taken to be alternative to the standard adiabatic mode, as we shall see, they fail to reproduce the observed patterns of the Doppler oscillations. However, it can happen that subleading entropic modes are present together with a dominant adiabatic mode. In this circumstance the phenomenological viability of the scenario will depend upon the relative weight of the adiabatic and of the entropic contributions.

8.1 The general plasma with four components

The pre-decoupling plasma contains four components, namely, photons, baryons, neutrinos and CDM particles. Under the assumption that the dark-energy component is parametrized in terms a cosmological constant, there are no extra sources of inhomogeneity to be considered on top of the metric fluctuations which shall be treated, as in chapter 7, within the longitudinal coordinate system. Since the neutrinos are present in the game with their anisotropic stress, it will not be possible any longer to consider the case $\phi = \psi$. When all the four species are simultaneously present in the plasma, Eq. (7.51) can be written in Fourier space as

$$-k^2\psi - 3\mathcal{H}(\mathcal{H}\phi + \psi') = 4\pi G\, a^2[\delta\rho_\gamma + \delta\rho_\nu + \delta\rho_c + \delta\rho_b]. \qquad (8.2)$$

Defining as δ_ν, δ_γ, δ_b and δ_c the neutrino, photon, baryon and CDM density contrasts, the Hamiltonian constraint of Eq. (8.2) can also be rearranged as

$$-3\mathcal{H}(\mathcal{H}\phi + \psi') - k^2\psi = \frac{3}{2}\mathcal{H}^2[(R_\nu\delta_\nu + (1 - R_\nu)\delta_\gamma) + \Omega_b\delta_b + \Omega_c\delta_c], \qquad (8.3)$$

where, for N_ν species of massless neutrinos,

$$r_\nu = \frac{7}{8}N_\nu\left(\frac{4}{11}\right)^{4/3}, \qquad R_\nu = \frac{r_\nu}{1 + r_\nu}, \qquad R_\gamma = 1 - R_\nu, \qquad (8.4)$$

so that R_ν and R_γ represent the fractional contributions of photons and neutrinos to the total density at early times deep within the radiation-dominated epoch. Two technical remarks are in order:

- Eq. (2.65) has been used (in the case of vanishing spatial curvature) in Eq. (8.2) in order to eliminate the explicit dependence upon the total energy density of the background;
- Eq. (8.3) has been written under the assumption that the total energy density of the sources is dominated by radiation since this is

the regime where the initial conditions for the numerical integration are customarily set.[a]

Taking now into account the four components of the plasma, from the momentum constraint of Eq. (7.52), and from Eq. (7.55) the following pair of equations can be derived:

$$k^2(\mathcal{H}\phi + \psi') = \frac{3}{2}\mathcal{H}^2\left[\frac{4}{3}(R_\nu\theta_\nu + R_\gamma\theta_\gamma) + \theta_b\Omega_b + \theta_c\Omega_c\right], \quad (8.5)$$

$$\psi'' + (2\psi' + \phi')\mathcal{H} + (2\mathcal{H}' + \mathcal{H}^2)\phi - \frac{k^2}{3}(\phi - \psi)$$
$$= \frac{\mathcal{H}^2}{2}(R_\nu\delta_\nu + \delta_\gamma R_\gamma), \quad (8.6)$$

where, following Eq. (7.54), the divergence of the (total) peculiar velocity field has been separated for the different species, i.e.

$$(p + \rho)\theta = \frac{4}{3}\rho_\nu\theta_\nu + \frac{4}{3}\rho_\gamma\theta_\gamma + \rho_c\theta_c + \rho_b\theta_b. \quad (8.7)$$

Note, once more, that in Eqs (8.5)–(8.6), Eq. (2.65) has been used in order to eliminate the explicit dependence upon the (total) energy and pressure densities.

Finally, according to Eqs. (7.56)–(7.57) the neutrino anisotropic stress fixes the difference between the two longitudinal fluctuations of the geometry. Recalling Eqs. (7.56) and (7.57) we will have

$$\nabla^4(\phi - \psi) = 12\pi G a^2 \partial_i \partial^j \Pi^i_j. \quad (8.8)$$

For temperatures smaller than $T \simeq$ MeV the only collisionless species of the plasma are neutrinos (which will be assumed to be massless, for simplicity). Neutrinos will then provide, in the absence of large-scale magnetic fields, the dominant contribution to the anisotropic stress.[b] The term at the right hand side of Eq. (8.8) can then be parametrized, for future convenience, as

$$\partial_i \partial^j \Pi^i_j = (p_\nu + \rho_\nu)\nabla^2 \sigma_\nu. \quad (8.9)$$

Hence, using Eqs. (2.65) and (2.66), Eq. (8.8) can then be written as

$$k^2(\phi - \psi) = -6\mathcal{H}^2 R_\nu \sigma_\nu, \quad (8.10)$$

---

[a]Needless to say that it is equally possible to drop this assumption and to write the equations in their full generality across the radiation-matter transition using the form of the scale factor given, for instance, in Eq. (2.93).
[b]Indeed, large-scale magnetic fields represent a rather plausible source of anisotropic stress that has been considered in the literature [253–255].

The evolution equations for the various species will now be discussed. The CDM and neutrino components are only coupled to the fluctuations of the geometry. Their evolution equations are then given respectively by

$$\theta'_c + \mathcal{H}\theta_c = k^2\phi, \tag{8.11}$$

$$\delta'_c = 3\psi' - \theta_c. \tag{8.12}$$

and by

$$\delta'_\nu = -\frac{4}{3}\theta_\nu + 4\psi', \tag{8.13}$$

$$\theta'_\nu = \frac{k^2}{4}\delta_\nu - k^2\sigma_\nu + k^2\phi, \tag{8.14}$$

$$\sigma'_\nu = \frac{4}{15}\theta_\nu - \frac{3}{10}k\mathcal{F}_{\nu 3}. \tag{8.15}$$

Equations (8.13) and (8.14) are directly obtained from Eqs. (7.59) and (7.60) in the case $w_\nu = 1/3$ and $\sigma_\nu \neq 0$. Equation (8.15) is not obtainable in the fluid approximation and the full Boltzmann hierarchy has to be introduced. The quantity $\mathcal{F}_{\nu 3}$ introduced in Eq. (8.15), is the octupole term of the neutrino phase space distribution. For a derivation of Eqs. (8.13), (8.14) and (8.15) see Eqs. (9.31), (9.32) and (9.33) in chapter 9.

8.2 CDM component

Equations (8.11) and (8.12) are the easiest to solve. Indeed, Eq. (8.11) can be easily rearranged as $(\theta_c a)' = k^2 \phi a$ and the solution will then be, formally,

$$\theta_c(k,\tau) = \frac{k^2}{a(\tau)} \int^\tau \phi(k,\tau')a(\tau')d\tau'. \tag{8.16}$$

Equation (8.12) can also be usefully recasted in the form $(\delta_c - 3\psi)' = -\theta_c$ whose solution is

$$\delta_c(k,\tau) = 3\psi(k,\tau) - k^2 \int^\tau \frac{d\tau'}{a(\tau')} \int^{\tau'} \phi(k,\tau'')a(\tau'')d\tau''. \tag{8.17}$$

Owing to its simplicity the CDM system can be solved analytically under different useful approximations. Here, again, we will refer to Fig. 7.1 which illustrates, through a simple cartoon, the relevant scales arising in the matter-radiation transition. Consider, indeed, the scales that are still outside the horizon *before* equality. For these scales, as discussed already in chapter 7, the adiabaticity condition implies that

$$\delta_c(k,\tau) = \frac{3}{4}\delta_r(k,\tau) = -2\psi_r(k) \tag{8.18}$$

where $\delta_{\rm r}(k,\tau)$ denotes, indifferently, the density contrast of photons and neutrinos and where $\psi_{\rm r}(k)$ parametrizes the primeval adiabatic mode that has been introduced in Eq. (7.72). Now, Eq. (7.72) is *exact* before matter radiation equality. So it holds both in the limit $k\tau \ll 1$ and in the limit $k\tau \gg 1$. Let us therefore consider Eq. (7.72) for $\tau \ll \tau_{\rm eq}$ and in the case $k\tau > 1$. This is the situation contemplated by the dashed line in Fig. 7.1 and corresponding to the case when the given physical wavelength reenters the Hubble radius prior to equality. In this limit the Hamiltonian constraint of Eq. (8.2) is dominated by the Laplacian, i.e.[c]

$$-k^2\psi \simeq \frac{3}{2}\mathcal{H}^2\left[\delta_{\rm r} + \frac{\rho_{\rm c}}{\rho_{\rm r}}\delta_{\rm c}\right], \qquad k\tau \gg 1. \tag{8.19}$$

Equation (7.72) then allows us to solve for $\delta_{\rm r}(k,\tau)$, i.e.

$$\psi(k,\tau) = -9\frac{\psi_{\rm r}(k)}{(k\tau)^2}\left[\cos\left(\frac{k\tau}{\sqrt{3}}\right) - \frac{\sqrt{3}}{(k\tau)}\sin\left(\frac{k\tau}{\sqrt{3}}\right)\right], \tag{8.20}$$

$$\delta_{\rm r}(k,\tau) = 6\psi_{\rm r}(k)\left[\cos\left(\frac{k\tau}{\sqrt{3}}\right) - \frac{\sqrt{3}}{(k\tau)}\sin\left(\frac{k\tau}{\sqrt{3}}\right)\right], \tag{8.21}$$

where, in comparison with Eq. (7.72), we traded y for $k\tau/\sqrt{3}$. It should be appreciated that, in the limit $k\tau \gg 1$ (i.e. when the dashed line of Fig. 7.1 is below the full (thick) line for $a < a_{\rm eq}$) the second term appearing in the two square brackets of Eqs. (8.20) and (8.21) (and going as a sine) is negligible in comparison with the first term (which goes as a cosine).

Let us now compute $\delta_{\rm c}(k,\tau)$ and $\theta_{\rm c}(k,\tau)$ before equality. For this calculation we need, certainly, Eqs. (8.16) and (8.17) but also Eq. (7.72). By changing integration variable in Eq. (8.16) and (8.17), and by assuming $\psi(k,\tau) = \phi(k,\tau)$ we can easily write

$$\theta_{\rm c}(k,y) = \frac{\sqrt{3}k}{y}\int^y w\psi(w,k)dw, \tag{8.22}$$

$$\psi(k,y) = 3\psi(y,k) - 3\int^y \frac{dw}{w}\int^w z\psi(k,z)dz, \tag{8.23}$$

where $y = k\tau/\sqrt{3}$, $w = k\tau'/\sqrt{3}$ and $z = k\tau''/\sqrt{3}$. Using Eq. (7.72) inside

---

[c]Equation (8.19) will be first solved in the case when the CDM density contrast is subleading in comparison with the radiation density contrast and in the case when the Universe is effectively dominated by radiation.

Eqs. (8.22) and (8.23) and doing the integrals as indicated we will have

$$\theta_c(k,y) = -3\sqrt{3}k\,\psi_r(k)\frac{\sin y}{y^2} + C_1(k), \tag{8.24}$$

$$\delta_c(k,y) = 9\psi_r(k)\left[C_1(k)\ln y + C_2(k) + \text{Ci}(y) - \frac{y\cos y - \sin y}{y^3}\right],$$

$$\text{Ci}(y) = -\int_y^\infty \frac{\cos t}{t}dt, \tag{8.25}$$

where $\text{Ci}(y)$ is the cosine integral function of y [215, 216]. As expected, in Eqs. (8.24) and (8.25) two integrations constants appear, i.e. $C_1(k)$ and $C_2(k)$. They can be determined by requiring that $\delta_c(k,y)$ matches the value of the CDM density contrast which is deduced both from the Hamiltonian constraint and from the adiabaticity condition. Indeed, for $y \gg 1$ and for $y \ll 1$, Eq. (8.25) implies, respectively:

$$\delta_c(k,y) \simeq 9\psi_r(k)\left[C_1(k)\ln y + C_2(k) + \mathcal{O}\left(\frac{1}{y}\right)\right], \tag{8.26}$$

$$\delta_c(k,y) \simeq \psi_k(k)[-6 + 9C_2(k) + 9\gamma$$
$$+ 9(1 + C_1(k))\ln y + \mathcal{O}(y^2)], \tag{8.27}$$

where $\gamma = 0.577216$ is the Euler-Mascheroni constant. We then have Eq. (8.27) that must equal to $-(3/2)\psi_r(k)$ and this implies, as it can be immediately checked, that

$$C_1(k) = -1, \qquad C_2(k) = \frac{1}{2} - \gamma. \tag{8.28}$$

The two conditions of Eq. (8.28) demand, when inserted in Eq. (8.26), that

$$\delta_c(k,\tau) \simeq -9\psi_r(k)\left[\ln\left(\frac{k\tau}{\sqrt{3}}\right) + \gamma - \frac{1}{2} + \mathcal{O}\left(\frac{\sqrt{3}}{k\tau}\right)\right]. \tag{8.29}$$

Equation (8.29) stipulates that, after crossing the Hubble radius (but *before equality*), the CDM density contrast is *constant* (up to logarithmic corrections). This observation will be rather relevant for the computations that will follow.

Owing to the slow growth of $\delta_c(k,\tau)$, the Hamiltonian constraint in the form (8.19) can be used to determine the metric fluctuation *after* equality (i.e. for $\tau > \tau_{eq}$) and prior to recombination (i.e. $\tau \leq \tau_{rec}$). Indeed, using Eq. (8.29) at the right hand side of Eq. (8.19) we can easily compute an approximate expression of $\psi(k,\tau_{eq})$ which turns out to be, after simple algebra:

$$\psi(k,\tau_{eq}) \simeq \frac{27}{2(k\tau_{eq})^2}\ln\left(\frac{k\tau_{eq}}{\sqrt{3}}\right). \tag{8.30}$$

This constant value of the metric fluctuation is preserved until recombination and decoupling. Recall, in fact, the results of chapter 7. Equation (7.73) gives the evolution of $\psi(k,\tau)$ during matter. Since, during matter domination, the total sound speed is effectively vanishing, the corresponding equation will not have Laplacians. Consequently the solution of Eq. (7.73) (reported in Eq. (7.74)) will hold both inside and outside the Hubble radius. Since this equation has a dominant constant mode, this justifies the statement that $\psi(k,\tau)$, for $\tau \geq \tau_{\text{eq}}$, freezes at its value at equality even for modes $k\tau > 1$.

While the statement of constancy of the metric fluctuation implied by Eq. (8.30) is correctly captured by the physical considerations we presented so far, the numerical value of the coefficient appearing at the right hand side of Eq. (8.30) is not so accurate. To determine the coefficient more appropriately it is useful to derive a slightly different expression for the evolution of δ_c. Indeed, let us assume that $\psi' \simeq 0$ to a good approximation and let us insert θ_c (determined from Eq. (8.12)) inside Eq. (8.11). The resulting expression will be:

$$\delta_c'' + \mathcal{H}\delta_c' + k^2\psi = 0. \tag{8.31}$$

We can now use Eq. (8.19) to eliminate $k^2\psi$ and the result will be

$$\delta_c'' + \mathcal{H}\delta_c' - 4\pi G a^2 \rho_c \left[\delta_c + \frac{\rho_r}{\rho_c}\delta_r\right] = 0, \tag{8.32}$$

We will now drop the contribution of the radiation density contrast but we will keep the radiation component of the background. The experienced reader will recognize Eq. (8.32) as the standard from of the Meszaros equation which is customarily derived in the nonrelativistic approximation and for typical wavenumbers $k \ll k_J$ where k_J is the Jeans scale. Equation (8.32) can be solved by passing from the conformal time to the auxiliary variable $\alpha = a/a_{\text{eq}}$. Recall, in this context, that from Eq. (2.65)

$$4\pi a^2 \rho_c = \frac{3}{2}\mathcal{H}^2 \frac{1}{1+\frac{\rho_r}{\rho_c}} \equiv \frac{3}{2}\mathcal{H}^2 \frac{\alpha}{\alpha+1}. \tag{8.33}$$

Furthermore, by expressing the derivatives with respect to τ (i.e. the conformal time coordinate) as derivatives with respect to α we will have that:

$$\delta_c' = \alpha\mathcal{H}\frac{d\delta_c}{d\alpha}, \qquad \delta_c'' = \alpha(\mathcal{H}^2 + \mathcal{H}')\frac{d\delta_c}{d\alpha} + \alpha^2\mathcal{H}^2\frac{d^2\delta_c}{d\alpha^2}. \tag{8.34}$$

Using Eqs. (8.33) and (8.34) inside Eq. (8.35) we do get the standard form of the equation:

$$\frac{d^2\delta_c}{d\alpha^2} + \frac{3\alpha+2}{2\alpha(\alpha+1)}\frac{d\delta_c}{d\alpha} - \frac{3}{2\alpha(\alpha+1)}\delta_c = 0, \tag{8.35}$$

if we recall that, from Eqs. (2.65) and (2.66),

$$2 + \frac{\mathcal{H}^2}{\mathcal{H}'} = \frac{2 + 3\alpha}{2(1+\alpha)}. \tag{8.36}$$

Equation (8.35) will have, of course, two linearly independent solutions. Since it is a second-order equation we can always find the second solution from the equation of the Wronskian, *provided* we are able to guess the first (linearly independent) solution. This can be easily done. Suppose, indeed, that the second derivative of δ_c with respect to α vanishes. Then the corresponding solution can be obtained by setting to zero the second derivative with respect to α in Eq. (8.35). The resulting expression will become:

$$\frac{d\delta_c}{d\alpha} = \frac{3}{3\alpha + 2}\delta_c \tag{8.37}$$

whose solution is clearly

$$\delta_c^{(1)}(\alpha) = \left(1 + \frac{3}{2}\alpha\right). \tag{8.38}$$

From the evolution equation of the Wronskian we can then deduce, after simple algebra, the second linearly independent solution which is:

$$\delta_c^{(2)}(\alpha) = \left[\left(1 + \frac{3}{2}\alpha\right)\ln\left(\frac{\sqrt{\alpha+1}+1}{\sqrt{\alpha+1}-1}\right) - 3\sqrt{\alpha+1}\right]. \tag{8.39}$$

Since this is a general technical trick, let us do the explicit calculation leading to Eq. (8.39). For sake of simplicity let us name, just for this derivation $\delta_c^{(1)} = \delta$ and Δ the second (linearly independent) solution which will turn out to be $\delta_c^{(2)}$. The Wronskian and its first derivative will be, by definition,

$$W(\alpha) = \delta\frac{d\Delta}{d\alpha} - \Delta\frac{d\delta}{d\alpha}, \tag{8.40}$$

$$\frac{dW}{d\alpha} = \delta\frac{d^2\Delta}{d\alpha^2} - \Delta\frac{d^2\delta}{d\alpha^2}. \tag{8.41}$$

Both δ and Δ satisfy Eq. (8.35). Therefore Eq. (8.35) written, respectively, for δ and Δ can be combined as

$$\Delta\left[\frac{d^2\delta}{d\alpha^2} + \frac{3\alpha+2}{2\alpha(\alpha+1)}\frac{d\delta}{d\alpha} - \frac{3}{2\alpha(\alpha+1)}\delta\right]$$
$$- \delta\left[\frac{d^2\Delta}{d\alpha^2} + \frac{3\alpha+2}{2\alpha(\alpha+1)}\frac{d\Delta}{d\alpha} - \frac{3}{2\alpha(\alpha+1)}\Delta\right] = 0. \tag{8.42}$$

By simplifying the terms containing $\delta\Delta$ the remaining terms, recalling Eqs. (8.40) and (8.41) we will reconstruct the evolution equation of the Wronskian:

$$\frac{dW}{d\alpha} + \frac{3\alpha+2}{2\alpha(\alpha+1)}W = 0. \tag{8.43}$$

Now Eq. (8.43) can be solved. Then, the obtained result will be inserted into Eq. (8.40). This will allow to determine Δ as a function of δ (which is known and given by Eq. (8.38)) and of W. The solution of Eq. (8.43) can be obtained with elementary methods and it is:

$$W(\alpha) = \frac{1}{\alpha\sqrt{\alpha+1}}. \tag{8.44}$$

Inserting Eq. (8.44) into Eq. (8.40) and recalling that $\delta = (2+3\alpha)/2$, we will have, with simple algebra, Δ that is determined by the following first-order relation:

$$\frac{d}{d\alpha}\left[\frac{\Delta}{3\alpha+2}\right] = \frac{2}{\alpha\sqrt{\alpha+1}(3\alpha+2)^2}. \tag{8.45}$$

To determine Δ from Eq. (8.45) the following elementary integral can be easily solved:

$$\int \frac{d\alpha}{\alpha(3\alpha+2)^2\sqrt{\alpha+1}} = \frac{3\sqrt{\alpha+1}}{2(3\alpha+2)} - \frac{1}{4}\ln\left[\frac{\sqrt{\alpha+1}+1}{\sqrt{\alpha+1}-1}\right]. \tag{8.46}$$

But then, Δ is determined to be

$$\Delta(\alpha) = 3\sqrt{\alpha+1} - \left(1+\frac{3}{2}\alpha\right)\ln\left[\frac{\sqrt{\alpha+1}-1}{\sqrt{\alpha+1}-1}\right]. \tag{8.47}$$

Equation (8.47) coincides with Eq. (8.39) up to a sign, i.e. $\Delta(\alpha) = -\delta_c^{(2)}$. This is not a crucial difference since the minus sign can simply be reabsorbed in the integration constant that parametrizes the appropriate part of the general solution of Eq. (8.35). The general solution of Eq. (8.35) will then become:

$$\delta_c(k,\alpha) = D_1(k)\delta_c^{(1)}(\alpha) + D_2(k)\delta_c^{(2)}(\alpha). \tag{8.48}$$

The two solutions $\delta_c^{(1)}$ and $\delta_c^{(2)}$ are illustrated in Fig. 8.1 as a function of $\alpha = a/a_{\text{eq}}$. As it will become clear in a moment, Fig. 8.1 exhibits an important aspect that will have an impact on the following steps of the present calculation. Namely the occurrence that, after equality, the dominant solution is always the one reported in Eq. (8.38). Of course, to make this qualitative statement more quantitative the two arbitrary constants appearing in the general solution (8.48) must be appropriately fixed.

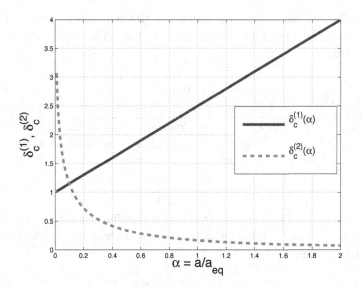

Fig. 8.1 The two linearly independent solutions of Eq. (8.35) are illustrated. It is clear from the graphic that for $a > a_{\rm eq}$ the dominant solution is the one denoted by $\delta_{\rm c}^{(1)}$ and reported in Eq. (8.38).

To obtain a correct estimate of the two integration constant in the intermediate regime (i.e. across the radiation-matter equality) the strategy is again rather simple. We can evaluate Eq. (8.48) in the limit $\alpha \ll 1$ and compare it with Eq. (8.29). The requirement that the two expressions coincide (as they should) will uniquely determine $D_1(k)$ and $D_2(k)$ in terms of $\psi_{\rm r}(k)$. By then inserting the obtained coefficients back into Eq. (8.48) a more accurate estimate of $\psi(k,\tau)$ can be obtained in the limit $\tau > \tau_{\rm eq}$.

Let us therefore determine $D_1(k)$ and $D_2(k)$. By expanding Eqs. (8.38) and (8.39) to leading order in α we can get the approximate form of Eq. (8.48) valid in the limit $\alpha = a/a_{\rm eq} \ll 1$. The result of this procedure is:

$$\delta_{\rm c}(k,\alpha) = D_1(k) - 3 D_2(k) - D_2(k) \ln\left(\frac{\alpha}{4}\right) + \mathcal{O}(\alpha) \qquad (8.49)$$

valid up to terms containing the first power of α. Now recall that $\alpha = (2x + x^2)$ where $x = \tau/\tau_1$. So, in the interesting limit we will have $\alpha \simeq 2(\tau/\tau_1)$.[d] By now demanding that Eq. (8.49) exactly matches Eq. (8.29)

---

[d] We remind the reader that, by definition, $\alpha = 1$ at $\tau_{\rm eq}$. Therefore we will have $\tau_{\rm eq}/\tau_1 = \sqrt{2} - 1$.

(which holds in the same limit) the following equality must hold

$$-9\psi_{\rm r}(k)\left[\left(\gamma - \frac{1}{2}\right) + \ln k - \frac{1}{2}\ln 3 + \ln\tau\right]$$
$$= D_1(k) - 3D_2(k) - D_2(k)\ln\tau + D_2(k)\ln(2\tau_1), \qquad (8.50)$$

which implies:

$$D_2(k) = 9\psi_{\rm r}(k), \qquad D_1(k) = 9\psi_{\rm r}(k)\left[\frac{7}{2} - \gamma - \ln\left(\frac{2k\tau_1}{\sqrt{3}}\right)\right]. \qquad (8.51)$$

By now inserting the obtained expression back into Eq. (8.49) we do get, to leading order in $(1/\alpha)$,

$$\delta_{\rm c}(k,\alpha) = D_1(k)\left[1 + \frac{3}{2}\alpha\right] + \mathcal{O}\left(\frac{1}{\alpha}\right), \qquad (8.52)$$

where the contribution of the second solution (multiplying $D_2(k)$) is negligible for $\alpha \gg 1$: indeed in this limit it is easy to find that

$$\delta_{\rm c}(k,\alpha) \simeq \frac{4}{15}\alpha^{-3/2} - \frac{6}{35}\alpha^{-5/2} + \mathcal{O}(\alpha^{-7/2}). \qquad (8.53)$$

Using Eq. (8.19) we shall obtain a more accurate estimate of the metric fluctuation prior to recombination

$$\psi(k,\tau_{\rm rec}) \simeq \frac{27}{2}\psi_{\rm r}(k)\left[\gamma - \frac{7}{2} + -\ln\left(\frac{2k\tau_1}{\sqrt{3}}\right)\right] \simeq \frac{\ln(0.15k\tau_{\rm eq})}{(0.272k\tau_{\rm eq})^2}. \qquad (8.54)$$

The second equality in the last equation can be simply obtained by noticing that τ_1 and $\tau_{\rm eq}$ (i.e. $\tau_1 = \tau_{\rm eq}/(\sqrt{2}-1)$) and that $[\gamma - 7/2 + \ln 2 - \ln\sqrt{3} - \ln(\sqrt{2}-1)] \simeq \ln 0.15$. The factor in the denominator simply arises because $27/2 \simeq (0.272)^{-2}$.

There exist both crude and more refined accounts of the evolution of metric fluctuations for $k\tau > 1$ and $\tau < \tau_{\rm eq}$. The crudest approximation one can imagine is to completely ignore the oscillatory contribution arising in Eq. (8.20) for the expression of $\psi(k,\tau)$ in the limit $k\tau > 1$ and for $\tau < \tau_{\rm eq}$. Under this approximation Eq. (8.20) implies $\psi \simeq \psi_{\rm r}(k)/(k^2\tau^2)$. Hence, ignoring oscillations,

$$\psi(k,\tau) = \frac{\psi_{\rm r}(k)}{1 + (k\tau)^2}, \qquad \tau < \tau_{\rm eq}. \qquad (8.55)$$

There exist also more refined ways of accounting of the value of the metric fluctuations for wavelengths shorter than the Hubble radius and involve the concept of transfer function. Equation (8.55) is a very crude example of transfer function. In more general terms we can say that

$$\psi(k,\tau_{\rm eq}) = T(k,k_{\rm eq})\psi_{\rm r}(k), \qquad k_{\rm eq} = \frac{1}{14}{\rm Mpc}^{-1} \qquad (8.56)$$

where the value of k_{eq} holds for the case of a spatially flat Universe. The function $T(k, k_{eq})$ is given by [271]:

$$T(k, k_e) = \frac{\ln[1 + 2.34\,q]}{2.34\,q}[1 + 3.89\,q + (1.61\,q)^2$$
$$+ (5.46\,q)^3 + (6.71\,q)^4]^{-1/4},$$
$$q = \frac{k}{14 k_{eq}} \tag{8.57}$$

The quantity q appearing in Eq. (8.57) is given, more generally, by $q = (k\mathrm{Mpc})/(\Gamma h_0)$ where k is measured in inverse Mpc so that q is ultimately dimensionless. The quantity Γ is the so called shape parameter. In the case of a spatially flat Universe where $\Omega_{b0} \ll \Omega_{M0}$ (as implied by observational data), we have $q \simeq k\mathrm{Mpc}$, i.e. more specifically

$$h_0 \Gamma = h_0^2 \Omega_{t0} e^{-\Omega_{b0} - \sqrt{2 h_0} \Omega_{b0}/\Omega_{t0}}. \tag{8.58}$$

Now, if we focus our attention on the case $k_{eq} < k < 11\,k_{eq}$ the transfer function of Eq. (8.57) can be approximated as

$$T(k) \simeq \frac{1}{4} \ln\left[\frac{14 k_{eq}}{k}\right]. \tag{8.59}$$

The selected interval of k/k_{eq} is the one relevant for the region off the first two Doppler oscillations of the angular power spectrum. The goodness of the approximation (8.59) can be understood by plotting (8.59) and (8.57) in the range $k_{eq} < k < 11 k_{eq}$. In Fig. 8.2, Eqs. (8.57) and (8.59) are illustrated, respectively, with the full line and with the dashed line.

8.3 Tight-coupling between photons and baryons

As discussed at the end of chapter 2, the strength of the Coulomb coupling (in comparison with the strength of Thompson scattering) justifies the consideration of a unique proton-electron component which will be the so-called *baryon fluid*.[e] The baryon fluid is however also coupled, through Thompson scattering, to the photons. Since the photon-electron cross section is larger than the photon-protron cross section, the momentum exchange between the two components will be dominated by *electrons*.[f] The

---

[e]This nomenclature is a bit dangerous since it may be misunderstood. Sometimes to avoid misunderstandings it is better to speak about baryon-lepton fluid. In what follows, for sake of simplicity, we will adhere to the common practice and talk about baryon fluid.

[f]This observation explains why the Thompson cross section for electrons will appear ubiquitously in our equations.

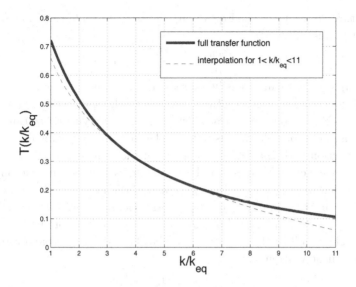

Fig. 8.2 The full transfer function and its interpolation.

evolution equation of the photon component can be written as:

$$\delta'_\gamma = -\frac{4}{3}\theta_\gamma + 4\psi', \tag{8.60}$$

$$\theta'_\gamma = \frac{k^2}{4}\delta_\gamma + k^2\phi + ax_e n_e \sigma_{\text{Th}}(\theta_b - \theta_\gamma). \tag{8.61}$$

For the baryon-lepton fluid, the two relevant equations are instead:

$$\delta'_b = 3\psi' - \theta_b, \tag{8.62}$$

$$\theta'_b = -\mathcal{H}\theta_b + k^2\phi + \frac{4}{3}\frac{\rho_\gamma}{\rho_b} a n_e x_e \sigma_{\text{Th}}(\theta_\gamma - \theta_b). \tag{8.63}$$

From Eqs. (8.61) and (8.63) it can be argued that the Thompson scattering terms drag the system to the final configuration where $\theta_\gamma \simeq \theta_b$. In fact, by taking the difference of Eqs. (8.61) and (8.63) the following equation can be easily obtained:

$$(\theta_\gamma - \theta_b)' + \Gamma_{\text{Th}}(\theta_\gamma - \theta_b) = J(\tau, \vec{x}), \tag{8.64}$$

where

$$\Gamma_{\text{Th}} = an_e x_e \sigma_{\text{Th}}\left(1 + \frac{4}{3}\frac{\rho_\gamma}{\rho_b}\right) \equiv an_e x_e \sigma_{\text{Th}}\left(\frac{R_b + 1}{R_b}\right), \tag{8.65}$$

$$J(\tau, \vec{x}) = \frac{k^2}{4}\delta_\gamma + \mathcal{H}\theta_b. \tag{8.66}$$

For future convenience, in Eq. (8.66) the baryon-to-photon ratio R_b has been introduced, i.e.

$$R_b(z) = \frac{3}{4}\frac{\rho_b}{\rho_\gamma} = \left(\frac{698.38}{z+1}\right)\left(\frac{h_0^2 \Omega_b}{0.023}\right). \tag{8.67}$$

From Eq. (8.64) it can be easily appreciated that any deviation of $(\theta_\gamma - \theta_b)$ swiftly decays away in spite of the strength of the source term $J(\tau, \vec{x})$. In fact, from Eq. (8.64), the characteristic time for the synchronization of the baryon and photon velocities is of the order of $(x_e n_e \sigma_{Th})^{-1}$, which is small in comparison with the expansion time. In the limit $\sigma_{Th} \to \infty$ the tight-coupling is exact and the photon-baryon velocity field is a unique physical entity which will be denoted by $\theta_{\gamma b}$. The evolution equation for $\theta_{\gamma b}$ can be easily obtained by summing up Eq. (8.61) and Eq. (8.63) with a relative weight (given by R_b) allowing the mutual cancellation of the scattering terms. The result of this procedure implies that the whole baryon-photon system can be written as

$$\delta'_\gamma = 4\psi' - \frac{4}{3}\theta_{\gamma b}, \tag{8.68}$$

$$\delta'_b = 3\psi' - \theta_{\gamma b}, \tag{8.69}$$

$$\theta'_{\gamma b} + \frac{\mathcal{H} R_b}{(1+R_b)}\theta_{\gamma b} = \frac{k^2 \delta_\gamma}{4(1+R_b)} + k^2 \phi. \tag{8.70}$$

The introduction of the baryon-photon velocity field also slightly modifies the form of the momentum constraint of Eq. (8.5) which now assumes the form:

$$k^2(\mathcal{H}\phi + \psi') = \frac{3}{2}\mathcal{H}^2\left[\frac{4}{3}R_\nu \theta_\nu + \frac{4}{3}R_\gamma(1+R_b)\theta_{\gamma b} + \theta_c \Omega_c\right]. \tag{8.71}$$

Equation (8.70) has been derived by adding and subtracting the evolution equations of the single species. Furthermore Eq. (8.70) does not include any dissipative effect (connected, for instance, with shear viscosity). Shear viscosity is important for the estimate of the Silk damping scale and appears in Eq. (8.70) in connection with an extra term going as $\nabla^2 \theta_{\gamma b}$. To derive this modification it is instructive to reconstruct Eq. (8.70) within a slightly different (but physically equivalent) framework.

8.4 Shear viscosity and silk damping

Consider, indeed, Eq. (D.51). In the longitudinal gauge (which is consistently used in this chapter), Eq. (D.51) can be written

$$(p+\rho)\theta' + \theta[(p'+\rho') + 4\mathcal{H}(p+\rho)]\nabla^2 \delta p + (p+\rho)\nabla^2 \phi = \frac{4}{3}\eta \nabla^2 \theta, \tag{8.72}$$

where η is nothing but the shear viscosity coefficient (see for instance [75]). In the photon-baryon system, Eq. (8.72) can be simply rewritten by recalling that

$$\theta \to \theta_{\gamma b}, \qquad (p+\rho) \to \left(\frac{4}{3}\rho_\gamma + \rho_b\right). \tag{8.73}$$

Using Eq. (8.73) into Eq. (8.72) we obtain

$$\left(\frac{4}{3}\rho_\gamma + \rho_b\right)\theta'_{\gamma b} + \mathcal{H}\rho_b\theta_{\gamma b} + \frac{\nabla^2 \delta\rho_\gamma}{3}$$
$$+ \left(\frac{4}{3}\rho_\gamma + \rho_b\right)\nabla^2\phi - \frac{4}{3}\eta\nabla^2\theta_{\gamma b} = 0 \tag{8.74}$$

where the relations $\rho'_b = -3\mathcal{H}\rho_b$ and $\rho'_\gamma = -4\mathcal{H}\rho_\gamma$ have been used. The shear viscosity coefficient η can be written in terms of the photon energy density and in terms of the Thompson mean free path λ_{Th} already defined in Eq. (2.116) of chapter 2:

$$\eta = \frac{4}{15}\rho_\gamma \lambda_{\text{Th}}. \tag{8.75}$$

Equation (8.74) can then be multiplied by $(3/4)\rho_\gamma^{-1}$ and the resulting expression will then be:

$$(1+R_b)\theta'_{\gamma b} + \mathcal{H}R_b\theta_{\gamma b} + \frac{1}{4}\nabla^2\delta_\gamma$$
$$+(1+R_b)\nabla^2\phi - \frac{\eta}{\rho_\gamma}\nabla^2\theta_{\gamma b} = 0. \tag{8.76}$$

Finally, dividing Eq. (8.76) by $(R_b + 1)$ and passing to Fourier space the wanted equation can be obtained:

$$\theta'_{\gamma b} + \frac{\mathcal{H}R_b}{(1+R_b)}\theta_{\gamma b} = \frac{k^2\delta_\gamma}{4(1+R_b)} + k^2\phi - \frac{4\lambda_{\text{Th}}}{15(1+R_b)}k^2\theta_{\gamma b}. \tag{8.77}$$

Equation (8.77) is exactly Eq. (8.70) but we also have an extra factor, i.e. the contribution of shear viscosity. It is possible now to obtain an equation for δ_γ which will play a major role in some of the considerations developed in the present chapter. From Eq. (8.68) we have, in fact, that

$$\theta_{\gamma b} = 3\psi' - \frac{3}{4}\delta'_\gamma. \tag{8.78}$$

By performing a (conformal) time derivation of both sides of Eq. (8.78) and by inserting the result into Eq. (8.77), the evolution equation for δ_γ becomes

$$\delta''_\gamma + \frac{\mathcal{H}R_b}{R_b+1}\delta'_\gamma + \frac{4}{5}\lambda_{\text{Th}} c_{\text{sb}}^2 k^2 \delta'_\gamma + c_{\text{sb}}^2 \delta_\gamma$$
$$= 4\left[\psi'' + \frac{\mathcal{H}R_b}{R_b+1}\psi' + \frac{4}{5}\lambda_{\text{Th}} c_{\text{sb}}^2 k^2 \psi' - \frac{k^2}{3}\phi\right], \tag{8.79}$$

where

$$c_{\rm sb} = \frac{1}{\sqrt{3(R_{\rm b}+1)}}, \qquad (8.80)$$

is the speed of sound in the baryon-photon system.

Equation (8.79) can be simplified and solved in several different ways. For the purposes of the present chapter it is useful to notice that the following equalities hold:

$$\frac{R_{\rm b}'}{R_{\rm b}} = \frac{\mathcal{H}R_{\rm b}}{1+R_{\rm b}} = -\frac{(c_{\rm sb}^2)'}{c_{\rm sb}^2}. \qquad (8.81)$$

Motivated by Eq. (8.81) it is then natural to define a new time variable which includes $c_{\rm sb}^2$. This new time variable[g] will be denoted by q and it is defined as $dq = c_{\rm sb}^2 d\tau$. It is easy to show that

$$\delta_\gamma' = c_{\rm sb}^2 \frac{d\delta_\gamma}{dq}, \qquad \delta_\gamma'' = (c_{\rm sb}^2)' \frac{d\delta_\gamma}{dq} + c_{\rm sb}^4 \frac{d^2\delta_\gamma}{dq^2},$$

$$\psi' = c_{\rm sb}^2 \frac{d\psi}{dq}, \qquad \psi'' = (c_{\rm sb}^2)' \frac{d\psi}{dq} + c_{\rm sb}^4 \frac{d^2\psi}{dq^2}. \qquad (8.82)$$

Using the results of Eq. (8.82), Eq. (8.79) can be written as

$$\frac{d^2\delta_\gamma}{dq^2} + \frac{4}{5}k^2\lambda_{\rm Th}k^2\frac{d\delta_\gamma}{dq} + \frac{k^2}{c_{\rm sb}^2}\delta_\gamma = 4\left[\frac{d^2\delta_\gamma}{dq^2} + \frac{4}{5}k^2\lambda_{\rm Th}k^2\frac{d\delta_\gamma}{dq} - \frac{k^2}{3c_{\rm sb}^4}\phi\right]. \qquad (8.83)$$

We shall get back to Eq. (8.83) after presenting some numerical estimates of the Doppler oscillations. Indeed, Eq. (8.83) may offer a pretty accurate analytical estimate of the numerical results for various classes of initial conditions. Prior to this analysis, however, it is important to formulate precisely the problem of the initial conditions both in the adiabatic and in the non-adiabatic case.

8.5 The adiabatic solution

We are now in the position to give an explicit example of solution of the whole generalized system of fluctuations in the case of the adiabatic mode. It is important to stress that the obtained solution will hold for wavelengths that are larger than the Hubble radius prior to equality, i.e. respectively,

---

[g]In chapter 10 and in the Appendix the variable q denotes the (scalar) normal modes of the scalar-tensor action perturbed to second order in the amplitude of the metric fluctuations. Here q is just a new time variable. There will not be potential ambiguities in this choice since the two quantities appear in conceptually distinct situations.

$|k\tau| \ll 1$ and $\tau < \tau_{\rm eq}$. In the cartoon of Fig. 7.1 the corresponding physical wavelengths will have slopes larger than the thick (full) line denoting, in the plot, the approximate evolution of the Hubble radius. The logic of the calculation will be the following:

- using the observation (see chapter 7) that the longitudinal fluctuations of the metric must be constant (in the limit $k\tau \ll 1$) for the adiabatic solution, the Hamiltonian constraint of Eq. (8.3) can be solved deep in the radiation-dominated stage;
- the adiabaticity condition will then be imposed by demanding that the entropy fluctuations vanish: this requirement will fix the relation between the various density contrasts of relativistic and non-relativistic species;
- the evolution equations of the peculiar velocities (i.e. Eqs. (8.11), (8.14) and (8.70)) will then be solved in the tight-coupling limit;
- the obtained solution for the peculiar velocities will have to satisfy the momentum constraint of Eq. (8.71) (see also Eqs. (8.5)–(8.7));
- the equations involving the neutrino anisotropic stress (i.e. Eqs. (8.10) and (8.15)) can also be solved: the result will determine the mismatch between the two longitudinal fluctuations of the metric, i.e. ψ and ϕ;
- finally, as a necessary cross-check, it is possible to verify that the solution obtained though the previous steps satisfies also the last equation, i.e. Eq. (8.6).

The logic followed in the determination of the adiabatic solution will persist also in the case of the non-adiabatic modes with the relevant difference that, in the case of the entropic modes, the adiabaticity condition *will not* be enforced for all the species as it will be explicitly discussed in the subsequent section.

In what follows the aforementioned steps will be addressed one by one. Consider first the Hamiltonian constraint of Eq. (8.3) deep in the radiation dominated epoch, i.e. for temperatures smaller than the temperature of the neutrino decoupling and temperatures larger than the equality temperature. In this case a solution can be found where the longitudinal fluctuations of the geometry are both constant in time, i.e.

$$\phi(k,\tau) = \phi_{\rm i}(k), \qquad \psi(k,\tau) = \psi_{\rm i}(k). \tag{8.84}$$

Equation (8.3) then implies, to the lowest order in $k\tau$ that the radiation density contrasts are also constant and given by

$$\delta_\gamma(k,\tau) \simeq \delta_\nu(k,\tau) \simeq -2\phi_{\rm i}(k). \tag{8.85}$$

Imposing now that the entropy fluctuations vanish we will also have

$$\delta_c(k,\tau) \simeq \delta_b(k,\tau) \simeq -\frac{3}{2}\phi_i(k) + \mathcal{O}(k^2\tau^2). \tag{8.86}$$

Direct integration of Eqs. (8.11), (8.14) and (8.70) implies, always to lowest order in $k\tau$ that

$$\theta_c(k,\tau) \simeq \theta_\nu(k,\tau) \simeq \theta_{\gamma b}(k,\tau) \simeq \phi_i(k)\frac{k^2\tau}{2}. \tag{8.87}$$

It can be checked that the momentum constraint of Eq. (8.71) is also satisfied in the radiation epoch. The relations of Eq. (8.87) express a general property of the adiabatic mode: to lowest order in $k\tau$, i.e. for typical wavelengths much larger than the Hubble radius at the corresponding epoch, the peculiar velocities of the various species are equal and much smaller than the density contrasts. Furthermore Eqs. (8.10) and (8.15) imply that the following two important relations:

$$\sigma_\nu(k,\tau) \simeq \frac{k^2\tau^2}{15}\phi_i(k), \tag{8.88}$$

$$\psi_i(k) = \left(1 + \frac{2}{5}R_\nu\right)\phi_i(k). \tag{8.89}$$

Recalling the definition of $\mathcal{R}$, it is also possible to relate the longitudinal fluctuations to the curvature fluctuations $\mathcal{R}$, i.e.

$$\psi_i(k) = -\frac{2(5+2R_\nu)}{15+4R_\nu}\mathcal{R}_*(k), \qquad \phi_i(k) = -\frac{10}{15+4R_\nu}\mathcal{R}_*(k). \tag{8.90}$$

Clearly, in the case $R_\nu = 0$ the relations of Eq. (8.90) reproduces the one already obtained and discussed in chapter 7. The initial conditions expressed through the Fourier components $\phi_i(k)$, $\psi_i(k)$ or $\mathcal{R}_*(k)$ are then computed by using the desired model of amplification. In particular, in the case of conventional inflationary models, the spectra of scalar fluctuations will be computed in chapter 10. Notice already that, in the case of the adiabatic mode, only one spectrum is necessary to set consistently the pre-decoupling initial conditions. Such a spectrum can be chosen to be either the one of $\phi_i(k)$ or the one of $\mathcal{R}_*(k)$. Indeed, as discussed in Eq. (7.104), the two variables are related via a simple numerical factor. As a final remark, it is useful to point out that the solution obtained for the adiabatic mode holds, in the present case, well before decoupling. In Eqs. (7.82), (7.83), (7.84) and (7.85) the solution has been derived, instead, during the matter epoch. These two solutions are physically different also because neutrinos (as well as baryons) have not been taken into account in chapter 7.

8.6 Pre-equality non-adiabatic initial conditions

It is now the moment for illustrating other non-adiabatic modes that may arise when the plasma possesses four different components. This generalization will also allow us to find the pre-equality form of the CDM-radiation non-adiabatic mode in the presence of a non-vanishing neutrino fraction, i.e. $r_\nu \neq 0$ and $R_\nu \neq 0$.

To make the treatment more symmetric it is practical to go back to the variable ζ already introduced in Eq. (7.120). Let us then introduce the following quantities

$$\zeta_\nu = -\psi + \frac{\delta_\nu}{4}, \qquad \zeta_\gamma = -\psi + \frac{\delta_\gamma}{4},$$
$$\zeta_c = -\psi + \frac{\delta_c}{3}, \qquad \zeta_b = -\psi + \frac{\delta_b}{3}, \qquad (8.91)$$

which are interpreted as the density contrasts for each independent fluid on uniform curvature hypersurfaces.

Equations (8.12), (8.13), (8.68) and (8.69) become, in terms of the variables introduced in (8.91),

$$\zeta'_\gamma = -\frac{\theta_{\gamma b}}{3}, \qquad \zeta'_\nu = -\frac{\theta_\nu}{3},$$
$$\zeta'_c = -\frac{\theta_c}{3}, \qquad \zeta'_b = -\frac{\theta_{\gamma b}}{3}. \qquad (8.92)$$

It is easy to show that

$$\zeta = \sum_i \frac{\rho'_i}{\rho'} \zeta_i, \qquad c_{st}^2 = \sum_i c_{si}^2 \frac{\rho'_i}{\rho'} \qquad (8.93)$$

where ζ_i and $c_i^2 = p'_i/\rho'_i$ and running index stands for each of the four components of the plasma. In this language, the adiabatic solution corresponds to the situation where

$$\zeta_\nu(k,\tau) = \zeta_\gamma(k,\tau) = \zeta_c(k,\tau) = \zeta_b(k,\tau) = \mathcal{R}_*(k). \qquad (8.94)$$

Equation (8.94) implies, in fact, that the entropy fluctuations of the mixture of fluids are vanishing. The entropy fluctuations are indeed defined as[h]

$$\mathcal{S}_{ij} = -3(\zeta_i - \zeta_j), \qquad \zeta_i = -\psi + \frac{\delta_i}{w_i + 1}, \qquad \zeta_j = -\psi + \frac{\delta_j}{w_j + 1}, \qquad (8.95)$$

where, as before, the indices run over the four components of the fluid. Using Eq. (8.94) into Eq. (8.95) the entropy fluctuations vanish for the adiabatic mode since $(\zeta_i - \zeta_j) = 0$.

---

[h] Equation (8.95) generalizes the analog relation (see Eqs. (7.88) and (7.90)) derived in the case of the CDM-radiation system in chapter 7.

The fluctuation of the total pressure can still be separated into an adiabatic and a non-adiabatic contribution as in Eq. (7.91), i.e. $\delta p = c_s^2 \delta\rho + \delta p_{\text{nad}}$ but now a generalized expression for the non-adiabatic pressure variation can be easily deduced by repeating the same calculation performed in chapter 7 with the difference that the plasma contains more than just two species. As easily arguable the final result is:

$$\delta p_{\text{nad}} = \frac{1}{6\mathcal{H}\rho_t'} \sum_{ij} \rho_i' \rho_j' (c_{si}^2 - c_{sj}^2) \mathcal{S}_{ij}, \qquad c_{st}^2 = \frac{p'}{\rho'}. \qquad (8.96)$$

In Eq. (8.96) c_{si}^2 and c_{sj}^2 are the sound speeds of each (generic) pair of fluids of the mixture. It is easy to show that Eq. (8.96) reproduces exactly the result of Eq. (7.99) if the mixture only consists of CDM and radiation. In this case the only non-vanishing contribution arises for $\mathcal{S}_{cr} = \mathcal{S}_*$. Substituting i $\to$ c and j $\to$ r in Eq. (8.96) and taking into account that $\mathcal{S}_{cr} = -\mathcal{S}_{rc}$ we get back to Eq. (7.99).

8.6.1 The CDM-radiation mode

We are now ready to compute the CDM-radiation mode for $\tau < \tau_{\text{eq}}$ (i.e. before equality) and for physical wavelengths which are, at the corresponding epoch, larger than the Hubble radius (i.e. $k\tau \ll 1$). The logic will be exactly the same as the one already outlined in the case of the adiabatic solution but with two important differences:

- the adiabaticity condition will not be enforced;
- the solution requires a more precise control of the background across the matter-radiation transition, so to be on the safe side the correct procedure will be to take the interpolating form of the scale factor discussed in Eq. (2.93) and insert it into the relevant equations; then each term of the equations can be expanded in the limit $x = |\tau/\tau_1| \ll 1$ and the wanted solution can then be obtained.

The first point translates immediately in the following two mathematical identities:

$$\frac{3}{4}\delta_\gamma(k,\tau) - \delta_c(k,\tau) = \mathcal{S}_*(k), \qquad \frac{3}{4}\delta_\nu(k,\tau) - \delta_c(k,\tau) = \mathcal{S}_*(k). \qquad (8.97)$$

Equation (8.97) means that the particular combination of the two density contrasts of radiation and CDM must lead to a quantity which is constant in time. Since radiation is composed by ν and γ Eq. (8.97) must hold independently for both species.

The second point listed above is merely technical and it means, at a practical level, that the interpolating form of the scale factor given in Eq. (2.93) will be introduced by employing the shorthand notations already exploited, for instance, in Eq. (7.98). The explicit form of the scale factor discussed in Eq. (2.93) implies that

$$\mathcal{H} = \frac{1}{\tau_1}\frac{2(x+1)}{x(x+2)},$$

$$\mathcal{H}' = -\frac{2}{\tau_1^2}\frac{x^2+2x+4}{x^2(x+2)^2}, \qquad (8.98)$$

$$\mathcal{H}^2 - \mathcal{H}' = \frac{1}{\tau_1^2}\frac{2(3x^2+6x+4)}{x^2(x+2)^2},$$

where $x = \tau/\tau_1$. Furthermore, since we have two matter species (i.e. baryon and CDM particles) we also have to modify slightly the notation of Eq. (7.98) and say that $\Omega_M(x) = \Omega_c(x) + \Omega_b(x)$. Since $\Omega_M(x) = 1 - \Omega_R(x)$, then we will also have

$$\Omega_c(x) = \overline{\Omega}_c \frac{x(x+2)}{(x+1)^2}, \qquad \Omega_b(x) = \overline{\Omega}_b \frac{x(x+2)}{(x+1)^2}, \qquad (8.99)$$

where $\overline{\Omega}_c$ and $\overline{\Omega}_b$ are two constants such that $\overline{\Omega}_c + \overline{\Omega}_b = 1$. The evolution of $\Omega_M(x)$ and $\Omega_R(x)$ are reported in Fig. 8.3 for the same background leading to Eq. (8.98).

The strategy will be to solve, first Eqs. (8.3), (8.5) and (8.6) with an ansatz of the type

$$\phi(k,\tau) = \phi_1(k)\left(\frac{\tau}{\tau_1}\right), \qquad \psi(k,\tau) = \psi_1(k)\left(\frac{\tau}{\tau_1}\right). \qquad (8.100)$$

where $\psi_1(k)$ and $\phi_1(k)$ will be determined from the consistency with all the other equations. This ansatz is simply motivated by the treatment of chapter 7 where it has been shown explicitly that, in the limit $\psi(k,\tau) = \phi(k,\tau)$, the longitudinal fluctuations of the geometry (as well as the curvature perturbations) must vanish in the limit $\tau \ll \tau_1$. This behaviour will now be different (since neutrinos are consistently included) but cannot be avoided. Based on the experience gained with the adiabatic mode we will then expect that $\phi_1(k) \neq \psi_1(k)$.

Let us now illustrate the procedure followed with the case of the Hamiltonian constraint. Inserting Eqs. (8.98) and (8.99) into Eq. (8.3) we get

$$\frac{12(x+1)^2}{x(x+2)^2}\phi_1 + \frac{6(x+1)}{x(x+2)}\psi_1 + \kappa^2\psi_1 x = -\frac{6(x+1)^2}{x^2(x+2)^2}\bigg[4\psi_1 x$$
$$- \mathcal{S}_* \frac{x(x+2)}{(x+1)^2}\overline{\Omega}_c + 3\psi_1 \frac{x^2(x+2)}{(x+1)^2}(\overline{\Omega}_b + \overline{\Omega}_b)\bigg], \qquad (8.101)$$

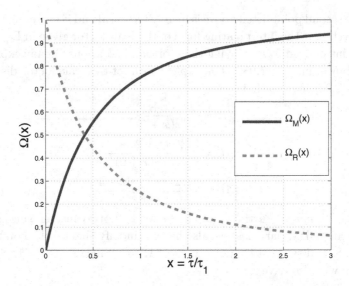

Fig. 8.3 The evolution of Ω_M (dashed curve) and Ω_R (full curve) are reported in the case when the radiation-matter transition is parametrized by Eq. (2.93) (see also the derived quantities reported in Eq. (8.98)).

where $\kappa = k\tau_1$. In Eq. (8.101) the condition (8.97) has been used. Furthermore, using the ansatz (8.100) we find that the evolution equations for the density contrasts are solved, provided

$$\delta_\gamma(k,\tau) = \delta_\nu(k,\tau) = 4\psi_1(k)\left(\frac{\tau}{\tau_1}\right),$$
$$\delta_c(k,\tau) = -\mathcal{S}_*(k) + 3\psi_1(k)\left(\frac{\tau}{\tau_1}\right), \quad (8.102)$$
$$\delta_b(k,\tau) = 3\psi_1(k)\left(\frac{\tau}{\tau_1}\right).$$

To solve Eq. (8.99) we have to use a double expansion, i.e. we have to expand for $x \ll 1$ (corresponding to the requirement that we are deep into the radiation epoch) and also for $k\tau = \kappa x \ll 1$ (corresponding to the requirement that the physical wavelengths are larger than the Hubble radius). Following this procedure three kinds of terms arise:

- terms containing κ^2 (which can be immediately dropped);
- terms going as $1/x$ (which should cancel will partially fix $\psi_1(k)$ and $\phi_1(k)$ as a function of $\mathcal{S}_*(k)$);

- terms which are constant and terms which are vanishing in the limit $x \to 0$.

The third class of terms necessarily requires higher orders in the ansatz. In other words, to get a match on those terms we will have to start with $\psi(k,\tau) = \psi_1(k)x + \psi_2(k)x^2$ and $\phi(k,\tau) = \phi_1(k)x + \phi_2(k)x^2$. The conditions stemming from the second class of terms mentioned above will imply the relation $\phi_1(k) + 3\psi_1(k) = \mathcal{S}_*(k)\overline{\Omega}_c$.

The evolution equations for the peculiar velocities can now be solved to the same order and they are:

$$\theta_c(k,\tau) = \frac{k^2\tau_1}{3}\phi_1(k)\left(\frac{\tau}{\tau_1}\right)^2,$$

$$\theta_{\gamma b}(k,\tau) = \frac{k^2\tau_1}{2}[\phi_1(k) + \psi_1(k)]\left(\frac{\tau}{\tau_1}\right)^2, \qquad (8.103)$$

$$\theta_\nu(k,\tau) = \frac{k^2\tau_1}{2}[\phi_1(k) + \psi_1(k)]\left(\frac{\tau}{\tau_1}\right)^2.$$

The evolution equation for the anisotropic stress will finally be solved provided

$$\mathcal{F}_{\nu 3}(k,\tau) = 0, \qquad \sigma_\nu(k,\tau) = \frac{k^2\tau_1^2}{6R_\nu}[\psi_1(k) - \phi_1(k)]\left(\frac{\tau}{\tau_1}\right)^3, \qquad (8.104)$$

with the conditions

$$\phi_1(k) + 3\psi_1(k) = \mathcal{S}_*(k)\overline{\Omega}_c,$$
$$\psi_1(k) - \phi_1(k) = \frac{2}{5}R_\nu[\phi_1(k) + \psi_1(k))], \qquad (8.105)$$

where the first condition is just the reminder of what obtained from the Hamiltonian constraint and the second equality stems from the evolution of the anisotropic stress. It is easy to see that the conditions given in Eq. (8.104) are solved provided:

$$\psi_1(k) = \frac{15 + 4R_\nu}{4(15 + 2R_\nu)}\mathcal{S}_*(k)\overline{\Omega}_c, \qquad \phi_1(k) = \frac{15 - 4R_\nu}{4(15 + 2R_\nu)}\mathcal{S}_*(k)\overline{\Omega}_c. \qquad (8.106)$$

We conclude the discussion on the CDM-radiation mode with two useful technical remarks. The first one deals with the form of the background geometry. Instead of taking the interpolating form of the scale factor described above and obtained in Eq. (2.93) we could have started at the beginning with a scale factor valid only in the radiation epoch like, for instance, $a(\tau) \sim \tau$. In this case the calculation follows the same steps, however, in the solution of the Hamilonian constraint a factor $1/2$ arises as a consequence of the fact that, from Eq. (2.93), the limit $\tau \ll \tau_1$ implies $a(\tau) \simeq 2 a_{\rm eq}(\tau/\tau_1) \simeq 2\tau$. So to make a long story short, the first relation in Eq. (8.105) will be modified as $(\phi_1 + 3\psi_1) = \mathcal{S}_*(k)\overline{\Omega}_{\rm c}/2$ since $\Omega_{\rm c}(\tau)$ vanishes, during radiation, as $a(\tau)$. This difference will propagate also to Eq. (8.106) and instead of having a factor 4 in the denominator there will be a factor 8. This is after all a minor point, however the interpolating parametrization of Eq. (2.93) is to be preferred since it allows, in practice, not only the evaluation of the lowest order of the solution but also the higher orders in the expansion.

The second remarks has to do with the constant factor $\overline{\Omega}_{\rm c}$ appearing in Eq. (8.106). In the limit $R_\nu \to 0$, $\psi(k,\tau) \simeq \mathcal{S}_*(k)\overline{\Omega}_{\rm c} a(\tau)$. In chapter 7 the same expression has been obtained but with $\overline{\Omega}_{\rm c} \to 1$. This was correct since there we only had CDM and no baryons. There exists the habit, in some literature, to include $\overline{\Omega}_{\rm c}$ in the definition of $\mathcal{S}_*(k)$. This is probably not the best convention one can imagine.

8.6.2 The baryon-entropy mode

The same algebra leading to the CDM-radiation mode also leads to the baryon-entropy mode. The important difference resides in the physical nature of the entropic contribution which arises, in this case, in the baryon-radiation system. In different words, Eqs. (8.97) and (8.100) respectively become

$$\frac{3}{4}\delta_\gamma(k,\tau) - \delta_{\rm b}(k,\tau) = \mathcal{S}_*(k), \tag{8.107}$$

$$\frac{3}{4}\delta_\nu(k,\tau) - \delta_{\rm b}(k,\tau) = \mathcal{S}_*(k), \tag{8.108}$$

$$\phi(k,\tau) = \overline{\phi}_1(k)\left(\frac{\tau}{\tau_1}\right), \qquad \psi(k,\tau) = \overline{\psi}_1(k)\left(\frac{\tau}{\tau_1}\right). \tag{8.109}$$

The solution of the evolution equations of the density contrasts will then give us, to leading order in $k\tau$ (or κx), the following conditions:

$$\delta_\gamma(k,\tau) = \delta_\nu(k,\tau) = 4\overline{\psi}_1(k)\left(\frac{\tau}{\tau_1}\right),$$

$$\delta_{\rm b}(k,\tau) = -\mathcal{S}_*(k) + 3\overline{\psi}_1(k)\left(\frac{\tau}{\tau_1}\right), \tag{8.110}$$

$$\delta_{\rm c}(k,\tau) = 3\overline{\psi}_1(k)\left(\frac{\tau}{\tau_1}\right).$$

The peculiar velocities can also be obtained by integrating the appropriate equations and the result is:

$$\theta_c(k,\tau) = \frac{k^2 \tau_1}{3} \overline{\phi}_1 \left(\frac{\tau}{\tau_1}\right)^2,$$

$$\theta_{\gamma b}(k,\tau) = \frac{k^2 \tau_1}{2}(\overline{\phi}_1 + \overline{\psi}_1)\left(\frac{\tau}{\tau_1}\right)^2, \quad (8.111)$$

$$\theta_\nu(k,\tau) = \frac{k^2 \tau_1}{2}(\overline{\phi}_1 + \overline{\psi}_1)\left(\frac{\tau}{\tau_1}\right)^2.$$

Also Eq. (8.104) translates easily to the case of baryon-entropy mode:

$$\mathcal{F}_{\nu 3}(k,\tau) = 0, \quad \sigma_\nu(k,\tau) = \frac{k^2 \tau_1^2}{6 R_\nu}(\overline{\psi}_1 - \overline{\phi}_1)\left(\frac{\tau}{\tau_1}\right)^3. \quad (8.112)$$

The most notable physical difference arises from Eqs. (8.105) and (8.106) which become, in this case:

$$\overline{\phi}_1(k) + 3\overline{\psi}_1(k) = \mathcal{S}_*(k)\overline{\Omega}_b,$$

$$[\overline{\psi}_1(k) - \overline{\phi}_1(k)] = \frac{2}{5} R_\nu(\overline{\phi}_1(k) + \overline{\psi}_1(k)), \quad (8.113)$$

$$\overline{\psi}_1(k) = \frac{15 + 4 R_\nu}{4(15 + 2 R_\nu)} \mathcal{S}_*(k)\overline{\Omega}_b, \quad (8.114)$$

$$\overline{\phi}_1(k) = \frac{15 - 4 R_\nu}{4(15 + 2 R_\nu)} \mathcal{S}_*(k)\overline{\Omega}_b. \quad (8.115)$$

Therefore, loosely speaking, we could say that the baryon-entropy mode is the analog of the CDM-radiation mode when the entropy fluctuation resides mainly in baryons. The nature of the result is, however, physically different. In the observed Universe we have evidence that the critical faction of CDM should be larger than the critical fraction in baryons. To make natural the baryon-entropy initial conditions (even in combination with a predominant adiabatic contribution) we should somehow assume an energy balance which is opposite to the observed one and we should posit (in contrast with the phenomenological evidence) that $\overline{\Omega}_b \gg \overline{\Omega}_c$.

8.6.3 The neutrino-entropy mode

The neutrino-entropy mode represents another interesting physical situation. In this case the entropy fluctuations arise in the neutrino-photon system. Consider, indeed, the circumstance where $\mathcal{S}_{\nu\gamma} \neq 0$. For notational convenience we will posit that $\mathcal{S}_{\nu\gamma}(k) = \tilde{\mathcal{S}}_*(k)$. Using the definition of $\mathcal{S}_{\nu\gamma}$ in terms of ζ_ν and ζ_γ (see Eq. (8.95)) we have

$$\mathcal{S}_{\nu\gamma}(k,\tau) = \frac{3}{4}[\delta_\gamma(k,\tau) - \delta_\nu(k,\tau)] = \tilde{\mathcal{S}}_*(k). \tag{8.116}$$

The inspection of the Hamiltonian constraint of Eq. (8.3) suggests that the wanted solution, to lowest order in $k\tau$, can be written as

$$\phi(k,\tau) = \phi_0(k), \qquad \psi = -\frac{\phi_0(k)}{2}. \tag{8.117}$$

So the longitudinal fluctuations of the metric are constant and the density contrasts are given, to the same order in $k\tau$, by

$$\delta_\gamma(k,\tau) = -2\phi_0(k) + \frac{4}{3}\tilde{\mathcal{S}}_*(k)R_\nu,$$

$$\delta_\nu(k,\tau) = -2\phi_0(k) - \frac{4}{3}\tilde{\mathcal{S}}_*(k)R_\gamma, \tag{8.118}$$

$$\delta_b(k,\tau) = \delta_c(k,\tau) = -\frac{3}{2}\phi_0(k).$$

The peculiar velocities and the neutrino anisotropic stress can be also computed by direct integration and the result is:

$$\theta_c(k,\tau) = \frac{k^2\tau}{2}\phi_0(k),$$

$$\theta_{\gamma b}(k,\tau) = \frac{k^2\tau}{4}\left[2\phi_0(k) + \frac{4}{3}\tilde{\mathcal{S}}_*(k)R_\nu\right],$$

$$\theta_\nu(k,\tau) = \frac{k^2\tau}{4}\left[2\phi_0(k) - \frac{4}{3}\tilde{\mathcal{S}}_*(k)R_\gamma\right], \tag{8.119}$$

$$\sigma_\nu(k,\tau) = -\frac{k^2\tau^2}{4R_\nu}\phi_0(k), \qquad \mathcal{F}_{\nu 3} = 0.$$

Finally, the consistency of Eqs. (8.10) and (8.15) implies a direct relation between $\phi_0(k)$ and $\tilde{\mathcal{S}}_*(k)$. The relation is:

$$\phi_0(k) = \frac{8R_\gamma R_\nu}{3(4R_\nu + 15)}\tilde{\mathcal{S}}_*(k). \tag{8.120}$$

An interesting feature of the neutrino-entropy mode is that the curvature perturbations on comoving orthogonal hypersurfaces[i] (i.e. $\mathcal{R}$) vanish for

---

[i]We refer the interested reader to chapter 11 where the physical and geometrical interpretation of $\mathcal{R}$ will be more deeply scrutinized. The definition of $\mathcal{R}$ in terms of longitudinal variables has however been already introduced in Eq. (7.101). This definition suffices for appreciating the statement that, during radiation, the neutrino-entropy mode leads to a vanishing value of $\mathcal{R}$ when the physical wavelengths are larger than the Hubble radius.

wavelengths larger than the Hubble radius before equality. This statement can be verified by using Eq. (8.117) in Eq. (7.101). Recalling that, for $\tau \ll \tau_1$, $\mathcal{H} = \tau^{-1}$, $\mathcal{R}(k) = -\psi_0 - \phi_0/2 \simeq 0$ (for $k\tau \ll 1$). In short, the most notable features of the neutrino-entropy mode can be summarized, in the longitudinal gauge, by saying that, to lowest order in $k\tau$:

- the longitudinal fluctuations are time-independent and the curvature perturbations vanish;
- the velocity fields are always negligible with respect to the density contrasts when the relevant wavelengths are larger than the Hubble radius;
- the density contrasts are time-independent.

These peculiar features are, of course, gauge-dependent. In a different gauge the properties of the neutrino-entropy mode may look different (see, in fact, chapter 11). In spite of this, the vanishing of the curvature perturbations is a gauge-invariant statement so the neutrino-entropy mode is really *isocurvature*. Indeed, as already mentioned, some authors talk about CDM isocurcature mode, baryon isocurvature mode and so on and so forth. This terminology is only appropriate at early times since, as we saw, for these non-adiabatic modes the spatial curvature vanish. We point out that this is not true, though, after equality and, in particular around recombination when the curvature fluctuations induced, for instance, by the CDM-radiation mode do not vanish.

There exists a further non-adiabatic mode that does not arise in the classification discussed so far. It is customarily called neutrino isocurvature velocity mode. This mode has peculiar features in the sense that it arises when the momentum constraint is vanishing at early times (see Eq. (8.71). At early times $(p_t + \rho_t)\theta_t = 0$ and, simultaneously, the total density is uniform. This requirement, once inserted into the evolution equations in the longitudinal gauge, entails a singularity in ϕ and ψ. The mode, however, is singular in the longitudinal description but perfectly regular in the synchronous gauge.[j]

We conclude this section with two references. They are reported in Refs. [272, 273] where a classification of the non-adiabatic fluctuations has been obtained in different (but related) frameworks.

---

[j]The evolution of the fluctuations in the synchronous coordinate system will be discussed in chapter 11.

8.7 Numerics in the tight-coupling approximation

It is useful to discuss some simplified example of numerical integration in the tight-coupling approximation. The idea will be to integrate numerically a set of equations where:

- baryons and photons are tightly coupled and the only relevant velocity fields are θ_c and $\theta_{\gamma b}$;
- neutrinos will be assumed to be absent;
- the evolution of the scalar modes will be implemented by means of $\mathcal{R}$ and ψ (i.e. by using Eq. (7.103)).

Since neutrinos are absent there is no source of anisotropic stress and the two longitudinal fluctuations of the metric are equal, i.e. $\phi = \psi$. Consequently, the system of equations to be solved becomes

$$\mathcal{R}' = \frac{k^2 c_s^2 \mathcal{H}}{\mathcal{H}^2 - \mathcal{H}'}\psi - \frac{\mathcal{H}}{p_t + \rho_t}\delta p_{\text{nad}}, \tag{8.121}$$

$$\psi' = -\left(2\mathcal{H} - \frac{\mathcal{H}'}{\mathcal{H}}\right)\psi - \left(\mathcal{H} - \frac{\mathcal{H}'}{\mathcal{H}}\right)\mathcal{R}, \tag{8.122}$$

$$\delta'_\gamma = 4\psi' - \frac{4}{3}\theta_{\gamma b}, \tag{8.123}$$

$$\theta'_{\gamma b} = -\frac{\mathcal{H} R_b}{R_b + 1}\theta_{\gamma b} + \frac{k^2}{4(1 + R_b)}\delta_\gamma + k^2\psi, \tag{8.124}$$

$$\delta'_c = 3\psi' - \theta_c, \tag{8.125}$$

$$\theta'_c = -\mathcal{H}\theta_c + k^2\psi. \tag{8.126}$$

Since the matter-radiation transition is crucial, the background will be parametrized by using Eq. (2.93). As already discussed in this chapter, from Eq. (2.93) the explicit expressions of $\mathcal{H}$ and $\mathcal{H}'$ easily follow and they have been already reported in Eq. (8.98) in terms of the rescaled time coordinate $x = \tau/\tau_1$. The results of chapter 7 (see, in particular, Eqs. (7.97)) can be used to express the barotropic index and the non-adiabatic pressure fluctuations in more explicit terms:

$$c_{\text{st}}^2 = \frac{4}{3}\frac{1}{3\alpha + 4}, \quad \delta p_{\text{nad}} = \rho_c c_{\text{st}}^2 \mathcal{S}_*, \quad \alpha(x) = \frac{a}{a_{\text{eq}}} = (x^2 + 2x). \tag{8.127}$$

With these specifications the evolution equations given in (8.121)–(8.126) become

$$\frac{d\mathcal{R}}{dx} = \frac{4}{3}\frac{x(x+1)(x+2)}{(3x^2+6x+4)^2}\kappa^2\psi - \frac{8(x+1)\mathcal{S}_*}{(3x^2+6x+4)^2}, \quad (8.128)$$

$$\frac{d\psi}{dx} = -\frac{3x^2+6x+4}{x(x+1)(x+2)}\mathcal{R} - \frac{5x^2+10x+6}{x(x+1)(x+2)}\psi, \quad (8.129)$$

$$\frac{d\delta_\gamma}{dx} = -\frac{4(3x^2+6x+4)}{x(x+1)(x+2)}\mathcal{R} - \frac{4(5x^2+10x+6)}{x(x+1)(x+2)}\psi - \frac{4}{3}\tilde{\theta}_{\gamma b}, \quad (8.130)$$

$$\frac{d\tilde{\theta}_{\gamma b}}{dx} = -\frac{2R_b}{R_b+1}\frac{(x+1)}{x(x+2)} + \frac{\kappa^2}{4(1+R_b)}\delta_\gamma + \kappa^2\psi, \quad (8.131)$$

$$\frac{d\delta_c}{dx} = -\frac{3(3x^2+6x+4)}{x(x+1)(x+2)}\mathcal{R} - \frac{3(5x^2+10x+6)}{x(x+1)(x+2)}\psi - \tilde{\theta}_c, \quad (8.132)$$

$$\frac{d\tilde{\theta}_c}{dx} = -\frac{2(x+1)}{x(x+2)}\tilde{\theta}_c + \kappa^2\psi. \quad (8.133)$$

In Eqs. (8.128)–(8.133) the following rescalings have been used (recall the role of τ_1 arising in Eq. (2.93)):

$$\kappa = k\tau_1, \qquad \tilde{\theta}_{\gamma b} = \tau_1\theta_{\gamma b}, \qquad \tilde{\theta}_c = \tau_1\theta_c. \quad (8.134)$$

The system of equations (8.128)–(8.133) can be readily integrated by giving initial conditions at $x_i \ll 1$. In the case of the adiabatic mode (which is the one contemplated by Eqs. (8.128)–(8.133) since we set $\delta p_{nad} = 0$) the initial conditions are:

$$\mathcal{R}(x_i) = \mathcal{R}_*, \qquad \psi(x_i) = -\frac{2}{3}\mathcal{R}_*,$$
$$\delta_\gamma(x_i) = -2\psi_*, \qquad \tilde{\theta}_{\gamma b}(x_i) = 0,$$
$$\delta_c(x_i) = \delta_b(x_i) = -\frac{3}{2}\psi_*, \qquad \tilde{\theta}_c(x_i) = 0, \quad (8.135)$$

and $\mathcal{S}_* = 0$ in Eq. (8.128). It can be shown by direct numerical integration that the system (8.128)–(8.133) gives a reasonable semi-quantitative description of the acoustic oscillations. To simplify initial conditions even further we can indeed assume a flat Harrison-Zeldovich spectrum and set $\mathcal{R}_* = 1$.

The same philosophy used to get to this simplified form can be used to integrate the full system. In this case, however, we would miss the important contribution of polarization since, to zeroth order in the tight-coupling expansion, the CMB is not polarized. In Figs. 8.4 and 8.5 the so-called Doppler (or Sakharov) oscillations are reported for fixed comoving wave-number k and as a function of the cosmic time coordinate. The

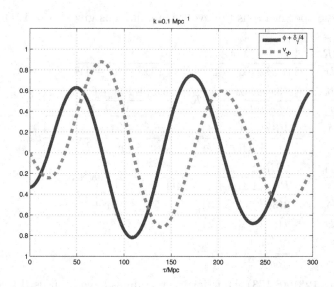

Fig. 8.4 The The (ordinary) SW and Doppler contributions are illustrated as a function of the conformal time at fixed comoving wave-number (in units of Mpc^{-1}). The initial conditions are adiabatic.

plots illustrate two different values of k in units of Mpc^{-1} in the case of adiabatic initial conditions (see Eq. (8.135)). In each plot the ordinary Sachs-Wolfe contribution and the Doppler contributions are illustrated, respectively, with full and dashed lines. To make the plot more clear we just report the Fourier mode and not the Fourier amplitude (which differs from the Fourier mode by a factor $k^{3/2}$). The quantity $v_{\gamma b}$ is simply $\theta_{\gamma b}/(\sqrt{3}k)$.

From Figs. 8.4 and 8.5 two general features emerge:

- in the adiabatic case the ordinary Sachs-Wolfe contribution oscillates as a cosine;
- in the adiabatic case the Doppler contribution (proportional to the peculiar velocity of the baryon-photon fluid) oscillates as a sine.

This rather naive observation has non-trivial consequences. In particular, the present discussion does not include polarization. However, the tight-coupling approximation can be made more accurate by going to higher orders. This will allow us to treat polarization (see chapter 9). Now, the Q Stokes parameter evaluated to first-order in the tight-coupling expansion will be proportional to the zeroth-order dipole. It is also useful to observe

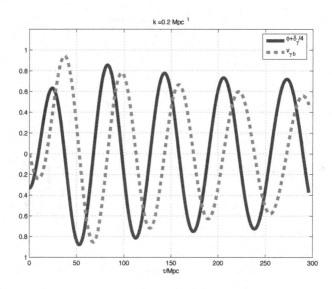

Fig. 8.5 The same quantities reported in Fig. 8.4 are illustrated but for a different value of the comoving wavenumber.

that in the units used in Figs. 8.4 and 8.5 the decoupling occurs, as discussed in connection with Eq. (2.93), for $\tau_{\rm dec} \simeq 284$ Mpc. The equality time is instead for $\tau_{\rm eq} \simeq 120$ Mpc. In Figs. 8.6 and 8.7 the ordinary Sachs-Wolfe contribution and the Doppler contribution are illustrated at fixed time (coinciding with $\tau_{\rm dec}$) and for different comoving wave-numbers. It is clear that the ordinary Sachs-Wolfe contribution gives a peak (the so-called Doppler peak) that corresponds to a mode of the order of the sound horizon at decoupling (see Eq. (9.151) in the following chapter). Note that because of the phase properties of the Sachs-Wolfe contribution there is a region where the ordinary Sachs-Wolfe contribution is quasi-flat. This is the so-called Sachs-Wolfe plateau. Recall that, for sake of simplicity, the curvature fluctuation has been normalized to 1 and the spectrum has been assumed scale-invariant. This is a rather crude approximation that has been adopted only for the purpose of illustration. Finally it should be remarked that diffusive effects, associated with Silk damping, have been completely neglected. This is a bad approximation for scales that are shorter than the scale of the first peak in the temperature autocorrelation. It should be however mentioned that there are semi-analytical ways of taking into account the Silk damping also in the framework of the tight-coupling expansion. In the tight-coupling expansion the Silk damping arises naturally when going

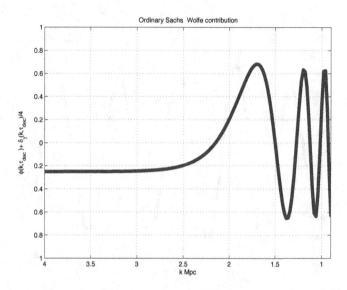

Fig. 8.6 The ordinary Sachs-Wolfe contribution is illustrated as a function of the comoving momentum at a fixed value of the conformal time. The initial conditions are adiabatic. On the horizontal axis, the base 10 logarithm of the comoving wave-number (in units of Mpc^{-1}) is reported.

to second-order in the small parameter that is used in the expansion and that corresponds, roughly, to the inverse of the photon mean free path (see chapter 9).

Let us now move to the case of the non-adiabatic initial conditions. In this case, as already discussed, the curvature fluctuations vanish in the limit $x \to 0$ and, in particular, the CDM-radiation non-adiabatic mode implies that

$$\mathcal{R}(x_i) = -\frac{\mathcal{S}_*}{3} x_i, \qquad \psi(x_i) = \frac{\mathcal{S}_*}{4} x_i,$$

$$\delta_\gamma(x_i) = \mathcal{S}_* x_i + \frac{4}{3}\mathcal{S}_*, \qquad \tilde{\theta}_{\gamma b}(x_i) = 0, \qquad (8.136)$$

$$\delta_c(x_i) = \delta_b(x_i) = \frac{3}{4}\mathcal{S}_* x_i, \qquad \tilde{\theta}_c(x_i) = 0.$$

In Figs. 8.8 and 8.9 the (ordinary) SW and Doppler contributions are reported for fixed comoving wave-numbers and as a function of the conformal time. Figs. 8.8 and 8.9 are the non-adiabatic counterpart of Figs. 8.4 and 8.5. It is clear that, in this case, the situation is reversed. While the adiabatic mode oscillates as cosine in the ordinary SW contribution, the non-adiabatic mode oscillates as a sine. Similarly, while the (adiabatic)

Improved Fluid Description of Pre-Decoupling Physics

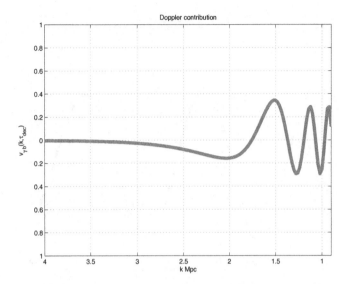

Fig. 8.7 The Doppler contribution is illustrated, for adiabatic initial conditions, as a function of the comoving momentum (and at a fixed value of the conformal time coordinate). On the horizontal axis the base 10 logarithm of the comoving wave-number is reported.

Doppler contribution oscillates as a sine, the non-adiabatic Doppler term oscillates as cosine. The different features of adiabatic and non-adiabatic contributions are even more evident in Fig. 8.10 and 8.11 which are the non-adiabatic counterpart of Figs. 8.6 and 8.7. Purely non-adiabatic initial conditions are excluded by current experimental data. However, a mixture of non-adiabatic and adiabatic initial conditions may be allowed as in the case of isocurvature modes induced by large-scale magnetic fields (see [253–255]).

8.7.1 *Interpretation of the numerical results*

The numerical results obtained so far can be corroborated by analytical solutions. In particular, for this purpose a useful tool is Eq. (8.83). Consider, in particular, the case where the longitudinal fluctuation of the metric is constant, i.e. $\phi(k) = \psi(k) = \phi_{\rm m}(k) = (9/10)\phi_i(k)$ (recall the discussion of chapter 7 on the super-Hubble evolution of curvature perturbations, i.e.

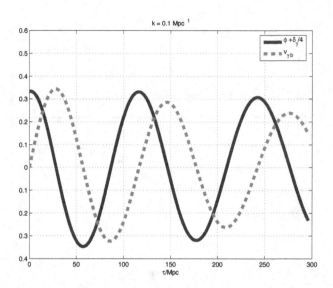

Fig. 8.8 The ordinary Sachs-Wolfe contribution is illustrated for non-adiabatic initial conditions (i.e. CDM-radiation mode). The comoving momentum is fixed.

Eqs. (7.75) and (7.104)). In this case Eq. (8.83) becomes:

$$\frac{d^2 \delta_\gamma}{dq^2} + \frac{4}{5} k^2 \lambda_{\mathrm{Th}} \frac{d\delta_\gamma}{dq} + \frac{k^2}{c_{\mathrm{sb}}^2} \delta_\gamma = -\frac{4k^2}{3 c_{\mathrm{sb}}^4} \phi_{\mathrm{m}}. \qquad (8.137)$$

The solution of the full equation will be given by the general solution of the homogeneous equation supplemented by a particular solution of the full equation. The particular solution can be easily found by inspection, i.e.

$$\overline{\delta}_\gamma(k) = -\frac{4}{3 c_{\mathrm{sb}}^2} \phi_{\mathrm{m}}(k). \qquad (8.138)$$

The general solution of the homogeneous equation is, in the WKB approximation,

$$\delta_\gamma(k,\tau) = \sqrt{c_{\mathrm{sb}}} \left[C_1(k) \cos\left(k \int c_{\mathrm{sb}} d\tau \right) \right.$$
$$\left. + C_2(k) \sin\left(k \int c_{\mathrm{sb}} d\tau \right) \right] e^{-\frac{k^2}{k_{\mathrm{D}}^2}}, \qquad (8.139)$$

where

$$k_{\mathrm{D}}^{-2} = \frac{2}{5} \int \frac{c_{\mathrm{sb}}^2}{(a/a_0) n_e x_e \sigma_{\mathrm{Th}}} d\tau. \qquad (8.140)$$

The integration constants $C_1(k)$ and $C_2(k)$ appearing in Eq. (8.138) can be determined by matching the solution to the value of the SW contribution

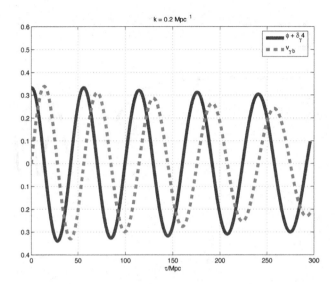

Fig. 8.9 The Doppler contribution is illustrated for the case of the CDM-radiation mode. As in Fig. 8.8 the comoving momentum is fixed while the velocity depends upon the conformal time coordinate.

when the relevant modes have wavelengths larger than the Hubble radius. In particular, in the adiabatic case we will have

$$\frac{\delta_\gamma(k,\tau)}{4} + \phi(k,\tau) \simeq \left(1 - \frac{1}{3c_{\rm sb}^2}\right)\phi_{\rm m}(k)$$
$$+ \frac{C_1(k)}{4}\sqrt{c_{\rm sb}}\cos\left(k\int c_{\rm sb}d\tau\right)e^{-\frac{k^2}{k_{\rm D}^2}} \quad (8.141)$$

showing that, indeed, for fixed comoving wave-number, the SW contribution oscillates like the cosine (see also Fig. 8.4). Furthermore, the baryon-photon velocity field will now be given by

$$\theta_{\gamma\rm b} \simeq -\frac{3}{4}\delta'_\gamma = -3\,k\,c_{\rm sb}^{3/2}C_1(k)\sin\left(k\int c_{\rm sb}d\tau\right)e^{-\frac{k^2}{k_{\rm D}^2}}, \quad (8.142)$$

showing that the Doppler contribution oscillates like a sine (see Figs. 8.4, 8.5 and 8.7). It is exactly the interplay of these two behaviours that will lead to the position and height of the Doppler peak in the temperature autocorrelations. The situation described for adiabatic initial conditions is clearly reversed when non-adiabatic initial conditions dominate. In the latter case the ordinary SW contribution goes as a sine while the Doppler term as a cosine (see Figs. 8.9 and 8.11). The baryon-to-photon ratio can

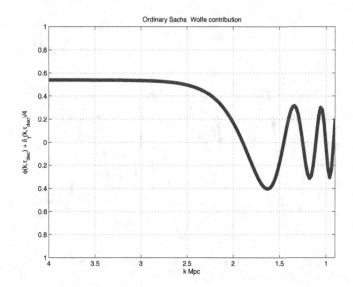

Fig. 8.10 The ordinary Sachs-Wolfe contribution is illustrated for a fixed value of the conformal time coordinate while. The initial conditions corresponds to the ones dictated by the CDM-radiation mode. On the horizontal axis the base 10 logarithm of the comoving wave-number (in units of Mpc^{-1}) is reported.

be referred to the redshift of recombination $z_{\text{rec}} \simeq 1050$. Thus we have, from Eq. (8.67)

$$R_{\text{rec}} = \frac{3}{4}\frac{\rho_{\text{b}}}{\rho_{\gamma}} = 0.669 \left(\frac{h_0^2 \Omega_{\text{b}}}{0.023}\right)\left(\frac{1051}{z_{\text{rec}} + 1}\right), \qquad (8.143)$$

where, by definition, $R_{\text{rec}} = R_{\text{b}}(\tau_{\text{rec}})$. It is now interesting, also in the light of forthcoming applications, to write the arguments of the cosine and sine appearing, respectively, in Eqs. (8.141) and (8.142) in a slightly different form by introducing the parameter γ which is implicitly defined by the following relation:

$$\gamma k \tau_0 = k \int_0^{\tau_{\text{rec}}} c_{\text{sb}}(\tau)d\tau, \qquad (8.144)$$

where τ_0 is the present value of the conformal time coordinate. Equation (8.144) implies, using Eq. (8.80) which defines $c_{\text{sb}}(\tau)$,

$$\gamma = \frac{1}{\sqrt{3}}\frac{1}{\tau_0}\int_0^{\tau_{\text{rec}}}\frac{d\tau}{\sqrt{1 + R_{\text{b}}(\tau)}}. \qquad (8.145)$$

To obtain the explicit expression of γ we have to write in explicit form $R_{\text{b}}(\tau)$ since, afterwards, the integral indicated in Eq. (8.145) has to be solved.

Improved Fluid Description of Pre-Decoupling Physics

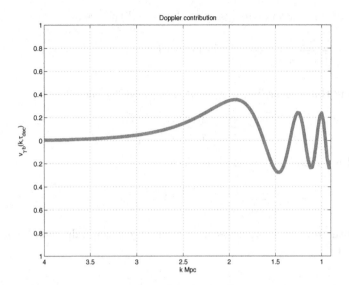

Fig. 8.11 The Doppler contribution is illustrated for the same initial conditions used to obtain Fig. 8.10. On the horizontal axis the base 10 logarithm of the comoving wave-number is reported.

There are two ways to express $R_b(\tau)$ which can be both useful, in slightly different ways. Recalling that, according to Eq. (2.93), $a(\tau) = a_{eq}(x^2 + 2x)$, Eq. (8.67) implies that $R_b(\tau) = \beta(x^2 + 2x)$ where the quantity denoted by β is simply given by[k]:

$$\beta = 698.38 \left(\frac{h_0^2 \Omega_{b0}}{0.023}\right) \frac{a_{eq}}{a_0} \simeq 0.216 \left(\frac{h_0^2 \Omega_{b0}}{0.023}\right) \left(\frac{h_0^2 \Omega_{M0}}{0.134}\right)^{-1}. \quad (8.146)$$

Looking together at Eqs. (8.143) and (8.146) it is also clear that:

$$\beta = R_{rec}\left(\frac{z_{rec}+1}{z_{eq}+1}\right), \qquad R_{rec} = 0.664\left(\frac{h_0^2 \Omega_{b0}}{0.023}\right)\left(\frac{1051}{z_{rec}+1}\right). \quad (8.147)$$

Equation (8.147) expresses the second (useful) form which allows to write $R_b(\tau)$ in explicit terms.

Let us therefore write γ from Eq. (8.145):

$$\gamma = \frac{1}{\sqrt{3}} \frac{\tau_1}{\tau_0} \int_0^y \frac{dx}{\sqrt{1 + \beta x(x+2)}} \quad (8.148)$$

where $y = \tau_{rec}/\tau_1$. The indefinite integral appearing in Eq. (8.148) can be

---

[k]Note that $a_0/a_{eq} = (z_{eq}+1)$ and $(z_{eq}+1)$ have been expressed by using Eq. (2.89).

easily obtained with elementary methods and the result is:
$$\int_0^y \frac{dx}{\sqrt{1+\beta x(x+2)}} = \left[\frac{1}{\sqrt{\beta}} \ln\left[2\sqrt{\beta}(x+1) + 2\sqrt{1+\beta x(x+2)}\right]\right]_0^y. \tag{8.149}$$

Recalling that, according to Eqs. (2.122), (2.123) and (2.124),
$$\frac{\tau_1}{\tau_0} = \frac{\tau_1}{\tau_{\text{rec}}} \frac{\tau_{\text{rec}}}{\tau_0} = \frac{\mathcal{T}_\Lambda}{\sqrt{z_{\text{eq}}}}, \tag{8.150}$$

we will have Eq. (8.148) that becomes:
$$\gamma = \frac{1}{\sqrt{3\beta}} \frac{\mathcal{T}_\Lambda}{\sqrt{z_{\text{eq}}}} \left[\frac{\beta(1+y) + \sqrt{1+\beta y(y+2)}}{\sqrt{\beta}+1}\right], \tag{8.151}$$

This is the first useful expression of γ whose explicit value can be obtained by simply recalling that:
$$y = \frac{\tau_{\text{rec}}}{\tau_1} = \sqrt{1 + \frac{z_{\text{eq}}+1}{z_{\text{rec}}+1}} - 1 \tag{8.152}$$

Recalling now Eq. (8.147) and using it into Eq. (8.151) we have a second useful expression for γ, namely:
$$\gamma = \frac{1}{\sqrt{3R_{\text{rec}}}} \frac{\mathcal{T}_\Lambda}{\sqrt{z_{\text{rec}}}} \ln\left\{\frac{\sqrt{[1+\frac{z_{\text{rec}}+1}{z_{\text{eq}}+1}]R_{\text{rec}}} + \sqrt{1+R_{\text{rec}}}}{1+\sqrt{R_{\text{rec}}\frac{z_{\text{rec}}+1}{z_{\text{eq}}+1}}}\right\}. \tag{8.153}$$

where, in the pre-factor, $z_{\text{eq}}+1$ has been approximated by z_{eq}. The value of γ can now be plotted for different choices of the cosmological parameters. In Figs. 8.12 and 8.13 the variation of γ is illustrated. From Figs. 8.12 and 8.13 it appears that γ decreases when the critical fraction of baryons and the redshift to recombination increase. On the contrary, an increase in Ω_{M0} entails an increase in γ. These observations will turn out to be useful to gain some intuition on the dependence of temperature autocorrelations upon the cosmological parameters.

8.7.2 Numerical estimates of diffusion damping

It is interesting, as a preparation for the subsequent applications

- to estimate the Silk damping scale as defined in Eq. (8.140);
- to estimate the phases of the oscillations as they emerge from Eq. (8.141).

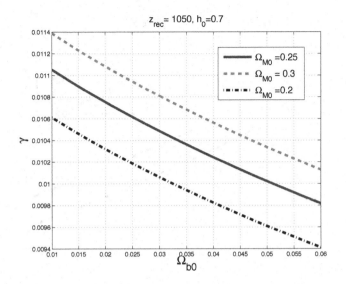

Fig. 8.12 The variation of γ as a function of Ω_{b0}.

Consider then Eq. (8.140) which can be also written as
$$\frac{1}{k_D^2(\tau)} = \frac{2}{5} \frac{1}{\eta_b n_{\gamma 0} \sigma_{\text{Th}} x_e} \int_0^\tau \left(\frac{a(\tau)}{a_0}\right)^2 c_{\text{sb}}^2(\tau'). \tag{8.154}$$

Concerning Eq. (8.154) in comparison with Eq. (8.140) we have to remark the following:

- the electron concentration n_e can be always expressed in terms of the baryonic charge since the Universe is assumed to be neutral, i.e. $n_e(\tau) = n_B(\tau) = \eta_b n_\gamma(\tau)$;
- the photon concentration at the time τ is related to the present electron concentration as $n_\gamma(\tau) = n_{\gamma 0}(a_0/a)^3$;
- the diffusion scale is time-dependent.

Using then the explicit form of the scale factor across radiation-matter equality, Eq. (8.154) can also be written, for $\tau \simeq \tau_{\text{rec}}$, as
$$\frac{1}{k_D^2} = \frac{2}{15} \frac{\tau_1}{\eta_b n_{\gamma 0} \sigma_{\text{Th}}} (1 + z_{\text{eq}})^{-2} \int_0^{\tau_{\text{rec}}/\tau_1} \frac{(x^2 + 2x)^2}{1 + \beta(x^2 + 2x)} dx, \tag{8.155}$$

where we used the fact that $(a_{\text{eq}}/a_0)^2 = (1 + z_{\text{eq}})^{-2}$; Eq. (8.146) has been also used to obtain the explicit form of the sound speed. Recall, finally, that according to the parametrization of Eq. (2.93), $x = \tau/\tau_1$. Equation (8.155) can now be evaluated in two different cases:

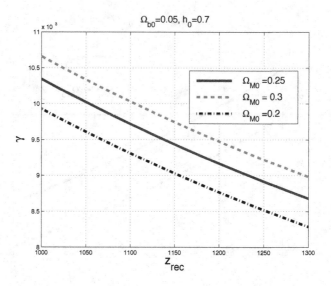

Fig. 8.13 The variation of γ as a function of $z_{\rm rec}$.

- the first case corresponds to set $\beta = 0$; in this case the integral over x becomes simple and, effectively, this assumption amounts to set $c_{\rm sb} = 1/\sqrt{3}$;
- the second case corresponds to keep $\beta \neq 0$.

Let us first proceed along the first case and then briefly discuss the second case.

If $\beta = 0$ we will have that the integral appearing in Eq. (8.155) can be approximated as follows:

$$\int_0^{\tau_{\rm rec}/\tau_1} (x^2 + 2x)^2 = \left[\frac{x^5}{5} + \frac{4}{3}x^3 + x^4 \right]_0^{\tau_{\rm rec}/\tau_1} \simeq \frac{1}{5}\left(\frac{\tau_{\rm rec}}{\tau_1} \right)^5, \quad (8.156)$$

which is justified since $\tau_{\rm rec}/\tau_1 > 1$. Consequently, using Eq. (8.156) inside Eq. (8.155) (and dividing both sides by $\tau_{\rm rec}^2$) the following expression can be easily obtained:

$$\frac{1}{k_{\rm D}^2 \tau_{\rm rec}^2} \simeq \frac{2}{75} \frac{1}{\eta_{\rm b} n_\gamma x_{\rm e} \sigma_{\rm Th} \tau_1} \left(\frac{\tau_{\rm rec}}{\tau_1} \right)^3. \quad (8.157)$$

Let us therefore plug the appropriate numerical values into Eq. (8.157). The values of the various constants are[1]:

$$\eta_{\rm b} = 6.27 \times 10^{-10} \left(\frac{h_0^2 \Omega_{\rm b0}}{0.023} \right),$$

$$\sigma_{\rm Th} = 6.65 \times 10^{-25} \, {\rm cm}^2,$$

$$n_\gamma = 411 \, {\rm cm}^{-3},$$

$$\left(\frac{a_{\rm eq}}{a_0} \right) = \frac{1}{3228.91} \left(\frac{h_0^2 \Omega_{\rm M0}}{0.134} \right)^{-1} \quad (8.158)$$

$$\frac{\tau_{\rm rec}}{\tau_1} \simeq \sqrt{\frac{z_{\rm eq}+1}{z_{\rm rec}+1}},$$

$$\tau_1 = 288.25 \left(\frac{h_0^2 \Omega_{\rm M0}}{0.134} \right)^{-1} {\rm Mpc}.$$

Notice that the value of τ_1 has been already computed in Eq. (2.94) but we found it practical to recall it explicitly. Moreover, the exact expression of $\tau_{\rm rec}/\tau_1$ is

$$\frac{\tau_{\rm rec}}{\tau_1} = \sqrt{1 + \frac{z_{\rm eq}+1}{z_{\rm rec}+1}} - 1, \quad (8.159)$$

which reduces to the estimate of Eq. (8.158) since $z_{\rm eq}/z_{\rm rec} > 1$. With the numerical values listed in Eq. (8.158) it is easy to find that

$$\frac{1}{k_{\rm D}\tau_{\rm rec}} = 9.63 \times 10^{-3} \left(\frac{h_0^2 \Omega_{\rm b0}}{0.023} \right)^{-1/2} \left(\frac{h_0^2 \Omega_{\rm M0}}{0.134} \right)^{1/4} \left(\frac{1050}{z_{\rm rec}} \right)^{3/4}. \quad (8.160)$$

Recalling the numerical value of $\tau_{\rm rec}$ in Mpc we do find that, depending on the values of the cosmological parameters $k_{\rm D}^{-1}$ is around 3 to 5 Mpc.

What happens now if $\beta \neq 0$? If $\beta \neq 0$ the new result will differ from the old (i.e. $\beta = 0$) result by the multiplicative factor

$$\mathcal{Q}(y,\beta) = \frac{5}{y^5} \int_0^y \frac{(x^2+2x)^2}{1+\beta(x^2+2x)} dx, \quad (8.161)$$

where, as in Eq. (8.148), the notation $y = \tau_{\rm rec}/\tau_1$ has been used. The indefinite integral appearing in Eq. (8.161) can be solved with elementary methods and the result is

$$\frac{(2+x)[\beta(x^2+x-2)-3]}{3\beta^2} + \frac{\arctan\left[\frac{\sqrt{\beta}(1+x)}{\sqrt{1-\beta}} \right]}{\beta^{5/2}\sqrt{1-\beta}}. \quad (8.162)$$

---
[1] We recall that 1 Mpc = 3.086×10^{24} cm.

The quantity $\mathcal{Q}(y,\beta)$ can then be precisely computed. It turns out that this quantity is rather sensitive to the estimate of y. For instance, if we take $y = \tau_{\text{rec}}/\tau_1$ we get $\mathcal{Q} \simeq 1.5$ which is pretty good[m] since the approximation of setting $\beta = 0$ would give $\mathcal{Q} = 1$. If, however, we estimate y according to the exact expression of Eq. (8.159) we get $\mathcal{Q} \simeq 5.11$. This shows that the presented estimate of the diffusion damping is larger by a factor $\sqrt{5.11} \simeq 2.26$ than what is reported in Eq. (8.160). This boils down, ultimately, to the occurrence that 9.63×10^{-3} must be replaced by 2.17×10^{-2}.

So we can conclude that the approximation of setting $\beta = 0$ underestimates (roughly by a factor of 2) the estimate of diffusion damping based on shear viscosity. This result could have been expected not so much for numerical reasons but simply because, in the present treatment, diffusion damping is just added and not computed. It will be shown in chapter 9 that diffusion damping arises naturally when going to second order in the tight-coupling expansion. To get to that point, however, it will be crucial to solve also the first-order and to consistently include the polarization. It will be actually shown that polarization affects, in a rather subtle way, the determination of the diffusion damping scale.

---

[m] For this estimate we have chosen $\beta = 0.66$ which is the value implied by Eqs. (8.143) and (8.146).

Chapter 9

Kinetic Hierarchies

The effect of metric inhomogeneities on the properties of the radiation field will now be analyzed using the radiative transfer (or radiative transport) equations. A classical preliminary reference is the textbook of Chandrasekar [274] (see in particular chapter 1 in light of the calculation of the collision term of Thompson scattering that is quite relevant here for us). Another recent reference is [275]. In broad terms the radiative transfer equations describe the evolution of the Stokes parameters of the radiation field through some layer of matter which could be, for instance, the stellar atmosphere or, in the present case, the primeval plasma around decoupling.

Radiative transfer equations have a further complication with respect to the flat space case: the collisionless part of the Boltzmann equation is modified by the inhomogeneities of the geometry. These inhomogeneities induce a direct coupling of the Boltzmann equation to the perturbed Einstein equations. An interesting system of equations naturally emerges: the Einstein-Boltzmann system of equations which becomes, when appropriately truncated, the multi-fluid treatment developed in chapters 7 and 8. In the multi-fluid approach the Einstein equations are coupled to a set of fluid equations for the density contrasts and for the peculiar velocities. These are, indeed, the first two terms (i.e. the monopole and the dipole) in the Boltzmann hierarchy. Truncated Boltzmann hierarchies are a useful tool for the analysis of initial conditions, but their limitations have been already emphasized in connection with the description of collisionless particles (see chapter 8).

While the general conventions established in the previous chapters will be consistently enforced, further conventions (related to the specific definition of the brightness perturbations) will emerge.[a] Denoting by Δ a

---
[a]See Eqs. (9.59)–(9.62) below for the definition of brightness perturbation.

brightness perturbation (related generically to one of the four Stokes parameters of the radiation field), the expansion of Δ in terms of Legendre polynomials will be written, in this paper, as

$$\Delta(\vec{k},\hat{n},\tau) = \sum_{\ell}(-i)^{\ell}\,(2\ell+1)\,\Delta_{\ell}(\vec{k},\tau)\,P_{\ell}(\hat{k}\cdot\hat{n}). \qquad (9.1)$$

Concerning Eq. (9.1) the following specifications are in order:

- $\vec{k}$ is the momentum of the Fourier expansion, $\hat{k}$ its direction;
- $\hat{n}$ is the direction of the photon momentum;
- $P_{\ell}(\hat{k}\cdot\hat{n})$ are the Legendre polynomials [215, 216].

The same expansion will be consistently employed for the momentum averaged phase-space density perturbation (see below Eq. (9.20)). This quantity will be also called, for short, reduced phase-space density and it is related to the brightness perturbation by a numerical factor. In the literature there exist different (and sometimes mutually confusing) conventions:

- the conventions of [276, 277, 279] (see also [280]) are such that the factor $(2\ell+1)$ *is not* included in the expansion;
- in [276–278] the metric fluctuations are parametrized in terms of the Bardeen potential while in [279] the treatment follows the conformally Newtonian gauge;
- in [197] the conventions are the same as the ones of Eq. (9.1) but the metric convention is mostly plus (i.e. $-,+,+,+$) and the definition of the longitudinal degrees of freedom is inverted (i.e. Ref. [197] calls ψ what we call ϕ and viceversa);
- in [281–283] (see also [284, 285]) the expansion of the brightness perturbation is different with respect to Eq. (9.1) since the authors *do not* include the factor $(-i)^{\ell}$ in the expansion; in the latter case the collision terms are modified by a sign difference in the dipole terms (involving a mismatch of $(-i)^2$ with respect to the conventions fixed by Eq. (9.1)).

9.1 Collisionless Boltzmann equation

If the space-time would be homogeneous the position variables x^i and the conjugate momenta P_j could constitute a practical set of pivot variables for the analysis of Boltzmann equation in curved backgrounds. However, since, in the present case, the space-time is not fully homogeneous, metric perturbations do affect the definition of conjugate momenta. Hence, for practical reasons, the approach usually followed is to write the Boltzmann equations in terms of the proper momenta, i.e. the momentum measured by an observer at a fixed value of the spatial coordinate. Consider, for simplicity, the case of massless particles (like photons or massless neutrinos). Their mass-shell condition can be written, in a curved background, as

$$g_{\alpha\beta}P^\alpha P^\beta = 0, \qquad (9.2)$$

where $g_{\alpha\beta}$ is now the full metric tensor (i.e. background plus inhomogeneities). Equation (9.2) implies, with simple algebra, that:

$$g_{00}P^0 P^0 = -g_{ij}P^i P^j \equiv \delta_{ij}p^i p^j, \qquad (9.3)$$

where the second equality is the definition of the physical three momentum p_i. Recalling that, to first order and in the longitudinal gauge,

$$g_{00} = a^2(1+2\phi), \qquad g_{ij} = -a^2(1-2\psi)\delta_{ij}, \qquad (9.4)$$

then the relation between the conjugate momenta and the physical three-momenta can be easily obtained by expanding the obtained expressions for small ϕ and ψ. The result is simply:

$$\begin{aligned} P^0 &= \frac{p}{a}(1-\phi) = \frac{q}{a^2}(1-\phi), \\ P_0 &= ap(1+\phi) = q(1+\phi), \\ P^i &= \frac{p^i}{a}(1+\psi) = \frac{q^i}{a^2}(1+\psi), \\ P_i &= -ap_i(1-\psi) = -q_i(1-\psi). \end{aligned} \qquad (9.5)$$

The vector q_i defined in Eq. (9.5) is nothing but the comoving three-momentum, i.e. $p_i a = q_i$, while $q = pa$ is the modulus of the comoving three-momentum. The generalization of Eq. (9.5) is trivial since, in the massive case, the mass-shell condition implies that $g_{\alpha\beta}P^\alpha P^\beta = m^2$ and, for instance $P^0 = \sqrt{q^2 + m^2 a^2}(1-\phi)$. In terms of the modulus and direction of the comoving three-momentum [286], i.e.

$$q_i = qn_i, \qquad n_i n^i = n_i n_j \delta^{ij} = 1, \qquad (9.6)$$

the Boltzmann equation can be written as

$$\frac{Df}{D\tau} = \frac{\partial f}{\partial \tau} + \frac{\partial f}{\partial x^i}\frac{dx^i}{d\tau} + \frac{\partial f}{\partial q}\frac{dq}{d\tau} + \frac{\partial f}{\partial n_i}\frac{dn^i}{d\tau} = \mathcal{C}_{\text{coll}}, \qquad (9.7)$$

where a generic collision term, $\mathcal{C}_{\text{coll}}$ has been included for future convenience. Eq. (9.7) can now be perturbed around a configuration of local thermodynamic (or kinetic) equilibrium[b] by writing

$$f(x^i, q, n_j, \tau) = f_0(q)[1 + f^{(1)}(x^i, q, n_j, \tau)], \qquad (9.8)$$

where $f_0(q)$ is the Bose-Einstein (or Fermi-Dirac in the case of fermionic degrees of freedom) distribution. Notice that $f_0(q)$ does not depend on n^i but only on q. Inserting Eq. (9.8) into Eq. (9.7) the first-order form of the perturbed Boltzmann equation can be readily obtained

$$f_0(q) \frac{\partial f^{(1)}}{\partial \tau} + f_0(q) \frac{\partial f^{(1)}}{\partial x^i} n^i + \frac{\partial f_0}{\partial q} \frac{dq}{d\tau} = \mathcal{C}_{\text{coll}}, \qquad (9.9)$$

by appreciating that a pair of terms

$$\frac{\partial f^{(1)}}{\partial q} \frac{dq}{d\tau}, \qquad \frac{\partial f^{(1)}}{\partial n_i} \frac{dn^i}{d\tau}, \qquad (9.10)$$

are of higher order (i.e. $\mathcal{O}(\psi^2)$) and have been neglected to first-order. Dividing by f_0, Eq. (9.9) can also be written as

$$\frac{\partial f^{(1)}}{\partial \tau} + \frac{\partial f^{(1)}}{\partial x^i} n^i + \frac{\partial \ln f_0}{\partial q} \frac{dq}{d\tau} = \frac{1}{f_0} \mathcal{C}_{\text{coll}}. \qquad (9.11)$$

Notice that in Eq. (9.9)–(9.11) the generalization of known special relativistic expressions

$$\frac{dx^i}{d\tau} = \frac{P^i}{P^0} = \frac{q^i}{q} = n^i, \qquad (9.12)$$

has been used. To complete the derivation, $dq/d\tau$ must be written in explicit terms. The geodesic equation gives essentially the first time derivative of the conjugate momentum, i.e.

$$\frac{dP^\mu}{d\lambda} = P^0 \frac{dP^\mu}{d\tau} = -\Gamma^\mu_{\alpha\beta} P^\alpha P^\beta, \qquad (9.13)$$

where λ is, as usual, the affine parameter; $\Gamma^\mu_{\alpha\beta}$ denotes the full Christoffel connection (background plus fulctuations). Using the values of the perturbed connections in the longitudinal gauge Eq. (9.13) becomes:

$$\frac{dP^i}{d\tau} = -\partial^i \phi P^0 + 2\psi' P^i - 2\mathcal{H} P^i - \frac{P^j P^k}{P^0} [\partial^i \psi \delta_{jk} - \partial_k \psi \delta^i_j - \partial_j \psi \delta^i_k]. \qquad (9.14)$$

Recalling now that $q = q_i n^i$, the explicit form of $dq/d\tau$ can be obtained from Eq. (9.14) and by taking into account the results of Eq. (9.5). The result is:

$$\frac{dq}{d\tau} = \left[\frac{\partial P^i}{\partial \tau} a^2 (1-\psi) + 2\mathcal{H} a^2 (1-\psi) P^i - a^2 \psi' P^i \right] n_i - P^i a^2 \partial_j \psi n^j n_i. \qquad (9.15)$$

---

[b]See Appendix B for a definition of thermodynamic, kinetic and chemical equulibria.

Inserting now Eq. (9.14) into Eq. (9.15) the explicit form of $dq/d\tau$ becomes[c]

$$\frac{dq}{d\tau} = q\psi' - qn_i\partial^i\phi. \qquad (9.16)$$

Finally, using Eq. (9.16) into Eq. (9.11) to eliminate $dq/d\tau$ the final form of the Boltzmann equation for massless particles becomes:

$$\frac{\partial f^{(1)}}{\partial \tau} + n^i \frac{\partial f^{(1)}}{\partial x^i} + \frac{\partial \ln f_0}{\partial \ln q}[\psi' - n_i\partial^i\phi] = \frac{1}{f_0}\mathcal{C}_{\text{coll}}, \qquad (9.17)$$

which can be also written, going to Fourier space, as

$$\frac{\partial f^{(1)}}{\partial \tau} + ik\mu f^{(1)} + \frac{\partial \ln f_0}{\partial \ln q}[\psi' - ik\mu\phi] = \frac{1}{f_0}\mathcal{C}_{\text{coll}}, \qquad (9.18)$$

where we have denoted, according to the standard notation, k as the Fourier mode and $\mu = \hat{k} \cdot \hat{n}$ as the projection of the Fourier mode along the direction of the photon momentum.[d] Clearly, given the axial symmetry of the problem it will be natural to identify the direction of $\vec{k}$ with the $\hat{z}$ direction in which case $\mu = \cos\theta$. The result obtained so far can be easily generalized to the case of massive particles

$$\frac{\partial f^{(1)}}{\partial \tau} + i\alpha(q,m)k\mu f^{(1)} + \frac{\partial \ln f_0}{\partial \ln q}[\psi' - i\alpha(q,m)k\mu\phi] = \frac{1}{f_0}\mathcal{C}_{\text{coll}}, \qquad (9.19)$$

where $\alpha(q,m) = q/\sqrt{q^2 + m^2a^2}$ and where, the appropriate mass dependence now has to appear in the equilibrium distribution $f_0(q)$.

9.2 Boltzmann hierarchy for massless neutrinos

The Boltzmann equations derived in Eqs. (9.18) and (9.19) are general. In the following, two relevant cases will be discussed, namely, the case of massless neutrinos and the case of photons. In order to proceed further with the case of massless neutrinos let us define the reduced phase-space distribution as

$$\mathcal{F}_\nu(\vec{k}, \hat{n}, \tau) = \frac{\int q^3 dq f_0 f^{(1)}}{\int q^3 dq f_0}. \qquad (9.20)$$

---

[c]To derive Eq. (9.16) from Eq. (9.15), the factors P^i and P^0 appearing at the right hand side Eq. (9.14) have to be replaced with their first-order expression in terms of the comoving-three momentum q^i (and q) as previously discussed in Eqs. (9.5).

[d]Notice that here there may be, in principle, a clash of notations since, in chapter 6 we denoted with μ the normal modes for the tensor action; in the present section q and μ denote, on the contrary the comoving three-momentum and the cosine between the Fourier mode and the photon direction. The two sets of variables never appear together and there should not be confusion.

Equation (9.18) becomes, in the absence of collision term,
$$\frac{\partial \mathcal{F}_\nu}{\partial \tau} + ik\mu\mathcal{F}_\nu = 4(\psi' - ik\mu\phi). \tag{9.21}$$

The factor 4 arising in Eq. (9.21) follows from the explicit expression of the equilibrium Fermi-Dirac distribution and observing that integration by parts implies
$$\int_0^\infty q^3 dq \frac{\partial f_0}{\partial \ln q} = -4 \int_0^\infty q^3 dq f_0. \tag{9.22}$$

The reduced phase-space distribution of Eq. (9.20) can be expanded in series of Legendre polynomials as defined in Eq. (9.1)
$$\mathcal{F}_\nu(\vec{k}, \hat{n}, \tau) = \sum_\ell (-i)^\ell (2\ell+1) \mathcal{F}_{\nu\ell}(\vec{k}, \tau) P_\ell(\mu). \tag{9.23}$$

Equation (9.23) will now be inserted into Eq. (9.21). The orthonormality relation for Legendre polynomials [215, 216],
$$\int_{-1}^1 P_\ell(\mu) P_{\ell'}(\mu) d\mu = \frac{2}{2\ell+1} \delta_{\ell\ell'}, \tag{9.24}$$

together with the well-known recurrence relation
$$(\ell+1)P_{\ell+1}(\mu) = (2\ell+1)\mu P_\ell(\mu) - \ell P_{\ell-1}(\mu), \tag{9.25}$$

allows us to get a hierarchy of differential equations coupling together the various multipoles. After having multiplied each of the terms of Eq. (9.21) by μ, integration of the obtained quantity will be performed over μ (varying between -1 and 1); in formulas:
$$\int_{-1}^1 P_{\ell'}(\mu) \mathcal{F}_\nu d\mu = 2(-i)^{\ell'} \mathcal{F}_{\nu\ell'}, \tag{9.26}$$

$$ik \int_{-1}^1 \mu P_{\ell'}(\mu) \mathcal{F}_\nu d\mu = 2ik\left[(-i)^{\ell'+1}\frac{\ell'+1}{2\ell'+1}\mathcal{F}_{\nu(\ell'+1)}\right.$$
$$\left. + (-i)^{\ell'-1}\frac{\ell'}{2\ell'+1}\mathcal{F}_{\nu(\ell'-1)}\right], \tag{9.27}$$

$$4\int_{-1}^1 \psi' P_{\ell'}(\mu) d\mu = 8\psi' \delta_{\ell'0}, \tag{9.28}$$

$$-4i\phi \int_{-1}^1 \mu P_{\ell'}(\mu) d\mu = -\frac{8}{3} ik\phi \delta_{\ell'1}. \tag{9.29}$$

Equation (9.27) follows from the relation
$$\int_{-1}^1 \mu P_\ell(\mu) P_{\ell'}(\mu) d\mu = \frac{2}{2\ell+1}\left[\frac{\ell'+1}{2\ell'+1}\delta_{\ell,\ell'+1} + \frac{\ell'}{2\ell'+1}\delta_{\ell,\ell'-1}\right], \tag{9.30}$$

that can be easily derived using Eqs. (9.25) and (9.24). Inserting Eqs. (9.26)–(9.29) into Eq. (9.21) the first example of Boltzmann hierarchy can be recovered:

$$\mathcal{F}'_{\nu 0} = -k\mathcal{F}_{\nu 1} + 4\psi', \tag{9.31}$$

$$\mathcal{F}'_{\nu 1} = \frac{k}{3}[\mathcal{F}_{\nu 0} - 2\mathcal{F}_{\nu 2}] + \frac{4}{3}k\phi, \tag{9.32}$$

$$\mathcal{F}'_{\nu \ell} = \frac{k}{2\ell+1}[\ell\mathcal{F}_{\nu,(\ell-1)} - (\ell+1)\mathcal{F}_{\nu(\ell+1)}]. \tag{9.33}$$

Equation (9.33) holds for $\ell \geq 2$. Eqs. (9.31) and (9.32) are nothing but the evolution equations for the density contrast and for the neutrino velocity field. This aspect can be easily appreciated by computing, in explicit terms, the components of the energy-momentum tensor as a function of the reduced neutrino phase-space density. In general terms, the energy-momentum tensor can be written, in the kinetic approach, as

$$T^\nu_\mu = -\int \frac{d^3P}{\sqrt{-g}} \frac{P_\mu P^\nu}{P^0} f(x^i, P_j, \tau). \tag{9.34}$$

Let us now verify that Eq. (9.34) indeed reproduces correctly the definitions of the various multipoles as we know them. Notice, first of all, that according to Eq. (9.5),

$$\frac{d^3P}{\sqrt{-g}} = \frac{dP_1 dP_2 dP_3}{\sqrt{-g}} = -\frac{d^3q(1-\psi)^3}{a^4(1+2\phi)^{1/2}(1-2\psi)^{3/2}}. \tag{9.35}$$

Consider, therefore, the (00) component of Eq. (9.34), i.e.

$$T^0_0 = \frac{1}{a^4}\int d^3q(1-\phi^2)(1-9\psi^2)[f_0(q) + f^{(1)}(x_i,q,n_j,\tau)] \tag{9.36}$$

where we used that $P_0 = q(1+\phi)$. Therefore, to lowest order and for the neutrinos we will have:

$$\rho_\nu = \frac{1}{a^4}\int d^3q\, q f_0(q), \tag{9.37}$$

i.e. the homogeneous energy density. Using the first-order phase space density, the density contrast, the peculiar velocity field and the fact that the neutrino anisotropic stress are connected, respectively, to the monopole, dipole and quadrupole moments of the (reduced) phase-space distribution:

$$\delta_\nu = \frac{1}{4\pi}\int d\Omega \mathcal{F}_\nu(\vec{k},\hat{n},\tau) = \mathcal{F}_{\nu 0}, \tag{9.38}$$

$$\theta_\nu = \frac{3i}{16\pi}\int d\Omega(\vec{k}\cdot\hat{n})\mathcal{F}_\nu(\vec{k},\hat{n},\tau) = \frac{3}{4}k\mathcal{F}_{\nu 1}, \tag{9.39}$$

$$\sigma_\nu = -\frac{3}{16\pi}\int d\Omega\left[(\vec{k}\cdot\hat{n})^2 - \frac{1}{3}\right]\mathcal{F}_\nu(\vec{k},\hat{n},\tau) = \frac{\mathcal{F}_{\nu 2}}{2}. \tag{9.40}$$

Inserting Eqs. (9.38) and (9.40) into Eqs. (9.31)–(9.33), the system following from the perturbation of the covariant conservation equations can be easily recovered

$$\delta'_\nu = -\frac{4}{3}\theta_\nu + 4\psi', \tag{9.41}$$

$$\theta'_\nu = \frac{k^2}{4}\delta_\nu - k^2\sigma_\nu + k^2\phi, \tag{9.42}$$

$$\sigma'_\nu = \frac{4}{15}\theta_\nu - \frac{3}{10}k\mathcal{F}_{\nu 3}, \tag{9.43}$$

For the adiabatic mode, after neutrino decoupling, $\mathcal{F}_{\nu 3} = 0$. The problem of dealing with neutrinos while setting initial conditions for the evolution of the CMB anisotropies can be now fully understood. The fluid approximation implies that the dynamics of neutrinos can be initially described, after neutrino decoupling, by the evolution of the monopole and dipole of the neutrino phase space distribution. However, in order to have an accurate description of the initial conditions one should solve an infinite hierarchy of equations for the time derivatives of higher order moments of the neutrino distribution function. Similar remarks will hold, with due differences, for the photons.

Equations (9.31)–(9.33) hold for massless neutrinos but a similar hierarchy can be derived also in the case of the photons or, more classically, in the case of the brightness perturbations of the radiation field to be discussed below. The spatial gradients of the longitudinal fluctuations of the metric are sources of the equations for the lowest multipoles, i.e. Eqs. (9.31) and (9.32). For $\ell > 2$, each multipole is coupled to the preceding (i.e. $(\ell - 1)$) and to the following (i.e. $(\ell + 1)$) multipoles. To solve numerically the hierarchy one could truncate the system at a certain $\ell_{\max}$. This is, however, not the best way of dealing with the problem since [197] the effect of the truncation could be an unphysical reflection of power down through the lower (i.e. $\ell < \ell_{\max}$) multipole moments. This problem can be efficiently addressed with the method of line-of-sight integration (to be discussed later in this section) that is also rather effective in the derivation of approximate expressions, for instance, of the polarization power spectrum. The method of line-of-sight integration is the one used, for instance, in CMBFAST [289, 290].

9.3 Brightness perturbations of the radiation field

Unlike neutrinos, photons are a collisional species, so the generic collision term appearing in Eq. (9.19) has to be introduced. With this warning in mind, all the results derived so far can be simply translated to the case of photons (collisionless part of Boltzmann equation, relations between the moments of the reduced phase-space and the components of the energy-momentum tensor...) provided the Fermi-Dirac equilibrium distribution is replaced by the Bose-Einstein distribution.

Thompson scattering leads to a collision term that depends both on the baryon velocity field[e] and on the direction cosine μ [274]. The collision term is different for the brightness function describing the fluctuations of the total intensity of the radiation field (related to the Stokes parameter I) and for the brightness functions describing the degree of polarization of the scattered radiation (related to the Stokes parameters U and V).

The conventions for the Stokes parameters and their well known properties will now be summarized: they can be found in standard electrodynamics textbooks [287] (see also [288, 292, 293] for phenomenological introduction to the problem of CMB polarization and [284] for a more theoretical perspective). Consider, for simplicity, a monochromatic radiation field decomposed according to its linear polarizations and travelling along the z axis:

$$\vec{E} = [E_1 \hat{e}_x + E_2 \hat{e}_y] e^{i(kz-\omega t)}. \tag{9.44}$$

The decomposition according to circular polarizations can be written as:

$$\vec{E} = [\hat{\epsilon}_+ E_+ + \hat{\epsilon}_- E_-] e^{i(kz-\omega t)}, \tag{9.45}$$

where

$$\hat{\epsilon}_+ = \frac{1}{\sqrt{2}} (\hat{e}_x + i\hat{e}_y), \tag{9.46}$$

$$\hat{\epsilon}_- = \frac{1}{\sqrt{2}} (\hat{e}_x - i\hat{e}_y). \tag{9.47}$$

Eq. (9.46) is conventionally defined to be a *positive* helicity, while Eq. (9.47) is the *negative* helicity. Recalling that E_1 and E_2 can be written as

$$E_1 = E_x e^{i\delta_x}, \qquad E_2 = E_y e^{i\delta_y}, \tag{9.48}$$

---

[e]Since the electron-ion collisions are sufficiently rapid, it is normally assumed, in analytical estimates of CMB effects, that electrons and ions are in kinetic equilibrium at a common temperature $T_{\rm eb}$.

the polarization properties of the radiation field can be described in terms of 4 real numbers given by the projections of the radiation field over the linear and circular polarization unit vectors, i.e.

$$(\hat{e}_x \cdot \vec{E}), \quad (\hat{e}_y \cdot \vec{E}), \quad (\hat{\epsilon}_+ \cdot \vec{E}), \quad (\hat{\epsilon}_- \cdot \vec{E}). \tag{9.49}$$

The four Stokes parameters then become, in the linear polarization basis

$$I = |\hat{e}_x \cdot \vec{E}|^2 + |\hat{e}_y \cdot \vec{E}|^2 = E_x^2 + E_y^2, \tag{9.50}$$

$$Q = |\hat{e}_x \cdot \vec{E}|^2 - |\hat{e}_y \cdot \vec{E}|^2 = E_x^2 - E_y^2, \tag{9.51}$$

$$U = 2\operatorname{Re}[(\hat{e}_x \cdot \vec{E})^*(\hat{e}_y \cdot \vec{E})] = 2E_x E_y \cos(\delta_y - \delta_x), \tag{9.52}$$

$$V = 2\operatorname{Im}[(\hat{e}_x \cdot \vec{E})^*(\hat{e}_y \cdot \vec{E})] = 2E_x E_y \sin(\delta_y - \delta_x). \tag{9.53}$$

Stokes parameters are not all invariant under rotations. Consider a two-dimensional (clock-wise) rotation of the coordinate system, namely

$$\hat{e}'_x = \cos\varphi \hat{e}_x + \sin\varphi \hat{e}_y,$$
$$\hat{e}'_y = -\sin\varphi \hat{e}_x + \cos\varphi \hat{e}_y. \tag{9.54}$$

Inserting Eq. (9.54) into Eqs. (9.50)–(9.53) it can be easily shown that $I' = I$ and $V' = V$ where the prime denotes the expression of the Stokes parameter in the rotated coordinate system. However, the remaining two parameters mix, i.e.

$$Q' = \cos 2\varphi Q + \sin 2\varphi U,$$
$$U' = -\sin 2\varphi Q + \cos 2\varphi U. \tag{9.55}$$

From the last expression it can be easily shown that the polarization degree P is invariant

$$P = \sqrt{Q^2 + U^2} = \sqrt{Q'^2 + U'^2}, \tag{9.56}$$

while $U/Q = \tan 2\alpha$ transform as $U'/Q' = \tan 2(\alpha - \varphi)$.

Stokes parameters are not independent (i.e. it holds that $I^2 = Q^2 + U^2 + V^2$), they only depend on the difference of the phases (i.e. $(\delta_x - \delta_y)$) but not on their sum (see Eqs. (9.50)–(9.53)). Hence the polarization tensor of the electromagnetic field can be written in matrix notation as

$$\rho = \begin{pmatrix} I+Q & U-iV \\ U+iV & I-Q \end{pmatrix} \equiv \begin{pmatrix} E_x^2 & E_x E_y e^{-i\Delta} \\ E_x E_y e^{i\Delta} & E_y^2 \end{pmatrix}, \tag{9.57}$$

where $\Delta = (\delta_y - \delta_x)$. If the radiation field would be treated in a second quantization approach, Eq. (9.57) can be promoted to the status of density matrix of the radiation field [284].

9.4 Evolution equations for the brightness perturbations

The evolution equations for the brightness functions will now be derived. Consider Eq. (9.18) written again, this time, in the case of photons. As in the case of neutrinos we can define a reduced phase space distribution $\mathcal{F}_\gamma$, just changing ν with γ in Eq. (9.20) and using the Bose-Einstein instead of the Fermi-Dirac equlibrium distribution. The (reduced) photon phase-space density describes the fluctuations of the intensity of the radiation field (related to the Stokes parameter I); a second reduced phase-space distribution, be it $\mathcal{G}_\gamma$, can be defined for the difference of the two intensities (related to the stokes parameter Q). The equations for $\mathcal{F}_\gamma$ and $\mathcal{G}_\gamma$ can be written as

$$\frac{\partial \mathcal{F}_\gamma}{\partial \tau} + ik\mu \mathcal{F}_\gamma - 4(\psi' - ik\mu\phi) = \mathcal{C}_I,$$

$$\frac{\partial \mathcal{G}_\gamma}{\partial \tau} + ik\mu \mathcal{G}_\gamma = \mathcal{C}_Q. \tag{9.58}$$

The collision terms for these two equations are different [295, 296] and can be obtained following the derivation reported in the chapter 1 of Ref. [274] or by following the derivation of Bond (with different notations) in the Appendix C of Ref. [46] (see from p. 638). Another way of deriving the collision terms for the evolution equations of the brightness perturbations is by employing the total angular momentum method [294] that will be quickly discussed in connection with CMB polarization.

Before writing the explicit form of the equations, including the collision terms, it is useful to pass directly to the brightness perturbations. For the fluctuations of the total intensity of the radiation field the brightness perturbations is simply given by

$$f(x^i, q, n_j, \tau) = f_0\left(\frac{q}{1 + \Delta_\mathrm{I}}\right). \tag{9.59}$$

Recalling now that, by definition,

$$f_0\left(\frac{q}{1 + \Delta_\mathrm{I}}\right) = f_0(q) + \frac{\partial f_0}{\partial q}[q(1 - \Delta_\mathrm{I}) - q], \tag{9.60}$$

the perturbed phase-space distribution and the brightness perturbation must satisfy:

$$f_0(q)[1 + f^{(1)}(x^i, q, n_j, \tau)] = f_0(q)\left[1 - \Delta_\mathrm{I}(x^i, q, n_j, \tau)\frac{\partial \ln f_0}{\partial \ln q}\right], \tag{9.61}$$

that also implies

$$\Delta_\mathrm{I} = -f^{(1)}\left(\frac{\partial \ln f_0}{\partial \ln q}\right)^{-1}, \quad \mathcal{F}_\gamma = -\Delta_\mathrm{I}\frac{\int q^3 dq f_0 \frac{\partial f_0}{\partial \ln q}}{\int q^3 dq f_0} = 4\Delta_\mathrm{I}, \tag{9.62}$$

where the second equality follows from integration by parts as in Eq. (9.22).
The Boltzmann equations for the perturbation of the brightness are then

$$\Delta_I' + ik\mu(\Delta_I + \phi) = \psi' + \epsilon'\left[-\Delta_I + \Delta_{I0} + \mu v_b - \frac{1}{2}P_2(\mu)S_Q\right], \quad (9.63)$$

$$\Delta_Q' + ik\mu\Delta_Q = \epsilon'\left\{-\Delta_Q + \frac{1}{2}[1 - P_2(\mu)]S_Q\right\}, \quad (9.64)$$

$$\Delta_U' + ik\mu\Delta_U = -\epsilon'\Delta_U, \quad (9.65)$$

$$\Delta_V' + ik\mu\Delta_V = -\epsilon'\left[\Delta_V + \frac{3}{2}\,i\mu\,\Delta_{V1}\right], \quad (9.66)$$

where we defined, for notational convenience and for homogeneity with the notations of other authors [281]

$$v_b = \frac{\theta_b}{ik} \quad (9.67)$$

and

$$S_Q = \Delta_{I2} + \Delta_{Q0} + \Delta_{Q2}. \quad (9.68)$$

In Eqs. (9.64)–(9.65), $P_2(\mu) = (3\mu^2 - 1)/2$ is the Legendre polynomial of second order, which appears in the collision operator of the Boltzmann equation for the photons due to the directional nature of Thompson scattering. Eq. (9.66) is somehow decoupled from the system. So if, initially, $\Delta_V = 0$ it will also vanish at later times. In Eqs. (9.63)–(9.65) the function ϵ' denotes the differential optical depth for Thompson scattering[f]

$$\epsilon' = x_e n_e \sigma_{Th}\frac{a}{a_0} = \frac{x_e n_e \sigma_{Th}}{z+1}, \quad (9.69)$$

having denoted with x_e the ionization fraction and $z = a_0/a - 1$ the redshift. Defining with τ_0 the time at which the signal is received, the optical depth will then be

$$\epsilon(\tau, \tau_0) = \int_\tau^{\tau_0} x_e n_e \sigma_{Th}\frac{a(\tau)}{a_0}d\tau. \quad (9.70)$$

There are two important limiting cases:

---

[f]Notice that, in comparison with Eq. (2.111), the ionization fraction has been taken out from the definition of electron density. This notation is often used in this context even if the notation used in chapter 2 can be also employed. Notice also that, conventionally, the differential optical depth is denoted by τ'. This notation would be highly ambiguous in the present case since τ denotes, according to the notations established in this book, the conformal time coordinate. This is the reason why the differential optical depth will be denoted by ϵ'.

- in the optically thin limit $\epsilon \ll 1$, the absorption along the ray path is negligible so that the emergent radiation is simply the sum of the contributions along the ray path;
- in the opposite case $\epsilon \gg 1$ the plasma is said to be optically thick.

To close the system the evolution of the baryon velocity field can be rewritten as

$$v_b' + \mathcal{H}v_b + ik\phi + \frac{\epsilon'}{R_b}\left(3i\Delta_{I1} + v_b\right) = 0, \tag{9.71}$$

where $R_b(z)$ has been already defined in Eq. (8.67). At the decoupling epoch occurring for $z_{\rm dec} \simeq 1100$, $R_b(z_{\rm dec}) \sim 7/11$ for a typical baryonic content of $h_0^2 \Omega_{b0} \sim 0.023$. Notice that the photon velocity field has been eliminated, in Eq. (9.71) with the corresponding expression involving the monopole of the brightness perturbation.

As pointed out in Eq. (9.56), while Q and U change under rotations, the degree of linear polarization is invariant. Thus, it is sometimes useful to combine Eqs. (9.63) and (9.64). The result of this combination is

$$\Delta_P' + (ik\mu + \epsilon')\Delta_P = \frac{3}{4}\epsilon'(1-\mu^2)S_P,$$
$$S_P = \Delta_{I2} + \Delta_{P0} + \Delta_{P2}. \tag{9.72}$$

With the same notations Eq. (9.63) can be written as

$$\Delta_I' + (ik\mu + \epsilon')\Delta_I = \psi' - ik\mu\phi + \epsilon'[\Delta_{I0} + \mu v_b - \frac{1}{2}P_2(\mu)S_P]. \tag{9.73}$$

By adding a ϕ' and $\epsilon'\phi$ both at the left and right hand sides of Eq. (9.73), the equation for the temperature fluctuations can also be written as:

$$(\Delta_I + \phi)' + (ik\mu + \epsilon')(\Delta_I + \phi) = (\psi' + \phi')$$
$$+\epsilon'[(\Delta_{I0} + \phi) + \mu v_b - \frac{1}{2}P_2(\mu)S_P]. \tag{9.74}$$

This form of the equation is relevant in order to find formal solutions of the evolution of the brightness perturbations (see below the discussion of the line of sight integrals).

9.4.1 Visibility function

An important function appearing ubiquitously in various subsequent expressions is the so-called *visibility function*, $\mathcal{K}(\tau)$, giving the probability that a CMB photon was last scattered between τ and $\tau + d\tau$; the definition of $\mathcal{K}(\tau)$ is

$$\mathcal{K}(\tau) = \epsilon' e^{-\epsilon(\tau,\tau_0)}, \tag{9.75}$$

usually denoted by $g(\tau)$ in the literature. The function $\mathcal{K}(\tau)$ is a rather important quantity since it is sensitive to the whole ionization history of the Universe. The visibility function is strongly peaked around the decoupling time $\tau_{\rm rec}$ and can be approximated, for analytical purposes, by a Gaussian with appropriate variance [297]. A relevant limit is the so-called sudden decoupling limit where the visibility function can be approximated by a Dirac delta function and its integral, i.e. the optical depth, can be approximated by a step function; in formulae:

$$\mathcal{K}(\tau) \simeq \delta(\tau - \tau_{\rm rec}), \qquad e^{-\epsilon(\tau,\tau_0)} \simeq \theta(\tau - \tau_{\rm rec}). \qquad (9.76)$$

This approximation will be used below for different applications and it is justified since the free electron density diminishes suddenly at decoupling. In spite of this occurrence there are compelling indications that, at some epoch after decoupling, the Universe was reionized.

Instead of using a visibility function defined in the sudden approximation, as in Eq. (9.76), it is rather useful, as already mentioned, to model the visibility function at recombination with a Gaussian. Within this parametrization, for semianalytical purposes we have[g]

$$\mathcal{K}(\tau_{\rm rec}) = e^{-\sigma_{\rm rec}^2 \tau_{\rm rec}^2 k^2},$$
$$\sigma_{\rm rec} = \frac{1}{\alpha_1 \mathcal{H}_{\rm rec} \tau_{\rm rec}}, \qquad (9.77)$$

where $\alpha_1 \simeq 33.55$. The value of α_1 can increase up to 36 and it is just a phenomenological parameter. In Eq. (9.77) the quantity $\mathcal{H}\tau$ can be made more explicit by using Eq. (2.93). Indeed, using Eq. (2.93) it is immediate to show that

$$\mathcal{H}\tau = \frac{2(x+1)}{x+2}, \qquad x = \frac{\tau}{\tau_1}. \qquad (9.78)$$

Using the smooth interpolation of Eq. (2.93) for the evolution of the background it is also immediate to show that

$$\frac{\tau_{\rm rec}}{\tau_1} = \sqrt{\frac{z_{\rm eq}+1}{z_{\rm rec}+1}+1}-1. \qquad (9.79)$$

Recall that, indeed, $a = a_{\rm eq}(x^2 + 2x)$ and that, by definition of redshift, $1 + z = a_0/a$. Consequently,

$$\mathcal{H}_{\rm rec}\tau_{\rm rec} = 2\frac{\sqrt{\frac{z_{\rm eq}+1}{z_{\rm rec}+1}+1}}{1+\sqrt{\frac{z_{\rm eq}+1}{z_{\rm rec}+1}+1}} \qquad (9.80)$$

---

[g]The discussion of the specific features of the process of recombination is beyond the scope of this book and can be found, for instance, in other dedicated publications such as [298].

and

$$\frac{1}{\mathcal{H}_{rec}\tau_{rec}} = \frac{1}{2}\left(1 + \frac{1}{\sqrt{1+\frac{z_{eq}+1}{z_{rec}+1}}}\right) \simeq \frac{1}{2}\left(1 + \frac{1}{\sqrt{1+\frac{z_{eq}}{z_{rec}}}}\right). \tag{9.81}$$

For typical values $z_{eq} \simeq 3228$ and $z_{rec} \simeq 1100$ we get $(\mathcal{H}_{rec}\tau_{rec})^{-1} \simeq 0.75$ (which can decrease down to 0.74 if $z_{rec} \simeq 1050$). Thus, for typical values of z_{rec} and z_{eq} we have $\sigma_{rec} \simeq 0.04$.

With a little bit of effort, using the formulae derived now, it is also possible to find the expression of the visibility function in multipole space. Let us perform the derivation and recall, on a side, the results already obtained in the final part of chapter 2 and, in particular, Eq. (2.125). We are interested in writing Eq. (9.77) as

$$\mathcal{K}(\tau) = e^{-k^2 \tau_{rec}^2 \sigma_{rec}^2} = e^{-(\ell/\ell_t)^2 w^2}, \tag{9.82}$$

where, by definition,

$$w = \frac{k\tau_0}{\ell}, \qquad \ell_t^2 = \frac{1}{\sigma_{rec}^2}\left(\frac{\tau_0}{\tau_{rec}}\right)^2. \tag{9.83}$$

Using Eq. (2.125) inside the definition of ℓ_t^2 we obtain

$$\ell_t^2 = \frac{\alpha_1^2 z_{eq}}{T_\Lambda^2} \frac{\left(1 + z_{eq}/z_{rec}\right)}{\left(\sqrt{1+z_{eq}/z_{rec}}+1\right)^2 \left(\sqrt{1+z_{eq}/z_{rec}}-1\right)^2}. \tag{9.84}$$

After simple algebra, taking the square root of Eq. (9.84) we obtain, for ℓ_t:

$$\ell_t = \frac{1112}{T_\Lambda}\sqrt{1+\frac{z_{eq}}{z_{rec}}}\sqrt{\frac{z_{rec}}{1100}}. \tag{9.85}$$

In Fig. 9.1 the base-10 logarithm of ℓ_t is reported for different values of z_{rec}.

9.5 Line of sight integrals

Equations (9.72) and (9.73)–(9.74) can be formally written as

$$\mathcal{M}' + (ik\mu + \epsilon')\mathcal{M} = \mathcal{N}(\vec{k},\mu,\tau), \tag{9.86}$$

where, by definition, $\mathcal{M} = \mathcal{M}(\vec{k},\mu,\tau)$, are appropriate functions changing from case to case and $\mathcal{N}(\vec{k},\mu,\tau)$ is a source term which also depends on

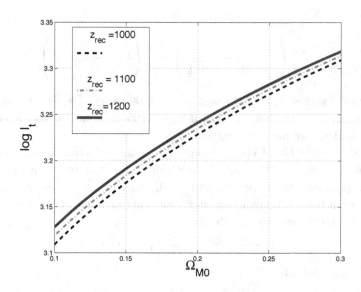

Fig. 9.1 The base-10 logarithm of ℓ_t is illustrated for different values of z_{rec} and as a function of Ω_{M0}.

the specific equation to be integrated. The formal solution of the class of equations parametrized in the form (9.86) can be written as

$$\mathcal{M}(\vec{k},\mu,\tau_0) = e^{-A(\vec{k},\mu,\tau_0)} \int_0^{\tau_0} e^{A(\vec{k},\mu,\tau)} \mathcal{N}(\vec{k},\tau) d\tau, \qquad (9.87)$$

where the boundary term for $\tau \to 0$ can be dropped since it is unobservable [276, 282]. The function $A(\vec{k},\mu,\tau)$ determines the solution of the homogeneous equations and it is:

$$A(\vec{k},\mu,\tau) = \int_0^\tau (ik\mu + \epsilon') d\tau = ik\mu\tau + \int_0^\tau x_e n_e \sigma_{Th} \frac{a}{a_0} d\tau. \qquad (9.88)$$

In the following few examples of this technique will be given. The first example we ought to discuss deals with the solution of Eqs. (9.72) and (9.74). Consider, indeed, Eq. (9.72) and compare it with Eq. (9.86). By identifying the appropriate terms we will have:

$$\mathcal{M}(\vec{k},\mu,\tau_0) \equiv \Delta_P(\vec{k},\mu,\tau_0),$$
$$\mathcal{N}(\vec{k},\mu,\tau_0) \equiv \frac{3}{4}(1-\mu^2) S_P. \qquad (9.89)$$

Thus, using the results of Eqs. (9.87) and (9.88), the solution of Eq. (9.72) can be formally written as

$$\Delta_P(\vec{k},\mu,\tau_0) = \frac{3}{4} \int_0^{\tau_0} \mathcal{K}(\tau) e^{-ik\mu\Delta\tau}(1-\mu^2) S_P(k,\tau) d\tau, \qquad (9.90)$$

where $\epsilon(\tau,\tau_0)$ is the optical depth already introduced in Eq. (9.70) and $\Delta\tau = (\tau_0 - \tau)$ is the (conformal time) increment between the reception of the signal (at τ_0) and the emission (taking place for $\tau \simeq \tau_{\rm rec}$). Indeed notice that, according to Eq. (9.87)

$$A(\vec{k},\mu,\tau) - A(\vec{k},\mu,\tau_0) = ik\mu(\tau - \tau_0) + \int_0^\tau x_e n_e \sigma_{\rm Th} \frac{a}{a_0}$$
$$- \int_0^{\tau_0} x_e n_e \sigma_{\rm Th} \frac{a}{a_0} \equiv -ik\mu\Delta\tau - \epsilon(\tau,\tau_0). \quad (9.91)$$

In Eq. (9.90) the visibility function $\mathcal{K}(\tau)$, already defined in Eq. (9.75), has been explicitly introduced. Consider now Eq. (9.74) and compare it with Eq. (9.86). By appropriately identifying the corresponding terms we get, in this case,

$$\mathcal{M}(\vec{k},\mu,\tau_0) \equiv \Delta_{\rm I}(\vec{k},\mu,\tau_0) + \phi(\vec{k},\tau),$$
$$\mathcal{N}(\vec{k},\mu,\tau_0) \equiv (\psi' + \phi') + \epsilon'[(\Delta_{\rm I0} + \phi) + \mu v_{\rm b} - \tfrac{1}{2}P_2(\mu)S_{\rm P}]. \quad (9.92)$$

Using again Eqs. (9.87) and (9.88) the formal result of the integration of Boltzmann equation can be written as:

$$(\Delta_{\rm I} + \phi)(\vec{k},\mu,\tau_0) = \int_0^{\tau_0} d\tau\, e^{-ik\mu\Delta\tau - \epsilon(\tau,\tau_0)}(\phi' + \psi')$$
$$+ \int_0^{\tau_0} d\tau\, \mathcal{K}(\tau) e^{-ik\mu\Delta\tau} \left[(\Delta_{\rm I0} + \phi + \mu v_{\rm b}) - \frac{1}{2}P_2(\mu)S_{\rm P}(k,\tau)\right], \quad (9.93)$$

Equations (9.93) and (9.90) are called for short line of sight integral solutions. There are at least two important applications of Eqs. (9.93) and (9.90). The first one is numerical and will only be quickly described. The second one is analytical and will be exploited both in the present section and in the following.

The formal solution of Eq. (9.72) can be written in a different form if the term μ^2 is integrated by parts (notice, in fact, that the μ also enters the exponential). The boundary terms arising as a result of the integration by parts can be dropped because they are vanishing in the limit $\tau \to 0$ and are irrelevant for $\tau = \tau_0$ (since only an unobservable monopole is induced). The result the integration by parts of the μ^2 term in Eq. (9.90) can be expressed as

$$\Delta_{\rm P}(\vec{k},\mu,\tau_0) = \int_0^{\tau_0} e^{-ik\mu\Delta\tau} \mathcal{N}_{\rm P}(k,\tau)\, d\tau, \quad (9.94)$$

$$\mathcal{N}_{\rm P}(\vec{k},\tau) = \frac{3}{4k^2}[\mathcal{K}(S_{\rm P}'' + k^2 S_{\rm P}) + 2\mathcal{K}' S_{\rm P}' + S_{\rm P}\mathcal{K}''], \quad (9.95)$$

where, as usual $\Delta\tau = (\tau_0 - \tau)$. The same exercise can be performed in the case of Eq. (9.73). Before giving the general result, let us just integrate by parts the term $-ik\mu\phi$ appearing at the right hand side of Eq. (9.73). The result of this manipulation is

$$\Delta_I(\vec{k},\mu,\tau_0) = \int_0^{\tau_0} e^{ik\mu(\tau-\tau_0)-\epsilon(\tau,\tau_0)}(\psi' + \phi') \, d\tau$$

$$+ \int_0^{\tau_0} \mathcal{K}(\tau)e^{ik\mu(\tau-\tau_0)} \, d\tau \left[\Delta_{I0} + \phi + \mu v_b - \frac{1}{2}P_2(\mu)S_P\right]. \quad (9.96)$$

Let us now exploit the sudden decay approximation illustrated around Eq. (9.76) and assume that the (Gaussian) visibility function $\mathcal{K}(\tau)$ is indeed a Dirac delta function centered around τ_{rec} (consequently the optical depth $\epsilon(\tau,\tau_0)$ will be a step function). Then Eq. (9.96) becomes:

$$\Delta_I(\vec{k},\mu,\tau_0) = \int_{\tau_{\text{rec}}}^{\tau_0} e^{ik\mu(\tau-\tau_0)}[\psi' + \phi']d\tau$$

$$+ e^{ik\mu(\tau_{\text{rec}}-\tau_0)}[\Delta_{I0} + \phi + \mu v_b]_{\tau_{\text{rec}}}, \quad (9.97)$$

where the term S_P has been neglected since it is subleading at large scales. Equation (9.97) is exactly (the Fourier space version of) Eq. (7.35) already derived with a different chain of arguments and we can directly recognize the integrated SW term (first term at the right hand side), the ordinary SW effect (proportional to [h] $(\Delta_{I0} + \phi)$) and the Doppler term receiving contribution from the peculiar velocity of the observer and of the emitter.

If all the μ dependent terms appearing in Eq. (9.73) are integrated by parts the result will be

$$\Delta_I(\vec{k},\mu,\tau_0) = \int_0^{\tau_0} e^{-ik\mu\Delta\tau-\epsilon(\tau,\tau_0)}(\psi' + \phi') \, d\tau$$

$$+ \int_0^{\tau_0} e^{-ik\mu\Delta\tau}\mathcal{N}_I(k,\tau)d\tau, \quad (9.98)$$

$$\mathcal{N}_I(k,\tau) = \left\{\mathcal{K}(\tau)\left[\Delta_{I,0} + \frac{S_P}{4} + \phi + \frac{i}{k}v_b' + \frac{3}{4k^2}S_P''\right]\right.$$

$$\left. + \mathcal{K}'\left[\frac{i}{k}v_b + \frac{3}{2k^2}S_P'\right] + \frac{3}{4k^2}\mathcal{K}''S_P\right\}. \quad (9.99)$$

There is another standard manipulation, analogous to the ones already discussed, that is often required in practical calculations. The idea is to use the line of sight solution and deduce the expression of the multipole moments in one shot. Consider, as an example, the line of sight solution for

---

[h]Recall, in fact, that because of the relation between brightness perturbations and the energy-momentum tensor, i.e. Eqs. (9.38) and (9.62), $4\Delta_{I0} = \delta_\gamma$.

the polarization. Recalling the definitions of the expansion in multipoles we can safely write

$$2(-i)^\ell \Delta_{\text{P}\ell} = \frac{3}{4} \int_0^{\tau_0} d\tau \, \mathcal{K}(\tau) S_{\text{P}}(k,\tau) \mathcal{D}_\ell(k) \tag{9.100}$$

where

$$\mathcal{D}_\ell(k) = \int_{-1}^{1} (1-\mu^2) e^{-ik\Delta\tau} P_\ell(\mu) d\mu. \tag{9.101}$$

Using the expansion of the plane wave in terms of Legendre polynomials, Eq. (9.101) can also be written as

$$\mathcal{D}_\ell(k) = \sum_{\ell'} (2\ell'+1) j_{\ell'}(k\Delta\tau)(-i)^{\ell'} \int_{-1}^{1} (1-\mu^2) P_{\ell'}(\mu) P_\ell(\mu) d\mu, \tag{9.102}$$

where

$$j_\ell(k\Delta\tau) = \sqrt{\frac{\pi}{2k\Delta\tau}} J_{\ell+1/2}(k\Delta\tau), \tag{9.103}$$

are the spherical Bessel functions. Let us now use extensively the recurrence relation for the Legendre polynomials already introduced in Eq. (9.25) and let us compute explicitly the integral over μ appearing in Eq. (9.102). We will have:

$$\int_{-1}^{1} (1-\mu^2) P_{\ell'}(\mu) P_\ell(\mu) = \frac{2}{2\ell+1} \delta_{\ell\ell'} - \int_{-1}^{1} \mu^2 P_{\ell'}(\mu) P_\ell(\mu) d\mu, \tag{9.104}$$

where, to integrate the first term, the orthogonality condition for $P_\ell(\mu)$ has been used (see Eq. (9.24)). To integrate the second term we first recall that, according to Eq. (9.25),

$$\mu^2 P_\ell = \frac{\ell+1}{2\ell+1} \mu P_{\ell+1} + \frac{\ell}{2\ell+1} \mu P_{\ell-1}. \tag{9.105}$$

The same recurrence relation allows us to eliminate the two terms at the right hand side of Eq. (9.105) going as $\mu P_{\ell+1}$ and as $\mu P_{\ell-1}$. Indeed, from Eq. (9.25), by shifting $\ell \to \ell+1$ and $\ell \to \ell-1$ we obtain, respectively:

$$\mu P_{\ell+1} = \frac{\ell+2}{2\ell+3} P_{\ell+2} + \frac{\ell+1}{2\ell+3} P_\ell, \tag{9.106}$$

$$\mu P_{\ell-1} = \frac{\ell}{2\ell-1} P_\ell + \frac{\ell-1}{2\ell-1} P_{\ell-2}. \tag{9.107}$$

Inserting Eqs. (9.106) and (9.107) into Eq. (9.105) we simply get

$$\mu^2 P_\ell = \frac{(\ell+1)(\ell+2)}{(2\ell+1)(2\ell+3)} P_{\ell+2} + \frac{\ell(\ell-1)}{(2\ell+1)(2\ell-1)} P_{\ell-2} + \left[\frac{(\ell+1)^2}{(2\ell+1)(2\ell+3)} + \frac{\ell^2}{(2\ell+1)(2\ell-1)} \right] P_\ell. \tag{9.108}$$

Finally, using Eq. (9.108) inside Eq. (9.102) and (9.100) the following result can be easily obtained:

$$\Delta_{\text{P}\ell}(k,\tau) = \frac{3}{4}\int_0^{\tau_0} \mathcal{K}(\tau)S_{\text{P}}(k,\tau)[c_\ell j_\ell(k\Delta\tau)$$
$$+ c_{\ell-2}j_{\ell-2}(k\Delta\tau) + c_{\ell+2}j_{\ell+2}(k\Delta\tau)], \qquad (9.109)$$

where

$$c_\ell = \frac{2(\ell^2 + \ell - 1)}{(2\ell - 1)(2\ell + 3)},$$
$$c_{\ell+2} = \frac{(\ell + 1)(\ell + 2)}{(2\ell + 1)(2\ell + 3)}, \qquad (9.110)$$
$$c_{\ell-2} = \frac{\ell(\ell - 1)}{(2\ell + 1)(2\ell - 1)}.$$

With the same kind of manipulation it is not difficult to obtain the expression for the multipoles of the brightness perturbation of the intensity, namely:

$$\Delta_{\text{I}\ell}(k,\tau_0) = \int_0^{\tau_0} e^{-\epsilon(\tau,\tau_0)}(\phi' + \psi')j_\ell(k\Delta\tau)d\tau$$
$$+ \int_0^{\tau_0} \mathcal{K}(\tau)(\Delta_{\text{I}0} + \phi)j_\ell(k\Delta\tau)d\tau$$
$$+ i\int_0^{\tau_0} \mathcal{K}(\tau)v_{\text{b}}\left[\frac{\ell}{2\ell+1}j_{\ell-1}(k\Delta\tau) - \frac{\ell+1}{2\ell+1}j_{\ell+1}(k\Delta\tau)\right]d\tau$$
$$- \frac{1}{4}\int_0^{\tau_0} \mathcal{K}(\tau)S_{\text{P}}(k,\tau)[\bar{c}_{\ell+2}j_{\ell+2}(k\Delta\tau)$$
$$+ \bar{c}_{\ell-2}j_{\ell-2}(k\Delta\tau) + \bar{c}_\ell j_\ell(k\Delta\tau)]d\tau \qquad (9.111)$$

where the coefficients are now given by:

$$\bar{c}_\ell = -\frac{2(\ell^2 + \ell - 1)}{(2\ell - 1)(2\ell + 3)},$$
$$\bar{c}_{\ell-2} = \frac{3\ell(\ell - 1)}{(2\ell + 1)(2\ell - 1)}, \qquad (9.112)$$
$$\bar{c}_{\ell+2} = \frac{3(\ell + 1)(\ell + 2)}{(2\ell + 3)(2\ell + 1)}.$$

In some of the previous examples the visibility function has been evaluated in the sudden decoupling approximation. This is because, as already discovered in chapter 7, the thickness of the last scattering surface does not crucially affect the estimate of the Sachs-Wolfe plateau which is the one obtained in Eq. (9.97). Finite thickness effects are however of upmost importance when going to smaller angular scales and this will be one of the themes of the forthcoming topics of the present chapter. Before going to smaller angular scales (the scales compatible with the celebrated Doppler peak) it is appropriate to address the tight-coupling expansion which serves as a useful bridge between the fluid and the kinetic discussion. This will be the topic covered by the following section. In the last part of the present section it will be shown that the calculation already studied in chapter 7 indeed applies with exactly the same steps also in the case when the temperature anisotropy is derived directly from the Boltzmann equation in the sudden recombination (or decoupling) limit.

9.5.1 Angular power spectrum and observables

Equations (9.97) together with the results derived in chapter 7 for the initial conditions of the metric fluctuations after equality allow the estimate of the angular power spectra in the region of the Sachs-Wolfe plateau. This calculation has been already performed in detail at the end of chapter 7 using the fluid approach. The two fluid treatment and the kinetic approach lead to the same results in the sudden decoupling limit and for $\ell < 30$. To emphasize the latter statement the results already obtained in the case of the adiabatic mode will be here rederived. The main (formal) difference is that here we will express the final result in terms of the spectrum of ψ and not in terms of the spectrum of ζ (this has been already done in chapter 7). This is just for pedagogical reasons since the two spectra, in the adiabatic case, are merely related via a numerical factor.

In the adiabatic case Eq. (9.97) (or Eq. (7.35)) has vanishing integrated contribution and vanishing Doppler contribution at large scales (as discussed in section 4). Then, using the result of Eq. (7.86), the adiabatic contribution to the temperature fluctuations can be written as

$$\Delta_{\mathrm{I}}^{\mathrm{ad}}(\vec{k},\tau_0) = e^{-ik\mu\tau_0}[\Delta_{\mathrm{I}0} + \phi]_{\tau_{\mathrm{rec}}} \simeq e^{-ik\mu\tau_0}\frac{1}{3}\psi_{\mathrm{m}}^{\mathrm{ad}}(\vec{k}), \quad (9.113)$$

noticing that, in the argument of the plane wave τ_{rec} can be dropped since $\tau_{\mathrm{rec}} \ll \tau_0$. The plane wave appearing in Eq. (9.113) can now be expanded in series of Legendre polynomials and, as a result,

$$\Delta_{\mathrm{I},\ell}^{\mathrm{ad}}(\vec{k},\tau_0) = \frac{j_\ell(k\tau_0)}{3}\psi_{\mathrm{m}}^{\mathrm{ad}}(\vec{k}), \quad (9.114)$$

where $j_\ell(k\tau_0)$ are the spherical Bessel functions of Eq. (9.103) but with a slightly different argument. Assuming now that $\psi_{\mathrm{m}}^{\mathrm{ad}}(\vec{k})$ are the Fourier components of a Gaussian and isotropic random field then

$$\langle \psi_{\mathrm{m}}^{\mathrm{ad}}(\vec{k}) \psi_{\mathrm{m}}^{\mathrm{ad}}(\vec{k}') \rangle = \frac{2\pi^2}{k^3} \mathcal{P}_\psi^{\mathrm{ad}}(k) \delta^{(3)}(\vec{k} - \vec{k}'),$$
$$\mathcal{P}_\psi^{\mathrm{ad}}(k) = \frac{k^3}{2\pi^2} |\psi_{\mathrm{m}}^{\mathrm{ad}}(k)|^2, \qquad (9.115)$$

where $\mathcal{P}_\psi^{\mathrm{ad}}(k)$ is the power spectrum of the longitudinal fluctuations of the metric after equality. Finally, Eq. (9.114) can be inserted into Eq. (3.27) and from Eq. (9.115) (together with the orthogonality relation of spherical harmonics) the C_ℓ turn out to be:

$$C_\ell^{(\mathrm{ad})} = \frac{4\pi}{9} \int_0^\infty \frac{dk}{k} \mathcal{P}_\psi^{\mathrm{ad}}(k) j_\ell(k\tau_0)^2. \qquad (9.116)$$

To perform the integral it is customarily assumed that the power spectrum of adiabatic fluctuations has a power-law dependence characterized by a single spectral index n

$$\mathcal{P}_\psi^{\mathrm{ad}}(k) = \frac{k^3}{2\pi^2} |\psi_k|^2 = A_{\mathrm{ad}} \left(\frac{k}{k_{\mathrm{p}}}\right)^{n-1}. \qquad (9.117)$$

Notice that k_{p} is a typical pivot scale which is conventional since the whole dependence on the parameters of the model is encoded in A_{ad} and n. For instance, the WMAP collaboration [37, 299], chooses to normalize A at[i]

$$k_{\mathrm{p}} = k_1 = 0.05 \text{ Mpc}^{-1}, \qquad (9.118)$$

while the scalar-tensor ratio (defined in section 6) is evaluated at a scale

$$k_0 = 0.002 \text{ Mpc}^{-1} \equiv 6.481 \times 10^{-28} \text{ cm}^{-1} = 1.943 \times 10^{-17} \text{ Hz}, \qquad (9.119)$$

recalling that 1 Mpc $= 3.085 \times 10^{24}$ cm.

Inserting Eq. (9.117) into Eq. (9.116) and recalling the explicit form of the spherical Bessel functions in terms of ordinary Bessel functions

$$C_\ell^{(\mathrm{ad})} = \frac{2\pi^2}{9} (\tau_0 k_{\mathrm{p}})^{1-n} A_{\mathrm{ad}} \int_0^\infty dy\, y^{n-3} J_{\ell+1/2}^2(y), \qquad (9.120)$$

where $y = k\tau_0$. The integral appearing in Eq. (9.120) can be performed for $-3 < n < 3$ with the result

$$\int_0^\infty dy\, y^{n-3} J_{\ell+1/2}^2(y) = \frac{1}{2\sqrt{\pi}} \frac{\Gamma(3-n/2)\Gamma(\ell+n/2-1/2)}{\Gamma(4-n/2)\Gamma(5/2+\ell-n/2)}. \qquad (9.121)$$

---

[i]There exist different possibilities for the actual value of the pivot scale especially in the case when the adiabatic mode is present together with isocurvature modes. For a lucid discussion see Ref. [266].

To get the standard form of the C_ℓ we use the duplication formula for the Γ function, in our case namely

$$\Gamma\left(\frac{3-n}{2}\right) = \frac{\sqrt{2\pi}\,\Gamma(3-n)}{2^{5/2-n}\Gamma(4-n/2)}. \tag{9.122}$$

Insert now Eq. (9.122) into Eq. (9.121); inserting then Eq. (9.121) into Eq. (9.120) we get

$$C_\ell^{(\mathrm{ad})} = \frac{\pi^2}{36} A_{\mathrm{ad}} \mathcal{Z}(n,\ell)$$
$$\mathcal{Z}(n,\ell) = (\tau_0\, k_{\mathrm{p}})^{1-n}\, 2^n \frac{\Gamma(3-n)\Gamma(\ell+n/2-1/2)}{\Gamma^2(4-n/2)\Gamma(5/2+\ell-n/2)}, \tag{9.123}$$

where the function $\mathcal{Z}(n,\ell)$ has been introduced in analogy with the notation employed in chapter 7 (see, in particular, Eqs. (7.148) and (7.153)). As already emphasized in chapter 7, for the approximations made in the evaluation of the Sachs-Wolfe plateau, Eq. (9.123) holds at large angular scales, i.e. $\ell < 30$.

It is now interesting to pause for a moment and to compare the results of Eq. (9.123) with their counterpart discussed in Eqs. (7.148) and (7.153). Indeed the two results are the same but they are a bit disguised. In the present calculation the amplitude A_{ad} refers to the spectrum of ψ, while in Eq. (7.36) the amplitude $\mathcal{A}_{\mathcal{R}}$ refers to the spectrum of $\mathcal{R}$. Having established this fact the two results are fully equivalent. Indeed, consider Eq. (7.148) in the limit when $\mathcal{A}_{\mathcal{S}} = \mathcal{A}_{\mathcal{RS}} = 0$. In this limit the Sachs-Wolfe plateau is simply expressed in terms of the adiabatic mode which is also the case contemplated by Eq. (9.123). Consider then, all together, Eqs. (7.148), (7.153) and (9.123). It emerges, after taking into account carefully the numerical factors, that

$$\frac{A_{\mathrm{ad}}}{36} = \frac{\mathcal{A}_{\mathcal{R}}}{100}, \tag{9.124}$$

which can be also expressed, after the appropriate simplifications, as

$$\mathcal{A}_{\mathcal{R}} = \left(-\frac{5}{3}\right)^2 A_{\mathrm{ad}}. \tag{9.125}$$

Recalling then Eq. (7.104) we can say that this result was expected since, during the pre-decoupling phase (i.e. when the background is already dominated by dusty matter)

$$\mathcal{R} = -\frac{5}{3}\psi, \qquad \mathcal{P}_{\mathcal{R}}^{(\mathrm{ad})} = \frac{25}{9}\mathcal{P}_\psi^{(\mathrm{ad})}. \tag{9.126}$$

But this is exactly the result reported in Eq. (9.125).

Consider, finally, the specific case of adiabatic fluctuations with Harrison-Zeldovich, i.e. the case $n = 1$ in eq. (9.123). In this case

$$\frac{\ell(\ell+1)}{2\pi} C_\ell^{(\text{ad})} = \frac{A_{\text{ad}}}{9}. \quad (9.127)$$

If the fluctuations were of purely adiabatic nature, then large-scale anisotropy experiments (see Fig. 1.3) imply[j] $A \sim 9 \times 10^{-10}$. Up to now the large angular scale anisotropies have been treated. In the following, the analysis of the smaller angular scales will be introduced within the tight-coupling approximation.

9.6 Tight-coupling expansion

If tight-coupling is exact, photons and baryons are synchronized so well that the photon phase-space distribution is isotropic in the baryon rest frame. In other words since the typical time-scale between two collisions is set by $\tau_c \sim 1/\epsilon'$, the scattering rate is rapid enough to equilibrate the photon-baryon fluid. Since the photon distribution is isotropic, the resulting radiation is not polarized. The idea is then to tailor a systematic expansion in $\tau_c \sim 1/\epsilon'$ or, more precisely, in $k\tau_c \ll 1$ and $\tau_c \mathcal{H} \ll 1$. Recall the expansion of the brightness perturbations in terms of Legendre polynomials:

$$\begin{aligned}
\Delta_{\text{I}}(\vec{k}, \hat{n}, \tau) &= \sum_\ell (-i)^\ell (2\ell+1) \Delta_{\text{I}\ell}(\vec{k}, \tau) P_\ell(\mu), \\
\Delta_{\text{Q}}(\vec{k}, \hat{n}, \tau) &= \sum_\ell (-i)^\ell (2\ell+1) \Delta_{\text{Q}\ell}(\vec{k}, \tau) P_\ell(\mu),
\end{aligned} \quad (9.128)$$

$\Delta_{\text{I}\ell}$ and $\Delta_{\text{Q}\ell}$ being the ℓ-th multipole of the brightness function Δ_{I} and Δ_{Q}. The strategy is now to expand Eqs. (9.63) and (9.64) in powers of the small parameter τ_c. Before doing the expansion, it is useful to derive the hierarchy for the brightness functions in full analogy with what has been discussed in this chapter for the case of the neutrino phase-space distribution. To this

---

[j]To understand fully the quantitative features of Fig. 1.3 it should be borne in mind that sometimes the C_ℓ are given not in absolute units (as implied in Eq. (9.127) but they are multiplied by the CMB temperature. To facilitate the conversion recall that the CMB temperature is $T_0 = 2.725 \times 10^6$ μK. For instance the WMAP collaboration normalizes the power spectrum of the curvature fluctuations at the pivot scale k_{p} as $\mathcal{P}_{\mathcal{R}}^{(\text{ad})} = (25/9) \times (800\pi^2/T_0^2) \times \tilde{A}$ where $\tilde{A}$ is not the A defined here but it can be easily related to it.

aim, each side of Eqs. (9.63)–(9.64) and (9.71) will be multiplied by the various Legendre polynomials and the integration over μ will be performed. Noticing that, from the orthonormality relation for Legendre polynomials (i.e. Eq. (9.24)),

$$\int_{-1}^{1} P_\ell(\mu)\Delta_{\rm I} d\mu = 2(-i)^\ell \Delta_{{\rm I}\ell},$$
$$\int_{-1}^{1} P_\ell(\mu)\Delta_{\rm Q} d\mu = 2(-i)^\ell \Delta_{{\rm Q}\ell}, \qquad (9.129)$$

and recalling that

$$P_0(\mu) = 1, \quad P_1(\mu) = \mu,$$
$$P_2(\mu) = \frac{1}{2}(3\mu^2 - 1), \quad P_3(\mu) = \frac{1}{2}(5\mu^3 - 3\mu), \qquad (9.130)$$

Eqs. (9.63)–(9.64) and (9.71) allow the determination of the first three sets of equations for the hierarchy of the brightness perturbations. More specifically, multiplying Eqs. (9.63)–(9.64) and (9.71) by $P_0(\mu)$ and integrating over μ, the following relations can be obtained

$$\Delta'_{\rm I0} + k\Delta_{\rm I1} = \psi', \qquad (9.131)$$

$$\Delta'_{\rm Q0} + k\Delta_{\rm Q1} = \frac{\epsilon'}{2}[\Delta_{\rm Q2} + \Delta_{\rm I2} - \Delta_{\rm Q0}], \qquad (9.132)$$

$$v'_b + \mathcal{H}v_b = -ik\phi - \frac{\epsilon'}{R_b}(3i\Delta_{\rm I1} + v_b). \qquad (9.133)$$

If Eqs. (9.63)–(9.64) and (9.71) are multiplied by $P_1(\mu)$, both at right and left-hand sides, the integration over μ of the various terms implies, using Eq. (9.129):

$$-\Delta'_{\rm I1} - \frac{2}{3}k\Delta_{\rm I2} + \frac{k}{3}\Delta_{\rm I0} = -\frac{k}{3}\phi + \epsilon'\left[\Delta_{\rm I1} + \frac{1}{3i}v_b\right], \qquad (9.134)$$

$$-\Delta'_{\rm Q1} - \frac{2}{3}k\Delta_{\rm Q2} + \frac{k}{3}\Delta_{\rm Q0} = \epsilon'\Delta_{\rm Q1}, \qquad (9.135)$$

$$v'_b + \mathcal{H}v_b = -ik\phi - \frac{\epsilon'}{R_b}(3i\Delta_{\rm I1} + v_b). \qquad (9.136)$$

The same procedure, using $P_2(\mu)$, leads to:

$$-\Delta'_{\rm I2} - \frac{3}{5}k\Delta_{\rm I3} + \frac{2}{5}k\Delta_{\rm I1} = \epsilon'\left[\frac{9}{10}\Delta_{\rm I2} - \frac{1}{10}(\Delta_{\rm Q0} + \Delta_{\rm Q2})\right], \qquad (9.137)$$

$$-\Delta'_{\rm Q2} - \frac{3}{5}k\Delta_{\rm Q3} + \frac{2}{5}k\Delta_{\rm Q1} = \epsilon'\left[\frac{9}{10}\Delta_{\rm Q2} - \frac{1}{10}(\Delta_{\rm Q0} + \Delta_{\rm I2})\right], \qquad (9.138)$$

$$v'_b + \mathcal{H}v_b = -ik\phi - \frac{\epsilon'}{R_b}\left(3i\Delta_{\rm I1} + v_b\right). \qquad (9.139)$$

For $\ell \geq 3$ the hierarchy of the brightness can be determined in general terms by using the recurrence relation for the Legendre polynomials reported in Eq. (9.25):

$$\Delta'_{I\ell} + \epsilon' \Delta_{I\ell} = \frac{k}{2\ell+1}[\ell \Delta_{I(\ell-1)} - (\ell+1)\Delta_{I(\ell+1)}],$$
$$\Delta'_{Q\ell} + \epsilon' \Delta_{Q\ell} = \frac{k}{2\ell+1}[\ell \Delta_{Q(\ell-1)} - (\ell+1)\Delta_{Q(\ell+1)}]. \tag{9.140}$$

9.7 Zeroth order in tight-coupling: acoustic oscillations

We are now ready to compute the evolution of the various terms to a given order in the tight-coupling expansion parameter $\tau_c = |1/\epsilon'|$. After expanding the various moments of the brightness function and the velocity field in τ_c

$$\Delta_{I\ell} = \overline{\Delta}_{I\ell} + \tau_c \delta_{I\ell},$$
$$\Delta_{Q\ell} = \overline{\Delta}_{Q\ell} + \tau_c \delta_{Q\ell}, \tag{9.141}$$
$$v_b = \overline{v}_b + \tau_c \delta_{v_b},$$

the obtained expressions can be inserted into Eqs. (9.131)–(9.136) and the evolution of the various moments of the brightness function can be found order by order. To zeroth order in the tight-coupling approximation, the evolution equation for the baryon velocity field, i.e. Eq. (9.133), leads to:

$$\overline{v}_b = -3i\overline{\Delta}_{I1}, \tag{9.142}$$

while Eqs. (9.132) and (9.135) lead, respectively, to

$$\overline{\Delta}_{Q0} = \overline{\Delta}_{I2} + \overline{\Delta}_{Q2}, \qquad \overline{\Delta}_{Q1} = 0. \tag{9.143}$$

Finally Eqs. (9.137) and (9.138) imply

$$9\overline{\Delta}_{I2} = \overline{\Delta}_{Q0} + \overline{\Delta}_{Q2}, \qquad 9\overline{\Delta}_{Q2} = \overline{\Delta}_{Q0} + \overline{\Delta}_{I2}. \tag{9.144}$$

Taking together the four conditions expressed by Eqs. (9.143) and (9.144) we have, to zeroth order in the tight-coupling approximation:

$$\overline{\Delta}_{Q\ell} = 0, \quad \ell \geq 0, \qquad \overline{\Delta}_{I\ell} = 0, \quad \ell \geq 2. \tag{9.145}$$

Hence, to zeroth order in the tight-coupling, the relevant equations are

$$\overline{v}_b = -3i\overline{\Delta}_{I1}, \tag{9.146}$$
$$\overline{\Delta}'_{I0} + k\overline{\Delta}_{I1} = \psi'. \tag{9.147}$$

This means, as anticipated, that to zeroth order in the tight-coupling expansion the CMB is not polarized since Δ_Q is vanishing.

A decoupled evolution equation for the monopole can be derived. Summing up Eq. (9.134) (multiplied by $3i$) and Eq. (9.136) (multiplied by R_b) we get, to zeroth order in the tight-coupling expansion:

$$R_b \bar{v}_b' - 3i\overline{\Delta}_{I1}' + ik\phi(R_b + 1) - 2ik\overline{\Delta}_{I2} + ik\overline{\Delta}_{I0} + R_b \mathcal{H}\bar{v}_b = 0. \qquad (9.148)$$

Recalling now Eq. (9.146) to eliminate $\bar{v}_b$ from Eq. (9.148), the following equation can be obtained

$$(R_b + 1)\overline{\Delta}_{I1}' + \mathcal{H}R_b \overline{\Delta}_{I1} - \frac{k}{3}\overline{\Delta}_{I0} = 0. \qquad (9.149)$$

Finally, the dipole term can be eliminated from Eq. (9.149) using Eq. (9.147). By doing so, Eq. (9.149) leads to the wanted decoupled equation for the monopole:

$$\overline{\Delta}_{I0}'' + \frac{R_b'}{R_b + 1}\overline{\Delta}_{I0}' + k^2 c_{sb}^2 \overline{\Delta}_{I0} = \left[\psi'' + \frac{R_b'}{R_b + 1}\psi' - \frac{k^2}{3}\phi\right], \qquad (9.150)$$

where c_{sb} has been already defined in Eq. (8.80) and it is the sound of the baryon-photon system. The term $k^2 c_{sb}^2 \overline{\Delta}_{I0}$ is the photon pressure. Defining, from Eq. (8.80), the sound horizon as

$$r_s(\tau) = \int_0^\tau c_{sb}(\tau')d\tau', \qquad (9.151)$$

the photon pressure cannot be neglected for modes $kr_s(\tau) \geq 1$. At the right hand side of Eq. (9.150) several forcing terms appear. The term ψ'' dominates, if present, on super-horizon scales and causes a dilation effect on $\overline{\Delta}_{I0}$. The term containing $k^2\phi$ leads to the adiabatic growth of the photon-baryon fluctuations and becomes important for $k\tau \simeq 1$. In Eq. (9.150) the damping term arises from the redshifting of the baryon momentum in an expanding Universe, while photon pressure provides the restoring force which is weakly suppressed by the additional inertia of the baryons. It is finally worth noticing that all the formalism developed in this section is nothing but an extension of the fluid treatment proposed in chapters 7 and 8. This aspect becomes immediately evident by comparing Eqs. (8.79) and (9.150). Equations (8.79) and (9.150) are indeed the same equation since $\overline{\Delta}_{I0} = \delta_\gamma/4$.

9.7.1 Solutions of the evolution of monopole and dipole

Equation (9.150) can be solved under different approximations (see for instance [276]). The first brutal approximation would be to set $R'_b = R_b = 0$, implying the the role of the baryons in the acoustic oscillations is totally neglected. As a consequence, in this case, $c_{sb} \equiv 1/\sqrt{3}$ which is nothing but the sound speed discussed in Eqs. (7.82)–(7.85) for the fluid analysis of the adiabatic mode. In the case of the adiabatic mode, neglecting neutrino anisotropic stress, $\psi = \phi = \psi_m$ and $\psi' = 0$. Hence, the solution for the monopole and the dipole to zeroth order in the tight-coupling expansion follows by solving Eq. (9.150) and by inserting the obtained result into Eqs. (9.146) and (9.147), i.e.

$$\overline{\Delta}_{I0}(k,\tau) = \frac{\psi}{3}[\cos(kc_{sb}\tau) - 3],$$
$$\overline{\Delta}_{I1}(k,\tau) = \frac{\psi_m}{3} kc_{sb} \sin(kc_{sb}\tau), \qquad (9.152)$$

which is exactly the solution discussed in section 4 if we recall Eq. (9.146) and the definition (9.67).

If $R'_b = 0$ but $R_b \neq 0$, then the solution of Eqs. (9.146)–(9.147) and (9.150) becomes, in the case of the adiabatic mode,

$$\overline{\Delta}_{I0}(k,\tau) = \frac{\psi_m}{3}(R_b + 1)[\cos(kc_{sb}\tau) - 3],$$
$$\overline{\Delta}_{I1}(k,\tau) = \frac{\psi_m}{3}\sqrt{\frac{R_b + 1}{3}} \sin(kc_{sb}\tau). \qquad (9.153)$$

Equation (9.153) shows that the presence of the baryons increases the amplitude of the monopole by a factor R_b. This phenomenon can be verified also in the case of generic time-dependent R_b. In the case of $R_b \neq 0$ the shift in the monopole term is $(R_b + 1)$ with respect to the case $R_b = 0$. This phenomenon produces a modulation of the height of the acoustic peak that depends on the baryon content of the model.

Consider now the possibility of setting directly initial conditions for the Boltzmann hierarchy during the radiation dominated epoch. During the radiation dominated epoch and for modes which are outside the horizon, the initial conditions for the monopole and the dipole are fixed as

$$\Delta_{I0}(k,\tau) = -\frac{\phi_0}{2} - \frac{525 + 188R_\nu + 16R_\nu^2}{180(25 + 2R_\nu)}\phi_0 k^2\tau^2,$$
$$\Delta_{I1}(k,\tau) = \frac{\phi_0}{6}k\tau - \frac{65 + 16R_\nu}{108(25 + 2R_\nu)}\phi_0 k^3\tau^3 \qquad (9.154)$$

where ϕ_0 is the constant value of ϕ during radiation. The constant value of ψ, i.e. ψ_0 will be related to ϕ_0 through R_ν, i.e. the fractional contribution of the neutrinos to the total density. It is useful to observe that in terms of the quantity $\Delta_0 = (\overline{\Delta}_{I0} - \psi)$, Eq. (9.150) becomes

$$\Delta_0'' + k^2 c_{\rm sb}^2 \Delta_0 = -k^2 \left[\frac{\phi}{3} + c_{\rm sb}^2 \psi\right]. \tag{9.155}$$

The initial conditions for Δ_0 are easily obtained from its definition in terms of Δ_{I0} and ψ. The same strategy can be applied to more realistic cases, such as the one where the scale factor interpolates between a radiation-dominated phase and a matter-dominated phase. In this case the solution of Eq. (9.150) will be more complicated but always analytically tractable. Equation (9.150) can indeed be solved in general terms. The general solution of the homogeneous equation is simply given, in the WKB approximation, as

$$\overline{\Delta}_{I0}(k,\tau) = (R_{\rm b} + 1)^{-1/4}[A \cos kr_{\rm s} + B \sin kr_{\rm s}]. \tag{9.156}$$

For adiabatic fluctuations, $k^2 \phi$ contributes primarily to the cosine. The reason is that, in this case, ψ is constant until the moment of Jeans scale crossing at which moment it begins to decay. Non-adiabatic fluctuations, on the contrary, have vanishing gravitational potential at early times and their monopole oscillates like a sine. Consequently, the peaks in the temperature power spectrum will be located, for adiabatic fluctuations, at a scale k_n such that $k_n r_{\rm s}(\tau_*) = n\pi$. Notice that, according to Eq. (9.147) the dipole, will be anticorrelated with the monopole. So if the monopole is cosinusoidal, the dipole will be instead sinusoidal. Hence the "zeros" of the cosine (as opposed to the maxima) will be filled by the monopole. The solution of Eq. (9.150) can then be obtained by supplementing the general solution of the homogeneous equation (9.156) with a particular solution of the inhomogeneous equations that can be found easily with the usual Green's function methods [276]. The amplitude of the monopole term shifts as $(1 + R_{\rm b})^{-1/4}$. Recalling the definition of $R_{\rm b}$ introduced in chapter 8, it can be argued that the height of the Doppler peak is weakly sensitive to $h_0^2 \Omega_{\rm b0}$ in the ΛCDM model where $\Omega_{\rm b0} \ll \Omega_{\rm M0}$ and $R_{\rm b}(\tau_{\rm rec}) < 1$.

9.7.2 Estimate of the sound horizon at decoupling

In ℓ space the position of the peaks for adiabatic and isocurvature modes is given, respectively, by

$$\ell^{(n)} = n\,\pi \frac{\overline{D}_A(z_{\text{dec}})}{r_s(\tau_{\text{rec}})}, \tag{9.157}$$

$$\ell^{(n)} = \left(n + \frac{1}{2}\right)\pi \frac{\overline{D}_A(z_{\text{dec}})}{r_s(\tau_{\text{rec}})}, \tag{9.158}$$

where $\overline{D}_A(z_{\text{dec}})$ is the (comoving) angular diameter distance to decoupling defined in Appendix A (see in particular Eq. (A.23) and (A.34)). We will now be interested in estimating (rather roughly) the position of the first peak, i.e.

$$\ell_A = \pi \frac{\overline{D}_A(z_{\text{dec}})}{r_s(\tau_{\text{rec}})}, \tag{9.159}$$

where the subscript A stands for acoustic. The first thing we have to do is to estimate the sound horizon at decoupling. From Eq. (9.151) we have

$$r_s(\tau_{\text{dec}}) = \int_0^{\tau_{\text{dec}}} \frac{d\tau}{\sqrt{3(1 + R_b(\tau))}}. \tag{9.160}$$

Equation (9.160) can also be written as

$$r_s(\tau_{\text{dec}}) = \int_0^{\alpha_{\text{dec}}} \frac{d\alpha}{\alpha\dot{\alpha}} c_{sb}(\alpha), \tag{9.161}$$

where, following the notation of Eqs. (2.80) and (2.81), $\alpha = (a/a_0)$. Indeed, recalling Eq. (2.80) we can write

$$\alpha\dot{\alpha} = H_0\sqrt{\Omega_{\Lambda 0}\alpha^4 + (1 - \Omega_{\Lambda 0} - \Omega_{M0})\alpha^2 + \Omega_{M0}\alpha + \Omega_{R0}}$$
$$\simeq H_0\sqrt{\Omega_{M0}\alpha + \Omega_{R0}}, \tag{9.162}$$

where, in the first equality we assumed that the spatial curvature vanishes; the second equality follows from the first since the contribution of matter and dark energy are subleading in the range of integration. Using Eq. (9.162) into Eq. (9.161) we can then write

$$\left(\frac{r_s(\tau_{\text{dec}})}{\text{Mpc}}\right) = \frac{2998}{\sqrt{h_0^2 \Omega_{R0}}} \int_0^{\alpha_{\text{dec}}} \frac{d\alpha}{\sqrt{(1 + \beta_1\alpha)(1 + \beta_2\alpha)}}, \tag{9.163}$$

where

$$\beta_1 = \frac{h_0^2 \Omega_{M0}}{h_0^2 \Omega_{R0}}, \qquad \beta_2 = \frac{3}{4}\frac{h_0^2 \Omega_{b0}}{h_0^2 \Omega_{\gamma 0}}. \tag{9.164}$$

Recall that, according to Eqs. (2.78), $h_0^2\Omega_{\gamma 0} = 2.47 \times 10^{-5}$ and $h_0^2\Omega_{R0} = 4.15 \times 10^{-5}$. From Eqs. (9.163) and (9.164) it is apparent that the sound horizon at decoupling depends on both Ω_{b0} and Ω_{M0}. If we increase either Ω_{b0} or Ω_{M0} the sound horizon gets smaller. The integral appearing in Eq. (9.163) can be done analytically and the result is:

$$\left(\frac{r_s(\tau_{\rm dec})}{\rm Mpc}\right) = \frac{2998}{\sqrt{1+z_{\rm dec}}} \frac{2}{\sqrt{3\,h_0^2\Omega_{M0}c_1}} \ln\left[\frac{\sqrt{1+c_1}+\sqrt{c_1+c_1c_2}}{1+\sqrt{c_1c_2}}\right], \tag{9.165}$$

where

$$\begin{aligned} c_1 &= \beta_2\alpha_{\rm dec} = 27.6\,h_0^2\Omega_{b0}\left(\frac{1100}{1+z_{\rm dec}}\right), \\ c_2 &= \frac{1}{\beta_1\alpha_{\rm dec}} = \frac{0.045}{h_0^2\Omega_{M0}}\left(\frac{1+z_{\rm dec}}{1100}\right). \end{aligned} \tag{9.166}$$

With our fiducial values of the parameters, $r_s(\tau_{\rm dec}) \simeq 150$ Mpc. Recalling now that the comoving angular diameter distance to decoupling is estimated in Appendix A (see in particular Eq. (A.36)) the multipole corresponding to the sound horizon at decoupling can be estimated as $\ell_A \sim 300$. The sound horizon has been a bit underestimated with our approximation. For $r_s(\tau_{\rm dec}) \sim 200$, $\ell_A \sim 220$, which is around the measured value of the first Doppler peak in the temperature autocorrelation (see Fig. 1.3).

It is difficult to obtain general analytic formulas for the position of the peaks. Degeneracies among the parameters may appear [300]. In [301] a semi-analytical expression for the integral giving the angular diameter distance has been derived for various cases of practical interest. Once the evolution of the lowest multipoles is known, the obtained expressions can be used in the integral solutions of the Boltzmann equation and the angular power spectrum can be computed analytically. Recently, Weinberg in a series of papers [302–304] computed the temperature fluctuations in terms of a pair of generalized form factors related, respectively, to the monopole and the dipole. This set of calculations were conducted in the synchronous gauge (see also [305–307] for earlier work on this subject; see also [308, 309]). Reference [310] also presents analytical estimates for the angular power spectrum exhibiting explicit dependence on the cosmological parameters in the case of the concordance model.

9.8 First order in tight-coupling: polarization

To first order in the tight-coupling limit, the relevant equations can be written down by keeping all terms of order τ_c and by using the first-order relations to simplify the expressions. From Eq. (9.135) the condition $\delta_{Q1} = 0$ can be derived. From Eqs. (9.132) and (9.137)–(9.138), the following remaining conditions are obtained respectively:

$$-\delta_{Q0} + \delta_{I2} + \delta_{Q2} = 0, \tag{9.167}$$

$$\frac{9}{10}\delta_{I2} - \frac{1}{10}[\delta_{Q0} + \delta_{Q2}] = \frac{2}{5}k\overline{\Delta}_{I1}, \tag{9.168}$$

$$\frac{9}{10}\delta_{Q2} - \frac{1}{10}[\delta_{Q0} + \delta_{I2}] = 0. \tag{9.169}$$

Equations (9.167)–(9.169) are a set of algebraic conditions implying that the relations to be satisfied are:

$$\delta_{Q0} = \frac{5}{4}\delta_{I2}, \tag{9.170}$$

$$\delta_{Q2} = \frac{1}{4}\delta_{I2}, \tag{9.171}$$

$$\delta_{I2} = \frac{8}{15}k\overline{\Delta}_{I1}. \tag{9.172}$$

Recalling the original form of the expansion of the quadrupole as defined in Eq. (9.141), Eq. (9.172) can also be written as

$$\Delta_{I2} = \tau_c \delta_{I2} = \frac{8}{15}k\tau_c\overline{\Delta}_{I1}, \tag{9.173}$$

since to zeroth order the quadrupole vanishes and the first non-vanishing effect comes from the first-order quadrupole whose value is determined from the zeroth-order monopole.

9.8.1 *Improved estimates of polarization*

From Eqs. (9.170) and (9.171), the quadrupole moment of Δ_Q is proportional to the quadrupole of Δ_I, which is, in turn, proportional to the dipole evaluated to first order in τ_c. But Δ_Q measures exactly the degree of linear polarization of the radiation field. So, to first order in the tight-coupling expansion, the CMB is linearly polarized. Notice that the same derivation performed in the case of the equation for Δ_Q can be more correctly carried on in the case of the evolution equation of Δ_P with the same result [282]. Using the definition of S_P (i.e. Eq. (9.72)), and recalling Eqs. (9.170)–(9.172), we have the source term of Eq. (9.90) that can be approximated

as
$$S_{\rm P}(k,\tau) \simeq \frac{4}{3} k\tau_{\rm c} \overline{\Delta}_{\rm I1}(k,\tau). \tag{9.174}$$

Since $\tau_{\rm c}$ grows very rapidly during recombination, to have quantitative estimates of the effect we have to know the evolution of $S_{\rm P}$ with better accuracy [282]. With this goal in mind, let us go back to the (exact) system describing the coupled evolution of the various multipoles and, in particular, to Eqs. (9.132) and (9.137)–(9.138). Taking the definition of $S_{\rm P}$ (or $S_{\rm Q}$) and performing a first time derivative we have

$$S_{\rm P}' = \Delta_{\rm I2}' + \Delta_{\rm P2}' + \Delta_{\rm P0}'. \tag{9.175}$$

Then Eqs. (9.132), (9.137) and (9.138) can be used to reconstruct the same combination appearing in Eq. (9.175). Equations (9.132), (9.137) and (9.138) can be written, respectively, as

$$\Delta_{\rm P0}' - \frac{\epsilon'}{2}[\Delta_{\rm P2} + \Delta_{\rm I2} - \Delta_{\rm P0}] = -k\Delta_{\rm Q1}, \tag{9.176}$$

$$\Delta_{\rm I2}' + \epsilon'\left[\frac{9}{10}\Delta_{\rm I2} - \frac{1}{10}(\Delta_{\rm P0} + \Delta_{\rm P2})\right] = -\frac{3}{5}k\Delta_{\rm I3} + \frac{2}{5}k\Delta_{\rm I1}, \tag{9.177}$$

$$\Delta_{\rm P2}' + \epsilon'\left[\frac{9}{10}\Delta_{\rm P2} - \frac{1}{10}(\Delta_{\rm P0} + \Delta_{\rm I2})\right] = -\frac{3}{5}k\Delta_{\rm P3} + \frac{2}{5}k\Delta_{\rm P1}. \tag{9.178}$$

Summing up Eqs. (9.176), (9.177) and (9.178) and recalling the definitions of $S_{\rm P}$ and $S_{\rm P}'$ the following equation can be easily obtained:

$$S_{\rm P}' + \frac{3}{10}\epsilon' S_{\rm P} = k\left[\frac{2}{5}\Delta_{\rm I1} - \frac{3}{5}\left(\Delta_{\rm P1} + \Delta_{\rm P3} + \Delta_{\rm I3}\right)\right]. \tag{9.179}$$

Clearly this equation is more accurate than the bare tight-coupling expansion. So now the logic will be:

- to evaluate the source terms in Eq. (9.179) by using the results valid to first-order in the tight coupling expansion;
- to integrate the resulting equation;
- to obtain, as a result of the previous step, an improved estimate of $S_{\rm P}$.

This is the program which will now be implemented by first solving Eq. (9.179). By evaluating the right hand side of Eq. (9.179) we obtain:

$$S_{\rm P}' + \frac{3}{10}\epsilon' S_{\rm P} = k\frac{2}{5}\overline{\Delta}_{\rm I1}(k,\tau). \tag{9.180}$$

This result is certainly accurate since, to first order, the octupole vanish while the dipole of the polarization is also vanishing. To understand the meaning of the present derivation, just compare Eq. (9.180) to Eq. (9.174) which was obtained in the bare tight-coupling approximation. Equation (9.180) can now be solved formally as

$$S_P(k,\tau) = \frac{2}{5} k e^{3\epsilon(\tau,\tau_0)/10} \int_0^\tau d\tau' \overline{\Delta}_{I1}(k,\tau') e^{-3\epsilon(\tau',\tau_0)/10}. \qquad (9.181)$$

It is relevant to note that, according to our definitions,

$$\frac{d\epsilon}{d\tau} = -\frac{a}{a_0} n_e \sigma_{Th}, \qquad \epsilon(\tau,\tau_0) = \int_\tau^{\tau_0} \frac{a}{a_0} n_e \sigma_{Th} d\tau. \qquad (9.182)$$

According to Eq. (9.182) we have, in the present notations, $d\epsilon/d\tau = -\epsilon'$ (a crucial minus sign appears as a consequence of the definition of $\epsilon(\tau,\tau_0)$ where τ is in the lower limit of integration). This odd occurrence is pretty common in CMB physics and it is related to the way the visibility function is defined. Some authors like to insert a minus sign directly in the definition of the visibility function $\mathcal{K}(\tau)$. We will proceed quietly by following our conventions and we will bear in mind the caveat expressed by Eq. (9.182). The integral appearing in Eq. (9.181) is over the variable τ'. Let us then transform the integration variable from τ' to

$$y = \frac{\epsilon(\tau,\tau_0)}{\epsilon(\tau',\tau_0)}, \qquad \frac{dy}{d\tau'} = -\frac{\epsilon(\tau,\tau_0)}{\epsilon^2(\tau',\tau_0)} \frac{d\epsilon}{d\tau'}. \qquad (9.183)$$

The quantity $d\epsilon/d\tau'$ can be estimated from the width of the visibility function and it can be written as

$$\frac{d\epsilon}{d\tau'} = \frac{\epsilon(\tau',\tau_0)}{\sigma_{rec}}, \qquad \frac{dy}{d\tau'} = -\frac{y}{\sigma_{rec}}. \qquad (9.184)$$

where the second relation in Eq. (9.184) is a consequence of Eq. (9.183). Equation (9.181) can therefore be rearranged as

$$S_P(k,\tau) = \frac{2}{5} k \overline{\Delta}_{I1}(\tau_{rec}) \sigma_{rec} e^{3\epsilon/10} \int_1^\infty e^{-3\epsilon y/10} \frac{dy}{y}. \qquad (9.185)$$

Concerning Eq. (9.185) few comments are in order:

- the limits of integration have been changed according to the change of variables of Eq. (9.183); this occurrence can be intuitively understood since for $\tau \to \tau_{dec}$ the mean free path tends to infinity (notice, indeed, that y defined in Eq. (9.183) is a normalized version of the mean free path);
- the dipole of the intensity has been evaluated directly at recombination so it can be taken out of the integral.

The result expressed by Eq. (9.185) can be used to evaluate the brightness perturbation of the polarization. Indeed, integrating along the line of sight and using Eq. (9.90), Eq. (9.185) implies:

$$\Delta_{\rm P}(k,\mu,\tau) = -\frac{3}{10}k(1-\mu^2)\overline{\Delta}_{\rm I1}(\tau_{\rm rec})e^{ik\mu(\tau_{\rm rec}-\tau_0)}\sigma_{\rm rec}\mathcal{I}_1, \qquad (9.186)$$

where

$$\mathcal{I}_1 = \int_0^\infty d\epsilon\, e^{-7\epsilon/10}\int_1^\infty \frac{dy}{y}e^{-3\epsilon y/10} \simeq 1.71996. \qquad (9.187)$$

The result appearing in the second equality of Eq. (9.187) is just the result of the numerical integration which leads to a convergent result over the domain of integration. Collecting the numerical factors and rearranging the terms of Eq. (9.186) the following result can easily be obtained:

$$\Delta_{\rm P} \simeq -(1-\mu^2)e^{ik\mu(\tau_{\rm rec}-\tau_0)}\mathcal{D}(k),$$
$$\mathcal{D}(k) \simeq 0.515\, k\, \sigma_{\rm rec}\overline{\Delta}_{\rm I1}(\tau_{\rm rec}). \qquad (9.188)$$

Equation (9.188) allows us to estimate with reasonable accuracy the angular power spectrum of the cross-correlation between temperature and polarization (see below), for instance, in the case of the adiabatic mode.

9.8.2 *Polarization power spectra*

While the derivation of the polarization dependence of Thompson scattering has been conducted within the framework of the tight-coupling approximation, it is useful to recall here that these properties follow directly from the polarization dependence of Thompson scattering whose differential cross-section can be written as

$$\frac{d\sigma}{d\Omega} = r_0^2|\epsilon^{(\alpha)}\cdot\epsilon^{(\alpha')}|^2 \equiv \frac{3\sigma_{\rm Th}}{8\pi}|\epsilon^{(\alpha)}\cdot\epsilon^{(\alpha')}|^2. \qquad (9.189)$$

where $\epsilon^{(\alpha)}$ is the incident polarization and $\epsilon^{(\alpha')}$ is the scattered polarization; r_0 is the classical radius of the electron and $\sigma_{\rm Th}$ is, as usual, the total Thompson cross-section.

Suppose that the incident radiation is not polarized, i.e. $U = V = Q = 0$; then we can write

$$Q = \mathcal{I}_x - \mathcal{I}_y = 0, \qquad \mathcal{I}_x = \mathcal{I}_y = \frac{\mathcal{I}}{2}. \qquad (9.190)$$

Defining the incoming and outgoing polarization vectors as

$$\epsilon_x = (1,\ 0,\ 0), \qquad \epsilon_y = (0,\ 1,\ 0), \qquad \hat{k} = (0,\ 0,\ 1).$$
$$\epsilon'_x = (-\sin\varphi,\ -\cos\varphi,\ 0), \qquad (9.191)$$
$$\epsilon'_y = (\cos\vartheta\cos\varphi,\ -\cos\vartheta\sin\varphi,\ -\sin\vartheta),$$

the explicit form of the scattered amplitudes will be:

$$\mathcal{I}'_x = \frac{3\sigma_{Th}}{8\pi}\left[|\epsilon_x \cdot \epsilon'_x|^2 \mathcal{I}_x + |\epsilon_y \cdot \epsilon'_x|^2 \mathcal{I}_y\right] = \frac{3\sigma_{Th}}{16\pi}\mathcal{I},$$
$$\mathcal{I}'_y = \frac{3\sigma_{Th}}{8\pi}\left[|\epsilon_x \cdot \epsilon'_y|^2 \mathcal{I}_x + |\epsilon_y \cdot \epsilon'_y|^2 \mathcal{I}_y\right] = \frac{3\sigma_{Th}}{16\pi}\mathcal{I}\cos^2\vartheta.$$
(9.192)

Recalling the definition of Stokes parameters:

$$I' = \mathcal{I}'_x + \mathcal{I}'_y = \frac{3}{16\pi}\sigma_{Th}\mathcal{I}(1+\cos^2\vartheta),$$
$$Q' = \mathcal{I}'_x - \mathcal{I}'_y = \frac{3}{16\pi}\sigma_{Th}\mathcal{I}\sin^2\vartheta.$$
(9.193)

Even if $U' = 0$ the obtained Q and U must be rotated to a common coordinate system:

$$Q' = \cos 2\varphi Q, \qquad U' = -\sin 2\varphi Q. \qquad (9.194)$$

So the final expressions for the Stokes parameters of the scattered radiation are:

$$I' = \frac{3}{16\pi}\sigma_{Th}\mathcal{I}(1+\cos^2\vartheta),$$
$$Q' = \frac{3}{16\pi}\sigma_{Th}\mathcal{I}\sin^2\vartheta\cos 2\varphi,$$
$$U' = -\frac{3}{16\pi}\sigma_{Th}\mathcal{I}\sin^2\vartheta\sin 2\varphi.$$
(9.195)

We can now expand the incident intensity in spherical harmonics:

$$\mathcal{I}(\theta,\varphi) = \sum_{\ell m} a_{\ell m} Y_{\ell m}(\vartheta,\varphi). \qquad (9.196)$$

So, for instance, Q' will be

$$Q' = \frac{3}{16\pi}\sigma_{Th}\int \sum_{\ell m} Y_{\ell m}(\vartheta,\varphi)a_{\ell m}\sin^2\vartheta\cos 2\varphi d\Omega. \qquad (9.197)$$

By inserting the explicit form of the spherical harmonics into Eq. (9.197) it can be easily shown that $Q' \neq 0$ provided the term $a_{22} \neq 0$ in the expansion of Eq. (9.196). The analysis performed in terms of Δ_Q and Δ_U can also be expressed in terms of the E and B modes. Recalling, in fact, that under clockwise rotations of φ the quantities $\mathcal{M}_\pm = \Delta_Q \pm i\Delta_U$ transform as

$$\tilde{\mathcal{M}}_\pm = e^{\mp 2i\varphi}\mathcal{M}_\pm \qquad (9.198)$$

the combinations

$$\mathcal{M}_\pm(\hat{n}) = \sum_{\ell m} a_{\pm 2,\ell m} \, {}_{\pm 2}\mathcal{Y}_{\ell m}(\hat{n}) \qquad (9.199)$$

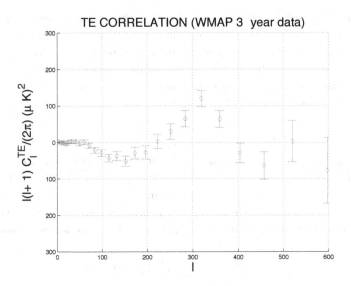

Fig. 9.2 The anticorrelation between the intensity fluctuations and the fluctuations in the degree of linear polarization is illustrated as presented in the WMAP 3-year data [39, 40].

can be expanded in terms of the spin-2 spherical harmonics, i.e. $\pm_2 \mathcal{Y}_\ell^m(\hat{n})$, with

$$a_{\pm 2,\ell m} = \int d\hat{n} \;\; \pm_2 \mathcal{Y}_{\ell m}^*(\hat{n}) \; (\Delta_Q \pm i\Delta_U)(\hat{n}). \tag{9.200}$$

The "electric" and "magnetic" components of polarization are eigenstates of parity and may be defined as

$$\begin{aligned} a_{\ell m}^{\rm E} &= -\frac{1}{2}(a_{2,\ell m} + a_{-2,\ell m}), \\ a_{\ell m}^{\rm B} &= \frac{i}{2}(a_{2,\ell m} - a_{-2,\ell m}). \end{aligned} \tag{9.201}$$

These newly defined variables are expanded in terms of ordinary spherical harmonics, $Y_{\ell m}(\hat{n})$,

$$\mathrm{E}(\hat{n}) = \sum_{\ell m} a_{\ell m}^{\rm E} Y_{\ell m}(\hat{n}), \qquad \mathrm{B}(\hat{n}) = \sum_{\ell m} a_{\ell m}^{\rm B} Y_{\ell m}(\hat{n}). \tag{9.202}$$

In connection with the spin-2 spherical harmonics appearing in Eqs. (9.199) and (9.200) it is relevant to mention here that a generic spin-s spherical harmonic is a known concept in the quantum mechanical theory of angular momentum (see, for instance, [314]). A typical quantum mechanical problem is to look for the representations of the operator specifying three-dimensional rotations, i.e. $\hat{R}$; this problem is usually approached within the so-called Wigner matrix elements, i.e. $\mathcal{D}^{(j)}_{mm'}(R) = \langle j, m'|\hat{R}|j, m\rangle$ where j denotes the eigenvalue of J^2 and m denotes the eigenvalue of J_z. Now, if we replace $m' \to -s$, $j \to \ell$, we have the definition of spin-s spherical harmonics in terms of the $\mathcal{D}^{(\ell)}_{-s,m}(\alpha, \beta, 0)$, i.e.

$$_s\mathcal{Y}_{\ell m}(\alpha, \beta) = \sqrt{\frac{2\ell+1}{4\pi}} \mathcal{D}^{(\ell)}_{-s,m}(\alpha, \beta, 0), \tag{9.203}$$

where α, β and γ (set to zero in the above definition) are the Euler angles defined as in [314]. If $s = 0$, $\mathcal{D}^{(\ell)}_{0,m}(\alpha, \beta, 0) = \sqrt{(2\ell+1)/4\pi}Y_{\ell m}(\alpha, \beta)$ where $Y_{\ell m}(\alpha, \beta)$ are the ordinary spherical harmonics.

To define properly the cross power spectra we also recall the expansion of the intensity fluctuations of the radiation field, in terms of spherical harmonics:

$$\Delta_I(\hat{n}) = \sum_{\ell m} a^T_{\ell m} Y_{\ell m}(\hat{n}), \tag{9.204}$$

where we wrote explicitly $a^T_{\ell m}$ since the fluctuations in the intensity of the radiation field, i.e. Δ_I are nothing but the fluctuations in the CMB temperature.

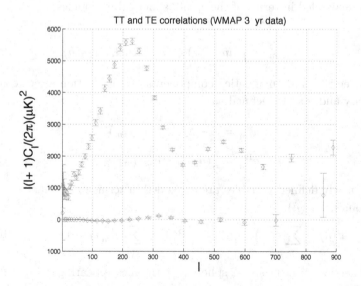

Fig. 9.3 The TE correlations and the TT correlations are reported on the same scale to display their relative magnitude. The lower plot in this figure corresponds to the data points already reported in Fig. 9.2.

Under parity inversion, the components appearing in Eqs. (9.202) and (9.204) transform as

$$a_{\ell m}^{\rm E} \to (-1)^\ell\, a_{\ell m}^{\rm E}, \qquad a_{\ell m}^{\rm B} \to (-1)^{\ell+1}\, a_{\ell m}^{\rm B}, \qquad a_{\ell m}^{\rm T} \to (-1)^\ell\, a_{\ell m}^{\rm T}. \tag{9.205}$$

Therefore, the E-modes have the same parity of the T-modes which have, in turn, the same parity of spherical harmonics, i.e. $(-1)^\ell$. On the contrary, the B-modes have $(-1)^{\ell+1}$ parity.

The existence of linear polarization allows for 6 different cross power spectra to be determined, in principle, from data that measure the full temperature and polarization anisotropy information. The cross power spectra can be defined in terms of the spectral functions $C_\ell^{X,Y}$ where X and Y stand for E, B or T depending on the cross-correlation one is interested in:

$$C_\ell^{X,Y} = \frac{1}{2\pi^2} \int k^2\, dk \sum_{m=-\ell}^{\ell} \frac{(a_{\ell m}^X)^* a_{\ell m}^Y}{(2\ell+1)}. \tag{9.206}$$

Therefore, if we are interested in the TE correlations we just have to set $X = {\rm T}$ and $Y = {\rm E}$ and use the relevant expansions given above. In the following, we will denote the correlations as TT, EE, BB, TB and so on. This notation refers to the definition given in Eq. (9.206).

Let us now see how the definition (9.206) works. Suppose we are interested in the TT correlations, i.e. the usual and well known temperature correlations. From Eq. (9.206) we will have

$$C_\ell^{\rm TT} = \frac{1}{2\pi^2} \int k^2\, dk \sum_{m=-\ell}^{\ell} \frac{[a_{\ell m}^{\rm T}(k)]^* a_{\ell m}^{\rm T}(k)}{(2\ell+1)}. \tag{9.207}$$

Now, from Eq. (9.204), using the orthogonality of spherical harmonics, we have

$$a_{\ell m}^{\rm T}(k) = \int d\hat{n}\, Y_{\ell m}(\hat{n}) \Delta_{\rm I}(\vec{k}, \hat{n}). \tag{9.208}$$

Inserting Eq. (9.208) into Eq. (9.207) and recalling the expansion of $\Delta_{\rm I}(\vec{k}, \hat{n})$ in terms of Legendre polynomials we get

$$C_\ell^{\rm TT} = \frac{2}{\pi} \int dk\, k^2 |\Delta_{{\rm I}\,\ell}|^2. \tag{9.209}$$

To get to Eq. (9.209) the following two identities have been used, i.e.

$$\int d\hat{n}\, P_{\ell'}(\hat{k} \cdot \hat{n}) Y_{\ell m}^*(\hat{n}) = \frac{4\pi}{(2\ell+1)} \delta_{\ell\ell'},$$

$$\sum_{m=-\ell}^{\ell} Y_{\ell m}^*(\hat{k}) Y_{\ell m}(\hat{n}) = \frac{2\ell+1}{4\pi} P_\ell(\hat{k} \cdot \hat{n}). \tag{9.210}$$

The second identity in Eq. (9.210) has already been exploited in chapter 3.

In similar ways, different expressions for the other correlations may be obtained. Notice that Eq. (9.209) is a consequence of the specific conventions adopted in the expansion of the brightness perturbations in Legendre polynomials. In particular, note that the a factor $(2\ell + 1)$ is included in the expansion. It must be clearly stated that this is matter of convention. Consequently, the cross correlations will inherit extra-terms that are simply a consequence of the different conventions adopted. So, for instance, in Eq. (9.209) and in the conventions of [276], a factor $(2\ell + 1)^2$ typically appears in the denominator.

Recalling the connection between the Wigner matrix elements and the spin-s spherical harmonics, i.e. Eq. (9.203), it is possible to show that, under complex conjugation,

$$a^*_{\pm 2,\ell m} = (-1)^m a_{\mp 2,\ell m}, \qquad \left(a^{\mathrm{T,E,B}}_{\ell,m}\right)^* = (-1)^m a^{\mathrm{T,E,B}}_{\ell,-m}, \qquad (9.211)$$

where the second equality follows from the first one by using Eq. (9.201). It is then possible to show that while the TB and EB correlators are parity-odd, all the other correlators (i.e. TT, BB, EE, TE) are parity even.

Scalar perturbations generate only the E mode [315]. While the scalar fluctuations only generate an E mode, tensor fluctuations also generate a B mode. The Boltzmann equation for the tensor modes can be easily derived following Refs. [316, 317] (see also [284, 318]).

Consider now, specifically, the adiabatic mode. While the temperature fluctuation Δ_{I} oscillates like $\cos(kc_{\mathrm{s,b}}\tau_{\mathrm{rec}})$, the polarization is proportional to the dipole and oscillates like the sine of the same argument. The correlation function of the temperature and polarization, i.e. $\langle \Delta_{\mathrm{I}}\Delta_{\mathrm{P}}\rangle$ will then be proportional to $\sin(kc_{\mathrm{s,b}}\tau_{\mathrm{rec}})\cos(kc_{\mathrm{s,b}}\tau_{\mathrm{rec}})$. An analytical prediction for this quantity can be inferred from Eq. (9.188) (see [282]). The spectrum of the cross correlation must then have a peak for $kc_{\mathrm{s,b}}\tau_{\mathrm{rec}} \sim 3\pi/4$, corresponding to $\ell \sim 150$. This is the result suggested by Figs. 9.2 and 9.3 (see also [37] reporting the measurement of the WMAP collaboration). Note that in Fig. 9.2 the TE spectrum has been reported alone while in Fig. 9.3 the TT and TE spectra have been illustrated together to emphasize the relative weight of the two power spectra.

We should mention here that a rather effective method to treat on equal footing the scalar, vector and tensor radiative transfer equations is the total angular momentum method [294, 319, 320]. Within this approach, the collision terms couple only the quadrupole moments of the distributions and each moment corresponds directly to observable patterns in the microwave sky. In this language the analysis of the polarization of the radiation field becomes somehow more transparent.

We conclude this section by defining the relation between the quantity called Δ_P (see for instance Eq. (9.72) and the power spectrum for the E-modes. Recall that the polarization depends effectively on the two polar angles ϑ and φ. Defining as $\vec{\ell}$ the vector conjugate to $\vec{\vartheta}$ we can Fourier transform the Stokes parameters and pass from the ϑ space to the $\vec{\ell}$ space. Recalling the transformation properties for the fluctuations of the brightness perturbations already introduced in Eq. (9.55) we can write, in $\vec{\ell}$ space

$$E(\vec{\ell}) = \overline{Q}(\vec{\ell}) \cos 2\varphi_\ell + \overline{U}(\vec{\ell}) \sin 2\varphi_\ell,$$
$$B(\vec{\ell}) = -\overline{Q}(\vec{\ell}) \sin 2\varphi_\ell + \overline{U}(\vec{\ell}) \cos 2\varphi_\ell, \qquad (9.212)$$

where $\overline{Q}(\vec{\ell}) = \Delta_{P\ll} \cos 2\varphi_\ell$ and $\overline{U}(\vec{\ell}) = \Delta_{P\ll} \sin 2\varphi_\ell$. It then follows from Eq. (9.212) that the angular power spectrum of B vanishes exactly, while the angular power spectrum of E does not vanish and, in particular, $C^{EE}(\ell) \simeq C^{PP}(\ell)$ and $C^{ET}(\ell) \simeq C^{PT}(\ell)$ (where both relations hold, strictly speaking, in the large ℓ limit, i.e. $\ell \gg 1$). Thus we will have

$$C_\ell^{EE} \simeq \frac{2}{\pi} \int k^3 d\ln k |\Delta_P(k, \tau_0)|^2,$$
$$C_\ell^{ET} \simeq \frac{2}{\pi} \int k^3 d\ln k |\Delta_P(k, \tau_0)\Delta_I(k, \tau_0)| \qquad (9.213)$$

9.9 Second order in tight-coupling: diffusion damping

The results of the tight-coupling expansion hold for $k\tau_c \ll 1$. Thus the present approximation scheme breaks down, strictly speaking, for wave-numbers $k > \tau_c^{-1}$. Equation (9.150) holds to zeroth-order in the tight-coupling expansion, i.e. it can only be applied on scales much larger than the photon mean free path. By comparing the rate of the Universe expansion with the rate of dissipation, we can estimate that $\tau_c k^2 \sim \tau^{-1}$ defines approximately the scale above which the wave-numbers will experience damping. From these considerations the typical damping scale can be approximated by

$$\frac{1}{k_D^2} = \frac{2}{5} \int_0^\tau \frac{c_{sb}^2}{(a/a_0)n_e x_e \sigma_{Th}} d\tau'. \qquad (9.214)$$

This is the result that has been already obtained in chapter 8 (see, in particular, Eq. (8.140)). The effect of diffusion is to damp the photon and baryon oscillations exponentially by the time of last scattering on comoving scales smaller than 3 Mpc. For an experimental evidence of this effect see [34] and references therein. In chapter 8 the diffusion damping has been estimated by including, as accurately as possible, the contribution of shear viscosity in the evolution equation of the baryon-photon velocity field. In that context, therefore, the diffusion damping has been introduced in some extrinsic approach, i.e. by simply adding the viscosity appropriately related to the photon mean free path. There is a more intrinsic way of discussing the diffusion damping and it has to do with the dispersion relations and, in particular, with the presence of an imaginary part in these relations. This analysis can be carried on, consistently, in the tight-coupling expansion. As we shall see below, this result arises to *second order* in the tight-coupling expansion. This statement already implies an important physical consideration, namely the occurrence that the polarization of the CMB (which arises to first order in the tight-coupling expansion) is essential for the correctness of the result. To perform the derivation we need:

- the evolution equation of v_b (see, for instance, Eq. (9.133));
- the evolution equation for Δ_{I1} (i.e. Eq. (9.134));
- the zeroth-order relation between v_b and the photon dipole (i.e. Eq. (9.146));
- the zeroth-order relation between the dipole and the monopole (i.e. Eq. (9.147));
- the evolution equation of the quadrupole (i.e. Eq. (9.137)) and the first-order relation between the quadrupole end the first-order dipole (i.e. Eq. (9.173)).

We can start with the evolution equation for v_b. This equation allows us to determine the precise relation between v_b and Δ_{I1} up to second order terms in the tight-coupling expansion parameter. Then we can express the monopole and dipole in terms of the dipole. Inserting everything into Eq. (9.137) the wanted dispersion relation can be obtained. Physics helps greatly in expediently achieving the correct result. Indeed, diffusion damping is a short-scale effect. This observation implies that:

- the metric fluctuations, over the diffusion damping scale, can be neglected;
- the expansion rate can be neglected.

Therefore, the terms $\mathcal{O}(\psi)$, $\mathcal{O}(\phi)$ and $\mathcal{O}(\mathcal{H}v_b)$ can be neglected in the corresponding equations.

With these warnings in mind let us take the generalized Laplace transform of the various terms. So, for instance,

$$v_{\rm b}(k,\tau) = v_{\rm b}(k,\omega)e^{i\int^\tau \omega(\tau')d\tau'},$$
$$\Delta_{\rm I1}(k,\tau) = \Delta_{\rm I1}(k,\omega)e^{i\int^\tau \omega(\tau')d\tau'}. \quad (9.215)$$

Consider then Eq. (9.133) which we can also write, neglecting the Hubble rate, as

$$v_{\rm b}(k,\omega) = -\frac{3i\Delta_{\rm I1}(k,\omega)}{1 + i\frac{\omega R_{\rm b}}{\epsilon'}}. \quad (9.216)$$

In what follows we will not write explicitly the eigenvalues in the argument of each function so, for instance, the shorthand notation $v_{\rm b}(k,\omega) \to v_{\rm b}$ will be used for $v_{\rm b}$ and analogously for the other functions appearing in the game (i.e. $\Delta_{\rm I1}$, $\Delta_{\rm I2}$ and so on and so forth). The right hand side of Eq. (9.216) can be expanded in powers of $1/|\epsilon'|$ with the obvious result that

$$v_{\rm b} \simeq -3i\Delta_{\rm I1}\left[1 - \frac{i\omega R_{\rm b}}{\epsilon'} - \left(\frac{\omega R_{\rm b}}{\epsilon'}\right)^2\right]. \quad (9.217)$$

Inserting then Eq. (9.217) into Eq. (9.137) the following relation can be easily derived:

$$-i\omega\Delta_{\rm I1} - \frac{2}{3}k\Delta_{\rm I2} + \frac{k}{3}\Delta_{\rm I0} = \epsilon'\Delta_{\rm I1}\left[\frac{i\omega R_{\rm b}}{\epsilon'} + \left(\frac{\omega R_{\rm b}}{\epsilon'}\right)^2\right]. \quad (9.218)$$

Now the trick is to eliminate $\Delta_{\rm I0}$ and $\Delta_{\rm I2}$ by using, respectively, the first-order and the second-order tight-coupling relations, i.e. from Eqs. (9.147) and (9.173):

$$\overline{\Delta}_{\rm I0} = -\frac{k}{i\omega}\overline{\Delta}_{\rm I1}, \qquad \Delta_{\rm I2} = \frac{8}{15}\frac{k}{\epsilon'}\overline{\Delta}_{\rm I1}. \quad (9.219)$$

Inserting Eq. (9.219) into Eq. (9.218) we will therefore get:

$$\left[-i\omega - \frac{16}{45}\frac{k^2}{\epsilon'} - \frac{k^2}{3i\omega}\right]\overline{\Delta}_{\rm I1} = \overline{\Delta}_{\rm I1}\epsilon'\left[\frac{i\omega R_{\rm b}}{\epsilon'} + \left(\frac{\omega R_{\rm b}}{\epsilon'}\right)^2\right]. \quad (9.220)$$

The dependence on the first-order dipole can therefore be simplified and, after simple algebra, the dispersion relation becomes:

$$\omega^2(R_{\rm b} + 1) - \frac{k^2}{3} - \frac{16}{45}\frac{k^2}{\epsilon'}i\omega = i\frac{\omega^3 R_{\rm b}}{\epsilon'} \quad (9.221)$$

To solve Eq. (9.221) the frequency can be written as $\omega = \overline{\omega} + \delta\omega$. The terms containing $\delta\omega$ can be thought of as being of higher order with respect to $\overline{\omega}$. Using this decomposition it is easy to obtain the following pair of relations

$$\overline{\omega} = \frac{k}{3\sqrt{R_{\rm b}+1}}, \qquad (9.222)$$

$$\delta\omega = \frac{i}{2(R_{\rm b}+1)}\left[\frac{16}{45}\frac{k^2}{\epsilon'} + \frac{R_{\rm b}^2\overline{\omega}}{\epsilon'}\right]. \qquad (9.223)$$

Equation (9.222) is the trivial dispersion relation which can be obtained in the lowest order of the tight-coupling expansion (see chapter 8 and also Eq. (9.150)). In Eq. (9.150) the relation (9.222) has been dubbed as $\overline{\omega} = kc_{\rm sb}$ where $c_{\rm sb}$ has been already defined in Eq. (8.80). Equation (9.223) defines the imaginary part of the frequency which leads, ultimately, to the diffusion damping. Inserting Eq. (9.222) into Eq. (9.223), $\delta\omega$ is determined to be

$$\delta\omega = \frac{i}{2(R_{\rm b}+1)}\left[\frac{16}{45}\frac{k^2}{\epsilon'} + \frac{R_{\rm b}^2}{3(R_{\rm b}+1)\epsilon'}\right]. \qquad (9.224)$$

Consider, finally, the expression of the velocity or of the monopole which can be obtained at a given order in the tight-coupling expansion. Such an expression will contain an oscillatory contribution and a damping contribution of the kind

$$e^{\pm i\int^\tau \overline{\omega}(\tau')d\tau'} e^{-k^2\int^\tau \delta\omega(\tau')d\tau'}. \qquad (9.225)$$

The oscillatory contribution in Eq. (9.225) is exactly the same factor arising in the evolution equation of the monopole (see Eq. (9.150)) and it has been previously parametrized in terms of the sound horizon since, by definition of the sound horizon (see Eq. (9.151)),

$$\int^\tau \overline{\omega}(\tau')d\tau' \equiv kr_{\rm s}(\tau). \qquad (9.226)$$

The damping factor appearing in Eq. (9.225) can also be written as $\exp\left(-k^2/k_{\rm D}^2\right)$ where $k_{\rm D}$ will now be the tight-coupling estimate of the diffusion damping which turns out to be, by identifying the appropriate terms,

$$\frac{1}{k_{\rm D}^2} = \int_0^\tau \frac{d\tau'}{6(R_{\rm b}+1)\epsilon'}\left[\frac{16}{15} + \frac{R_{\rm b}^2}{R_{\rm b}+1}\right]. \qquad (9.227)$$

The factor 16/15 arises since the polarization fluctuations are taken consistently into account in the derivation. This difference is physically relevant. Grossly speaking we can indeed say that

- more polarization implies more anisotropy (and vice versa);
- more polarization implies a faster damping by diffusion.

The accurate estimate of the diffusion damping scale is a necessary ingredient for the semi-analytical evaluation of the temperature and polarization autocorrelations.

Let us therefore complete the estimate provided by Eq. (9.227). The first observation we make is that the calculation of the diffusion damping reported in chapter 8 (and based on Eq. (9.214)) seems, at least naively, to underestimate the effect at least by a factor of 2. This statement can be easily justified by looking at Eq. (9.214) (deduced from the inclusion of shear viscosity in the zeroth order tight-coupling approximation) and by comparing it to the new estimate provided by Eq. (9.227). Let us therefore write both expressions in the same notations by recalling that $\epsilon' = (a/a_0)x_e n_e \sigma_{Th}$ as well as the explicit expression of c_{sb} in terms of the baryon-to-photon ratio R_b. Equations (9.214) and (9.227) can then be written as

$$\frac{1}{\overline{k}_D^2} = \frac{2}{15} \int_0^\tau \frac{d\tau'}{(R_b+1)\epsilon'}, \tag{9.228}$$

$$\frac{1}{k_D^2} = \frac{8}{45} \int_0^\tau \frac{d\tau'}{(R_b+1)\epsilon'} + \int_0^\tau \frac{R_b^2}{6(R_b+1)^2 \epsilon'}, \tag{9.229}$$

where we denoted with $\overline{k}_D$ the old estimate of the diffusion scale based on shear viscosity and that was extensively discussed in Eq. (8.160) of chapter 8. It is not difficult to see by looking at Eqs. (9.228) and (9.229) together that the new estimate of the diffusion damping is related to the old one by the following simple relation

$$\frac{1}{k_D^2} = \frac{4}{\overline{k}_D^2} + \int_0^\tau \frac{R_b^2}{6(R_b+1)^2 \epsilon'} d\tau', \tag{9.230}$$

which implies that, at least, $k_D^{-1} \simeq 2\overline{k}_D^{-1}$. We say at least because this consideration excludes the second integral appearing at the right hand side of Eq. (9.230). In slightly different words we can reach the same conclusion by taking the limit $R_b \to 0$ (corresponding to $c_{sb} \to 1/\sqrt{3}$). In this case the previous statement is exact. For the typical parameters of Eq. (8.160) $1/\overline{k}_D$ turned out to be of the order of 6 to 8 Mpc (recall the factor of 2 discussed at the end of chapter 8) the new estimates increases the old one in the range of 10–15 Mpc.

To corroborate this impression let us do the complete estimate of k_D^{-2} as it emerges from Eqs. (9.227) and (9.230). If we divide by τ_{rec}^2 both sides of Eq. (9.230) we can write

$$\frac{1}{k_D^2 \tau_{rec}^2} = \frac{4}{\overline{k}_D^2 \tau_{rec}^2} + \frac{1}{\tau_1} \left(\frac{\tau_1^2}{\tau_{rec}}\right)^2 \int_0^y \frac{R_b^2}{6(R_b+1)^2 \epsilon'} dx, \tag{9.231}$$

where, as in chapter 8, $x = \tau'/\tau_1$ and $y = \tau_{rec}/\tau_1$. The integral at the right hand side of Eq. (9.231) can be performed analytically since

$$\frac{1}{\tau_1}\left(\frac{\tau_1}{\tau_{rec}}\right)^2 \int_0^y \frac{R_b^2}{6(R_b^2+1)^2 \epsilon'} dx$$
$$= \frac{\beta^2}{6\tau_1}\left(\frac{\tau_1}{\tau_{rec}}\right)^2 \frac{1}{n_{\gamma 0}\eta_b \sigma_{Th}} \left(\frac{a_{eq}}{a_0}\right) \int_0^y \frac{x^4(x+2)^4}{[\beta(x^2+2x)\beta+1]^2} dx, \quad (9.232)$$

where we recalled, according to the notations established in Eq. (8.146), that $R_b = \beta(x^2 + 2x)$. The result of the explicit integration is not so inspiring but it can be expanded in powers of β (which is always smaller than one for the physical range of the parameters). It is then possible to derive the following equation

$$\frac{1}{k_D^2 \tau_{rec}^2} \simeq \frac{1}{\overline{k}_D^2 \tau_{rec}^2}\left(4 + \frac{25}{36}\beta^2 y^2\right). \quad (9.233)$$

Consider now a typical choice of the parameters. For $h_0^2 \Omega_{M0} \simeq 0.134$ and for $h_0^2 \Omega_{b0} \simeq 0.023$, $y \simeq 1.75$ and $\beta \simeq 0.669$. This implies that the estimate of diffusion damping derived on the basis of the second order in the tight-coupling expansion is always between 2 and $\sqrt{5}$ times larger than the estimate based on shear viscosity and reported in Eq. (8.160).

9.10 Semi-analytical approach to Doppler oscillations

It will now be shown how it is possible to obtain reasonable semi-analytical results on the Doppler oscillations by working in the framework of the tight-coupling expansion and by parametrizing appropriately both the finite thickness effects and the diffusive effects connected with Silk damping. It should be borne in mind that, for very large length-scales (i.e. $\ell < 30$) the analytical calculation of the SW plateau suffices to impose the large-scale normalization. For smaller angular scales the Bessel functions that appear ubiquitously in this kind of calculations can be expanded, not for large argument but for large order[k] (i.e. $\ell \gg 1$). This guiding observation will allow us, in turn, to reduce the problem of computing the temperature autocorrelations. It will suffice to compute numerically four integrals. The encouraging results indeed suggest that this approximation scheme captures the basic physics of the Doppler regime.

---

[k]This approach has been also exploited in [302–304]. The present discussion, however, does not aim at a full analytical evaluation of the temperature autocorrelations, since the resulting integrals, as we shall see, will be estimated numerically.

The main physical assumption of the present evaluation will be that the dominant source of inhomogeneities is provided, at an early time, by the adiabatic initial conditions. The main technical tool for the analysis is the tight-coupling expansion and the logic will therefore be to estimate the line of sight integrals by using the results obtained in the previous sections of this chapter. For immediate convenience we recall that the C_ℓ are defined as (see, for instance, Eqs. (3.25) and (3.27)):

$$\langle a_{\ell m} a^*_{\ell' m'} \rangle = C_\ell \delta_{\ell \ell'} \delta_{m m'}, \tag{9.234}$$

where, as we saw in chapter 3,

$$a_{\ell m} = \frac{4\pi(-i)^\ell}{(2\pi)^{3/2}} \int d^3 k Y^*_{\ell m}(\hat{k}) \Delta_{I\ell}(\vec{k}, \tau_0). \tag{9.235}$$

Using Eqs. (9.234) and (9.235) it is immediate to obtain the following expression

$$C_\ell \delta_{\ell \ell'} \delta_{m m'} = \frac{2}{\pi} \int d^3 k \, d^3 p \, \langle \Delta_{I\ell}(\vec{k}, \tau_0) \Delta^*_{I\ell'}(\vec{p}, \tau_0) \rangle Y_{\ell m}(\hat{k}) Y^*_{\ell' m'}(\hat{p}), \tag{9.236}$$

where the angular brackets denote ensemble average. From the line of sight integrals obtained in this chapter (see, in particular, Eq. (9.96)) the brightness perturbations can then be expressed, to lowest order in the tight-coupling expansion, as

$$\Delta_I(\vec{k}, \tau_0) = \int_0^{\tau_0} d\tau \, \mathcal{K}(\tau) [\overline{\Delta}_{I0} + \psi + \mu \bar{v}_b] e^{-i\mu x_0}, \qquad x_0 = k \tau_0. \tag{9.237}$$

The terms containing the time derivatives of the metric fluctuations will be neglected since, for the adiabatic mode, the time dependence is not crucial for intermediate angular scales and for the range of spectral index close to the Harrison-Zeldovich benchmark value. Integrating by parts Eq. (9.237) we can also write:

$$\Delta_I(\vec{k}, \tau_0) = \int_0^{\tau_0} d\tau \, \mathcal{K}(\tau) [\overline{\Delta}_{I0} + \psi + i\bar{v}_b \frac{d}{dx_0}] e^{-i\mu x_0}. \tag{9.238}$$

Both sides of Eq. (9.238) can now be expanded in series of Legendre polynomials,

$$\Delta_I(\vec{k}, \tau_0) = \sum_\ell (-i)^\ell (2\ell+1) P_\ell(\mu) \Delta_{I\ell}(\vec{k}, \tau_0),$$
$$e^{-i\mu x_0} = \sum_\ell (-i)^\ell (2\ell+1) P_\ell(\mu) j_\ell(x_0). \tag{9.239}$$

By now approximating the visibility function with a Gaussian (see Eq. (9.77)) the brightness perturbation can be written as:

$$\Delta_{I\ell}(k, \tau_0) = \mathcal{K}(\tau_{\text{rec}}) \left[(\overline{\Delta}_{I0} + \psi)_{\tau_{\text{rec}}} j_\ell(x_0) + i\bar{v}_b \frac{dj_\ell(x_0)}{dx_0} \right], \tag{9.240}$$

Recall now the tight-coupling relations derived before in this chapter (see Eqs. (9.146) and (9.147)) which we rewrite in the case $\psi' = 0$:

$$\overline{v}_b = -3i\overline{\Delta}_{I1}, \qquad \overline{\Delta}'_{I0} = -k\overline{\Delta}_{I1}. \tag{9.241}$$

Inserting Eq. (9.241) into Eq. (9.240) and using the Gaussian form of the visibility function discussed in Eq. (9.77) we have:

$$\Delta_{I\ell}(k,\tau_0) = \mathcal{K}(\tau_{\rm rec})\left[(\overline{\Delta}_{I0} + \psi)_{\tau_{\rm rec}} j_\ell(x_0) - 3\overline{\Delta}'_{I0}\frac{dj_\ell(x_0)}{dx_0}\right], \tag{9.242}$$

where, as previously discussed, $\mathcal{K}(\tau_{\rm rec}) = \exp[-\sigma^2\tau_{\rm rec}^2 k^2]$. Thus, the C_ℓ can be expressed as

$$C_\ell = \frac{2}{\pi}\int k^3 d\ln k |\Delta_{I\ell}(k,\tau_0)|^2. \tag{9.243}$$

where

$$|\Delta_{I\ell}(k,\tau_0)|^2 = \frac{e^{-2\sigma^2\tau_{\rm rec}^2 k^2}}{k^2}\left\{|\overline{\Delta}_{I0} + \psi|^2 j_\ell^2 + 9|\overline{\Delta}'_{I0}|^2\left(\frac{dj_\ell}{dx_0}\right)^2\right\}. \tag{9.244}$$

Equations (9.243) and (9.244) then imply:

$$C_\ell = \frac{2}{\pi}\int dk e^{-2\sigma^2\tau_{\rm rec}^2 k^2}\left\{|\overline{\Delta}_{I0} + \psi|^2 j_\ell^2 + 9|\overline{\Delta}'_{I0}|^2\left(\frac{dj_\ell}{dx_0}\right)^2\right\}. \tag{9.245}$$

Recalling the recurrence relations of Bessel functions [215, 216]

$$\left(\frac{dj_\ell}{dx_0}\right)^2 = \left[1 - \frac{\ell(\ell+1)}{x_0^2}\right]j_\ell(x_0)^2 + \frac{(x_0 j_\ell^2(x_0))''}{2x_0}, \tag{9.246}$$

Eq. (9.245) can be rearranged as

$$C_\ell = \frac{2}{\pi}\int dk e^{-2\sigma^2\tau_{\rm rec}^2 k^2}\left\{|\overline{\Delta}_{I0} + \psi|^2 + 9|\overline{\Delta}'_{I0}|^2\left[1 - \frac{\ell(\ell+1)}{x_0^2}\right]\right\}j_\ell(x_0)^2, \tag{9.247}$$

where the second term at the right hand side of Eq. (9.246) can be neglected since it will turn out to be of higher order.

The photon density contrast can be determined under the assumption that the entropic contribution is absent, as stressed at the beginning of this section. Thus, if only the adiabatic mode is present, the tight-coupling approximation leads to the following solution

$$\overline{\Delta}_{I0}(k,\tau) = -\frac{1}{3c_{\rm sb}^2}\psi(k,\tau) + \sqrt{c_{\rm sb}}\,A_1(k)\cos k r_s(\tau)\, e^{-\frac{k^2}{k_{\rm D}^2}}, \tag{9.248}$$

where $\psi(k,\tau)$ is assumed to be slowly varying in time and where $r_s(\tau)$ is the sound horizon. The constant $A_1(k)$ can be determined by matching the solution to the large-scale (i.e. super-Hubble) behaviour of the fluctuations, i.e.

$$\overline{\Delta}_{\text{I}0}(k,\tau) + \psi(k,\tau) \to \frac{\psi_{\text{m}}(k)}{3} = -\frac{\zeta_*(k)}{5}, \qquad (9.249)$$

where ψ_{m} denotes the value of $\psi(k)$ after equality and for $k\tau < 1$. From the solution of the evolution equation of $\overline{\Delta}_{\text{I}0}$, $\overline{v}_{\text{b}}$ can also be easily obtained. The final result can be expressed, for the present purposes, as

$$\left[\overline{\Delta}_{\text{I}0} + \psi\right] = \mathcal{L}_\zeta(k,\tau) + \mathcal{M}_\zeta(k,\tau)\mathcal{D}_\zeta(k)\sqrt{c_{\text{sb}}}\cos kr_{\text{s}}, \qquad (9.250)$$

$$\overline{\Delta}'_{\text{I}0} = k\, c_{\text{sb}}^{3/2} \mathcal{M}_\zeta(k,\tau)\mathcal{D}_\zeta(k)\sin kr_{\text{s}}. \qquad (9.251)$$

The functions $\mathcal{L}_\zeta(k,\tau)$ and $\mathcal{M}_\zeta(k,\tau)$ are directly related to the curvature perturbations and can be determined by interpolating the large-scale behaviour with the small-scale solutions. In the present case they can be written as[1]

$$\mathcal{L}_\zeta(k,\tau) = -\frac{\zeta_*(k)}{6}\left(1 - \frac{1}{3c_{\text{sb}}^2}\right)\ln\left[\frac{14}{w\ell}\frac{\tau_0}{\tau_{\text{eq}}}\right], \qquad (9.252)$$

$$\mathcal{M}_\zeta(k,\tau) = -\frac{6}{25}\zeta_*(k)\ln\left[\frac{14}{25}w\ell\frac{\tau_{\text{eq}}}{\tau_0}\right]. \qquad (9.253)$$

In Eqs. (9.252) and (9.253) the variable $w = k\tau_0/\ell$ has been introduced. This way of writing may seem, at the moment, obscure. However, the variable w will appear as an integration variable in the angular power spectrum, so it is practical, as early as possible, to express the integrands directly in terms of w. The function $\mathcal{D}_\zeta(k)$ encode the informations related to the diffusivity wave-number:

$$\mathcal{D}_\zeta(k) = e^{-\frac{k^2}{k_{\text{D}}^2}} \equiv e^{-\frac{\ell^2}{\ell_{\text{D}}^2}w^2}. \qquad (9.254)$$

As introduced before k_{D} is the thermal diffusivity scale (i.e. shear viscosity). For the purpose of simplifying the integrals to be evaluated numerically it is practical to introduce the following rescaled quantities:

$$\mathcal{L}_\zeta(k,\tau) = \zeta_*(k) L_\zeta(k,\tau), \qquad \mathcal{M}_\zeta(k,\tau) = \zeta_*(k) M_\zeta(k,\tau). \quad (9.255)$$

After some algebra the angular power spectrum can be written as the sum of four integrals, i.e.

$$C_\ell = \mathcal{U}_1(\ell) + \mathcal{U}_2(\ell) + \mathcal{U}_3(\ell) + \mathcal{U}_4(\ell), \qquad (9.256)$$

---

[1] Recall at this point the discussion on the transfer function for the adiabatic mode of chapter 8 and, in particular, Eqs. (8.57) and (8.59) as well as Fig. 8.2.

where:

$$\mathcal{U}_1(\ell) = 4\pi \int_0^\infty \frac{dw}{w} \overline{\mathcal{U}}_1(\ell, w) \mathcal{K}^2(\ell, \ell_t, w) j_\ell^2(\ell w), \tag{9.257}$$

$$\mathcal{U}_2(\ell) = 2\pi c_{\text{sb}} \int_0^\infty \frac{dw}{w^3} [w^2 + 9c_{\text{sb}}^2(w^2 - 1)]$$
$$\times \overline{\mathcal{U}}_2(\ell, w) \mathcal{K}^2(\ell, \ell_t, w) j_\ell^2(\ell w), \tag{9.258}$$

$$\mathcal{U}_3(\ell) = 2\pi c_{\text{sb}} \int_0^\infty \frac{dw}{w^3} [w^2 - 9c_{\text{sb}}^2(w^2 - 1)]$$
$$\times \overline{\mathcal{U}}_3(\ell, w) \cos(2\ell\gamma w) \mathcal{K}^2(\ell, \ell_t, w) j_\ell^2(\ell w), \tag{9.259}$$

$$\mathcal{U}_4(\ell) = 8\pi \sqrt{c_{\text{sb}}} \int_0^\infty \frac{dw}{w} \cos(\ell\gamma w)$$
$$\times \overline{\mathcal{U}}_4(\ell, w) \mathcal{K}^2(\ell, \ell_t, w) j_\ell^2(\ell w), \tag{9.260}$$

where $j_\ell(y)$ denote the spherical Bessel functions of the first kind [215, 216] which are related to the ordinary Bessel functions of the first kind as $j_\ell(y) = \sqrt{\pi/(2y)} J_{\ell+1/2}(y)$. The various functions appearing in Eqs. (9.257), (9.258), (9.259) and (9.260) are:

$$\overline{\mathcal{U}}_1(\ell, w) = \mathcal{P}_\zeta(w, \ell) L_\zeta^2(\ell, w), \tag{9.261}$$
$$\overline{\mathcal{U}}_2(\ell, w) = \mathcal{P}_\zeta(w, \ell) M_\zeta^2(\ell, w) \mathcal{D}_\zeta^2(\ell, \ell_D, w), \tag{9.262}$$
$$\overline{\mathcal{U}}_3(\ell, w) = \mathcal{P}_\zeta(w, \ell) M_\zeta^2(\ell, w) \mathcal{D}_\zeta^2(\ell, \ell_D, w), \tag{9.263}$$
$$\overline{\mathcal{U}}_4(\ell, w) = \mathcal{P}_\zeta(w, \ell) L_\zeta(\ell, w) M_\zeta(\ell, w) \mathcal{D}_\zeta(\ell, \ell_D, w), \tag{9.264}$$

where ℓ_D denotes the typical diffusion damping scale already introduced in Eq. (9.254).

In Eqs. (9.259) and (9.260) the oscillatory terms arising originally in the full expression of the angular power spectrum have been simplified. The two oscillatory contributions in Eqs. (9.259) and (9.260) go, respectively, as $\cos(2\gamma\ell w)$ and as $\cos(\gamma\ell w)$. The definition of γ can be easily deduced from the original parametrization of the oscillatory contribution in Eq. (9.248). In fact we can write $k r_s(\tau_{\text{rec}}) = \gamma(\tau_{\text{rec}}) \ell w$. Recalling that $w = k\tau_0/\ell$, and defining, for notational convenience, $\gamma \equiv \gamma(\tau_{\text{rec}})$, the following expression for γ can be easily obtained

$$\gamma = \frac{1}{\tau_0} \int_0^{\tau_{\text{rec}}} c_{\text{sb}}(\tau) d\tau = \frac{\tau_1}{\sqrt{3}\tau_0} \int_0^{\tau_{\text{rec}}/\tau_1} \frac{dx}{\sqrt{1 + \beta(x^2 + 2x)}}. \tag{9.265}$$

Equation (9.265) can be expressed in more explicit terms and the result is the one already derived in chapter 8 (see Eq. (8.151) and discussion therein). At this point the spherical Bessel functions appearing in the above

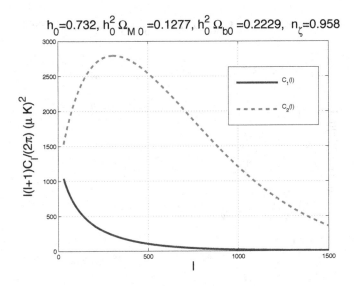

Fig. 9.4 The contribution of the integrals $C_1(\ell)$ and $C_2(\ell)$ appearing in Eqs. (9.268) and (9.269).

expressions can be evaluated in the limit of large ℓ with the result that the integrals can be made more explicit. In particular, focusing the attention on $j_\ell(\ell w)$, we have [215, 216]

$$j_\ell^2(\ell w) \simeq \frac{\cos^2[\beta(w,\ell)]}{\ell w \sqrt{\ell^2 w^2 - \ell^2}}, \quad w > 1. \tag{9.266}$$

The result expressed by Eq. (9.266) allows us to write the integrals of Eqs. (9.257), (9.258), (9.259) and (9.260) as

$$C_\ell = \frac{1}{\ell^2}[\mathcal{C}_1(\ell) + \mathcal{C}_2(\ell) + \mathcal{C}_3(\ell) + \mathcal{C}_4(\ell)], \tag{9.267}$$

where

$$\mathcal{C}_1(\ell) = \int_1^\infty I_1(w,\ell)\overline{\mathcal{U}}_1(\ell,w)\,dw, \tag{9.268}$$

$$\mathcal{C}_2(\ell) = \int_1^\infty I_2(w,\ell)\overline{\mathcal{U}}_2(\ell,w)\,dw, \tag{9.269}$$

$$\mathcal{C}_3(\ell) = \int_1^\infty I_3(w,\ell)\overline{\mathcal{U}}_3(\ell,w)\,dw, \tag{9.270}$$

$$\mathcal{C}_4(\ell) = \int_1^\infty I_4(w,\ell)\overline{\mathcal{U}}_4(\ell,w)\,dw, \tag{9.271}$$

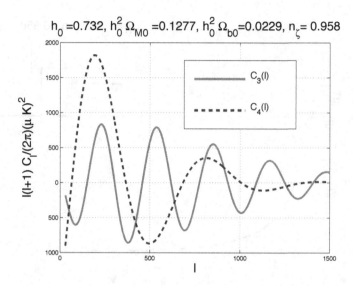

Fig. 9.5 The contribution of the integrals $C_3(\ell)$ and $C_4(\ell)$ appearing in Eqs. (9.270) and (9.271).

and where
$$I_1(w,\ell) = \frac{4\pi \cos^2[\beta(w,\ell)]}{w^2\sqrt{w^2-1}} e^{-2\frac{\ell^2}{\ell_t^2}w^2}, \tag{9.272}$$

$$I_2(w,\ell) = 2\pi c_{\text{sb}} \frac{w^2 + 9c_{\text{sb}}^2(w^2-1)}{w^4\sqrt{w^2-1}} \cos^2[\beta(w,\ell)] e^{-2\frac{\ell^2}{\ell_t^2}w^2}, \tag{9.273}$$

$$I_3(w,\ell) = 2\pi c_s \frac{w^2 - 9c_{\text{sb}}^2(w^2-1)}{w^4\sqrt{w^2-1}} \cos(2\ell\gamma w)$$
$$\times \cos^2[\beta(w,\ell)] e^{-2\frac{\ell^2}{\ell_t^2}w^2}, \tag{9.274}$$

$$I_4(w,\ell) = 8\pi\sqrt{c_{\text{sb}}} \frac{\cos(\ell\gamma w)}{w^2\sqrt{w^2-1}} \cos^2[\beta(w,\ell)] e^{-2\frac{\ell^2}{\ell_t^2}w^2}. \tag{9.275}$$

Concerning Eqs. (9.268)–(9.271) and (9.272)–(9.275) the following comments are in order:

- the lower limit of integration over w is 1 in Eqs. (9.268)–(9.271) since the asymptotic expansion of Bessel functions implies that $k\tau_0 \geq \ell$, i.e. $w \geq 1$;
- the obtained expressions valid for the angular power spectrum will be applicable for sufficiently large ℓ; in practice, as we shall see the obtained results are in good agreement with the data in the Doppler region;

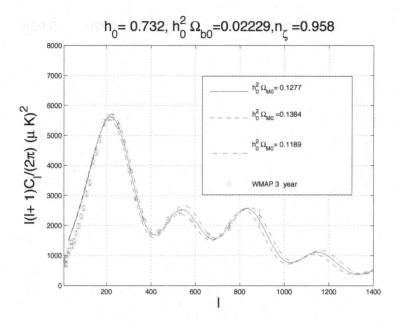

Fig. 9.6 The temperature autocorrelations for a fiducial set of cosmological parameters chosen within a concordance model. The variation of $h_0^2 \Omega_{M0}$ is illustrated.

- the function $\beta(w, \ell) = \ell\sqrt{w^2 - 1} - \ell \cos^{-1}(w^{-1}) - \frac{\pi}{4}$ leads to a rapidly oscillating argument whose effect will be to slow down the convergence of the numerical integration; it is practical, for the present purposes, to replace $\cos^2[\beta(w, \ell)]$ by its average (i.e. 1/2).

In the integrals (9.272), (9.273), (9.274) and (9.275) the scale ℓ_t stems from the finite thickness of the last scattering surface and it is defined as

$$\ell_t^2 = \frac{1}{\sigma^2}\left(\frac{\tau_0}{\tau_{\rm rec}}\right)^2, \qquad \sigma_{\rm rec} = 1.49 \times 10^{-2}\frac{\sqrt{z_{\rm rec} + z_{\rm eq}} + \sqrt{z_{\rm rec}}}{\sqrt{z_{\rm rec} + z_{\rm eq}}}. \qquad (9.276)$$

Furthermore, within the present approximations, one can also define the Silk damping scale ℓ_S as

$$\frac{\ell_t^2}{\ell_S^2} = \frac{\sigma^2 k_D^2 \tau_{\rm rec}^2}{\sigma^2 k_D^2 \tau_{\rm rec}^2 + 1}, \qquad (9.277)$$

To simplify further the obtained expressions we can also change variable in some of the integrals. Consider, as an example, the integrals appearing

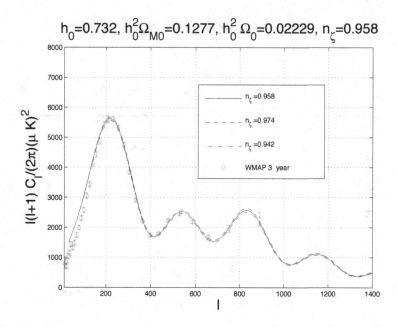

Fig. 9.7 The impact of the variation of n_ζ within the limits set by the concordance model.

in the expression of $C_1(\ell)$ (see Eq. (9.268)). Changing the variable of integration as $w = y^2 + 1$ we will have

$$C_1(\ell) = \int_0^\infty \overline{I}_1(y,\ell)\overline{\mathcal{U}}_1(\ell,y)dy, \qquad (9.278)$$

where, in explicit terms and after the change of variables,

$$\overline{I}_1(y,\ell) = \frac{4\pi}{(y^2+1)^2\sqrt{y^2+2}}e^{-2\frac{\ell^2}{\ell_t^2}(y^2+1)^2}, \qquad (9.279)$$

and

$$\overline{\mathcal{U}}_1(\ell,y) = \mathcal{A}_\zeta \left(\frac{k_0}{k_p}\right)^{n-1} \ell^{n-1} L_\zeta^2(\ell,y)(y^2+1)^{n-1}. \qquad (9.280)$$

With similar manipulations it is possible to also transform all the other integrands appearing in Eqs. (9.269), (9.270) and (9.271).

So far the necessary ingredients for the estimate of the temperature autocorrelations have been sorted out. In particular the angular power spectrum has been computed semi-analytically in the two relevant regions, i.e. the Sachs-Wolfe regime (corresponding to large angular scales and

$\ell \leq 30$) and the Doppler region, i.e. $\ell > 100$. Furthermore, for the nature of the approximations made we do not expect the greatest accuracy of the algorithm in the intermediate region (i.e. $30 < \ell < 100$). Indeed, it was recognized (in a different semi-analytical approach [276, 277]) that it is somehow necessary to smooth the joining of the two regimes by assuming an interpolating form of the metric fluctuations that depends upon two fitting parameters. We prefer here to stress that this method is inaccurate in the matching regime since the spherical Bessel functions have been approximated for large ℓ. Therefore, the comparison with experimental data should be preferentially conducted, for the present purposes, in the Doppler region.

Before plunging into the discussion, it is appropriate to comment on the choice of the cosmological parameters that will be employed throughout this section. The WMAP 3-year [39] data have been combined, so far, with various sets of data. These data sets include the 2dF Galaxy Redshift Survey [62], the combination of Boomerang and ACBAR data [63, 64], and the combination of CBI and VSA data [65, 66]. Furthermore the WMAP 3-year data can also be combined with the Hubble Space Telescope Key Project (HSTKP) data [67] as well as with the Sloan Digital Sky Survey (SDSS) [68, 69] data. Finally, the WMAP 3-year data can also be usefully combined with the weak lensing data [70, 71] and with the observations of type Ia supernovae[m] (SNIa).

Each of the data sets mentioned in the previous paragraph can be analyzed within different frameworks. The minimal ΛCDM model with no cut-off in the primordial spectrum of the adiabatic mode and with vanishing contribution of tensor modes is the simplest concordance framework. This is the one that has been adopted in this paper. Diverse completions of this minimal model are possible: they include the addition of the tensor modes, a sharp cut-off in the spectrum and so on and so forth.

All these sets of data (combined with different theoretical models) lead necessarily to slightly different determinations of the relevant cosmological parameters. To have an idea of the range of variations of the parameters the following examples are useful[n]:

- the WMAP 3-year data alone [39] (in a ΛCDM framework) seem to favour a slightly smaller value $h_0^2 \Omega_{M0} = 0.127$;

---

[m]In particular the data of the Supernova Legacy Survey (SNLS) [72] and the so-called Supernova "Gold Sample" (SNGS) [73, 74].

[n]The values quoted for all the cosmological observables always refer to the case of a spatially flat Universe where the semi-analytical calculation has been performed.

- if the WMAP 3-year data are combined with the "gold" sample of SNIa [73] (see also [74]) the favoured value $h_0^2\Omega_{M0}$ is of the order of 0.134; if the WMAP 3-year data are combined with *all* the data sets $h_0^2\Omega_{M0} = 0.1324$.
- similarly, if the WMAP data alone are considered, the preferred value of $h_0^2\Omega_{b0}$ is 0.02229 while this value decreases to 0.02186 if the WMAP data are combined with all the other data sets.

The aforementioned list of statements refers to the case of a pure ΛCDM model. If, for instance, tensors are included, then the WMAP 3-year data combined with CBI and VSA increase by a little the value of $h_0^2\Omega_{b0}$ which becomes in this case, closer to 0.023. While it might be interesting to study the same problems in non-minimal ΛCDM scenarios, here the logic will be to take a best fit model to the WMAP data alone, compare it with the numerical scheme proposed in this section, and consequently, assess the accuracy of the semi-analytical method.

Consider, therefore, the case of a ΛCDM model with no tensors. In Figs. 9.4 and 9.5 the contribution of each of the integrals appearing in Eq. (9.267) is illustrated. The analytical form of these integrals has been derived in Eqs. (9.268), (9.269), (9.270) and (9.271). In Fig. 9.4 the separate contributions of $\ell(\ell+1)C_1(\ell)/(2\pi)$ and of $\ell(\ell+1)C_2(\ell)/(2\pi)$ have been reported for a fiducial set of parameters[o] (i.e. $n_\zeta = 0.958$, $h_0^2\Omega_{M0} = 0.1277$ and $h_0^2\Omega_{b0} = 0.0229$). This fiducial set of parameters corresponds to the best fit of the WMAP 3-year data alone [39]. As mentioned the pivot wave-number is $k_p = 0.002\,\mathrm{Mpc}^{-1}$. This is also the choice made by the WMAP team. In the plot of Fig. 9.5 the separate contributions of $\ell(\ell+1)C_3(\ell)/(2\pi)$ and of $\ell(\ell+1)C_4(\ell)/(2\pi)$ is illustrated for the same fiducial set of parameters (which is also described at the top of the plot). The various contributions are expressed in units of $(\mu K)^2$ (i.e. $1\mu K = 10^{-6}K$) which are the appropriate ones for comparison with the data. The normalization of the calculation is set by evaluating (analytically) the large-scale contribution for $\ell < 30$ and by comparing it, in this region, with the WMAP 3-year data release.

By summing up the four separate contributions illustrated in Fig. 9.4, Eq. (9.267) allows to determine, for a given choice of cosmological parameters, the full temperature autocorrelations. The results are reported in

---

[o] The scalar spectral index of the adiabatic mode is denoted, in this section, by n_ζ to emphasize that the Sachs-Wolfe normalization has been enforced using the spectrum of curvature perturbations and not the spectrum of the longitudinal fluctuations.

Figs. 9.6 and 9.7. In the plot at the left of Fig. 9.6 the critical fractions of matter and baryons, as well as h_0, are all fixed. The only quantity allowed to vary from one curve to the other is the scalar spectral index of curvature perturbations, i.e. n_ζ. The full line denotes the pivot case $n_\zeta = 0.958$ (corresponding to the central value for the spectral index as determined according to the WMAP data alone). The dashed and dotted-dash lines correspond, respectively, to $n_\zeta = 0.974$ and $n_\zeta = 0.942$ (which define the allowed range of n_ζ since $n_\zeta = 0.958 \pm 0.016$ [39]).

As we have already stressed, the regime $\ell < 100$ is only reasonably reproduced while the most interesting region, for the present purposes, is rather accurate (as the comparison with the WMAP data shows). The region of very large ℓ (i.e. $\ell > 1200$) is also beyond the treatment of diffusive effects adopted in the present approach. In Figure 9.7 the adiabatic spectral index is fixed (i.e. $n_\zeta = 0.958$) while the total (present) fraction of non-relativistic matter is allowed to vary (h_0 and $h_0^2 \Omega_{b0}$ are, again, kept fixed). It can be observed that, according to Fig. 9.7, the amplitude of the first peak increases as the total (dusty) matter fraction decreases.

Chapter 10

Early Initial Conditions?

The presentation of the CMB anisotropies has been developed, in the last three chapters, through a bottom-up approach. It has been argued that the properties of the temperature and polarization autocorrelations are determined from a set of initial conditions that can include either an adiabatic mode, or a non-adiabatic mode, or both (see, in particular, the final part of chapter 7).

The most general set of initial conditions required to integrate the Einstein-Boltzmann hierarchy is formed by one adiabatic mode and by four non-adiabatic modes.[a] In this approach, the initial conditions are set prior to equality but after neutrino decoupling. The wording *initial conditions* of CMB anisotropies may also have, in the present literature, a complementary (but crucially different) meaning. If we are to believe a compelling and not erroneous picture of the thermodynamic history of the Universe, also for temperatures much larger than 200 GeV, then initial conditions for the fluctuations of the FRW metric may be assigned not prior to decoupling but much earlier, for instance during a stage of inflationary expansion.

While the logic summarized in the previous paragraph is the one followed in chapters 7, 8 and 9, there is also a second (complementary) approach that has been partially explored already in chapter 6 when computing, in a simplified situation, the spectrum of relic gravitons produced thanks to the sudden transition form the de Sitter stage of expansion to the radiation-dominated stage of expansion. In Fig. 6.2 the spectrum of relic gravitons has been discussed for models making diverse assumptions about both the thermodynamic history of the Universe and about the nature of the laws of gravity at short distances. There is then no surprise that the

---

[a]Different gauge descriptions of the adiabatic and non-adiabatic modes can also be found in chapter 11.

obtained phenomenological signatures for the graviton spectra are indeed different.

It is rather understandable, therefore, that this second approach necessarily demands the adoption of a specific model of evolution of the background geometry valid well before the moment when weak interactions fall out of thermal equilibrium. For this purpose the standard approach is to suppose that the Universe underwent a phase of accelerated expansion that was replaced by a stage dominated by radiation. This oversimplified picture can be matched, below temperatures of the order of 1 MeV, to the firmer model of evolution that has been discussed in the previous chapters. In this sense the relevant plots that illustrate the overall evolution of the Hubble radius are the ones reported in Figs. 4.3 and 4.4. Concerning this general lore various observations are in order:

- if the duration of inflation is minimal (see Fig 4.3) the initial conditions may not be necessarily quantum mechanical;
- if the duration of inflation is nonminimal (see Fig. 4.4) the initial conditions for the scalar and tensor modes are, most likely, of quantum mechanical origin;
- the initial inflationary stage may be realized either through a single scalar degree of freedom or by means of a collection of scalar fields;
- in the case when more than one field is present, non-adiabatic modes are expected ab initio;
- if inflation is realized through a single scalar field, there is still the possibility that various other (spectator) fields are present during inflation and they may modify the evolution of curvature perturbations.

It is legitimate to stress that while we believe that we have a clear and verified picture of the history of the Universe for $T < 1$ MeV, the same confidence may not be justified in the case of the early stages of the life of the Universe. If initial conditions for the scalar and tensor modes are set during the inflationary epoch, the tacit assumption is that we do know pretty well the evolution of our Universe between the $H_{\rm inf} \simeq 10^{-5} M_{\rm P} \simeq 10^{15}$ GeV and $H_{\rm BBN} \simeq 10^{-31}$ GeV (which is, according to Eq. (B.44) the curvature scale corresponding to a temperature $T \simeq$ MeV). In this section we are going to assume that only a single scalar degree of freedom drives inflation and we will ask the question of what are the curvature fluctuations induced by the fluctuations of the inflaton. On a more technical ground, it is appropriate to mention that similar mathematical developments are required in the treatment of inhomogeneities in quintessence models driven by a single scalar field.

10.1 Minimally coupled scalar field

In diverse situations it is important to compute the fluctuations induced by a single (minimally coupled) scalar degree of freedom. This exercise is therefore technically relevant. The first step in this direction will be to write down the fluctuations of the energy-momentum tensor. We will do this in the conformally Newtonian gauge[b] and we will then learn how to translate the obtained result in any other gauge needed for the resolution of physical problems. Consider therefore the energy-momentum tensor of a scalar field φ characterized by a potential $V(\varphi)$. To first-order in the amplitude of the (scalar) metric fluctuations we will have:

$$\delta_s T_\mu^\nu = \delta_s g^{\nu\alpha}\partial_\alpha\varphi\partial_\mu\varphi + 2\overline{g}^{\nu\alpha}\partial_\alpha\varphi\partial_\mu\chi$$
$$-\delta_\mu^\nu\left[\frac{1}{2}\delta_s g^{\alpha\beta}\partial_\alpha\varphi\partial_\beta\varphi + \overline{g}^{\alpha\beta}\partial_\alpha\chi\partial_\beta\varphi - \frac{\partial V}{\partial\varphi}\right], \quad (10.1)$$

where $\chi = \delta\varphi$ is the first-order fluctuation of the scalar field in the conformally Newtonian gauge and $\overline{g}_{\alpha\beta}$ denotes the background metric while $\delta_s g_{\alpha\beta}$ are its first-order fluctuations. In the conformally Newtonian gauge, we recall, the only nonvanishing components are $\delta_s g_{00} = 2a^2\phi$ and $\delta_s g_{ij} = 2a^2\psi\delta_{ij}$. Using the results of Appendix C (in particular Eqs. (C.25), (C.28), (C.29) and (C.30)), component by component, Eq. (10.1) will give

$$\delta T_0^0 = \delta_s\rho_\varphi, \qquad \delta T_i^j = -\delta_s p_\varphi \delta_i^j, \qquad \delta T_0^i = \varphi'\partial^i\chi, \quad (10.2)$$

where

$$\delta_s\rho_\varphi = \frac{1}{a^2}\left[-\phi\varphi'^2 + \chi'\varphi' + a^2\frac{\partial V}{\partial\varphi}\chi\right],$$
$$\delta_s p_\varphi = \frac{1}{a^2}\left[-\phi\varphi'^2 + \chi'\varphi' - a^2\frac{\partial V}{\partial\varphi}\chi\right]. \quad (10.3)$$

The anisotropic stress arises, in the case of a single scalar field, only to second order in the amplitude of the fluctuations. This means that the anisotropic stress of a minimally coupled scalar field will contain two derivatives of χ so it will be, in stenographic notation, of $\mathcal{O}(|\partial\chi|^2)$ (see also Eqs. (5.50) and (5.52)). Since we are here perturbing

---

[b] As already emphasized in the previous chapters the longitudinal and conformally Newtonian are just used here as synonyms.

Einstein equations to first order, this contribution will be neglected. The perturbed Einstein equations are then easily written. In particular, by using the explicit fluctuations of the Einstein tensors reported in Eqs. (C.10), (C.11) and (C.12) of Appendix C we have:

$$\nabla^2 \psi - 3\mathcal{H}(\mathcal{H}\phi + \psi') = 4\pi G \left[-\phi\varphi'^2 + \chi'\varphi' + a^2 \frac{\partial V}{\partial \varphi}\chi \right], \quad (10.4)$$

$$\psi'' + \mathcal{H}(\phi' + 2\psi') + (\mathcal{H}^2 + 2\mathcal{H}')\phi + \frac{1}{3}\nabla^2(\phi - \psi)$$
$$= 4\pi G \left[-\phi\varphi'^2 + \chi'\varphi' - a^2 \frac{\partial V}{\partial \varphi}\chi \right], \quad (10.5)$$

$$\mathcal{H}\phi + \psi' = 4\pi G \varphi' \chi, \quad (10.6)$$

together with the condition $\phi = \psi$ since, as already remarked, the perturbed energy-momentum tensor of a single scalar degree of freedom does not possess, to first-order, an anisotropic stress. To equations (10.4), (10.5) and (10.6) it is sometimes practical to add the perturbed Klein-Gordon equation (see Eqs. (C.36), (C.38) and (C.39) of the Appendix C):

$$\chi'' + 2\mathcal{H}\chi' - \nabla^2 \chi + \frac{\partial^2 V}{\partial \varphi^2} a^2 \chi + 2\phi \frac{\partial V}{\partial \varphi} a^2 - \varphi'(\phi' + 3\psi') = 0. \quad (10.7)$$

10.1.1 Gauge-invariant description

Noticing that the gauge variation of the scalar field fluctuation reads

$$\chi \to \tilde{\chi} = \chi - \epsilon_0 \varphi', \quad (10.8)$$

the gauge-invariant generalization of ϕ, ψ and χ is given by the following three quantities[c]:

$$\Psi = \psi - \mathcal{H}(B - E'), \quad (10.9)$$
$$\Phi = \phi + \mathcal{H}(B - E') + (B - E')', \quad (10.10)$$
$$X = \chi + \varphi'(B - E'). \quad (10.11)$$

Sometimes Φ and Ψ are called Bardeen potentials [194]. An interesting property of the conformally Newtonian gauge is that, in terms of Ψ, Φ and X, the evolution equations have the same form they would have in terms

[c]While the discussion developed throughout this book is self-contained, the interested reader might also want to consult the interesting review of Ref. [321] which is based on the longitudinal variables and their gauge-invariant generalizations. As already mentioned in chapters 7 and 8 the perspective adopted in the present book is not to stick to a single gauge but rather to profit the gauge freedom in order to simplify the derivations. This approach will be even more clearly spelled out in chapter 11.

of ψ, ϕ and χ. So the corresponding evolution equations for Ψ, Φ and X can be obtained from Eqs. (10.4), (10.5) and (10.6) by replacing

$$\psi \to \Psi, \qquad \phi \to \Phi, \qquad \chi \to X. \tag{10.12}$$

The result of this trivial manipulation is given by

$$\nabla^2 \Psi - 3\mathcal{H}(\mathcal{H}\Phi + \Psi') = 4\pi G a^2 \delta^{(gi)} \rho_\varphi, \tag{10.13}$$

$$\Psi'' + \mathcal{H}(\Phi' + 2\Psi') + (\mathcal{H}^2 + 2\mathcal{H}')\Phi + \frac{1}{3}\nabla^2(\Phi - \Psi)$$
$$= 4\pi G a^2 \delta^{(gi)} \rho_\varphi, \tag{10.14}$$

$$\mathcal{H}\Phi + \Psi' = 4\pi G \varphi' X, \tag{10.15}$$

where the gauge-invariant energy density and pressure fluctuations arise from Eq. (10.3)

$$\begin{aligned}\delta^{(gi)} \rho_\varphi &= \frac{1}{a^2}\left[-\Phi\varphi'^2 + X'\varphi' + a^2 \frac{\partial V}{\partial \varphi} X\right], \\ \delta^{(gi)} p_\varphi &= \frac{1}{a^2}\left[-\Phi\varphi'^2 + X'\varphi' - a^2 \frac{\partial V}{\partial \varphi} X\right].\end{aligned} \tag{10.16}$$

Suppose we are interested in the evolution equations of the fluctuations in a gauge that is totally different from the conformally Newtonian gauge. The solution of this problem is very simple. Take the evolution equations written in explicitly gauge-invariant terms. Then express the gauge-invariant quantities in the gauge you like. Finally substitute the expressions of the gauge-invariant quantities (now expressed in a specific gauge) back into the gauge-invariant equations. You will obtain very quickly the evolution equations in the gauge you like. Let us give an example of this procedure. Consider, for instance, the so-called uniform field gauge, i.e. the gauge where the scalar field is homogeneous (see also chapter 11). Such a gauge is defined by the following two conditions:

$$\tilde{\chi} = 0, \qquad \tilde{E} = 0. \tag{10.17}$$

If we start from a generic gauge, we get to the uniform field gauge by fixing the relevant gauge parameters, i.e. ϵ and ϵ_0 to the following values:

$$\epsilon_0 = \frac{\chi}{\varphi'}, \qquad \epsilon = E. \tag{10.18}$$

These two conditions can be obtained from Eqs. (10.8) and (6.21) by imposing, respectively, $\tilde{\chi} = 0$ and $\tilde{E} = 0$. Notice that this gauge fixing eliminates completely the gauge freedom since ϵ_0 and ϵ are not determined up to arbitrary constants. In the uniform field gauge, the gauge-invariant quantities introduced in Eqs. (10.9), (10.10) and (10.11) assume the following form

$$\Phi = \phi + \mathcal{H}B + B', \qquad \Psi = \psi - \mathcal{H}B, \qquad X = \varphi' B. \tag{10.19}$$

It is now easy to get all the evolution equations. Consider, for instance, the following two equations, i.e.

$$\mathcal{H}(\mathcal{H}\Phi + \Psi') = 4\pi G\varphi' X, \qquad \Phi = \Psi, \qquad (10.20)$$

which are, respectively, the gauge-invariant form of the momentum constraint and the gauge-invariant condition stemming from the off-diagonal terms of the perturbed Einstein equations. Let us now use Eq. (10.19) in Eq. (10.20). The following pair of equations can be quickly obtained:

$$\psi' + \mathcal{H}\phi = 0, \qquad \phi = \psi - (B' + 2\mathcal{H}B). \qquad (10.21)$$

With a similar procedure the other relevant equations can also be transformed in the uniform field gauge. Notice that, depending on the problem at hand, gauge-dependent calculations may become much shorter than fully gauge-invariant treatments. This will be the guiding theme of the considerations collected in chapter 11.

10.1.2 Curvature perturbations and scalar normal modes

Among all the gauge-invariant quantities some combinations have a special status. For instance the so-called curvature perturbations [199, 200] or the gauge-invariant density constrast [198, 199] (see also [33] and references therein):

$$\mathcal{R} = -\left(\Psi + \frac{\mathcal{H}(\mathcal{H}\Phi + \Psi')}{\mathcal{H}^2 - \mathcal{H}'}\right), \qquad (10.22)$$

$$\zeta = -\left(\Psi + \mathcal{H}\frac{\delta^{(\mathrm{gi})}\rho_\varphi}{\rho'_\varphi}\right), \qquad (10.23)$$

where ρ_φ denotes the (background) energy density of the scalar field

$$\rho_\varphi = \left(\frac{\varphi'^2}{2a^2} + V\right), \qquad (10.24)$$

and where $\delta^{(\mathrm{gi})}\rho_\varphi$ has been introduced in Eq. (10.16). The variable $\mathcal{R}$, already discussed in chapter 7, has a simple interpretation in the comoving orthogonal gauge where it is exactly the fluctuation of the spatial curvature. In similar terms, ζ is nothing but the density contrast on hyeprsurfaces where the curvature is uniform. These two variables are clearly connected. In fact, putting Eqs. (10.22) and (10.23) into Eq. (10.16) we get to the following simple expression:

$$\mathcal{R} = \zeta - \frac{\nabla^2 \Psi}{12\pi G \varphi'^2}. \qquad (10.25)$$

The message of Eq. (10.25) is very important. It tells us that when the wavelength of the fluctuations exceeds the Hubble radius, $\mathcal{R} \simeq \zeta + \mathcal{O}(k^2\tau^2)$. In other words, as long as $k\tau \ll 1$ (i.e. when the relevant wavelength is larger than the Hubble radius), the variables ζ and $\mathcal{R}$ have the same spectrum. The variable $\mathcal{R}$ also has a special status since the scalar-tensor action perturbed to second order in the amplitude of the metric and scalar field fluctuations assumes the following simple form [322, 323] (see also [33]):

$$S_{\mathrm{s}} = \delta_{\mathrm{s}}^{(2)} S = \frac{1}{2} \int d^4x z^2 \eta^{\alpha\beta} \partial_\alpha \mathcal{R} \partial_\beta \mathcal{R}, \qquad (10.26)$$

where

$$z = \frac{a\varphi'}{\mathcal{H}}. \qquad (10.27)$$

The Euler-Lagrange equations derived from Eq. (10.26) imply

$$\mathcal{R}'' + 2\frac{z''}{z}\mathcal{R}' - \nabla^2 \mathcal{R} = 0. \qquad (10.28)$$

Equation (10.28) is derived in Appendix C (see, in particular, the algebra prior to Eq. (C.53)). The canonical normal mode that can be easily read off from Eqs. (10.26) and (10.28) is then $q = -z\mathcal{R}$. In terms of q the action (10.26) then becomes, up to total derivatives,

$$S_{\mathrm{s}} = \frac{1}{2} \int d^4x \left[q'^2 + \frac{z''}{z} q^2 - \partial_i q \partial^i q \right]. \qquad (10.29)$$

The canonical action (10.29) is exactly of the same form as of the one discussed in Eq. (6.61) for the tensor modes of the geometry. What changes is essentially the form of the pump field which is, in the case of the tensor modes, $a''/a = \mathcal{H}^2 + \mathcal{H}'$. In the case of the scalar modes the pump field is z''/z instead. From Eq. (10.29) the quantum theory of the scalar modes can be easily developed in full analogy with what has been done in the case of the tensor modes. In particular the canonical normal modes q and their conjugate momenta (i.e. q') can be promoted to the status of quantum mechanical operators obeying equal-time commutation relations. Recalling the notations of chapter 6 (see, in particular, Eq. (6.72)) the field operator will now be expanded as

$$\hat{q}(\vec{x}, \tau) = \frac{1}{2(2\pi)^{3/2}} \int d^3k \left[\hat{q}_{\vec{k}} e^{-i\vec{k}\cdot\vec{x}} + \hat{q}_{\vec{k}}^\dagger e^{i\vec{k}\cdot\vec{x}} \right], \qquad (10.30)$$

and analogously for the conjugate momentum. Also the phenomenon of super-adiabatic amplification (discussed in chapter 6) is simply translated in the context of the scalar modes since the operators $\hat{q}$ obey now, in Fourier space, a Schrödinger-like equation in the Heisenberg representation:

$$\hat{q}_{\vec{k}}'' + \left[k^2 - \frac{z''}{z}\right]\hat{q}_{\vec{k}} = 0. \quad (10.31)$$

Exactly as in the case of tensors[d] (see, for instance Eqs. (6.41) and (6.77)) Eq. (10.31) admits two physical regimes: the oscillating regime (i.e. $k^2 \gg |z''/z|$) and the super-adiabatic regime (i.e. $k^2 \ll |z''/z|$) where the field operators are amplified and, in a more correct terminology, scalar phonons may be copiously produced. The overall simplicity of these results must not be misunderstood. The perfect analogy between scalar and tensor modes only holds in the case of a *single* scalar field. Already in the case of two scalar degrees of freedom the generalization of these results is much more cumbersome.[e] In the case of scalar fields and fluid variables, furthermore, the perfect mirroring between tensors and scalars is somehow lost.

10.2 Spectral relations

10.2.1 *Some slow-roll algebra*

As a simple exercise the spectral relations (typical of single-field inflationary models) will now be derived. The logic of the derivation will be to connect the spectral slopes and amplitudes of the scalar and of the tensor modes to the slow-roll parameters introduced in chapter 5:

$$\epsilon = -\frac{\dot{H}}{H^2} = \frac{\overline{M}_{\rm P}^2}{2}\left(\frac{V_{,\varphi}}{V}\right)^2,$$

$$\eta = \frac{\ddot{\varphi}}{H\dot{\varphi}} = \epsilon - \overline{\eta}, \qquad \overline{\eta} = \overline{M}_{\rm P}^2 \frac{V_{,\varphi\varphi}}{V}, \quad (10.32)$$

where the terms $V_{,\varphi}$ and $V_{,\varphi\varphi}$ denote, respectively, the first and second derivatives of the potential with respect to φ. The slow-roll parameters affect the definition of the conformal time coordinate τ. In fact, by definition

$$\tau = \int \frac{dt}{a(t)} = -\frac{1}{aH} + \epsilon\int \frac{da}{a^2 H} \quad (10.33)$$

---

[d]The analog of Eq. (10.31) in the case of an irrotational fluid supporting scalar fluctuations has been studied long ago by Lukash in Ref. [322]. In that context it has been argued that it described the production of phonons in isotropic Universes. The quantum normalization has been also discussed

[e]Concerning this point see chapter 11 and the related discussion of the off-diagonal gauge.

where the second equality follows after integration by parts assuming that ϵ is constant (as it happens in the case when the potential, at least locally, can be approximated with a monomial in φ). Since

$$\int \frac{dt}{a} = \int \frac{da}{a^2 H}, \qquad (10.34)$$

Eq. (10.33) allows us to express aH in terms of τ:

$$aH = -\frac{1}{\tau(1-\epsilon)}. \qquad (10.35)$$

Using these observations, the pump fields of the scalar and tensor modes of the geometry can be expressed solely in terms of the slow-roll parameters. In particular, in the case of the tensor modes it is easy to derive the following chain of equality on the basis of the relation between cosmic and conformal time, and using Eq. (10.32):

$$\frac{a''}{a} = \mathcal{H}^2 + \mathcal{H}' = a^2 H^2 (2 + \frac{\dot{H}}{H^2}) = a^2 H^2 (2 - \epsilon), \qquad (10.36)$$

Inserting Eq. (10.35) into Eq. (10.36) we will also have, quite simply

$$\frac{a''}{a} = \frac{2-\epsilon}{\tau^2(1-\epsilon)^2}. \qquad (10.37)$$

The evolution equation for the tensor mode functions is

$$f_k'' + \left[k^2 - \frac{a''}{a}\right] f_k = 0, \qquad (10.38)$$

whose solution is

$$f_k(\tau) = \frac{\mathcal{N}}{\sqrt{2k}} \sqrt{-k\tau} H_\nu^{(1)}(-k\tau), \qquad \mathcal{N} = \sqrt{\frac{\pi}{2}} e^{i\pi(2\nu+1)/4}, \qquad (10.39)$$

where $H_\nu^{(1)}(-k\tau)$ are the Hankel functions [215, 216] already encountered in chapter 6. In Eq. (10.39) the relation of ν tp ϵ is determined from the relation

$$\nu^2 - \frac{1}{4} = \frac{2-\epsilon}{(1-\epsilon)^2}, \qquad (10.40)$$

which implies

$$\nu = \frac{3-\epsilon}{2(1-\epsilon)}. \qquad (10.41)$$

The same algebra allows us to determine the relation of the scalar pump field with the slow-roll parameters. In particular, the scalar pump field is

$$\frac{z''}{z} = \left(\frac{z'}{z}\right)^2 + \left(\frac{z'}{z}\right)'. \qquad (10.42)$$

and the corresponding evolution equation for the mode functions follows from Eq. (10.31) and can be written as

$$\tilde{f}_k'' + \left[k^2 - \frac{z''}{z}\right]\tilde{f}_k = 0. \tag{10.43}$$

Recalling now the explicit expression of z, i.e.

$$z = \frac{a\varphi'}{\mathcal{H}} = \frac{a\dot{\varphi}}{H}, \tag{10.44}$$

we will have

$$\frac{\dot{z}}{z} = H + \frac{\ddot{\varphi}}{\dot{\varphi}} - \frac{\dot{H}}{H}. \tag{10.45}$$

But using Eq. (10.44), Eq. (10.42) can be expressed as

$$\frac{z''}{z} = a^2\left[\left(\frac{\dot{z}}{z}\right)^2 + H\frac{\dot{z}}{z} + \frac{\partial}{\partial t}\left(\frac{\dot{z}}{z}\right)\right]. \tag{10.46}$$

Using Eq. (10.45) inside Eq. (10.46) we get an expression that is the scalar analog of Eq. (10.36):

$$\frac{z''}{z} = a^2 H^2 (2 + 2\epsilon + 3\eta + \epsilon\eta + \eta^2). \tag{10.47}$$

Notice that the explicit derivatives appearing in Eq. (10.46) lead to two kinds of terms:

- the terms of the first kind can be immediately written in terms of the slow-roll parameters;
- the second kind of terms involve three (time) derivatives either of the scalar field or of the Hubble parameter.

In these two cases we can still say, from the definitions of ϵ and η, that

$$\frac{\partial^3 \varphi}{\partial t^3} = \eta \dot{H}\dot{\varphi} + \eta H \ddot{\varphi}, \qquad \ddot{H} = -2\epsilon H \dot{H}. \tag{10.48}$$

Again, using Eq. (10.35) inside Eq. (10.47) we get

$$\frac{z''}{z} = \frac{(2 + 2\epsilon + 3\eta + \epsilon\eta + \eta^2)}{(1-\epsilon)^2 \tau^2}, \tag{10.49}$$

implying that the relation of the Bessel index $\tilde{\nu}$ to the slow-roll parameters is now determined from [215, 216]

$$\tilde{\nu}^2 - \frac{1}{4} = \frac{2 + 2\epsilon + 3\eta + \epsilon\eta + \eta^2}{(1-\epsilon)^2}. \tag{10.50}$$

By solving Eq. (10.50) with respect to $\tilde{\nu}$ the following simple expression can be readily obtained:
$$\tilde{\nu} = \frac{3 + \epsilon + 2\eta}{2(1 - \epsilon)}. \tag{10.51}$$
Consequently, the solution of Eq. (10.43) will be, formally, the same as the one of the tensors (see Eq. (10.39)) but with a Bessel index $\tilde{\nu}$ instead of ν:
$$\tilde{f}_k(\tau) = \frac{\tilde{\mathcal{N}}}{\sqrt{2k}} \sqrt{-k\tau} H_{\tilde{\nu}}^{(1)}(-k\tau), \quad \tilde{\mathcal{N}} = \sqrt{\frac{\pi}{2}} e^{i\pi(2\tilde{\nu}+1)/4}, \tag{10.52}$$
Of course, this formal analogy should not be misunderstood: the difference in the Bessel index will entail, necessarily, a different behaviour in the small argument limit of Hankel functions [215, 216] and, ultimately, slightly different spectra whose essential features will be the subject of the remaining part of the present section.

10.2.2 Tensor power spectra

The tensor power-spectrum, in a given model, is the Fourier transform of the two-point function of the corresponding fluctuations (either tensor or scalar, in the examples discussed along the present section). Consider, therefore, the two-point function of the tensor modes of the geometry computed in the operator formalism:
$$\langle 0 | \hat{h}_i^j(\vec{x}, \tau) \hat{h}_j^i(\vec{y}, \tau) | 0 \rangle = \frac{8\ell_P^2}{a^2} \langle 0 | \hat{\mu}(\vec{x}, \tau) \hat{\mu}(\vec{y}, \tau) | 0 \rangle, \tag{10.53}$$
where μ are the operators corresponding to the canonical normal modes of the tensor problem discussed in chapter 6. By now evaluating the expectation value we obtain
$$\langle 0 | \hat{h}_i^j(\vec{x}, \tau) \hat{h}_j^i(\vec{y}, \tau) | 0 \rangle = \frac{8\ell_P^2}{a^2} \int \frac{d^3k}{(2\pi)^3} |f_k(\tau)|^2 e^{-i\vec{k}\cdot\vec{r}}. \tag{10.54}$$
By making explicit the phase-space integral in Eq. (10.54) we get
$$\langle 0 | \hat{h}_i^j(\vec{x}, \tau) \hat{h}_j^i(\vec{y}, \tau) | 0 \rangle = \int d\ln k \, \mathcal{P}_T(k) \frac{\sin kr}{kr}, \tag{10.55}$$
where $\mathcal{P}_T(k)$ is the tensor power spectrum, i.e.
$$\mathcal{P}_T(k) = \frac{4\ell_P^2}{a^2 \pi^2} k^3 |f_k(\tau)|^2. \tag{10.56}$$
But from Eq. (10.39) we have
$$|f_k(\tau)|^2 = \frac{|\mathcal{N}|^2}{2k} (-x) H_\nu^{(1)}(-x) H_\nu^{(2)}(-x) \simeq \frac{\Gamma^2(\nu)}{4\pi k} 2^{2\nu} (-x)^{1-2\nu}. \tag{10.57}$$

The second equality in Eq. (10.57) follows from the small argument limit of Hankel functions [215, 216]:

$$H_\nu^{(1)}(-x) \simeq -\frac{i}{\pi}\Gamma(\nu)\left(-\frac{x}{2}\right)^{-\nu}, \qquad (10.58)$$

for $|x| \ll 1$. The physical rationale for the small argument limit is that we are considering modes whose wavelengths are larger than the Hubble radius, i.e. $|x| = k\tau \ll 1$. Notice that, in the last equation, $\Gamma(\nu)$ denotes, as usual, the Euler Γ function [215, 216]. Equation (10.56) then gives the *super-Hubble* tensor power spectrum, i.e. the spectrum valid for those modes whose wavelength is larger than the Hubble radius at the decoupling, i.e.

$$\mathcal{P}_T(k) = \ell_P H^2 \frac{2^{2\nu}}{\pi^3}\Gamma^2(\nu)(1-\epsilon)^{2\nu-1}\left(\frac{k}{aH}\right)^{3-2\nu}. \qquad (10.59)$$

In Eq. (10.59), the term $(1-\epsilon)^{2\nu-1}$ arises by eliminating τ in favor of $(aH)^{-1}$ as dictated by Eq. (10.34). There are now different (but equivalent) ways of expressing the result of Eq. (10.59). Recalling that $\ell_P = \overline{M}_P^{-1}$, the spectrum (10.59) at horizon crossing (i.e. $k \simeq Ha$) can be expressed as

$$\mathcal{P}_T(k) = \frac{2^{2\nu}}{\pi^3}\Gamma^2(\nu)(1-\epsilon)^{2\nu-1}\left(\frac{H^2}{\overline{M}_P^2}\right)_{k\simeq aH}. \qquad (10.60)$$

The subscript arising in Eq. (10.60) refers to the moment at which a given wavelength crosses the Hubble radius is defined as the time at which

$$\frac{k}{\mathcal{H}} = \frac{k}{aH} \simeq k\tau \simeq 1. \qquad (10.61)$$

This condition is also called (somehow improperly, as discussed in chapter 4), horizon crossing. Notice once more that the equalities in Eq. (10.61) simply follow from the relation between the cosmic and the conformal time coordinate (see, for instance, Eq. (2.68)).

As discussed in chapter 5, the slow-roll parameters are much smaller than one during inflation and become of order 1 as inflation ends. Now the typical scales relevant for CMB anisotropies crossed the Hubble radius the first time (see Figs. 4.3 and 4.4) about 60 e-folds before the end of inflation (see Eqs. (5.119) and (5.120)) when the slow-roll parameters had to be, for consistency, sufficiently smaller than 1. It is therefore legitimate to expand the Bessel indices in powers of the slow-roll parameters and, from Eq. (10.41), we get:

$$\nu \simeq \frac{3}{2} + \epsilon + \mathcal{O}(\epsilon^2). \qquad (10.62)$$

Equation (10.60) can also be written as

$$\mathcal{P}_T(k) \simeq \frac{2}{3\pi^2}\left(\frac{V}{\overline{M}_P^4}\right)_{k\simeq aH}, \qquad (10.63)$$

where we used the slow-roll relation $3\overline{M}_P^2 H^2 \simeq V$ (see Eqs. (5.98) and (5.99)). Finally, expressing $\overline{M}_P$ in terms of M_P (see Eq. (5.59)),

$$\mathcal{P}_T(k) \simeq \frac{128}{3}\left(\frac{V}{M_P^4}\right)_{k\simeq aH}. \qquad (10.64)$$

The tensor spectral index n_T is then defined as

$$\mathcal{P}_T(k) \simeq k^{n_T}, \qquad n_T = \frac{d\ln\mathcal{P}_T}{d\ln k}. \qquad (10.65)$$

Taking now the spectrum in the parametrization of Eq. (10.64) we have, from Eq. (10.65),

$$n_T = \frac{V_{,\varphi}}{V}\frac{\partial\varphi}{\partial\ln k}. \qquad (10.66)$$

But at horizon crossing, $k = aH$. By taking the derivative with respect to φ of both sides of the latter relation we have:

$$\frac{\partial\ln k}{\partial\varphi} = \frac{1}{a}\frac{\partial a}{\partial\varphi} + \frac{1}{H}\frac{\partial H}{\partial\varphi}. \qquad (10.67)$$

The right hand side of Eq. (10.67) can then be rearranged by using the definitions of the slow-roll parameter ϵ (see Eq. (10.32)) and it is

$$\frac{\partial\ln k}{\partial\varphi} = -\frac{V}{V_{,\varphi}}\left(\frac{1-\epsilon}{\overline{M}_P^2}\right). \qquad (10.68)$$

Inserting Eq. (10.68) into Eq. (10.66) it is easy to obtain

$$n_T = -\left(\frac{V_{,\varphi}}{V}\right)^2\frac{\overline{M}_P^2}{1-\epsilon} \simeq -\frac{2\epsilon}{1-\epsilon} \simeq -2\epsilon + \mathcal{O}(\epsilon^2). \qquad (10.69)$$

Slightly different definitions for the slow-roll parameters and for the spectral indices exist in the literature. At the very end the results obtained with different sets of conventions must necessarily all agree. In the present discussion the conventions adopted are, for practical reasons, the same as the ones of the WMAP collaboration (see, for instance, [37, 38]).

10.2.3 Scalar power spectra

The scalar power spectrum is computed by considering the two-point function of the curvature perturbations on comoving orthogonal hypersurfaces. This choice is, to some extent, conventional and also dictated by practical reasons since the relation of the curvature perturbations $\mathcal{R}$ to the scalar normal mode q is rather simple, i.e. $z\mathcal{R} = -q$ (see Eqs. (10.26) and (10.28) as well as the algebra prior to Eq. (C.53) in Appendix C). Having said this, the scalar modes of the geometry can be parametrized in terms of the two-point function of any other gauge-invariant operator. Eventually, after the calculation, the spectrum of the curvature perturbations can always be obtained by means of (sometimes lengthy) algebraic manipulations. In the operator formalism the quantity to be computed is

$$\langle 0|\hat{\mathcal{R}}(\vec{x},\tau)\hat{\mathcal{R}}(\vec{y},\tau)|0\rangle = \int \mathcal{P}_{\mathcal{R}}(k)\frac{\sin kr}{kr} d\ln k, \tag{10.70}$$

where

$$\mathcal{P}_{\mathcal{R}}(k) = \frac{k^3}{2\pi^2 z^2}|\tilde{f}_k(\tau)|^2, \tag{10.71}$$

where the mode functions $\tilde{f}_k(\tau)$ are now the functions given in Eq. (10.52). By repeating exactly the same steps outlined in the tensor case, the scalar power spectrum can be written as

$$\mathcal{P}_{\mathcal{R}}(k) = \frac{2^{2\tilde{\nu}-3}}{\pi^3}\Gamma^2(\tilde{\nu})(1-\epsilon)^{1-2\tilde{\nu}}\left(\frac{k}{aH}\right)^{3-2\tilde{\nu}}\left(\frac{H^2}{\dot{\varphi}}\right)^2. \tag{10.72}$$

Recalling Eq. (10.51) and expanding in the limit $\epsilon \ll 1$ and $\eta \ll 1$ we have, in full analogy with Eq. (10.62),

$$\tilde{\nu} = \frac{3+\epsilon+2\eta}{2(1-\epsilon)} \simeq \frac{3}{2} + 2\epsilon + \eta + \mathcal{O}(\epsilon^2). \tag{10.73}$$

By comparing Eqs. (10.62) with (10.73) it appears clearly that the difference between ν and $\tilde{\nu}$ arises as a first-order correction that depends upon (both) ϵ and η. Equation (10.72) can then be written in various (equivalent) forms, for instance, evaluating the expression at horizon crossing and taking into account that $\tilde{\nu}$, to lowest order, is $3/2$. The result is:

$$\mathcal{P}_{\mathcal{R}}(k) = \frac{1}{4\pi^2}\left(\frac{H^2}{\dot{\varphi}}\right)^2_{k\simeq aH}. \tag{10.74}$$

Since, from the slow-roll equations,

$$\dot{\varphi}^2 = \frac{V_{,\varphi}}{9H^2}, \quad \frac{1}{2\pi^2}\frac{H^4}{\dot{\varphi}^2} = \frac{1}{12\pi^2}\frac{V}{\epsilon M_P^4}. \tag{10.75}$$

Hence, Eq. (10.74) becomes
$$\mathcal{P}_{\mathcal{R}}(k) = \frac{8}{3 M_P^4} \left(\frac{V}{\epsilon}\right)_{k \simeq aH}, \qquad (10.76)$$
where we used Eq. (10.75) in Eq. (10.74) and we recalled that $\overline{M}_P^2 = M_P^2/(8\pi)$. The scalar spectral index is now defined as
$$\mathcal{P}_{\mathcal{R}}(k) \simeq k^{n_s - 1}, \qquad n_s - 1 = \frac{d \ln \mathcal{P}_{\mathcal{R}}}{d \ln k} \qquad (10.77)$$
It should be stressed that, again, the definition of the spectral index is conventional and, in particular, it appears that while in the scalar case the exponent of the wave-number has been parametrized as $n_s - 1$, in the tensor case the analog quantity has been parametrized as n_T. Using the parametrization of the power spectrum given in Eq. (10.76) and recalling Eq. (10.68), Eq. (10.77) implies that
$$n_s - 1 = \frac{\dot{\varphi}}{H} \frac{1}{1-\epsilon} \left[\frac{V_{,\varphi}}{V} - \frac{\epsilon_{,\varphi}}{\epsilon}\right]. \qquad (10.78)$$
Since
$$\frac{\epsilon_{,\varphi}}{\epsilon} = 2\frac{V_{,\varphi\varphi}}{V_{,\varphi}} - 2\left(\frac{V_{,\varphi}}{V}\right), \qquad \frac{\dot{\varphi}}{H} = -\frac{V_{,\varphi}}{3H^2}, \qquad (10.79)$$
Eq. (10.78) implies that
$$n_s = 1 - 6\epsilon + 2\overline{\eta}. \qquad (10.80)$$
Equation (10.80) is the standard result for the scalar spectral index arising in single field inflationary models. Notice that the value of n_s is close to (but smaller than) the Harrison-Zeldovich value (i.e. $n_s = 1$).

10.2.4 Consistency relation

Therefore, we will have, in summary,
$$\mathcal{P}_T = \frac{128}{3}\left(\frac{V}{M_P^4}\right)_{k \sim aH}, \qquad \mathcal{P}_{\mathcal{R}} = \frac{8}{3}\left(\frac{V}{\epsilon M_P^4}\right)_{k \sim aH}, \qquad (10.81)$$
with
$$n_T = -2\epsilon, \qquad n_s = 1 - 6\epsilon + 2\overline{\eta}. \qquad (10.82)$$
For applications, the ratio between the tensor and the scalar spectrum is also defined as[f]
$$r = \frac{\mathcal{P}_T}{\mathcal{P}_{\mathcal{R}}}. \qquad (10.83)$$

---

[f]Finally the rationale for the introduction of r as a parameter of the ΛCDM model can be understood. The standard set of the cosmological parameters is briefly discussed in the last section of chapter 1.

Using Eq. (10.81) into Eq. (10.83) we obtain

$$r = 16\epsilon = -8n_T \qquad (10.84)$$

which is also known as consistency condition. Again, as previously remarked, the conventions underlying Eq. (10.84) are the same ones adopted from the WMAP collaboration [37, 38].

The scalar and tensor power spectra computed here represent the (single field) inflationary result for the CMB initial conditions. This exercise shows that the quantum-mechanically normalized inflationary perturbations lead, prior to decoupling, to a single adiabatic mode. Few final comments are in order:

- while the situation is rather simple when only one scalar degree of freedom is present, a more complicated system can easily be imagined when more than one scalar is present in the game;
- if many scalar fields are simultaneously present, the evolution of curvature perturbations may be more complex.

The approximate conservation of curvature perturbations for typical wavelengths larger than the Hubble radius holds, strictly speaking, only in the case of single field inflationary models. It is therefore the opinion of the author that the approach of setting early initial conditions (i.e. during inflation) should always be complemented and corroborated with a model-independent treatment of the late initial conditions. In this way we will not only understand which is the simplest model fitting the data but also, hopefully, the correct one.

10.3 Curvature perturbations and density contrasts

In this section we are going to anticipate some ideas that will be more thoroughly discussed in chapter 11. The main difference is that here the main attention will be concentrated on the case of a minimally coupled scalar degree of freedom and on the induced curvature inhomogeneities.

In the comoving gauge the three velocity of the fluid vanishes, i.e. $v_C = 0$ (where $\theta_C = \nabla^2 v_C$). Since the constant (conformal) time hypersurfaces should be orthogonal to the four-velocity, we will also have $B_C = 0$. In this gauge the curvature perturbation can be computed directly from the expressions of the perturbed Christoffel connections bearing in mind that, unlike the Appendix, we want to compute here the fluctuations in the spatial curvature, namely

$$\delta R_C^{(3)} = \delta_s \gamma^{ij} \overline{R}_{ij}^{(3)} + \overline{\gamma}^{ij} \delta_s R_{ij}^{(3)} \equiv \frac{4}{a^2} \nabla^2 \psi_C, \qquad (10.85)$$

where the subscript C refers to the fact that the calculation has been conducted on comoving hypersurfaces. The curvature fluctuations of the comoving gauge can be connected to the fluctuations in a different gauge characterized by a different value of the time coordinate, i.e. $\tau_C \to \tau = \tau_C + \epsilon_0$. Under this shift

$$\psi_C \to \psi = \psi_C + \mathcal{H}\epsilon_0, \qquad (10.86)$$
$$(v_C + B_C) \to B + v = (v_C + B_C) + \epsilon_0. \qquad (10.87)$$

Since in the comoving orthogonal gauge $B_C + v_C = 0$, Eqs. (10.86)–(10.87) imply, in the new coordinate system,

$$\psi_C = \psi - \mathcal{H}(v + B) \equiv (\Psi - \mathcal{H} V_g), \qquad (10.88)$$

where the second equality follows from the definitions of gauge-invariant fluctuations given in Eqs (10.10)–(10.9). From Eq. (10.88) it can be concluded that $\mathcal{R}$ corresponds to the curvature fluctuations of the spatial curvature on comoving orthogonal hypersurfaces.

Other quantities, defined in specific gauges, turn out to have a gauge-invariant interpretation. Take, for instance, the curvature fluctuations on constant density hypersurfaces, ψ_D. Under infinitesimal gauge transformations

$$\psi_D \to \psi = \psi_D + \mathcal{H}\epsilon_0,$$
$$\delta\rho_D \to \delta\rho = \delta\rho_D - \rho'\epsilon_0. \qquad (10.89)$$

But on constant density hypersurfaces $\delta\rho_D = 0$, by the definition of the uniform density gauge. Hence, from Eq. (10.89),

$$\psi_D = \psi + \mathcal{H}\frac{\delta\rho}{\rho'} \equiv \Psi + \mathcal{H}\frac{\delta\rho_g}{\rho'}, \qquad (10.90)$$

where, again, the second equality follows from the definitions of gauge-invariant fluctuations given in Eqs. (10.10) and (10.9) since the gauge-invariant velocity potential is simply given by $\partial^i V_g = \partial^i v + \partial^i E'$. Hence, the (gauge-invariant) curvature fluctuations on constant density hypersurfaces can be defined as

$$\zeta = -\left(\Psi + \mathcal{H}\frac{\delta\rho_g}{\rho'}\right). \qquad (10.91)$$

Equation (10.91) defines the gauge-invariant curvature fluctuations on constant density hypersurfaces. The last sentence seems to contain a contradiction, but such an expression simply means that ζ coincides with the curvature fluctuations in the uniform density gauge. In a different gauge (for instance the longitudinal gauge) ζ has the same value but does not coincide with the curvature fluctuations.

The values of ζ and $\mathcal{R}$ are equal up to terms proportional to the Laplacian of Ψ. This can be shown by using explicitly the definitions of $\mathcal{R}$ and ζ whose difference gives

$$\zeta - \mathcal{R} = -\mathcal{H}\left(V_g + \frac{\delta\rho_g}{\rho'}\right). \tag{10.92}$$

Recalling now that from the Hamiltonian and momentum constraints and from the conservation of the energy density of the background

$$V_g = -\frac{1}{4\pi G a^2(p+\rho)}(\mathcal{H}\Phi + \Psi'),$$
$$\frac{\delta\rho_g}{\rho'} = \frac{1}{4\pi G a^2(p+\rho)}[\nabla^2\Psi - 3\mathcal{H}(\mathcal{H}\Phi + \Psi')], \tag{10.93}$$

we obtain

$$\zeta - \mathcal{R} = \frac{\nabla^2\Psi}{12\pi G a^2(p+\rho)}. \tag{10.94}$$

The density contrast on comoving orthogonal hypersurfaces can also be defined as [194, 198]

$$\epsilon_m = \frac{\delta\rho_C}{\rho} \equiv \frac{\delta\rho + \rho'(v+B)}{\rho} \equiv \frac{\delta\rho_g - 3\mathcal{H}(p+\rho)V_g}{\rho}, \tag{10.95}$$

where the second equality follows from the first one by using the definitions of gauge-invariant fluctuations. Again, using the Hamiltonian and momentum constraints,

$$\epsilon_m = \frac{1}{4\pi G a^2 \rho}\nabla^2\Psi \tag{10.96}$$

that means, according to Eq. (10.94), that $(\zeta - \mathcal{R}) \propto \epsilon_m$.

10.4 Hamiltonians for the scalar problem

In light of possible applications, it is desirable to treat the evolution of the scalar fluctuations of the geometry in terms of a suitable variational principle. On this basis, Hamiltonians for the evolution of the fluctuations can be defined. By perturbing the action of the scalar fluctuations of the geometry, the final form of the action can be expressed in terms of the curvature fluctuations

$$S^{(1)}_{\text{scal}} = \frac{1}{2}\int d^4x\, z^2 \left[\mathcal{R}'^2 - (\partial_i\mathcal{R})^2\right]. \tag{10.97}$$

Defining now the canonical momentum $\pi_\mathcal{R} = z^2\mathcal{R}'$, the Hamiltonian related to the action (10.97) becomes

$$H^{(1)}_{\text{scal}}(\tau) = \frac{1}{2}\int d^3x \left[\frac{\pi_\mathcal{R}^2}{z^2} + z^2(\partial_i\mathcal{R})^2\right], \tag{10.98}$$

and the Hamilton equations

$$\pi'_\mathcal{R} = z^2\nabla^2\mathcal{R}, \qquad \mathcal{R}' = \frac{\pi_\mathcal{R}}{z^2}. \tag{10.99}$$

Equations (10.99) can be combined in a single second order equation so that Eq. (10.28) is recovered.

The canonically conjugate momentum, $\pi_\mathcal{R}$ is related to the density contrast on comoving hypersurfaces, namely, in the case of a single scalar field source [194, 199],

$$\epsilon_m = \frac{\delta\rho_\varphi + 3\mathcal{H}(\rho_\varphi + p_\varphi)V}{\rho_\varphi} = \frac{a^2\delta\rho_\varphi + 3\mathcal{H}\varphi'X}{a^2\rho_\varphi}, \tag{10.100}$$

where the second equality can be obtained using the fact that $\rho_\varphi + p_\varphi = \varphi'^2/a^2$ and that the effective "velocity" field in the case of a scalar field is $V = X/\varphi'$. Making use of Eq. (10.15) in Eq. (10.13), Eq. (10.100) can be expressed as

$$\epsilon_m = \frac{2M_P^2\nabla^2\Psi}{a^2\rho_\varphi} \equiv \frac{2}{3}\frac{\nabla^2\Psi}{\mathcal{H}^2}, \tag{10.101}$$

where the last equality follows from the background equations. From Eq. (10.101), it also follows that

$$\pi_\mathcal{R} = z^2\mathcal{R}' \equiv -6a^2\mathcal{H}\epsilon_m, \tag{10.102}$$

where Eq. (10.101) has been used. Hence, in this description, while the canonical field is the curvature fluctuations on comoving spatial hypersurfaces, the canonical momentum is the density contrast on the same hypersurfaces.

To bring the second-order action in the simple form of Eq. (10.97), various (non-covariant) total derivatives have been dropped. Hence, there is always the freedom of redefining the canonical fields through time-dependent functions of the background geometry. In particular, the action (10.97) can be rewritten in terms of the variable q defined in prior to Eq. (10.29). Then

$$S_{\text{scal}}^{(2)} = \frac{1}{2} \int d^4x \left[q'^2 - 2\frac{z'}{z}qq' - (\partial_i q)^2 + \left(\frac{z'}{z}\right)^2 q^2 \right], \qquad (10.103)$$

and its related Hamiltonian and canonical momentum are, respectively

$$H_{\text{scal}}^{(2)}(\tau) = \frac{1}{2} \int d^3x \left[\pi_q^2 + 2\pi_q q + (\partial_i q)^2 \right], \quad \text{and } \pi = q' - \frac{z'}{z}q. \qquad (10.104)$$

In Eq. (10.103) a further total derivative term can be dropped, leading to another action:

$$S_{\text{scal}}^{(3)} = \frac{1}{2} \int d^4x \left[q'^2 - (\partial_i q)^2 + \frac{z''}{z}q^2 \right], \qquad (10.105)$$

and another Hamiltonian

$$H^{(3)}(\tau)_{\text{scal}} = \frac{1}{2} \int d^3x \left[\tilde{\pi}_q^2 + (\partial_i q)^2 - \frac{z''}{z}q^2 \right], \qquad (10.106)$$

where $\tilde{\pi} = q'$. As in the case of the Hamiltonians for the tensor modes, Eqs. (10.98), (10.104) and (10.106) are all related by canonical transformations. Furthermore, notice that classically the scalar end tensor Hamiltonians have exactly the same form in the case of power-law inflation.

10.5 Trans-Planckian problems?

Consider, for simplicity, the case of power inflation parametrized in terms of the conformal time coordinate τ as $a(\tau) \sim (-\tau)^{-\beta}$ with $\beta > 0$. If the standard normalization prescription is interpreted in strict mathematical terms, then it will happen that if the initial normalization time τ_0 is sent to $-\infty$, a given physical frequency at the time τ_0, $\omega = k/a(\tau_0)$ will become much larger than the Planck mass, or as often emphasized, trans-Planckian. A common theme of various investigations in this direction of research is the observation that in the "trans-Planckian" regime the precise description of the evolution of the metric fluctuations could be foggy. These statements leave room for various proposals which can be summarized as follows:

- the dispersion relations can be modified by trans-Planckian effect;
- the indetermination relations are modified;
- trans-Planckian physics induces the presence of a new fundamental scale M smaller than the Planck mass; observable effects in the scalar and tensor power spectra can be expected.

The modifications invoked in the dispersion relations, as often acknowledged by the authors, always have some ad hoc feature. This possibility was extensively investigated, in a series of papers, by Brandenberger and Martin [325–327] as well as by other authors [328–334]. The observation may be summarized by saying that the wavenumber of the fluctuation is modified in such a way that $k^2 \to k_{\text{eff}}^2(k,\tau)$ [325–327]. Such modifications should be derived from a suitable variational principle [331] (but this is not completely obvious) and they can well be nonlinear [329]. Some effort has been made in order to justify the existence of modified dispersion relations in various frameworks ranging from analog modifications arising in black-hole physics [331] to quantum-Poincaré algebras applied to a cosmological setting [330], to non-commutative geometries [332].

The modifications of the indetermination relations are sometimes motivated by the scattering of superstrings at Planckian energies [335–338]. However, in that context the modifications in the dispersion relations occurs in critical dimensions (26 in the case of the bosonic string); the "position" operator is an impact parameter of two colliding strings. In the trans-Planckian context, on the contrary, the modifications of the indetermination relations are studied in a four-dimensional context. Tachyonic instabilities then occur (this is also a problem, in some cases, when dispersion relations are modified). This approach has been followed by various authors [339–342].

One possible approach to the problem would be to assume that there is a fundamental scale $M < M_{\text{P}}$ and that fluctuations should be normalized as soon as they "exit" from the physical regime characterized by the scale M. This aspect is partially illustrated in Fig. 10.1 where with the dotted line the new fundamental scale is reported. According to these type of proposals [343, 344, 360] the modes of the quantum field should be normalized (at a finite conformal time τ_0) as soon as the physical frequency (with dashed line in Fig. 10.1) crosses the dotted line. This crossing clearly occurs at different times for different comoving frequencies. In the case of power-law inflation, a given physical frequency $\omega(\tau) = k/a(\tau)$ will "cross" the scale M at a characteristic time $\tau_0(k)$ determined as

$$\omega(\tau_0) = \frac{k}{a(\tau_0)} \simeq M, \ \to \ \tau_0(k) = -\tau_1 \left(\frac{M}{k}\right)^{\frac{1}{\beta}} \simeq -\tau_1 \left(\frac{M}{k}\right)^{1-\frac{1}{p}}, \quad (10.107)$$

where $a(\tau) \sim (-\tau/\tau_1)^{-\beta}$. The second equality in the second equation follows from the relation between the cosmic and conformal times in power-law backgrounds (see, for instance, before Eq. (6.81)). In the expanding branch of de Sitter space-time $\beta = 1$ and $p \to \infty$. Consequently $k\tau_0$ is constant and roughly given by M/H where H is the Hubble rate during the de Sitter phase. Now the claim is that [343, 344] this prescription for setting the initial condition for quantum fluctuations produces observable corrections to the tensor power spectrum[g] of the form

$$\overline{\mathcal{P}}_{\mathrm{T}} = \mathcal{P}_{\mathrm{T}}\left[1 + c_\gamma\left(\frac{H}{M}\right)^\gamma + ...\right], \qquad (10.108)$$

where c_γ is a numerical constant; $\mathcal{P}_{\mathrm{T}}$ is the tensor power spectrum obtained through the standard normalization prescription discussed earlier in this section. Oscillating terms multiplying the ratio $(H/M)^\gamma$ have been neglected (see below, however). The ellipses in Eq. (10.108) stand for terms which are of higher order in H/M. Concerning the parametrization given in Eq. (10.108) two comments are in order:

- it is clear that if the prescription (10.107) is applied the correction to the power spectrum has to be in the form (10.108): the reason is that the power spectrum will now depend upon the new scale $x_0 = k\tau_0 \sim M/H \gg 1$ that becomes the argument of Hankel functions whose limit for $x_0 \gg 1$ must indeed produce a result of the type (10.108);
- the power γ is crucial for the possible observational relevance of trans-Planckian effects so it becomes crucial to understand *what* controls the power γ.

Some authors claim that γ can be 1 [343, 344]. In this case the ratio H/M will be larger than H/M_{P} that we know can be as large as 10^{-6}–10^{-5}. If $\gamma = 2$ or even $\gamma = 3$ the correction is not relevant. Different arguments have been put forward for values $\gamma > 1$ [346–350]. Notice that often the discussions of these effects take place in pure de Sitter space (where, as already pointed out, scalar modes would be absent). In this context different vacua can be defined. They are connected by unitary transformations and generically indicated as α vacua [351–355]. Flat space has a global time-like killing vector. This allows us to define a Hamiltonian whose minimization defines a "vacuum", i.e. the lowest energy eigenstate. In de Sitter space there is no global time-like killing vector. So a globally conserved energy cannot be defined. It is however still possible to find states which are invariant under the connected part of the isometry group

---

[g]Here we discuss the tensor case since, for *exact* (expanding) de Sitter space only the tensor modes are excited. However, the same discussion hold also for the scalar power spectra in the quasi-de Sitter space-time.

of de Sitter space, i.e. $SO(4,1)$. In the case of a free scalar field there is an infinite family of invariant states called α vacua since they can be distinguished by a single complex number α.

10.5.1 *Minimization of canonically related Hamiltonians*

In the following the roots of the corrections to the power spectrum will be understood in terms of a much more mundane feature of the theory of cosmological fluctuations, i.e. the possibility, already discussed in chapter 6, of defining different Hamiltonians related by canonical transformations. More specifically it will be demonstrated than the value of γ given in Eq. (10.108) depends on *which Hamiltionian* one wishes to have minimized at the moment when the given physical frequency cosses the new fundamental scale M as explained in Eq. (10.107). By averaging the energy-momentum pseudo-tensor of the fluctuations of the geometry over the state minimizing a given Hamiltonian at $\tau_0(k)$, the energetic content of the fluctuations can be estimated. The result of such a calculation must be *smaller* than the energy density of the background geometry. This chain of calculations has been presented recently in [213] and [214] and will be briefly reviewed here.

The first step is to show that, depending on the which Hamiltonian is minimized at $\tau_0(k)$ (defined as in Eq. (10.107)) a different power of γ is obtained in Eq. (10.108). In order to be specific consider the case of conventional power-law inflationary models. We will first derive the result for the cases of the Hamiltonians (10.98). Then we derive the results of the minimization of the other two Hamiltonians (i.e. Eqs. (10.104)-(10.106)). As already discussed in chapter 6 the scalar Hamiltonians have a direct counterpart in the case of the tensor modes. So we will be able to report the analog results holding in the tensor case.

Equation (10.98) implies that the canonical field is $\mathcal{R}$, i.e. the curvature perturbation. The canonical momentum is the density contrast as discussed in Eq. (10.102). The Hamiltonian operator will be

$$\hat{H}(\tau) = \frac{1}{2} \int d^3x \left[\frac{\hat{\pi}_\mathcal{R}^2}{z^2} + z^2 (\partial_i \hat{\mathcal{R}})^2 \right]. \tag{10.109}$$

Repeating the same procedure outlined earlier in this section, Eq. (10.109) can be written as

$$\hat{H}(\tau) = \frac{1}{4} \int d^3k \left[\frac{1}{z^2} (\hat{\pi}_{\vec{k}} \hat{\pi}_{\vec{k}}^\dagger + \hat{\pi}_{\vec{k}}^\dagger \hat{\pi}_{\vec{k}}) + k^2 z^2 (\hat{\mathcal{R}}_{\vec{k}} \hat{\mathcal{R}}_{\vec{k}}^\dagger + \hat{\mathcal{R}}_{\vec{k}}^\dagger \hat{\mathcal{R}}_{\vec{k}}) \right]. \tag{10.110}$$

The Hamiltonian (10.110) can be written at $\tau_0(k)$ as

$$\hat{H}(\tau_0) = \frac{1}{4}\int d^3k k\left[\hat{Q}^\dagger_{\vec{k}}\hat{Q}_{\vec{k}} + \hat{Q}_{\vec{k}}\hat{Q}^\dagger_{\vec{k}} + \hat{Q}^\dagger_{-\vec{k}}\hat{Q}_{-\vec{k}} + \hat{Q}_{-\vec{k}}\hat{Q}^\dagger_{-\vec{k}}\right], \quad (10.111)$$

where

$$\hat{Q}_{\vec{k}}(\tau_0) = \frac{1}{\sqrt{2k}}\left[\frac{\hat{\pi}_{\vec{k}}(\tau_0)}{z(\tau_0)} - i z(\tau_0) k \hat{\mathcal{R}}_{\vec{k}}(\tau_0)\right], \quad (10.112)$$

obeying $[\hat{Q}_{\vec{k}}, \hat{Q}^\dagger_{\vec{p}}] = \delta^{(3)}(\vec{k}-\vec{p})$. Consequently, the state minimizing (10.110) at τ_0 is the one annihilated by $\hat{Q}_{\vec{k}}$, i.e.

$$\hat{Q}_{\vec{k}}(\tau_0)|0^{(1)}\rangle = 0, \qquad \hat{Q}_{-\vec{k}}(\tau_0)|0^{(1)}\rangle = 0. \quad (10.113)$$

The specific relation between field operators dictated by (10.113) provides initial conditions for the Heisenberg equations

$$i\hat{\mathcal{R}}' = [\hat{\mathcal{R}}, \hat{H}], \qquad i\hat{\pi}'_\mathcal{R} = [\hat{\pi}_\mathcal{R}, \hat{H}]. \quad (10.114)$$

The full solution of this equation can be written as

$$\hat{\mathcal{R}}_{\vec{k}}(\tau) = \hat{a}_{\vec{k}}(\tau_0) f_k(\tau) + \hat{a}^\dagger_{-\vec{k}}(\tau_0) f^*_k(\tau), \quad (10.115)$$

$$\hat{\pi}_{\vec{k}}(\tau) = \hat{a}_{\vec{k}}(\tau_0) g_k(\tau) + \hat{a}^\dagger_{-\vec{k}}(\tau_0) g^*_k(\tau), \quad (10.116)$$

where, recalling the explicit solution of the equations in the case of the exponential potential (see chapter 5) and defining $x = k\tau$

$$f_k(\tau) = \frac{\sqrt{\pi}}{4}\frac{e^{\frac{i}{2}(\mu+1/2)\pi}}{z(\tau)\sqrt{k}}\sqrt{-x}H^{(1)}_\nu(-x), \qquad \nu = \frac{3p-1}{2(p-1)}$$

$$g_k(\tau) = -\frac{\sqrt{\pi}}{4}e^{\frac{i}{2}(\mu+1/2)\pi}z(\tau)\sqrt{k}\sqrt{-x}H^{(1)}_{\nu-1}(-x), \quad (10.117)$$

satisfy the Wronskian normalization condition

$$f_k(\tau)g^*_k(\tau) - f^*_k(\tau)g_k(\tau) = i. \quad (10.118)$$

Notice that the mode functions of Eq. (10.117) are different from the ones previously written earlier in this section. The creation and annihilation operators appearing in (6.79) are defined as

$$\hat{a}_{\vec{k}}(\tau_0) = \frac{1}{z_0\sqrt{2k}}\{[g^*_k(\tau_0) + ikz_0^2 f^*_k(\tau_0)]\hat{Q}_{\vec{k}}(\tau_0)$$
$$- [g^*_k(\tau_0) - ikz_0^2 f^*_k(\tau_0)]\hat{Q}^\dagger_{-\vec{k}}(\tau_0)\},$$
$$\hat{a}^\dagger_{-\vec{k}}(\tau_0) = \frac{1}{z_0\sqrt{2k}}\{[g_k(\tau_0) - ikz_0^2 f_k(\tau_0)]\hat{Q}^\dagger_{-\vec{k}}(\tau_0) \quad (10.119)$$
$$- [g_k(\tau_0) + ikz_0^2 f_k(\tau_0)]\hat{Q}_{\vec{k}}(\tau_0)\}.$$

So far two sets of creation and annihilation operators have been introduced: the operators $\hat{Q}_{\vec{k}}(\tau_0)$ and the operators $\hat{a}_{\vec{k}}(\tau_0)$. The state annihilated by $\hat{Q}_{\vec{k}}(\tau_0)$ minimizes the Hamiltonian at τ_0 while the state annihilated by $\hat{a}(\tau_0)$ *does not* minimize the Hamiltonian at τ_0. It is relevant to introduce these operators not so much for the calculation of the two-point function but for the subsequent applications to the back-reaction effects. In fact, in the standard approach to the initial value problem for the quantum mechanical fluctuations, the initial state is chosen to be the one annihilated by $\hat{a}_{\vec{k}}(\tau_0)$ for $\tau_0 \to -\infty$.

The Fourier transform of the two-point function,

$$\langle 0^{(1)}, \tau_0 | \hat{\mathcal{R}}(\vec{x}, \tau) \hat{\mathcal{R}}(\vec{y}, \tau) | \tau_0, 0^{(1)} \rangle = \int \frac{dk}{k} \overline{\mathcal{P}}_{\mathcal{R}} \frac{\sin kr}{kr}, \quad r = |\vec{x} - \vec{y}|, \quad (10.120)$$

can now be computed, and the result is

$$\overline{\mathcal{P}}_{\mathcal{R}} = \frac{k^2}{2\pi^2} \left\{ |f_k(\tau)|^2 \left[\frac{|g_k(\tau_0)|^2}{z(\tau_0)^2} + k^2 z(\tau_0)^2 |f_k(\tau_0)|^2 \right] \right.$$

$$- \frac{f_k(\tau)^2}{2} \left[\frac{g_k^*(\tau_0)^2}{z(\tau_0)^2} + k^2 z(\tau_0)^2 f_k^*(\tau_0)^2 \right]$$

$$\left. - \frac{f_k^*(\tau)^2}{2} \left[\frac{g_k(\tau_0)^2}{z(\tau_0)^2} + k^2 z(\tau_0)^2 f_k(\tau_0)^2 \right] \right\}. \quad (10.121)$$

The explicit form of $\overline{\mathcal{P}}_{\mathcal{R}}$ and $\overline{\mathcal{P}}_T$ can be obtained by inserting Eqs. (10.117) into Eq. (10.121). The results should be expanded for $|x| = k\tau \ll 1$ and for $|x_0| = k\tau_0 \gg 1$. While $|k\tau|$ measures how much a given mode is outside the horizon,

$$|x_0| = |k\tau_0| \simeq \frac{M}{H(t_0(k))} = \frac{M}{H_{\text{ex}}} \quad (10.122)$$

defines the moment at which the given mode crosses the scale M. Notice that Eq. (10.122) has exactly the same content of Eq. (10.107)

To have the explicit form of the power spectrum, Eq. (10.121) should be expanded for $k\tau \ll 1$ and $|k\tau_0| \gg 1$ and the result is

$$\overline{\mathcal{P}}_{\mathcal{R}}^{1/2} = \mathcal{P}_{\mathcal{R}}^{1/2} \left[1 + \frac{p}{2(p-1)} \frac{\sin[2x_0 + p\pi/(p-1)]}{x_0} \right], \quad (10.123)$$

where $\mathcal{P}_{\mathcal{R}}^{1/2}$ is the power spectrum of the scalar fluctutations already obtained within the standard lore.

The results obtained in the case of Eq. (10.98) can be also generalized to the case of the tensor modes. In this case the calculation is identical with the only difference that $z(\tau)$ is replaced by $a(\tau)$ and $\mathcal{R}$ is replaced by h. The result will be:

$$\overline{\mathcal{P}}_{\mathrm{T}}^{1/2} = \mathcal{P}_{\mathrm{T}}^{1/2}\left[1 + \frac{p}{2(p-1)}\frac{\sin\left[2x_0 + p\pi/(p-1)\right]}{x_0}\right], \qquad (10.124)$$

where, now $\mathcal{P}_{\mathrm{T}}^{1/2}$ is the same as the spectrum obtained in the case of the standard lore.

In Eqs. (10.123) and (10.124), on top of the standard (leading) terms there is a correction that goes, roughly, as $1/x_0 \sim H_{\mathrm{ex}}/M$ where, as discussed in Eq. (10.122), H_{ex} denotes the Hubble parameter evaluated at the moment the given scale crosses M. If $M \sim M_{\mathrm{P}}$, $H_{\mathrm{ex}}/M \sim 10^{-6}$. This is the correction that would apply in the scalar power spectrum if quantum mechanical initial conditions were assigned in such a way that the initial state minimizes (10.98).

Having discussed in detail the results for the case of (10.98), attention will now be turned to the case of Eqs. (10.104) and (10.106). In this case the whole procedure of canonical quantization can be repeated with the crucial difference that the state minimizing the quantum version of (10.104) will not be the same minimizing (10.98). If the Hamiltonian (10.104) is minimized the result is

$$\overline{\mathcal{P}}_{\mathcal{R},\mathrm{T}}^{1/2} = \mathcal{P}_{\mathcal{R},\mathrm{T}}^{1/2}\left[1 - \frac{p}{4(p-1)}\frac{\cos\left[2x_0 + p\pi/(p-1)\right]}{x_0^2}\right], \qquad (10.125)$$

where the subscript denotes either the scalar or the tensor power spectrum since the correction, in a power-law inflationary background, is the same for both scalars and tensors.

A comparison of Eqs. (10.124) and (10.125) shows two important facts. The first is that the leading term of the spectra is the same. This phenomenon simply reflects the occurrence that different Hamiltonians, connected by canonical transformations, must lead to the same evolution and to the same leading term in the power spectra. The second fact to be noticed is that the correction to the power spectrum goes as $1/x_0^2$ in the case of (10.125). This correction is then much smaller than the one appearing in (10.124). If $M \sim M_{\mathrm{P}}$ then the correction will be, in the de Sitter case, $\mathcal{O}(10^{-12})$, i.e. six orders of magnitude smaller than the correction appearing in Eq. (10.124).

Finally the case of the Hamiltonian (10.106) will be examined. Equation (10.106) can be minimized following the same procedure as already discussed in the case of Eqs. (10.98) and (10.104). However, again, the state minimizing the Hamiltonian of Eq. (10.106) will be different from

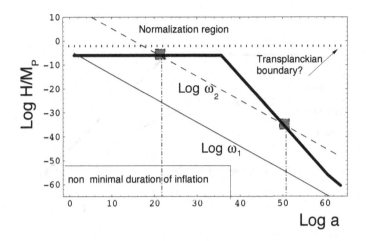

Fig. 10.1 The evolution of the Hubble rate in Planck units is illustrated in a schematic model of post-inflationary evolution. The evolution of two different physical frequencies is also illustrated in the case when the duration of inflation is non-minimal (about 80 efolds). With the dotted line some fundamental physical scale (smaller than the Planck mass) is also illustrated.

the states minimizing the Hamiltonians of Eqs. (10.98) and (10.104). The result for the power spectra will then be

$$\overline{\mathcal{P}}^{1/2}_{\mathcal{R},T} = \mathcal{P}^{1/2}_{\mathcal{R},T}\left[1 + \frac{p(2p-1)}{(p-1)^2}\frac{\sin[2x_0 + p\pi/(p-1)]}{4x_0^3}\right]. \tag{10.126}$$

In Eq. (10.126) the correction arising from the initial state goes as $1/x_0^3$ and again, if $M \sim M_{\rm P}$, it is $\mathcal{O}(10^{-18})$, i.e. 12 orders of magnitude smaller than in the case discussed in Eqs. (10.123) and (10.124).

10.5.2 Back-reaction effects

In order to select the correct Hamiltonian in a way compatible with the idea of assigning initial conditions when a given physical frequency crosses the scale M, it is desirable to address the issue of back-reaction effects. The energetic content of the quantum-mechanical state minimizing the given Hamiltonian should be estimated and compared with the energy density of the background geometry. The back-reaction effects of the different quantum-mechanical states minimizing the Hamiltonians will now be computed. Without loss of generality, attention will be focused on the tensor modes of the geometry. The advantage of discussing the gravitons

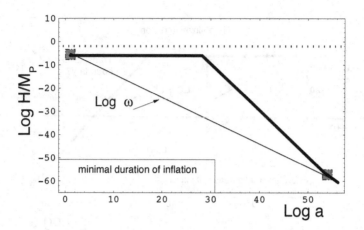

Fig. 10.2 The same quantities illustrated in Fig. 10.1 are reported in the situation when the flat plateau lasts for about 65 efolds.

is that they do not couple to the sources and, therefore, the form of the energy-momentum pseudo-tensor is simpler than in the case of the scalar modes [356, 357].

The appropriate energy-momentum tensor of the fluctuations of the geometry will be averaged over the state minimizing a given Hamiltonian at $\tau_0(k)$ and the result compared with the energy density of the background geometry, since the energy density of the perturbations cannot exceed that of the background geometry.

The energy density of the tensor inhomogeneities can be computed from the energy-momentum pseudo-tensor written, for simplicity, for one of the two polarizations:

$$\langle \hat{T}_0^0 \rangle = \frac{\mathcal{H}}{2a^2} \langle (\hat{h}'\hat{h} + \hat{h}\hat{h}') \rangle + \frac{1}{8a^2} \langle [\hat{h}'^2 + (\partial_i \hat{h})^2] \rangle, \qquad (10.127)$$

where $\langle ... \rangle$ denotes the expectation value with respect to a quantum mechanical state minimizing a given Hamiltonian and $\hat{h}$ denotes the field operator corresponding to a single tensor polarization of the geometry.

If, in Eq. (10.127), the average is taken with respect to the state minimizing the tensor analog of Eq. (10.98), then the energy density is, in the case of de Sitter space [214]:

$$\overline{\rho}_{\rm GW}^{(1)}(\tau, \tau_0) \sim \frac{H^4}{64\pi^2}\left[x_0^2 + \mathcal{O}\left(\frac{1}{x_0^2}\right)\right]$$

$$\simeq \frac{H^4}{64\pi^2}\left(\frac{M}{H}\right)^2\left[1 + \mathcal{O}\left(\frac{H^2}{M^2}\right)\right]. \qquad (10.128)$$

Since, as already discussed, $|x_0| = M/H \gg 1$, in the case of de Sitter space, the back-reaction effects related to the state minimizing the first Hamiltonian are then large. Recall, in fact, that the energy density of the background geometry is $\mathcal{O}(H^2 M_P^2)$. Hence, if $M \sim M_P$ the energy density of the fluctuations will be of the same order as that of the background geometry, which is not acceptable since if this is the case, inflation could not even start. Let us now turn our attention to the case of the state minimizing the tensor analog of Eq. (10.104). The expectation value of the energy-momentum pseudo-tensor will now be taken over the state minimizing the second class of Hamiltonians. Applying the same procedure as before, the averaged energy density is smaller than the result obtained in Eq. (10.128) by a factor $(H/M)^2$.

Finally, in the third and last case, we have to average the energy-momentum pseudo-tensor over the state minimizing the tensor analog of Eq. (10.106). In this case the result is [214]

$$\overline{\rho}_{\text{GW}}^{(3)}(\tau, \tau_0) \simeq \frac{27}{256\pi^2} H^4 \left(\frac{H}{M}\right)^2 \left[1 + \mathcal{O}\left(\frac{H^4}{M^4}\right)\right], \qquad (10.129)$$

i.e. even smaller than the result discussed in the case of the second Hamiltonian. In this case, the averaged energy density is much smaller than that of the background.

The results reported in [214] and summarized above suggest the following reflections:

- it is physically interesting to set quantum mechanical initial conditions as soon as a given physical frequency crosses a fundamental scale M;
- to compute "observables", it is crucial to determine which Hamiltonian is minimized as soon as the given physical frequency crosses a fundamental scale M;
- large (observable) effects seems to be ruled out if one accepts that the energy density of the fluctuations should be smaller than the energy density of the background geometry.

Similar considerations, but through different arguments, appear in [358–360]. It is here appropriate to mention that the authors of Ref. [361, 362] claim to reach more optimistic conclusions concerning the back-reaction effects associated with particular classes of trans-Planckian effects. In particular, in Ref. [361] the analysis has been performed just in terms of the scalar field fluctuation (that is not gauge-invariant for infinitesimal diffeomorphisms). It would be interesting to repeat the same analysis in terms of the tensor modes of the geometry.

10.6 How many adiabatic modes?

Concerning the the evolution of scalar fluctuations, a plausible question to ask is how many adiabatic solutions and how many non-adiabatic solutions are compatible, for instance, with the conservation of $\mathcal{R}$ and ζ discussed respectively in Eqs. (7.103) and (7.123).

This question, usually approached within the separate Universe picture [198] (see also [363] for a reintroduction of some of arguments given in [198]), was recently addressed by Weinberg in a series of papers [364–366]. The separate Universe picture amounts to stipulating that any portion of the Universe that is larger than H^{-1} (H being the Hubble rate) but smaller than the physical wavelength on the perturbation a/k will look like a separate unperturbed Universe. In the following the separate Universe picture will not be invoked but the way of reasoning put forward in Refs. [364–366] will be briefly outlined. The hypotheses of the theorem are the following:

- the theory of gravity is, for simplicity, of Einstein-Hilbert type;
- the background geometry is given by a FRW metric which we can take for simplicity to be spatially flat;
- that the Universe is always expanding (possibly with different rates);
- the sources of the geometry are relativistic fluids.

The last assumption is quite general and it allows for the presence of anisotropic stresses. From the mentioned assumptions it also follows that Eq. (7.103) is fully valid. This is true for two separate reasons:

- the first reason is that Eq. (7.103) is deduced within an Einsteinian theory of gravity;
- the second reason is that, since the Universe is assumed to be always expanding, the time-dependent coefficients of the terms at the right hand side of Eq. (7.103) are all non-singular.

Moreover, since fluids with different barotropic indices can be simultaneously present, the non-adiabatic fluctuation of the pressure density, $\delta p_{\rm nad}$, can be present.

The thesis of the theorem is that the evolution equations, in the longitudinal gauge description, always have a pair of physical solutions for which $\delta p_{\text{nad}} \to 0$ and $\mathcal{R}$ approaches a constant for $k \to 0$, where k denotes, as usual, the comoving wavenumber. In other words, following the usual terminology, there will always be at least a pair of adiabatic solutions with $\delta p_{\text{nad}} = 0$ and $\mathcal{R}$ constant in the limit $k \to 0$:

- one solution with $\mathcal{R} \neq 0$;
- the other with $\mathcal{R} = 0$.

In order to understand correctly the thesis of the theorem, the limit $k \to 0$ is crucial. The second point to be borne in mind is that there could also be other solutions with $\mathcal{R}' \neq 0$. These solutions, if allowed by the Hamiltonian constraint, will be non-adiabatic.

The theorem can be demonstrated in general terms, as illustrated in Ref. [364]. Here it will be illustrated in the simple case of a scalar field source. In this case, the evolution equations in the longitudinal gauge can be simply obtained from Eqs. (10.4), (10.5), (10.6) and (10.7). By definition of longitudinal coordinate system, the gauge freedom is completely specified. However, in the limit $k \to 0$ there is an extra accidental symmetry which is a remnant of the gauge symmetry and which allows us to find exact solutions in the limit of vanishing comoving momenta.

Consider, therefore, the aforementioned evolution equations in the limit $k \to 0$ (or, equivalently, the limit of vanishing spatial gradients). The key observation in order to prove the theorem in general terms is that these equations have an extra accidental symmetry for a particular class of coordinate transformations, i.e.

$$\tau \to \tilde{\tau} = \tau + \epsilon_0(\tau), \qquad x^i \to \tilde{x}^i = x^i - \lambda x^i, \qquad (10.130)$$

where the parameter λ is a space-time constant. This coordinate transformation clearly induces a transformation in the fluctuations defined in the longitudinal gauge, in particular, using Eq. (6.13), and recalling the definitions of longitudinal fluctuations, the transformations will be:

$$\phi \to \tilde{\phi} = \phi - \mathcal{H}\epsilon_0 - \epsilon_0', \qquad \psi \to \tilde{\psi} = \psi + \mathcal{H}\epsilon_0 - \lambda, \chi \to \tilde{\chi} = \chi - \varphi'\epsilon_0.$$

While the gauge transformation for ϕ is exactly the one derived in Eq. (6.18), the transformation for ψ derived on the basis of (10.130) differs from Eq. (6.19). There is no surprise for this occurrence: in fact we are here taking the limit $k \to 0$. Hence, the part of the perturbed metric proportional to $k_i k_j$ (which would be generically induced by a coordinate transformation) is not induced by the specific coordinate transformation given in Eq. (10.130). Therefore, from Eqs. (10.131) it is clear that a solution of the longitudinal gauge equations can be obtained, in the limit $k \to 0$, by setting

$$\phi = -\epsilon_0' - \mathcal{H}\epsilon_0, \qquad \psi = \mathcal{H}\epsilon_0 - \lambda,$$
$$\chi = -\epsilon_0 \varphi'. \tag{10.131}$$

This solution can be generalized to the case when the sources are more general, namely, for instance, with several fluids, several interacting scalar fields and so on.

The solution given in (10.131) would simply be a pure gauge. However, it can be extended to the case of non-zero k by looking at the conditions implied on ϵ_0 by the components of the perturbed Einstein equations which are trivially satisfied in the limit $k \to 0$: these are the off-diagonal terms in the (ij) equation, implying $\phi = \psi$ (in the absence of an anisotropic stress) and the momentum constraint of Eq. (10.6) implying $\psi' + \mathcal{H}\psi = 4\pi G \varphi' \chi$. The condition $\phi = \psi$ implies, according to Eq. (10.131), that ϵ_0 satisfies the following simple equation

$$\epsilon_0' + 2\mathcal{H}\epsilon_0 = \lambda, \tag{10.132}$$

whose solution can be written as $\epsilon_0 = \epsilon_0^{(1)} + \epsilon_0^{(2)}$ where

$$\epsilon_0^{(1)} = \frac{\lambda}{a^2(\tau)} \int^\tau a^2(\tau') d\tau', \qquad \epsilon_0^{(2)} = \frac{c_2}{a^2(\tau)}, \tag{10.133}$$

are, respectively, a particular solution of the inhomoheneous equation and the general solution of the homogeneous equation. The solution expressed by $\epsilon_0^{(2)}$ can also be understood from the freedom of shifting the lower limit of the integral appearing in $\epsilon_0^{(1)}$.

With this observation, the system of the longitudinal fluctuations can be solved. In order to appreciate the significance of the results to be obtained, it is useful to write down the system as a third order differential system in the variables ψ_k, χ_k and $\chi_k' = f_k$. Using Eqs. (10.6) and (10.7), the system to be solved is, in Fourier space,[h]

$$\psi_k' = -\mathcal{H}\psi_k + 4\pi G \varphi' \chi_k, \tag{10.134}$$
$$\chi_k' = f_k, \tag{10.135}$$
$$f_k' = -2\mathcal{H}f_k - k^2 \chi_k - \frac{\partial^2 W}{\partial \varphi^2} a^2 \chi_k - 2\psi_k \frac{\partial W}{\partial \varphi} a^2 + 4\psi_k' \varphi'. \tag{10.136}$$

---

[h]In the following the index denoting Fourier transformed quantities will be restored since no ambiguity can arise with different subscripts. This is important since it has to be stressed that the pure gauge solution in the limit $k \to 0$ can be lifted to the status of physical solution also for $k \neq 0$.

In this case the coefficients of the various terms appearing in the system are continuous in a neighborhood of $k = 0$. This condition guarantees that the would-be gauge mode can be lifted to the status of physical solution in the limit $k \to 0$. In the general case (when the sources are more complicated than a single scalar field) this condition may not be satisfied in general but it is certainly rather plausible given the typical forms of energy-momentum tensors customarily employed in cosmological model building.

After repeated use of the background equations, (and in particular of Eq. (5.80)), Eq. (10.136) implies for $k = 0$ that

$$\psi_k = \frac{1}{a}\left(\frac{\chi_k}{\varphi'}a\right)'. \tag{10.137}$$

Now, since

$$\chi_k^{(1)} = -\epsilon_0^{(1)}\varphi', \qquad \chi_k^{(2)} = -\epsilon_0^{(2)}\varphi', \tag{10.138}$$

we will also have

$$\psi_k^{(1)} = -\lambda + \lambda\mathcal{H}\int^\tau a^2(\tau')d\tau', \tag{10.139}$$

$$\psi_k^{(2)} = c_2\frac{\mathcal{H}}{a^2}. \tag{10.140}$$

Knowing what $\chi_k^{(1)}$, $\chi_k^{(2)}$ are, Eqs. (10.139) and (10.140) allow us to determine what $\mathcal{R}_k$ is in the limit $k \to 0$ from the by now familiar relation

$$\mathcal{R}_k^{(1,2)} = -\psi_k^{(1,2)} - \frac{\mathcal{H}}{\varphi'}\chi_k^{(1,2)}. \tag{10.141}$$

In particular, inserting Eqs. (10.138) and (10.139)–(10.140) into Eq. (10.141) we will have

$$\mathcal{R}_k^{(1)} = \lambda, \qquad \mathcal{R}_k^{(2)} = 0. \tag{10.142}$$

The result is: for $k \to 0$ there always two adiabatic modes, one with $\mathcal{R}$ constant and the other with $\mathcal{R}$ vanishing. Since the system of Eqs. (10.134)–(10.136) is a third-order system there will be 3 independent solutions. However, the Hamiltonian constraint eliminates one, the non-adiabatic, i.e. the one leading to $\mathcal{R}' \neq 0$ for $k \to 0$. If instead of one scalar field there are N_s scalar degrees of freedom the theorem applies as well. The system of Eqs. (10.134)–(10.136) can be generalized to this case with the result that the number of independent equations will be $(2N_s+1)$. Of these $(2N_s + 1)$, 2 are adiabatic and a third one is eliminated by the Hamiltonian constraint. Thus, the remaining $2(N_s - 1)$ are non-adiabatic modes.

This perspective was further developed in [366] (see also [365]) with the purpose of discussing the fate of non-adiabatic modes in multi-field inflationary models. As remarked in [367], in a related context, the conservation laws for $\mathcal{R}$ and its analog for the tensor modes of the geometry bears some similarities with the Goldstone theorem [368, 369] normally employed in relativistic quantum field theories. The modes for which $\mathcal{R}$ or h_{ij} are constant outside the horizon take the place here with the Goldstone bosons that become free particles for long-wavelength.

As a final side remark, it is interesting to mention that the derivation of the present results has been conducted in the longitudinal coordinate system. What would happen in a different coordinate system, for instance, the synchronous? The answer to this question is that different gauges may suggest that different quantities are constant outside the horizon [364]. However, this does not imply that different gauges are not equivalent as far as the evolution of physical quantities are concerned. The reason is that the limit $k \to 0$ may have different meanings in different gauges [364].

Part IV

Complements

Chapter 11

Surfing on the Gauges

The discussion of gravitational inhomogeneities has been conducted, so far, by exploiting a specific choice of coordinate system. In chapter 6 it has been noticed that while the tensor modes of the geometry are invariant under infinitesimal diffeomorphisms, the scalar and vector modes are not. As a consequence of this statement, the scalar modes have been treated in the conformally Newtonian (or longitudinal) gauge in chapters 7, 8 and 9. In chapter 10 it has been shown how it is possible to switch from a gauge-dependent to a gauge-invariant description and the general observation has been applied to the early evolution of scalar fluctuations generated by the inhomogeneities of the inflaton field. This theme has been further expanded in Appendix D.

There are no compelling reason why a coordinate system should be regarded as intrinsically superior to another coordinate system. Every coordinate system has its virtues and its possible drawbacks. Different classes of physical problems, ranging from classical mechanics to gauge theories, suggest there is nothing deep about the selection of a gauge. The only guiding principle in the choice of a gauge could be practical considerations. In a partical problem it can happen, for instance, that the evolution equations are more transparent (or more easily solvable) in a particular gauge. This does not mean that every problem should be treated in *the same* gauge. On the contrary, the possibility of changing the coordinate system at every moment should be seen as a powerful tool that allows us to cross-check the results of long calculations.

In the present chapter it will be shown why it is productive (and sometimes highly desirable) to surf on different gauge descriptions. Along this discussion the peculiar features of different coordinate systems will be highlighted. The gauge choices can be divided, generally speaking, into two broad classes:

- gauges where the arbitrariness of fixing the coordinate system is completely removed;
- gauges where the arbitrariness of fixing the coordinate system is not completely removed.

As already discussed in chapter 6, the gauge freedom is completely removed if the parameters of the gauge transformation (called for short gauge parameters) are fully determined. If the gauge parameters are determined only up to an integration constant (which is time-independent but space-dependent) then, the gauge freedom is not completely removed. The longitudinal gauge and the off-diagonal gauge (sometimes also called uniform curvature gauge) belong to first of the two aforementioned classes. The synchronous gauge and the comoving orthogonal gauge belong to the second class. In the first part of this chapter the most common gauge choices will be examined. In the second part of this chapter various gauge-invariant descriptions will be studied with the aim of stressing that not only gauge-dependent descriptions are dictated by the specific problem but also that gauge-invariant descriptions may change depending on the practical circumstances of a given calculation.

11.0.1 Generalities on scalar gauge transformations

The essential elements of the decomposition discussed in Eqs. (6.3), (6.4) and (6.5) will now be quickly reviewed with specific attention to the scalar modes of the geometry. Indeed, according to Eqs. (6.3), (6.4) and (6.5), the scalar fluctuations of the geometry can be parametrized as:

$$\delta_s g_{00} = 2a^2 \phi, \tag{11.1}$$

$$\delta_s g_{ij} = 2a^2(\psi \delta_{ij} - \partial_i \partial_j E), \tag{11.2}$$

$$\delta_s g_{0i} = -a^2 \partial_i B, \tag{11.3}$$

where, as in Appendix C and D, δ_s denotes a scalar fluctuation of the corresponding quantity. For infinitesimal coordinate transformations

$$\tau \to \tilde{\tau} = \tau + \epsilon_0, \tag{11.4}$$

$$x^i \to \tilde{x}^i = x^i + \partial^i \epsilon, \tag{11.5}$$

the four independent functions ϕ, ψ, B and E appearing in Eqs. (11.1), (11.2) and (11.3) transform as

$$\phi \to \tilde{\phi} = \phi - \mathcal{H}\epsilon_0 - \epsilon_0', \qquad (11.6)$$

$$\psi \to \tilde{\psi} = \psi + \mathcal{H}\epsilon_0, \qquad (11.7)$$

$$B \to \tilde{B} = B + \epsilon_0 - \epsilon', \qquad (11.8)$$

$$E \to \tilde{E} = E - \epsilon. \qquad (11.9)$$

In Eqs. (11.6), (11.7), (11.8) and (11.9) the tilde means that the corresponding quantities are evaluated in the tilded coordinate system as implied by Eqs. (11.4) and (11.5).

To fix (either completely or partially) the gauge parameters ϵ_0 and ϵ amounts to selecting a specific coordinate system. If this would be the end of the story, the choice of a gauge would not be so problematic. Not only must the scalar fluctuations of the geometry change under Eqs. (11.4) and (11.5); the scalar fluctuations of the sources must also transform accordingly. So the possible ways in which the gauge parameters can be posited increases. Consider, for instance, the situation when the sources are represented by a collection of perfect fluids. For each species λ, the gauge transformations (11.4) and (11.5) induce the following shift:

$$\delta_s \rho_\lambda \to \delta_s \tilde{\rho}_\lambda = \delta_s \rho_\lambda - \rho_\lambda' \epsilon_0, \qquad (11.10)$$

$$\delta_s p_\lambda \to \delta_s \tilde{p}_\lambda = \delta_s p_\lambda - p_\lambda' \epsilon_0, \qquad (11.11)$$

$$\theta_\lambda \to \tilde{\theta}_\lambda = \theta_\lambda + \nabla^2 \epsilon'. \qquad (11.12)$$

The quantities ρ_λ and p_λ are the background energy density and pressure of each single species. The gauge transformations for the global variables can be obtained by recalling that:

$$p_t = \sum_\lambda p_\lambda, \qquad \rho_t = \sum_\lambda \rho_\lambda, \qquad (p_t + \rho_t)\theta_t = \sum_\lambda (p_\lambda + \rho_\lambda)\theta_\lambda. \qquad (11.13)$$

Equations (11.10), (11.11) and (11.12) imply, together with Eq. (11.13), that

$$\delta_s \rho_t \to \delta_s \tilde{\rho}_t = \delta_s \rho_t - \rho_t' \epsilon_0, \qquad (11.14)$$

$$\delta_s p_t \to \delta_s \tilde{p}_t = \delta p_t - p_\lambda' \epsilon_0, \qquad (11.15)$$

$$\theta_t \to \tilde{\theta}_t = \theta_t + \nabla^2 \epsilon'. \qquad (11.16)$$

Using the covariant conservation equation for the background fluid density, i.e. $\rho_\lambda' = -3\mathcal{H}(w_\lambda+1)\rho_\lambda$, the gauge transformation for the density contrast, i.e. $\delta_{(\lambda)} = \delta\rho_{(\lambda)}/\rho_{(\lambda)}$, follows easily from Eq. (11.10):

$$\tilde{\delta}_{(\lambda)} = \delta_{(\lambda)} + 3\mathcal{H}(1 + w_{(\lambda)})\epsilon_0. \qquad (11.17)$$

Another possibility for the energy-momentum tensor of the fluid sources is represented by one (or more) scalar degrees of freedom. In this case the relevant gauge transformation has been already discussed and employed in chapter 10 (see, in particular, Eq. (10.8)). Defining as φ the scalar degree of freedom and as $\chi = \delta_s \varphi$ the corresponding fluctuation, the relevant gauge transformation induced on χ by Eqs. (11.4) and (11.5) reads:

$$\chi \to \tilde{\chi} = \chi - \epsilon_0 \varphi'. \tag{11.18}$$

Since the scalar sources also shift under infinitesimal coordinate transformations, the two gauge parameters can be fixed either from the sources or from the scalar degrees of freedom of the geometry. For instance we can choose the gauge where the fluctuations of the energy density vanish: this will be the *uniform density gauge*. Or we can choose the gauge where the fluctuations of the scalar field vanish: this will be the *uniform field gauge* and so on and so forth.

It is appropriate to mention also the gauge variation of another important quantity, namely the brightness perturbations which will be important for expressing the perturbed Boltzmann equation in different gauges. Recalling the expression of the four-momentum

$$P^\mu = \frac{dx^\mu}{d\lambda} = \frac{d\tau}{d\lambda}\frac{dx^\mu}{d\tau} = P^0 \frac{dx^\mu}{d\tau}, \tag{11.19}$$

from the condition $g_{\alpha\beta} P^\alpha P^\beta = 0$ we get the modulus of the comoving three-momentum that reads:

$$q = n_i q^i = n_i \frac{dx^i}{d\tau}. \tag{11.20}$$

By now, gauge-shifting the spatial coordinates as $x^i \to \tilde{x}^i = x^i + \epsilon^i$, the comoving three-momentum transforms as:

$$\tilde{q} = q[1 + n_i \partial^i \epsilon']. \tag{11.21}$$

The quantity appearing in the phase space distribution is a scalar under gauge transformations. Consequently, the ratio (q/T) must transform as a scalar under infinitesimal coordinate transformations:

$$\left(\frac{\tilde{q}}{\tilde{T}}\right) = \left(\frac{q}{T}\right) - \epsilon_0 \left(\frac{q}{T}\right)'. \tag{11.22}$$

The inhomogeneous temperature can then be separated into a homogeneous contribution (denoted by $T_0(\tau)$) and a fully inhomogeneous perturbation:

$$T(\vec{x}, n^i, \tau) = T_0(\tau) + \delta_s T(\vec{x}, n^i, \tau). \tag{11.23}$$

The quantity which must now be computed is the value of δT in the gauge-shifted frame. Inserting Eq. (11.23) into Eq. (11.22) the following relation can be easily obtained:

$$\frac{\tilde{q}}{T_0\left(1+\frac{\delta_s \tilde{T}}{T_0}\right)} = \frac{q}{T_0\left(1+\frac{\delta_s T}{T_0}\right)} + \epsilon_0 q \frac{T_0'}{T_0^2}. \tag{11.24}$$

Both sides of Eq. (11.24) can now be expanded for small values of the relative temperature fluctuations with the result that:

$$\frac{\tilde{q}}{T_0}\left(1-\frac{\delta_s \tilde{T}}{T_0}\right) = \frac{q}{T_0}\left(1-\frac{\delta_s T}{T_0}\right) + \epsilon_0 q \frac{T_0'}{T_0^2}. \tag{11.25}$$

We can now use Eq. (11.21) to eliminate $\tilde{q}$; the resulting relation will give directly the gauge-shift of the temperature under infinitesimal coordinate transformations:

$$\delta_s \tilde{T} = \delta T - \epsilon_0 T_0' + n_i \partial^i \epsilon' T_0, \tag{11.26}$$

which also implies, in the notation of chapter 9, that

$$\tilde{\Delta}_I = \Delta_I - \epsilon_0 T_0' + n_i \partial^i \epsilon' T_0. \tag{11.27}$$

The collisionless Boltzmann equation written in a given gauge can be easily transformed into another suitable gauge. In fully analog terms it is possible to obtain the gauge variation of the reduced phase space distribution both for photons and neutrinos. The connection between the brightness perturbation and the reduced phase space distribution has been derived in Eqs. (9.61) and (9.62). For instance, in the case of neutrinos the gauge transformed $\mathcal{F}_\nu$ is given by

$$\tilde{\mathcal{F}}_\nu = \mathcal{F}_\nu - 4\frac{T_0'}{T_0}\epsilon_0 + 4n_i \partial^i \epsilon', \tag{11.28}$$

where the factor of 4 comes from the definition of $\mathcal{F}_\nu$ versus $f^{(1)}$ and where $T_0'/T_0 = -\mathcal{H}$ (see Eqs. (9.61) and (9.62) for the relation between $\mathcal{F}_\nu$ and $f^{(1)}$). A relevant observation that applies both to Eq. (11.27) and (11.28) is that while the monopole and dipole are not invariant under infinitesimal coordinate transformations, the higher multipoles are automatically gauge invariant. This property can be exhibited by taking the various multipoles of both sides of Eq. (11.27). Even if this short calculation is rather trivial, let us discuss it in detail. Equation (11.28) reads, in Fourier space,

$$\tilde{\mathcal{F}}_\nu = \mathcal{F}_\nu - 4\frac{T_0'}{T_0}\epsilon_0 + 4ik\mu\epsilon'. \tag{11.29}$$

By definition of multipole moment we will have (recall that $\mu = \hat{k}\cdot\hat{n}$)

$$\int_{-1}^{1} P_\ell(\mu)\mathcal{F}_\nu(\vec{k},\hat{n},\tau)d\mu = 2(-i)^\ell \mathcal{F}_{\nu\ell}(k,\tau). \tag{11.30}$$

By now taking the multipole moments of both sides of Eq. (11.29), it can be easily found after straightforward integration over μ that while the monopole and dipole shift under gauge transformations as

$$\tilde{\mathcal{F}}_{\nu 0} = \mathcal{F}_{\nu 0} - 4\frac{T_0'}{T_0}\epsilon_0, \qquad \tilde{\mathcal{F}}_{\nu 1} = \mathcal{F}_{\nu 1} + \frac{4}{3}k\epsilon', \tag{11.31}$$

the higher multipoles *do not* shift. We said that this property is trivial since it can be directly inferred from the gauge-shift of the radiation density contrast and of the peculiar velocity. We actually learned in chapter 9 that the lowest multipoles of the Boltzmann hierarchy do correspond, respectively, to the density contrast (monopole), to the peculiar velocity (dipole) and to the anisotropic stress (quadrupole).

In the following, a selection of gauges will be specifically discussed. For the sake of comparison we will also discuss, very briefly, the longitudinal gauge which has been used for different applications in chapters 7, 8, 9 and 10. Of course, as repeatedly stressed at the beginning of this chapter, the selection of gauges which will be presented is by no means exhaustive: its purpose is solely to account for some of the most commonly employed choices of coordinate systems.

11.1 The longitudinal gauge

The longitudinal (often named conformally Newtonian) gauge is obtained by setting to zero the functions E and B appearing in the parametrization of Eqs. (11.1), (11.2) and (11.3). In the longitudinal coordinate system we will have $E_{\rm L} = 0$ and $B_{\rm L} = 0$. The perturbed Christoffel connections and the perturbed components of the Ricci and Einstein tensors can be easily obtained, in this gauge, by setting to zero E and B in the corresponding equations appearing in Appendix D. The resulting expressions will contain ϕ and ψ which should be understood as the longitudinal fluctuations of the metric *computed in the longitudinal gauge*. In this sense it would be convenient to name those quantities $\phi_{\rm L}$ and $\psi_{\rm L}$. Even if this notation is desirable it is in practice a bit redundant due to the possible proliferation of indices. We will try to avoid it as much as possible. In spite of this, in some circumstances, it will be practical to recall in which gauge the fluctuations are effectively computed. For instance it will be interesting to know is what ψ when computed in the comoving orthogonal gauge. In this case the convention adopted will be to write ψ_C, i.e. the value of ψ *computed in the comoving orthogonal gauge*. Similar situations will be encountered along the way.

If we are to reach the longitudinal gauge from a generic parametrization of the perturbed line element we will have to demand, according to Eqs. (11.8) and (11.9), that

$$B \to B_\mathrm{L} = B + \epsilon_0 - \epsilon', \tag{11.32}$$

$$E \to E_\mathrm{L} = E - \epsilon. \tag{11.33}$$

Since $E_\mathrm{L} = B_\mathrm{L} = 0$, Eqs. (11.32) and (11.33) imply that the two gauge parameters are fixed as

$$\epsilon = E, \qquad \epsilon_0 = E' - B. \tag{11.34}$$

As anticipated the longitudinal condition removes the gauge freedom completely since both ϵ and ϵ_0 do not depend upon any arbitrary constant. The evolution equations in the longitudinal gauge have been extensively studied in chapters 7, 8 and 9. Moreover, in chapter 10 the longitudinal gauge equations have been also written in the presence of a single scalar degree of freedom which is the relevant situation for the discussion of the problem of initial conditions deep within the inflationary epoch.

11.1.1 *Gauge-invariant generalizations*

Since the gauge freedom is completely specified, it is possible to construct a set of gauge-invariant variables whose main property will be to reduce to the longitudinal variables once the condition $E = B = 0$ is consistently imposed. Of course, these equations will still hold, by construction, also in other gauges where $E \neq 0$ or $B \neq 0$. The relevant set of gauge-invariant variables can be written as

$$\Phi = \phi + (B - E')' + \mathcal{H}(B - E'), \tag{11.35}$$

$$\Psi = \psi - \mathcal{H}(B - E'), \tag{11.36}$$

$$\delta\rho_\mathrm{g} = \delta_\mathrm{s}\rho_\mathrm{t} + \rho_\mathrm{t}'(B - E'), \tag{11.37}$$

$$\delta p_\mathrm{g} = \delta_\mathrm{s} p_\mathrm{t} + p_\mathrm{t}'(B - E'), \tag{11.38}$$

$$\Theta_\mathrm{g} = \theta + \nabla^2 E'. \tag{11.39}$$

Equations (11.35) and (11.36) have already been anticipated in chapter 10 (see Eqs. (10.10) and (10.9)) where the gauge-invariant fluctuation of the scalar degree of freedom was specifically treated (see Eq. (10.11)). We immediately verify, according to Eqs. (11.6), (11.7), (11.8) and (11.9), that for infinitesimal coordinates transformations, the variables defined in Eqs. (11.35)–(11.39) are left invariant. We also immediately verify that when $E = B = 0$, $\Phi \to \phi$, $\Psi \to \psi$, and so on. Namely, the set of gauge-invariant variables defined in Eqs. (11.35)–(11.39) reduces to the longitudinal gauge variables when the longitudinal gauge condition are posited. As a useful exercise it is appropriate to write down the gauge-invariant evolution equations in terms of the quantities defined in Eqs. (11.35)–(11.39). Inserting the definitions (11.35)–(11.39) into the general expressions of the perturbed Einstein and energy-momentum tensors (see Appendix D), the gauge-invariant forms of the Hamiltonian and momentum constraints becomes:

$$\nabla^2 \Psi - 3\mathcal{H}(\Psi' + \mathcal{H}\Phi) = 4\pi a^2 G \delta \rho_g, \tag{11.40}$$

$$\nabla^2 (\Psi' + \mathcal{H}\Phi) = -4\pi G a^2 (p_t + \rho_t) \Theta_g, \tag{11.41}$$

while the two remaining conditions stemming from the (i, j) components of the perturbed Einstein equations are[a]:

$$\Psi'' + \mathcal{H}(\Phi' + 2\Psi') + (\mathcal{H}^2 + 2\mathcal{H}')\Phi$$
$$+ \frac{1}{3}\nabla^2(\Phi - \Psi) = 4\pi G a^2 \delta p_g, \tag{11.42}$$

$$\nabla^2(\Phi - \Psi) = 12\pi G a^2 (p_t + \rho_t) \sigma_t. \tag{11.43}$$

Equation accounts for the presence of the anisotropic stress. We prefer to write the *total* anisotropic stress (instead of the neutrino contribution) since there might also be other species with anisotropic stress. In this case we will have, in analogy to what happens with the total velocity field,

$$(p_t + \rho_t)\sigma_t = \sum_\lambda (p_\lambda + \rho_\lambda)\sigma_\lambda, \tag{11.44}$$

where the index λ runs over all the species of the mixture.

Plugging Eqs. (11.35)–(11.39) into Eqs. (D.50) and (D.51) and using the evolution equations of the background (i.e. Eqs. (2.65), (2.66) and (2.67)), the gauge-invariant form of the perturbed covariant conservation equations can easily be obtained:

$$\delta \rho_g' - 3(p + \rho)\Psi' + (p_t + \rho_t)\Theta_g + 3\mathcal{H}(\delta\rho_g + \delta p_g) = 0, \tag{11.45}$$

$$(p_t + \rho_t)\Theta_g' + [(p_t' + \rho_t') + 4\mathcal{H}(p_t + \rho_t)]\Theta_g$$
$$+ \nabla^2 \delta p_g + (p_t + \rho_t)\nabla^2 \Phi - (p_t + \rho_t)\nabla^2 \sigma_t = 0. \tag{11.46}$$

---

[a]It is worth noticing that the set of gauge-invariant equations (11.40)–(11.41) and (11.42)–(11.43) are the fluid generalization of Eqs. (10.13), (10.14) and (10.15) that have been already introduced in chapter 10 when treating the fluctuations induced by a self-interacting scalar field.

Another expedient way of obtaining Eqs. (11.40)–(11.43) and (11.45)–(11.46) is to use the evolution equations written directly in the longitudinal gauge (see, for instance, chapter 7) and to use the observation that they must have the same form of the evolution equations for the corresponding gauge-invariant quantities. The two methods lead to the same results.

11.2 The synchronous gauge

The synchronous coordinate system is defined by setting $\phi = 0$ and $B = 0$ in Eqs. (11.1) and (11.3). Therefore, in the synchronous coordinate system $\phi_S = B_S = 0$. As previously seen in the context of the longitudinal gauge, if we start from a generic coordinate system, the synchronous gauge is recovered if we posit, according to Eqs. (11.6) and (11.8),

$$B \to B_S = B + \epsilon_0 - \epsilon', \qquad (11.47)$$

$$\phi \to \phi_S = \phi - \mathcal{H}\epsilon_0 - \epsilon'_0. \qquad (11.48)$$

Since, by construction, $\phi_S = B_S = 0$, the gauge parameters are determined this time as

$$\epsilon' = B + \epsilon_0, \qquad (a\epsilon_0)' = a\phi. \qquad (11.49)$$

The explicit solution of Eq. (11.49) gives the gauge parameters in terms of $B(\vec{x}, \tau)$ and $\phi(\vec{x}, \tau)$. Direct integration of Eq. (11.49) implies:

$$\epsilon_0(\vec{x}, \tau) = \frac{1}{a(\tau)} \int^\tau a(\tau') \phi(\vec{x}, \tau') d\tau' + \frac{C_1(\vec{x})}{a(\tau)}, \qquad (11.50)$$

$$\epsilon(\vec{x}, \tau) = \int^\tau B(\vec{x}, \tau') d\tau' + C_2(\vec{x})$$

$$+ C_1(\vec{x}) \int^\tau \frac{d\tau'}{a(\tau')} + \int^\tau \frac{d\tau''}{a(\tau'')} \int^{\tau''} a(\tau') \phi(\vec{x}, \tau') d\tau'. \qquad (11.51)$$

Equations (11.50) and (11.51) depend upon two arbitrary functions $C_1(\vec{x})$ and $C_2(\vec{x})$ signaling that, in the synchronous frame, the gauge freedom is not completely removed. According to Eq. (11.2), the synchronous fluctuation of the metric can then be written, in Fourier space, as

$$\delta_s g_{ij}(k, \tau) = 2a^2(\psi_S \delta_{ij} + k_i k_j E_S). \qquad (11.52)$$

Another commonly used parametrization for the perturbed line element in the synchronous gauge is given by

$$\delta_s g_{ij}(k, \tau) = a^2(\tau) \left[\hat{k}_i \hat{k}_j h(k, \tau) + 6\xi(k, \tau) \left(\hat{k}_i \hat{k}_j - \frac{1}{3} \delta_{ij} \right) \right]. \qquad (11.53)$$

Since the parametrizations of Eqs. (11.52) and (11.53) must coincide, ψ_S and E_S will just be a combination of ξ and h:

$$\psi_S = -\xi, \qquad E_S = \frac{(h + 6\xi)}{2k^2}. \tag{11.54}$$

The evolution equations in the synchronous gauge can be derived with two complementary strategies:

- In the first approach the expressions of Appendix D can be specialized to the case of the synchronous gauge by setting in the relevant expressions $\phi = B = 0$; the remaining equations will depend upon ψ_S, E_S as well as the relevant density contrasts and peculiar velocities computed in the synchronous gauge; then using Eq. (11.54) the variables ψ_S and E_S can be traded for h and ξ and the wanted system can be recovered;
- in a complementary perspective it is also possible to take the evolution equations written in another gauge and then work out the connection between the dynamical variables expressed in the two gauges; then by expressing the old system in terms of the synchronous variables the very same system of synchronous equations can be obtained.

Since in chapters 7, 8 and 9 the Einstein-Boltzmann hierarchy has already been written in the longitudinal frame it is possible to take the longitudinal equations and to transform them into the synchronous equations.

To spell out the relation between two different gauges it suffices to write the relation between the dynamical variables defined in the two coordinate systems. Suppose, therefore, for us to be in the longitudinal gauge where the variables are ϕ_L, ψ_L, δ_L and θ_L. The corresponding variables in the synchronous gauge will be given by

$$\xi(k,\tau) = -\psi_L(k,\tau) - \frac{\mathcal{H}}{a} \int^\tau a(\tau')\phi_L(k,\tau')d\tau',$$

$$h(k,\tau) = 6\psi_L(k,\tau) + 6\frac{\mathcal{H}}{a} \int^\tau a(\tau')\phi_L(k,\tau')d\tau'$$

$$- 2k^2 \int^\tau \frac{d\tau'}{a(\tau')} \int^{\tau'} a(\tau'')\phi_L(k,\tau'')d\tau'', \tag{11.55}$$

$$\delta_S(k,\tau) = \delta_L(k,\tau) + \frac{3\mathcal{H}(w_{(\lambda)} + 1)}{a} \int^\tau a(k,\tau')\phi_L(k,\tau')d\tau'$$

$$\theta_S(k,\tau) = \theta_L(k,\tau) - \frac{k^2}{a} \int^\tau a(\tau')\phi_L(k,\tau')d\tau'.$$

Eq. (11.55) follows from Eqs. (11.6)–(11.9) and (11.10)–(11.12) and from the conditions defining the longitudinal gauge. In Eq. (11.55), $w_{(\lambda)}$ refers to the barotropic index of each single species. Concerning Eq. (11.55) a comment is in order. Since in the synchronous gauge $\phi_S = 0$, the gauge parameters that drive the solution from the longitudinal to the synchronous system will simply be determined from the equations

$$(a\epsilon_0)' = a\phi_L, \qquad \epsilon' = \epsilon_0, \tag{11.56}$$

which are the analog of Eqs. (11.49) but written for the specific case of a transformation from longitudinal to synchronous coordinates. As discussed in Eqs. (11.50) and (11.51), Eq. (11.56) necessarily entails two integration constants that may depend, in Fourier space, upon the comoving wavenumber k. These two constants must be set to zero since they would induce a spurious solution which is sometimes named gauge mode, i.e. an unphysical solution that can be removed by a further coordinate transformation that does not drive the transformed variables outside the synchronous gauge.

Conversely, a solution defined in the synchronous frame can be transformed in the longitudinal frame. In this case the gauge parameters driving the solution from the synchronous to the longitudinal frame are determined from the relations

$$B_S \to B_L = B_S + \epsilon_0 - \epsilon', \qquad E_S \to E_L = E_S - \epsilon, \tag{11.57}$$

recalling that $E_L = B_S = B_L = 0$. Consequently, from Eq. (11.57), the gauge parameters are determined to be:

$$\epsilon = \frac{h + 6\xi}{2k^2}, \qquad \epsilon_0 = \frac{h' + 6\xi'}{2k^2}, \tag{11.58}$$

where we recall that, according to Eq. (11.54), $E_S = (h + 6\xi)/(2k^2)$. Using the form of the gauge parameters obtained in Eq. (11.58), a given solution in the synchronous frame can be transformed in the longitudinal gauge according to the following expressions:

$$\begin{aligned}
\phi_L &= -\frac{1}{2k^2}[(h + 6\xi)'' + \mathcal{H}(h + 6\xi)'], \\
\psi_L &= -\xi + \frac{\mathcal{H}}{2k^2}(h + 6\xi)', \\
\delta_L &= \delta_S + \frac{3\mathcal{H}(w_\lambda + 1)}{2k^2}(h + 6\xi)' \\
\theta_L &= \theta_S - \frac{1}{2}(h + 6\xi)'.
\end{aligned} \tag{11.59}$$

Let us now see how Eq. (11.55) works on a practical example, namely, the way in which the Hamiltonian constraint is transformed from the longitudinal to the synchronous gauge. In the longitudinal gauge (and in Fourier space) the Hamiltonian constraint reads (see Eq. (7.51))

$$-k^2 \psi_L - 3\mathcal{H}(\mathcal{H}\phi_L + \psi_L') = 4\pi G a^2 \delta_s^{(L)} \rho_t \tag{11.60}$$

where, by definition, $\delta_s^{(L)} \rho_t$ is the density contrast computed in the longitudinal gauge, i.e.

$$\delta_s^{(L)} \rho_t = \rho_\gamma \delta_\gamma^{(L)} + \rho_\nu \delta_\nu^{(L)} + \rho_c \delta_c^{(L)} + \rho_b \delta_b^{(L)}. \tag{11.61}$$

Applying Eq. (11.55) we can easily find, after simple algebra, that Eq. (11.60) translates into the following equation

$$k^2 \xi - \frac{\mathcal{H}}{2} h' = 4\pi G a^2 \delta_s^{(S)} \rho_t, \tag{11.62}$$

where now $\delta_s^{(S)} \rho_t$ is the total density contrast *computed in the synchronous gauge*, i.e.

$$\delta_s^{(S)} \rho_t = \rho_\gamma \delta_\gamma^{(S)} + \rho_\nu \delta_\nu^{(S)} + \rho_c \delta_c^{(S)} + \rho_b \delta_b^{(S)}. \tag{11.63}$$

Notice that Eq. (11.62) is obtained by using one of the background equations and, more specifically, Eq. (2.66) which stipulates that $(\mathcal{H}^2 - \mathcal{H}') = 4\pi G a^2 (p_t + \rho_t)$. Following a similar procedure all the other evolution equations (already obtained in the longitudinal gauge) can be easily transformed in the synchronous gauge.

11.2.1 Evolution equations in the synchronous gauge

The complete set of evolution equations in the synchronous gauge will now be presented[b]:

$$k^2 \xi - \frac{\mathcal{H}}{2} h' = 4\pi G a^2 \delta \rho_t, \tag{11.64}$$

$$h'' + 2\mathcal{H} h' - 2k^2 \xi = 24\pi G a^2 \delta p_t, \tag{11.65}$$

$$(h + 6\xi)'' + 2\mathcal{H}(h + 6\xi)' - 2k^2 \xi = 24\pi G a^2 (p_\nu + \rho_\nu) \sigma_\nu, \tag{11.66}$$

$$k^2 \xi' = -4\pi G a^2 (p_t + \rho_t) \theta_t. \tag{11.67}$$

---

[b]For the sake of simplicity various subscripts and superscripts will be omitted. In particular the superscript S (referring to the synchronous gauge) will be avoided. The notation will be shortened by writing $\delta_s^{(S)} \rho_t = \delta \rho_t$ (and analogously for the pressure and for the divergence of the total peculiar velocity): it must actually be clear, at this point of the discussion, that these variables will be the ones computed in the synchronous frame.

Equations (11.64), (11.65), (11.66) and (11.67) follow, respectively, from the (00), (ii), $(i \neq j)$ and $(0i)$ components of the perturbed Einstein equations. Indeed Eqs. (11.64)–(11.67) can be obtained directly from the general expressions reported in Appendix D by imposing ab initio the synchronous gauge condition. This exercise is a useful cross-check for the consistency of the whole procedure. The evolution equations for the different species can be obtained with similar methods. In particular for CDM particles we have:

$$\delta'_c = -\theta_c + \frac{h'}{2},$$
$$\theta'_c + \mathcal{H}\theta_c = 0.$$
(11.68)

The evolution equations for the photons are instead[c]:

$$\delta'_\gamma = -\frac{4}{3}\theta_\gamma + \frac{2}{3}h',$$
$$\theta'_\gamma = \frac{k^2}{4}\delta_\gamma + \epsilon'(\theta_b - \theta_\gamma).$$
(11.69)

The equations for the baryons will be:

$$\delta'_b = -\theta_b + \frac{h'}{2},$$
$$\theta'_b + \mathcal{H}\theta_b = \frac{4}{3}\frac{\rho_\gamma}{\rho_b}\epsilon'(\theta_\gamma - \theta_b).$$
(11.70)

Finally, for neutrinos we will have:

$$\delta'_\nu = -\frac{4}{3}\theta_\nu + \frac{2}{3}h', \qquad \theta'_\nu = -k^2\sigma_\nu + \frac{k^2}{4}\delta_\nu,$$
$$\sigma'_\nu = \frac{4}{15}\theta_\nu - \frac{3}{10}k\mathcal{F}_{\nu 3} - \frac{2}{15}h' - \frac{4}{5}\xi'.$$
(11.71)

Using the observation that, prior to decoupling, photons are tightly coupled to baryons, the relevant evolution equations for the baryon-photon system are given by:

$$\delta'_\gamma = -\frac{4}{3}\theta_{\gamma b} + \frac{2}{3}h', \qquad \delta'_b = -\theta_{\gamma b} + \frac{h'}{2},$$
$$\theta'_{\gamma b} + \frac{\mathcal{H}R_b}{R_b + 1}\theta_{\gamma b} = \frac{k^2}{4}\frac{\delta_\gamma}{R_b + 1},$$
(11.72)

---

[c]It should be mentioned, to avoid confusion, that the quantity ϵ' appearing in Eqs. (11.69) and (11.70) denotes (as in chapter 7) the differential optical depth. Therefore it should not be confused with the first time derivative of the gauge parameter which is also denoted by ϵ'. With this specification a potentially dangerous clash of notations is avoided.

where $\theta_{\gamma b}$ is the photon-baryon velocity field. A word of care must be spent for the transformation of the evolution equations of the neutrinos. As already mentioned the equations for the neutrinos are obtained by keeping the lowest multipoles of the Boltzmann hierarchy which arise from the collisionless Boltzmann equation. So we need to find the synchronous expression of the evolution equation for $\mathcal{F}_\nu$, i.e. the reduced neutrino phase space distribution. Consider, therefore, the equation for $\mathcal{F}_\nu$ that was derived in chapter 9 (see Eq. (9.21)):

$$\mathcal{F}'_L + ik\mu \mathcal{F}_L = 4\psi'_L - 4ik\mu\phi_L, \tag{11.73}$$

where the subscript ν referring to the neutrinos has been suppressed in favour of the subscript L stressing the fact that the reduced phase space distribution is expressed in the longitudinal coordinate system. The problem is then to determine the form of Eq. (11.73) in the synchronous gauge. Using Eq. (11.28) and the gauge transformations given in Eqs. (11.6) and (11.7) we have:

$$\begin{aligned}\mathcal{F}_S \to \mathcal{F}_L &= \mathcal{F}_S + 4\mathcal{H}\epsilon_0 + 4ik\mu\epsilon',\\ \psi_S \to \psi_L &= \psi_S + \mathcal{H}\epsilon_0,\\ \psi_S \to \phi_L &= -\mathcal{H}\epsilon_0 - \epsilon'_0,\end{aligned} \tag{11.74}$$

where we used the fact that $\phi_S = 0$ by definition of synchronous gauge. Inserting Eq. (11.74) into Eq. (11.73) the following equation is swiftly obtained:

$$\mathcal{F}'_S + ik\mu\mathcal{F}_S = -4\left[\xi' - \frac{\mu^2}{2}(h' + 6\xi')\right]. \tag{11.75}$$

To get to Eq. (11.75) the observation is that the relevant gauge parameters ϵ_0 and ϵ are given in Eq. (11.58) and that, in particular, $\epsilon'' = \epsilon'_0$.

The useful exercise will now be to derive the Boltzmann equation directly in the synchronous frame. Starting from the general form of the collisional Boltzmann equation for massless particles, i.e.

$$\frac{Df}{D\tau} = \frac{\partial f}{\partial \tau} + \frac{\partial f}{\partial x^i}\frac{dx^i}{d\tau} + \frac{\partial f}{\partial q^i}\frac{dq^i}{d\tau} = \mathcal{C}_{\text{coll}}, \tag{11.76}$$

whose explicit perturbation, according to the method treated in Eqs. (9.9) and (9.11), will be

$$f(x^i, P_j, \tau) = f_0(q)[1 + f^{(1)}(x^i, q_j, \tau)]. \tag{11.77}$$

To first order Eq. (11.76) will then become:

$$\frac{\partial f^{(1)}}{\partial \tau} + \frac{\partial f^{(1)}}{\partial x^i}n^i + \frac{\partial \ln f_0}{\partial q}\frac{dq}{d\tau} = \frac{1}{f_0}\mathcal{C}_{\text{coll}}. \tag{11.78}$$

The four-momenta can then be expressed as[d]

$$P^0 = \frac{p}{a}, \qquad P^0 = \frac{q}{a^2}, \qquad q^i = a^2\left(\delta^i_j - \frac{h^i_j}{2}\right)P^j. \qquad (11.79)$$

To write Eq. (11.78) in explicit terms the derivative of the comoving three-momentum should be computed and the result is:

$$\frac{dq^i}{d\tau} = 2\mathcal{H}a^2 P^j\left[\delta^i_j - \frac{h^i_j}{2}\right] + a^2\left[\delta^i_j - \frac{h^i_j}{2}\right]\frac{dP^j}{d\tau} - \frac{a^2}{2}P^j\frac{\partial h^i_j}{\partial \tau}. \qquad (11.80)$$

The explicit form of the time derivative of the conjugate momentum can then be derived by recalling that

$$\frac{dP^\mu}{d\lambda} = -\Gamma^\mu_{\alpha\beta}P^\alpha P^\beta. \qquad (11.81)$$

Since

$$\frac{dP^\mu}{d\lambda} = \frac{dP^\mu}{d\tau}\frac{d\tau}{d\lambda} = P^0\frac{dP^\mu}{d\tau}, \qquad (11.82)$$

we also easily get

$$\frac{dP^i}{d\tau} = -2\Gamma^i_{j0}P^j = -2\left(\mathcal{H}\delta^i_j - \frac{1}{2}h^{i\prime}_j\right)P^j. \qquad (11.83)$$

Using Eq. (11.83) in Eq. (11.80)

$$\frac{dq}{d\tau} = \frac{q}{2}h'_{ij}n^i n^j. \qquad (11.84)$$

Equation (11.78) can be written in explicit terms and the overall result of the previous manipulations will be:

$$\frac{\partial f^{(1)}}{\partial \tau} + \frac{\partial f^{(1)}}{\partial x^i}n^i + \frac{\partial \ln f_0}{\partial q}h'_{ij}n^i n^j = \frac{1}{f_0}\mathcal{C}_{\text{coll}}. \qquad (11.85)$$

According to the parametrization of the line element in the synchronous gauge (see Eq. (11.53)) we then have:

$$n^i n^j h'_{ij} = \mu^2(h' + 6\xi') - \frac{\xi'}{3}. \qquad (11.86)$$

Using Eq. (11.86) in Eq. (11.85) the evolution equation of $f^{(1)}$ becomes, finally:

$$\frac{\partial f^{(1)}}{\partial \tau} + ik\mu f^{(1)} - \frac{\partial \ln f_0}{\partial q}\left[\xi' - \frac{\mu^2}{2}(h' + 6\xi')\right] = \frac{1}{f_0}\mathcal{C}_{\text{coll}}. \qquad (11.87)$$

---

[d] In the following equations h_{ij} denotes the (global) scalar metric fluctuation in the synchronous gauge defined in Eq. (11.53). The quantity h_{ij} (only used with this meaning in the present section) should not be confused with the tensor fluctuation which has been introduced in chapter 6.

Defining the reduced neutrino phase space distribution as

$$\mathcal{F}_\nu(\vec{k},\hat{n},\tau) = \frac{\int q^3 f_0 f^{(1)}\, dq}{\int f^{(0)} q^3\, dq} = \sum_\ell (-i)^\ell (2\ell+1) \mathcal{F}_{\nu\ell} P_\ell(\mu), \qquad (11.88)$$

the collisionless Boltzmann equation for massless neutrinos then becomes, simply,

$$\mathcal{F}'_\nu + ik\mu \mathcal{F}_\nu = 4\left[-\xi' + \frac{\mu^2}{2}(h' + 6\xi')\right], \qquad (11.89)$$

which is exactly Eq. (11.73). Recalling the definition of the second Legendre polynomial Eq. (11.73) (or Eq. (11.89)) can also be expressed as:

$$\mathcal{F}'_\nu + ik\mu \mathcal{F}_\nu = \frac{2}{3}h' + \frac{4}{3}P_2(\mu)(h' + 6\xi'). \qquad (11.90)$$

Following the same procedure for the multipolar expansion discussed in chapter 9, Eq. (11.90) implies

$$\begin{aligned}
\mathcal{F}'_{\nu 0} + k\mathcal{F}_{\nu 1} &= \frac{2}{3}h', \\
\mathcal{F}'_{\nu 1} &= \frac{k}{3}\mathcal{F}_{\nu 0} - \frac{2}{3}k\mathcal{F}_{\nu 2}, \\
\mathcal{F}'_{\nu 2} &= \frac{2}{5}k\mathcal{F}_{\nu 1} - \frac{3}{5}k\mathcal{F}_{\nu 3} - \frac{4}{15}(h' + 6\xi'), \\
\mathcal{F}'_{\nu \ell} &= \frac{k}{2\ell+1}[\ell \mathcal{F}_{\nu(\ell-1)} - (\ell+1)\mathcal{F}_{\nu(\ell+1)}], \qquad \ell \geq 3.
\end{aligned} \qquad (11.91)$$

Recalling the connection of the multipole moments with the density contrast, the velocity field and the anisotropic stress, the first three relations of the previous hierarchy become:

$$\begin{aligned}
\delta'_\nu &= -\frac{4}{3}\theta_\nu + \frac{2}{3}h', \\
\theta'_\nu &= \frac{k^2}{4}\delta_\nu - k^2\sigma_\nu, \\
\sigma'_\nu &= \frac{4}{15}\theta_\nu - \frac{3}{10}k\mathcal{F}_{\nu 3} - \frac{2}{15}(h' + 6\xi'),
\end{aligned} \qquad (11.92)$$

which coincide with the fluid equations previously deduced (see Eq. (11.71)) by transforming the longitudinal equations in the synchronous frame.

Always on the theme of the gauge-transformed Boltzmann equations, it is appropriate to write the evolution equations for the brightness perturbations of the radiation field (see Eq. (11.27)) directly in the synchronous gauge:

$$\Delta_I' + ik\mu\Delta_I = -\left[\xi' - \frac{\mu^2}{2}(h' + 6\xi')\right] + \epsilon'\left[-\Delta_I + \Delta_{I0} + \mu v_b - \frac{1}{2}P_2(\mu)S_Q\right],$$

$$\Delta_Q' + ik\mu\Delta_Q = \epsilon'\left[-\Delta_Q + \frac{1}{2}(1 - P_2(\mu))S_Q\right],$$

$$\Delta_Q' + ik\mu\Delta_Q = -\epsilon'\Delta_Q,$$

$$v_b' + \mathcal{H}v_b + \frac{\epsilon'}{R_b}(3i\Delta_{I1} + v_b) = 0, \qquad (11.93)$$

where, as usual, $S_Q = \Delta_{I2} + \Delta_{Q0} + \Delta_{Q2}$. For the derivation of the results mentioned in this section it is useful to recall that the gauge-transformed dipoles and the gauge-transformed monopole are:

$$\tilde{\Delta}_{I1} = \Delta_{I1} - \frac{k}{3}\epsilon' = \Delta_{I1} - \frac{1}{6k}(h' + 6\xi'),$$

$$\tilde{v}_b = v_b + ik\epsilon' = v_b + \frac{i}{2k}(h' + 6\xi'), \qquad (11.94)$$

$$\tilde{\Delta}_{I0} = \Delta_{I0} + \mathcal{H}\epsilon_0 = \Delta_{I0} + \frac{\mathcal{H}}{2k^2}(h' + 6\xi').$$

As in chapter 9 we defined the baryon peculiar velocity as $v_b = \theta_b/(ik)$.

11.2.2 The adiabatic mode in the synchronous gauge

An interesting exercise will now be to derive the solution for the primordial adiabatic mode directly in the synchronous gauge. This derivation has already been discussed in the longitudinal gauge (see the discussion from Eq. (8.84) to Eq. (8.89)). The purpose of the present exercise is to look for a solution of the lowest multipoles of the Einstein-Boltzmann hierarchy to leading order in $k\tau$ and prior to matter-radiation equality. The interest of this derivation resides in the method rather than in the result that has been already obtained by working in the longitudinal gauge (see chapter 8). In the present chapter, the derivation will be performed directly in the synchronous coordinate system. In the longitudinal gauge the adiabaticity of the initial conditions imply the constancy of the metric fluctuations. In the synchronous gauge the situation is a bit different. Consider first the evolution equations of the CDM component, i.e. Eq. (11.68). The CDM velocity can be set to zero and it will remain zero later on; thus Eq. (11.68) implies that

$$\theta_c(k,\tau) = 0, \qquad \delta_c(k,\tau) \simeq \frac{h(k,\tau)}{2}. \qquad (11.95)$$

The adiabaticity condition then implies immediately that

$$\delta_\gamma(k,\tau) = \delta_\nu(k,\tau), \qquad \delta_c(k,\tau) = \delta_b(k,\tau), \qquad \delta_\gamma(k,\tau) = \frac{4}{3}\delta_c(k,\tau). \tag{11.96}$$

Using Eq. (11.95), Eq. (11.96) also implies that $\delta_\gamma \simeq (2/3)h$. Inspection of the Hamiltonian constraint of Eq. (11.64) then suggests the following parametrization for h and ξ:

$$h(k,\tau) = -C(k)k^2\tau^2, \qquad \xi(k,\tau) = -2C(k) + B(k)k^2\tau^2, \tag{11.97}$$

which indeed solves, to the lowest order in $k\tau$, Eq. (11.64). Notice that B and C appearing in Eq. (11.97) do not depend on τ but do depend on k, i.e. $B = B(k)$ and $C = C(k)$. The relation between $B(k)$ and $C(k)$ will be determined from the consistency with the other equations of the system. Within the parametrization provided by Eq. (11.97), Eq. (11.65) is immediately solved. Equations (11.96) and (11.97) then imply:

$$\delta_\gamma(k,\tau) = -\frac{2}{3}C(k)k^2\tau^2, \qquad \delta_\nu(k,\tau) = -\frac{2}{3}C(k)k^2\tau^2,$$
$$\delta_c(k,\tau) = -\frac{C(k)}{2}k^2\tau^2, \qquad \delta_b(k,\tau) = -\frac{C(k)}{2}k^2\tau^2. \tag{11.98}$$

The evolution of the neutrino anisotropic stress and of the remaining velocity fields is consistently solved by

$$\sigma_\nu(k,\tau) = A(k)k^2\tau^2, \qquad \theta_\nu(k,\tau) = -\frac{k^4\tau^3}{3}\left[A(k) + \frac{C(k)}{6}\right],$$
$$\theta_{\gamma b}(k,\tau) = -\frac{C(k)}{18}k^4\tau^3. \tag{11.99}$$

From the momentum constraint of Eq. (11.66) and from Eq. (11.71) the arbitrary constants $A(k)$, $B(k)$ and $C(k)$ must satisfy, respectively, the following pair of conditions:

$$18B(k) - C(k) = 6R_\nu A(k), \qquad 4C(k) - 24B(k) = 30A(k). \tag{11.100}$$

The solution of the two algebraic conditions of Eq. (11.100) leads to the determination of $A(k)$ and $B(k)$ in terms of $C(k)$:

$$B(k) = \frac{4R_\nu + 5}{6(4R_\nu + 15)}C(k), \qquad A(k) = \frac{4C(k)}{3(4R_\nu + 15)}. \tag{11.101}$$

With the help of Eq. (11.59) the synchronous solution can be transformed back into the longitudinal gauge. We then discover that what has been called $\phi_i(k)$ in Eq. (8.84) is connected with $C(k)$ through a numerical constant that depends upon the neutrino fraction. The complete result will then be:

$$\phi_i(k) = \frac{20C(k)}{4R_\nu + 15}, \qquad \psi_i(k) = \left(\frac{8R_\nu + 20}{4R_\nu + 15}\right)C(k), \qquad (11.102)$$

$$\delta_\gamma(k) = \delta_\nu(k) = -2\phi_i(k), \qquad \delta_c(k) = \delta_b(k) - \frac{3}{2}\phi_i(k), \qquad (11.103)$$

$$\sigma_\nu(k,\tau) = \frac{k^2\tau^2}{6R_\nu}[\psi_i(k) - \phi_i(k)], \qquad (11.104)$$

$$\theta_{\gamma b}(k,\tau) \simeq \theta_\nu(k,\tau) \simeq \theta_c(k,\tau) = \frac{k^2\tau}{2}\phi_i(k). \qquad (11.105)$$

The solution given by Eqs. (11.102), (11.103) and (11.105) is clearly consistent with the solution previously discussed (directly in the longitudinal gauge) and reported in chapter 8. Indeed, from Eq. (11.102) we immediately derive the relation between $\psi_i(k)$ and $\phi_i(k)$, namely $\psi_i(k) = (1 + 2R_\nu/5)\phi_i(k)$. The latter relation is exactly the one already obtained (directly in the longitudinal gauge) and reported in Eq. (8.89).

11.2.3 Entropic modes in the synchronous gauge

In chapter 8 the entropic modes have been described in the language of the longitudinal gauge. It is appropriate here to present the synchronous description of the entropic modes. We can either solve the equations directly in the synchronous gauge or transform the longitudinal solutions. As already discussed, these two procedure lead to the same result. The solutions reported in the remaining part of the present section are all approximate in the sense that they hold during the radiation-dominated epoch and to lowest order in $k\tau$.

Consider, to begin with, the CDM-radiation mode. In the synchronous frame the solution becomes:

$$h(k,\tau) = 2\mathcal{S}_*(k)\overline{\Omega}_c\left(\frac{\tau}{\tau_1}\right), \qquad \xi(k,\tau) = -\frac{\mathcal{S}_*(k)}{3}\overline{\Omega}_c\left(\frac{\tau}{\tau_1}\right), \qquad (11.106)$$

$$\delta_\gamma(k,\tau) = \delta_\nu(k,\tau) = \frac{4}{3}\mathcal{S}_*(k)\overline{\Omega}_c\left(\frac{\tau}{\tau_1}\right), \qquad (11.107)$$

$$\delta_c(k,\tau) = -\mathcal{S}_*(k) + \mathcal{S}_*(k)\overline{\Omega}_c\left(\frac{\tau}{\tau_1}\right), \qquad (11.108)$$

$$\delta_b(k,\tau) = \mathcal{S}_*(k)\overline{\Omega}_c\left(\frac{\tau}{\tau_1}\right), \qquad (11.109)$$

$$\theta_c(k,\tau) = 0, \qquad (11.110)$$

$$\theta_\nu(k,\tau) = \theta_{\gamma b}(k,\tau) = \frac{k^2\tau_1}{6}\mathcal{S}_*\overline{\Omega}_c\left(\frac{\tau}{\tau_1}\right)^2, \qquad (11.111)$$

$$\sigma_\nu(k,\tau) = \frac{k^2\tau_1}{6}\left(\frac{R_\nu}{15+4R_\nu}\right)\overline{\Omega}_c\mathcal{S}_*(k)\left(\frac{\tau}{\tau_1}\right)^3. \qquad (11.112)$$

As in the case of the discussion of chapter 8 the quantity $\mathcal{S}_*(k)$ denotes the entropy mode of the CDM-radiation system, i.e. according to the definition of Eq. (8.95)

$$\mathcal{S}_{c\gamma}(k,\tau) = \mathcal{S}_{c\nu}(k,\tau) = \mathcal{S}_*(k). \qquad (11.113)$$

The important point to stress is that *entropy fluctuations are gauge-invariant* so they have the same value no matter what gauge is used to compute them. Let us indeed consider their definition given in Eq. (8.95) in terms of variables defined in the longitudinal gauge. In terms of the longitudinal density contrasts, following Eq. (8.95), we can write

$$\mathcal{S}_{ij} = 3\left[\frac{\delta_j}{(w_j+1)} - \frac{\delta_i}{(w_i+1)}\right]. \qquad (11.114)$$

Let us then perform an infinitesimal shift of the time slicing as $\tau \to \tilde{\tau} = \tau + \epsilon_0$. The two density contrasts (pertaining to two generic fluid of the mixture) will transform (see Eq. (11.17)) as $\delta_i \to \tilde{\delta}_i = \delta_i + (w_i+1)\epsilon_0$) and as $\delta_j \to \tilde{\delta}_j = \delta_j + (w_j+1)\epsilon_0$. Consequently, from Eq. (11.114), we will have

$$\mathcal{S}_{ij} \to \tilde{\mathcal{S}}_{ij} = \mathcal{S}_{ij}, \qquad (11.115)$$

which shows that, indeed, $\mathcal{S}_{ij}$ is invariant under infinitesimal coordinate transformations. It may seem that this result depends upon the occurrence that the two generic fluids of the mixture have *constant* barotropic indices w_i and w_j. This is not the case. In fact, in the following sections it will be demonstrated that the quantities ζ_i and ζ_j (see Eq. (8.95)) have a simple gauge-invariant interpretation in terms of the curvature perturbations on the hypersurfaces where the energy density of the two generic species i and j are not perturbed. This result allows us to generalize the statement of Eq. (11.115) also to the situation where the barotropic index is not constant.

Let us now move on and discuss the synchronous gauge description of the baryon-entropy mode. In this case the synchronous gauge variables are simply given by:

$$h(k,\tau) = 2\mathcal{S}_*(k)\overline{\Omega}_{\rm b}\left(\frac{\tau}{\tau_1}\right), \qquad \xi(k,\tau) = -\frac{\mathcal{S}_*(k)}{3}\overline{\Omega}_{\rm b}\left(\frac{\tau}{\tau_1}\right), \qquad (11.116)$$

$$\delta_\gamma(k,\tau) = \delta_\nu(k,\tau) = \frac{4}{3}\mathcal{S}_*(k)\overline{\Omega}_{\rm b}\left(\frac{\tau}{\tau_1}\right), \qquad (11.117)$$

$$\delta_{\rm c}(k,\tau) = \mathcal{S}_*(k)\overline{\Omega}_{\rm b}\left(\frac{\tau}{\tau_1}\right), \qquad (11.118)$$

$$\delta_{\rm b}(k,\tau) = -\mathcal{S}_*(k) + \mathcal{S}_*(k)\overline{\Omega}_{\rm b}\left(\frac{\tau}{\tau_1}\right), \qquad (11.119)$$

$$\theta_{\rm c}(k,\tau) = 0, \qquad (11.120)$$

$$\theta_\nu(k,\tau) = \theta_{\gamma\rm b}(k,\tau) = \frac{k^2\tau_1}{6}\mathcal{S}_*\overline{\Omega}_{\rm b}\left(\frac{\tau}{\tau_1}\right)^2, \qquad (11.121)$$

$$\sigma_\nu(k,\tau) = \frac{k^2\tau_1}{6}\left(\frac{R_\nu}{15+4R_\nu}\right)\overline{\Omega}_{\rm b}\mathcal{S}_*(k)\left(\frac{\tau}{\tau_1}\right)^3. \qquad (11.122)$$

Equations (11.116)–(11.122) completely mirror the analog equations obtained in the case of the CDM-radiation mode. The only notable physical difference is that the entropic contribution now resides in the baryon sector. The notations have been chosen in accordance with the notations already established in the longitudinal gauge (see Eqs. (8.107)–(8.115)).

The solution corresponding to the neutrino-entropy mode reads, in the synchronous gauge,

$$\xi(k,\tau) = -\frac{2}{9}\frac{R_\gamma R_\nu}{(4R_\nu+15)}\tilde{\mathcal{S}}_*(k)k^2\tau^2, \qquad h(k,\tau) = 0 \qquad (11.123)$$

$$\delta_\gamma(k,\tau) = \frac{4}{3}R_\nu\tilde{\mathcal{S}}_*(k), \qquad \delta_\nu(k,\tau) = -\frac{4}{3}R_\gamma\tilde{\mathcal{S}}_*(k), \qquad (11.124)$$

$$\delta_{\rm b}(k,\tau) = \delta_{\rm c}(k,\tau) = 0, \qquad \theta_{\rm c} = 0 \qquad (11.125)$$

$$\theta_{\gamma\rm b} = R_\nu\frac{\tilde{\mathcal{S}}_*(k)}{3}k^2\tau, \qquad \theta_\nu = -R_\gamma\frac{\tilde{\mathcal{S}}_*(k)}{3}k^2\tau \qquad (11.126)$$

$$\sigma_\nu(k,\tau) = -\frac{2R_\gamma}{3(4R_\nu+15)}\tilde{\mathcal{S}}_*(k)k^2\tau^2. \qquad (11.127)$$

11.3 Comoving orthogonal hypersurfaces

There are two important variables that have been already treated and that have a natural interpretation in specific gauge descriptions. They have been introduced in chapter 7 in terms of the longitudinal gauge variables:

- the first variable is $\mathcal{R}$ (see Eq. (7.101));
- the second variable is ζ (see Eq. (7.120)).

As we saw (see Eq. (7.101)) the variable $\mathcal{R}$ arises directly as a simplifying variable allowing a direct integration for the evolution equations of ψ in the longitudinal gauge. This had been the main practical reason for its introduction in chapter 7. In similar terms the introduction of ζ has been rather useful for simplifying the discussion of non-adiabatic initial conditions (see Eq. (7.120) and discussion therein). In this section the physical and geometric interpretation of these two variables will be further scrutinized.

The variable $\mathcal{R}$ measures the curvature perturbations in the comoving orthogonal gauge. The comoving orthogonal gauge is defined as the coordinate system where $B_\mathrm{C} = 0$ and $\theta_\mathrm{C} = 0$. Since the comoving orthogonal gauge sets a condition on the velocity field there might be some subtleties connected with the physical nature of θ_C. If a single total fluid is present in the game the statement $\theta_\mathrm{C} = 0$ is not ambiguous. However, if there are many different fluids the situation could become more tricky. In the present case this subtlety will be left aside. By the condition $\theta_\mathrm{C} = 0$, we mean that either the velocity field vanishes (in the single fluid case) or the total velocity field vanishes (if many fluids are present in the mixture). To prove the statement that $\mathcal{R}$ measures the fluctuations of the spatial curvature in the comoving orthogonal gauge, let us compute the fluctuations of the spatial curvature when $B_\mathrm{C} = \theta_\mathrm{C} = 0$. In this coordinate system the curvature perturbation can be computed directly from the expressions of the perturbed Christoffel connections. From the equations reported in Appendix D (see, in particular, Eq. (D.1)) we have

$$\delta_\mathrm{s}\Gamma_{ij}^k = (\partial^k \psi_\mathrm{C} \delta_{ij} - \partial_i \psi_\mathrm{C} \delta_j^k - \partial_j \psi_\mathrm{C} \delta_i^k) + \partial_i \partial_j \partial^k E_\mathrm{C} \qquad (11.128)$$

The fluctuations of the spatial curvature are then defined as

$$\delta R_\mathrm{C}^{(3)} = \delta_\mathrm{s} \gamma^{ij} \overline{R}_{ij}^{(3)} + \overline{\gamma}^{ij} \delta_\mathrm{s} R_{ij}^{(3)}, \qquad (11.129)$$

where $\gamma_{ij} = a^2 \delta_{ij}$ and where $R_{ij}^{(3)} = 0$ since the background metric is spatially flat. By definition we will have

$$\delta_\mathrm{s} R_{ij}^{(3)} = \partial_k \delta_\mathrm{s} \Gamma_{ij}^k - \partial_j \delta_\mathrm{s} \Gamma_{ik}^k \equiv \nabla^2 \psi_\mathrm{C} \delta_{ij} + \partial_i \partial_j \psi_\mathrm{C}, \qquad (11.130)$$

where the second equality follows from Eq. (11.128) (notice that E_C disappears from the final expression). Thus, from Eq. (11.129)

$$\delta_s R^{(3)} \equiv \frac{4}{a^2}\nabla^2 \psi_C. \qquad (11.131)$$

The curvature fluctuations of the comoving gauge can be connected to the fluctuations in a different gauge characterized by a different value of the time coordinate, i.e. $\tau_C \to \tau = \tau_C + \epsilon_0$. Under this shift

$$\psi_C \to \psi = \psi_C + \mathcal{H}\epsilon_0, \qquad (11.132)$$

$$\left(B_C - \frac{\theta_C}{k^2}\right) \to \left(B - \frac{\theta}{k^2}\right) = \left(B_C - \frac{\theta_C}{k^2}\right) + \epsilon_0. \qquad (11.133)$$

Since in the comoving orthogonal gauge $B_C - \theta_C/k^2 = 0$, Eqs. (11.132)–(11.133) imply, in the new coordinate system,

$$\psi_C = \psi - \mathcal{H}\left(B - \frac{\theta}{k^2}\right). \qquad (11.134)$$

Equation (11.134) still looks very different from Eq. (7.101) (which was written, however, in the longitudinal gauge). Let us notice immediately that ψ_C is gauge-invariant. This statement may look, at first sight, a bit strange. The variable ψ_C is defined on comoving orthogonal hypersurfaces, how come it is also gauge-invariant? The answer to this question stems directly from Eq. (11.134). We can clearly see that ψ_C, when expressed in a generic coordinate system where neither θ nor B vanish, does not change under infinitesimal coordinate transformations. In fact, under a shift of the time coordinate, ψ transform through $\mathcal{H}\epsilon_0$ while the combination $(B - \theta/k^2)$ shifts as ϵ_0. Since ψ_C is gauge-invariant it can be expressed directly as a combination of gauge-invariant variables. Indeed, using Eqs. (11.36) and (11.39) it is easy to show that

$$\psi_C = \left(\Psi + \mathcal{H}\frac{\Theta_g}{k^2}\right) = \left(\Psi + \frac{\mathcal{H}(\mathcal{H}\Phi + \Psi')}{\mathcal{H}^2 - \mathcal{H}'}\right), \qquad (11.135)$$

where the second equality follows from the gauge-invariant momentum constraint of Eq. (11.41) which allows us to eliminate Θ_g in favour of a combination involving Φ and Ψ'. Notice also that in writing the second equality of Eq. (11.135), the background equations have been used (and, in particular, Eq. (2.66)). Since Eq. (11.135) is written in explicit gauge-invariant terms, it can be evaluated in any gauge. Let us choose, in particular, the longitudinal frame. In this coordinate system, as we saw before in this chapter, $\Psi \to \psi_L$ and $\Phi \to \phi_L$. But then is evident that $\mathcal{R}$ *is* ψ_C up to a sign, i.e. $\mathcal{R} = -\psi_C$. Consequently, from Eq. (11.135) it can be concluded that $\mathcal{R}$ corresponds to the curvature fluctuations of the spatial curvature on comoving orthogonal hypersurfaces.

11.4 Uniform density hypersurfaces

Let us now concentrate our attention on ζ. The physical interpretation of ζ can be twofold. It can be interpreted as the fluctuation of the spatial curvature on the hypersurfaces of constant density. But it can also be interpreted as the (total) density contrast on the hypersurfaces of constant spatial curvature. Let us start from the first possible interpretation. On the hypersurfaces of constant density the measure of curvature perturbations is provided by ψ_D. Under infinitesimal gauge transformations

$$\psi_D \to \psi = \psi_D + \mathcal{H}\epsilon_0,$$
$$\delta\rho_D \to \delta\rho = \delta\rho_D - \rho'\epsilon_0. \tag{11.136}$$

But on constant density hypersurfaces, $\delta\rho_D = 0$, by the definition of the uniform density gauge. Hence, from Eq. (11.136),

$$\psi_D = \psi + \mathcal{H}\frac{\delta\rho}{\rho'_t} \equiv \Psi + \mathcal{H}\frac{\delta\rho_g}{\rho'_t}, \tag{11.137}$$

where, again, the second equality follows from the definitions of gauge-invariant fluctuations given in Eqs. (11.36) and (11.37). Hence, the (gauge-invariant) curvature fluctuations on constant density hypersurfaces can be defined as

$$\zeta = -\left(\Psi + \mathcal{H}\frac{\delta\rho_g}{\rho'_t}\right). \tag{11.138}$$

Equation (11.138) defines the gauge-invariant curvature fluctuations on constant density hypersurfaces. Again, as in the case of curvature fluctuations on comoving orthogonal hypersurfaces, the last sentence seems to contain a contradiction, but such an expression simply means that ζ coincides with the curvature fluctuations in the uniform density gauge. In a different gauge (for instance the longitudinal gauge) ζ has the same value but does not coincide with the curvature fluctuations. In particular we might choose to evaluate ζ on the hypersurfaces of constant curvature. The hypersurfaces of constant curvature are the ones for which $\psi = 0$. We can immediately verify that, in this gauge, ζ equals $\delta\rho_t/[3(p_t + \rho_t)]$. This conclusion can be drawn, for instance, from Eq. (11.137) which implies, in the constant curvature gauge, that $\psi_D = -\delta\rho_t/[3(p_t + \rho_t)]$. But, by definition, $\zeta = -\psi_D$. This proves the statement that, indeed, ζ can be also interpreted as the total density contrast on uniform curvature hypersurfaces.

The values of ζ and $\mathcal{R}$ are equal up to terms proportional to the Laplacian of Ψ. This can be shown by using explicitly the definitions of $\mathcal{R}$ and ζ

whose difference gives

$$\zeta - \mathcal{R} = \mathcal{H}\left(\frac{\delta\rho_g}{\rho_t'} - \frac{\Theta_g}{k^2}\right). \tag{11.139}$$

Recalling now the form of the Hamiltonian and momentum constraints in gauge-invariant notations (see Eqs. (11.40) and (11.41)), and using the conservation of the energy density of the background (i.e. Eq. (2.67)), the total peculiar velocity and the total density contrast can be expressed as:

$$\begin{aligned}\frac{\Theta_g}{k^2} &= \frac{1}{4\pi G a^2 (p_t + \rho_t)}(\mathcal{H}\Phi + \Psi'), \\ \frac{\delta\rho_g}{\rho_t'} &= -\frac{1}{4\pi G a^2 (p_t + \rho_t)}[k^2 \Psi + 3\mathcal{H}(\mathcal{H}\Phi + \Psi')].\end{aligned} \tag{11.140}$$

Thus, from Eq. (11.139) we obtain

$$\zeta - \mathcal{R} = -\frac{k^2 \Psi}{12\pi G a^2 (p_t + \rho_t)}. \tag{11.141}$$

It is finally useful to mention another interesting quantity that often appears in the treatment of cosmological perturbations, namely the density contrast on comoving orthogonal hypersurfaces:

$$\epsilon_m = \frac{\delta\rho_C}{\rho_t} \equiv \frac{\delta\rho_t + \rho_t'(B - \theta/k^2)}{\rho_t} \equiv \frac{\delta\rho_g + 3\mathcal{H}(p + \rho)\Theta_g/k^2}{\rho_t}, \tag{11.142}$$

where the second equality follows from the first one by using the definitions of gauge-invariant fluctuations given in Eqs. (11.35)–(11.39). Again, using the Hamiltonian and momentum constraints,

$$\epsilon_m = -\frac{k^2}{4\pi G a^2 \rho_t}\Psi \tag{11.143}$$

that means, according to Eq. (11.141), that $(\zeta - \mathcal{R}) \propto \epsilon_m$, or more specifically,

$$\epsilon_m = \frac{3(p_t + \rho_t)}{\rho_t}(\zeta - \mathcal{R}). \tag{11.144}$$

In conclusion, the three gauge-invariant quantities discussed in the present and in the previous section can be summarized, in explicit gauge-invariant terms, as

$$\begin{aligned}\mathcal{R} &= -\Psi - \frac{\Theta_g}{k^2}, \qquad \zeta = -\Psi - \mathcal{H}\frac{\delta\rho_g}{\rho_t'}, \\ \epsilon_m &= \frac{\delta\rho_g}{\rho_t} + \frac{3\mathcal{H}(p_t + \rho_t)}{\rho_t}\frac{\Theta_g}{k^2}.\end{aligned} \tag{11.145}$$

They are not independent since, by virtue of the gauge-invariant Hamiltonian constraint of Eq. (11.40), Eq. (11.144) must always hold.

It is finally appropriate to remark that there exist physical situations where more than one fluid is present in the system. In this circumstance it is always possible to define ζ and $\mathcal{R}$ in terms of global variables. However, it is also productive, sometimes, to think in terms of the variables defining the separate dynamics of each fluid of the mixture. This is the case, for instance, in the treatment of non-adiabatic initial conditions of CMB anisotropies that have been introduced in chapter 8. Consider, therefore, a realistic pre-decoupling fluid where there are photons, neutrinos, CDM particles and baryons. It is possible, in this situation, to introduce the curvature fluctuations on the hypersurfaces where one of the species composing the mixture has constant energy density. These will be the hypersurfaces where the energy density of a given component of the plasma is unperturbed. Let us give an example and consider the curvature fluctuations computed on the hypersurface where the photon density contrast vanishes. On this hypersurface, by definition, $\delta\rho_{D\gamma} = 0$, meaning that (with this notation) this is the hypersurface where the energy density of the photons is unperturbed. Therefore, it will be possible to perform a gauge shift of the time coordinate as $\tau_{D\gamma} \to \tau = \tau_{D\gamma} + \epsilon_0$ which will induce the following shift on the two relevant dynamical variables:

$$\psi_{D\gamma} \to \psi = \psi_{D\gamma} + \mathcal{H}\epsilon_0,$$
$$\delta\rho_{D\gamma} \to \delta\rho_\gamma = \delta\rho_{D\gamma} - \rho'_\gamma \epsilon_0. \qquad (11.146)$$

Consequently, the new (gauge-invariant) variable arising from this observation will be:

$$\psi_{D\gamma} = \psi + \mathcal{H}\frac{\delta\rho_\gamma}{\rho'_\gamma} = \psi - \frac{\delta_\gamma}{4}. \qquad (11.147)$$

To be consistent with the previous conventions, therefore, we will define a new gauge-invariant variable $\zeta_\gamma = -\psi_{D\gamma}$ whose physical interpretation will be the curvature fluctuation on the hypersurfaces of constant *photon* energy density. Alternatively, the variable ζ_γ can be interpreted as the photon density contrast on the hypersurfaces of unperturbed spatial curvature (i.e. $\psi = 0$). The same kind of game can be played with all the species of the plasma so that, in the pre-decoupling system, the following set of gauge-invariant variables arises naturally

$$\zeta_\nu = -\Psi + \frac{\delta_\nu^{(g)}}{4}, \qquad \zeta_\gamma = -\Psi + \frac{\delta_\gamma^{(g)}}{4},$$
$$\zeta_c = -\Psi + \frac{\delta_c^{(g)}}{3}, \qquad \zeta_b = -\Psi + \frac{\delta_b^{(g)}}{3}. \qquad (11.148)$$

Equation (11.148) is the analog of Eq. (8.91) that has already been encountered in chapter 8. The difference is that, in Eq. (11.148), all the quantities are written in gauge-invariant terms: Ψ is the well known Bardeen potential and

$$\delta_\gamma^{(g)} = \delta_\gamma - 4\mathcal{H}(B - E'), \qquad \delta_\nu^{(g)} = \delta_\nu - 4\mathcal{H}(B - E'), \qquad (11.149)$$

$$\delta_b^{(g)} = \delta_b - 3\mathcal{H}(B - E'), \qquad \delta_c^{(g)} = \delta_c - 3\mathcal{H}(B - E'), \qquad (11.150)$$

are the gauge-invariant density contrasts for each single species of the plasma. Equations (11.149) and (11.150) define the various gauge-invariant density contrasts (characterized by the superscript g) in terms of the longitudinal gauge variables in full analogy with Eq. (11.37) that holds for the total density contrast of the mixture. The density contrasts of the various species, when evaluated on uniform curvature hypersurfaces, are clearly related to ζ, i.e. the total density contrast on uniform curvature hypersurfaces. It is easy to show that, indeed,

$$\zeta = \sum_i \frac{\rho_i'}{\rho_t'} \zeta_i, \qquad c_{st}^2 = \sum_i c_{si}^2 \frac{\rho_i'}{\rho_t'}, \qquad (11.151)$$

where the index i runs over the different species of the plasma. By gauge-invariance of the ζ_i and of ζ, the same relation holds also in any other coordinate system. Equation (11.151) has been already derived in chapter 8 (see Eq. (8.93)) in the course of a discussion conducted in the longitudinal frame.

11.5 The off-diagonal gauge

The last remark of the previous section has to do with the explicit form of $\mathcal{R}$ in the case where the fluctuations are provided by a self-interacting scalar degree of freedom. This theme has already been explored in chapter 10 but it is appropriate here to deepen the implication of those results. In the case of a single scalar degree of fredom the gauge-invariant generalization of the fluctuation of the scalar field is given by Eq. (10.11), i.e. $X = \chi + \varphi'(B - E')$. In terms of X the momentum constraint of Eq. (11.41) is given by Eq. (10.15), i.e. $\mathcal{H}\Phi + \Psi' = 4\pi G\varphi'X$. Thus, from Eq. (11.145) the gauge-invariant curvature fluctuation $\mathcal{R}$ becomes

$$\mathcal{R} = -\Psi - \frac{\mathcal{H}}{\varphi'}X = -\frac{q}{z}, \qquad (11.152)$$

where $z = a\varphi'/\mathcal{H}$ has already been introduced in Eqs. (10.27). The variable q defined by the second equality of Eq. (11.152) arises naturally as a canonical normal mode of the scalar-tensor action perturbed to second-order in the amplitude of the scalar fluctuations of the metric (see Eq. (10.29)) and plays a crucial role in the problem of the early initial conditions of CMB anisotropies (see, also, the final part of the Appendix C). By definition of Ψ and X it is also immediate for us to obtain, from Eq. (11.152) that

$$\mathcal{R} = -\psi - \frac{\mathcal{H}}{\varphi'}\chi. \tag{11.153}$$

From Eq. (11.153) it emerges that $\mathcal{R}$ has other two other potentially interesting interpretations:

- $\mathcal{R}$ can be interpreted as the curvature fluctuation on uniform field hypersurfaces, i.e. the hypersurfaces where the scalar field is unperturbed and $\chi = 0$;
- $\mathcal{R}$ can be interpreted as the scalar field fluctuation on uniform curvature hypersurfaces where ψ vanishes.

The second interpretation lead us directly to exploit the uniform curvature gauge which is called also *off diagonal gauge*. The rationale for this nomenclature is very simple and stems from the fact that the requirement of uniform curvature does not specify completely the coordinate system. If we insist that the gauge freedom should be completely removed, the natural choice is to select the new gauge by requiring that, not only $\psi_{od} = 0$ but also that $E_{od} = 0$. If we want to reach the off-diagonal gauge from another generic coordinate system we just have to posit the off-diagonal gauge conditions and determine accordingly the two gauge parameters. In fact, from Eqs. (11.7) and (11.9) we have

$$\psi \to \psi_{od} = \psi + \mathcal{H}\epsilon_0, \qquad E \to E_{od} = E - \epsilon. \tag{11.154}$$

Imposing now $E_{od} = \psi_{od} = 0$ the two gauge parameters ϵ_0 and ϵ are determined to be

$$\epsilon_0 = -\frac{\psi}{\mathcal{H}}, \qquad \epsilon = E. \tag{11.155}$$

It is clear that according to Eq. (11.155) the off-diagonal gauge fixes completely the coordinate system. In this sense the off-diagonal gauge belongs to the same class of the longitudinal gauge. In both gauges the gauge parameters are completely fixed. This is in contrast with what we saw happening in the synchronous gauge where two arbitrary space-dependent functions arise in the determination of the gauge parameters.

11.5.1 Evolution equations in the off-diagonal gauge

The observation that, in the off-diagonal gauge, $\psi_{\rm od} = 0$ implies directly (according to Eq. (11.152)) that $q = a\chi$, i.e. the canonical normal mode of the second-order action coincides in this gauge solely with the fluctuation of the scalar field. This observation suggests that the off-diagonal gauge is particularly suitable when the dominant source of the energy-momentum tensor is represented by one (or more) scalar fields. To illustrate the advantages of conducting explicit calculations in the off-diagonal gauge, we will analyze the evolution equations of the scalar fluctuations of the geometry when two interacting scalar fields are simultaneously present in the system. The matter part of the action will therefore be written as

$$S_{\varphi\sigma} = \int d^4x \sqrt{-g}\left[\frac{1}{2}g^{\alpha\beta}\partial_\alpha\varphi\partial_\beta\varphi + \frac{1}{2}g^{\alpha\beta}\partial_\alpha\sigma\partial_\beta\sigma - W(\varphi,\sigma)\right]. \quad (11.156)$$

The evolution equations of the conformally flat background geometry can be usefully written as:

$$\mathcal{H}^2 - \mathcal{H}' = 4\pi G(\sigma'^2 + \varphi'^2), \quad (11.157)$$

$$\varphi'' + 2\mathcal{H}\varphi' + a^2\frac{\partial W}{\partial \varphi} = 0, \quad (11.158)$$

$$\sigma'' + 2\mathcal{H}\sigma' + a^2\frac{\partial W}{\partial \sigma} = 0. \quad (11.159)$$

In the off-diagonal gauge the $(i \neq j)$, (00) and $(0i)$ components of the perturbed Einstein equations lead, respectively, to:

$$B' + 2\mathcal{H}B + \phi = 0, \quad (11.160)$$

$$\mathcal{H}\nabla^2 B + (\mathcal{H}' + 2\mathcal{H}^2)\phi = -4\pi G\bigg[\varphi'\chi'_\varphi + \sigma'\chi'_\sigma$$
$$+ \frac{\partial W}{\partial \varphi}a^2\chi_\varphi + \frac{\partial W}{\partial \sigma}a^2\chi_\sigma\bigg], \quad (11.161)$$

$$\mathcal{H}\phi = 4\pi G(\chi_\varphi\varphi' + \chi_\sigma\sigma'). \quad (11.162)$$

In Eqs. (11.161) and (11.162) the variables χ_φ and χ_σ denote the fluctuations of φ and σ. Equations (11.160), (11.161) and (11.162) must be supplemented by the two perturbed Klein-Gordon equations that can be easily obtained from

$$g^{\alpha\beta}\nabla_\alpha\nabla_\beta\varphi + \frac{\partial W}{\partial \varphi} = 0, \qquad g^{\alpha\beta}\nabla_\alpha\nabla_\beta\sigma + \frac{\partial W}{\partial \sigma} = 0. \quad (11.163)$$

Using the results reported in Appendix D (with the appropriate modifications implied by the presence of two scalar fields), the first-order form of

Eq. (11.163) is:

$$\chi_\varphi'' + 2\mathcal{H}\chi_\varphi' - \nabla^2\chi_\varphi + \frac{\partial^2 W}{\partial\varphi^2}a^2\chi_\varphi$$

$$+ \frac{\partial^2 W}{\partial\varphi\partial\sigma}a^2\chi_\sigma + 2\phi\frac{\partial W}{\partial\varphi}a^2 - \varphi'\phi' - \varphi'\nabla^2 B = 0, \qquad (11.164)$$

$$\chi_\sigma'' + 2\mathcal{H}\chi_\sigma' - \nabla^2\chi_\sigma + \frac{\partial^2 W}{\partial\sigma^2}a^2\chi_\sigma$$

$$+ \frac{\partial^2 W}{\partial\varphi\partial\sigma}a^2\chi_\varphi + 2\phi\frac{\partial W}{\partial\sigma}a^2 - \sigma'\phi' - \sigma'\nabla^2 B = 0. \qquad (11.165)$$

Equations (11.164) and (11.165) can be used to integrate the system in explicit terms. By inserting Eqs. (11.160), (11.161) and (11.162) into Eqs. (11.164) and (11.165) the dependence upon B and ϕ (i.e. the fluctuations of the metric) can be totally eliminated. Indeed, Eqs. (11.160), (11.161) and (11.162) imply

$$\phi = 4\pi G\left(\frac{\varphi'}{\mathcal{H}}\chi_\varphi + \frac{\sigma'}{\mathcal{H}}\chi_\sigma\right), \qquad (11.166)$$

$$\phi' = 4\pi G\left[\frac{\varphi'}{\mathcal{H}}\chi_\varphi' + \frac{\sigma'}{\mathcal{H}}\chi_\sigma' + \left(\frac{\varphi'}{\mathcal{H}}\right)'\chi_\varphi + \left(\frac{\sigma'}{\mathcal{H}}\right)'\chi_\sigma\right], \qquad (11.167)$$

$$\nabla^2 B = -4\pi G\bigg\{(\mathcal{H}' + 2\mathcal{H}^2)\left[\frac{\varphi'}{\mathcal{H}}\chi_\varphi + \frac{\sigma'}{\mathcal{H}}\chi_\sigma\right]$$

$$+ \frac{\varphi'}{\mathcal{H}}\chi_\varphi' + \frac{\sigma'}{\mathcal{H}}\chi_\sigma' + \frac{\partial W}{\partial\varphi}\frac{a^2}{\mathcal{H}}\chi_\varphi + \frac{\partial W}{\partial\sigma}\frac{a^2}{\mathcal{H}}\chi_\sigma\bigg\}, \qquad (11.168)$$

where Eq. (11.167) is just the derivative of Eq. (11.166) and Eq. (11.168) follows from Eq. (11.161) by using Eq. (11.166). Using now Eqs. (11.166), (11.167) and (11.168) to eliminate ϕ, ϕ' and $\nabla^2 B$ from Eqs. (11.164) and (11.165), the wanted result can be obtained, namely:

$$\chi_\varphi'' + 2\mathcal{H}\chi_\varphi' - \nabla^2\chi_\varphi + \mathcal{A}_{\varphi\varphi}\chi_\varphi + \mathcal{A}_{\varphi\sigma}\chi_\sigma = 0, \qquad (11.169)$$

$$\chi_\sigma'' + 2\mathcal{H}\chi_\sigma' - \nabla^2\chi_\sigma + \mathcal{A}_{\sigma\sigma}\chi_\sigma + \mathcal{A}_{\varphi\sigma}\chi_\varphi = 0, \qquad (11.170)$$

where

$$\mathcal{A}_{\varphi\varphi} = \frac{\partial^2 W}{\partial\varphi^2} + 4\pi G\left[4\frac{\partial W}{\partial\varphi}a^2\left(\frac{\varphi'}{\mathcal{H}}\right) + \varphi'^2\left(4 + 2\frac{\mathcal{H}'}{\mathcal{H}^2}\right)\right], \qquad (11.171)$$

$$\mathcal{A}_{\sigma\sigma} = \frac{\partial^2 W}{\partial\sigma^2} + 4\pi G\left[4\frac{\partial W}{\partial\sigma}a^2\left(\frac{\sigma'}{\mathcal{H}}\right) + \sigma'^2\left(4 + 2\frac{\mathcal{H}'}{\mathcal{H}^2}\right)\right], \qquad (11.172)$$

$$\mathcal{A}_{\varphi\sigma} = \frac{\partial^2 W}{\partial\sigma\partial\varphi} + 4\pi G\left[2\frac{\partial W}{\partial\sigma}a^2\left(\frac{\sigma'}{\mathcal{H}}\right) + 2\frac{\partial W}{\partial\varphi}a^2\left(\frac{\varphi'}{\mathcal{H}}\right)\right.$$

$$\left. + \varphi'\sigma'\left(4 + 2\frac{\mathcal{H}'}{\mathcal{H}^2}\right)\right], \qquad (11.173)$$

where $\mathcal{A}_{\varphi\sigma} = \mathcal{A}_{\sigma\varphi}$. Equations (11.169) and (11.170) cannot be decoupled. They can be rearranged in such a way that the scalar normal modes $q_\varphi = a\chi_\varphi$ and $q_\sigma = a\chi_\sigma$ arise naturally. In this case Eqs. (11.169) and (11.170) become

$$q_\varphi'' - \nabla^2 q_\varphi + \left[-\frac{a''}{a} + \mathcal{A}_{\varphi\varphi}\right] q_\varphi + \mathcal{A}_{\varphi\sigma} q_\sigma = 0, \qquad (11.174)$$

$$q_\sigma'' - \nabla^2 q_\sigma + \left[-\frac{a''}{a} + \mathcal{A}_{\varphi\varphi}\right] q_\sigma + \mathcal{A}_{\varphi\sigma} q_\varphi = 0. \qquad (11.175)$$

By solving Eqs. (11.174) and (11.175), all the other interesting quantities can be obtained. In the off-diagonal gauge the curvature perturbation $\mathcal{R}$ has a very simple expression, since

$$\mathcal{R} = -\frac{\mathcal{H}^2}{\mathcal{H}^2 - \mathcal{H}'}\phi. \qquad (11.176)$$

Equation (11.176) can be directly obtained, for instance, from the second identity of Eq. (11.135) by recalling that, in the off-diagonal gauge the two Bardeen potentials can be expressed as $\Psi = -\mathcal{H}B$ and $\Phi = \phi + B' + \mathcal{H}B$. Using Eq. (11.166) to eliminate ϕ, Eq. (11.176) can then be written as

$$\mathcal{R} = -\frac{1}{\varphi'^2 + \sigma'^2}\left[\frac{q_\varphi}{z_\varphi}\varphi'^2 + \frac{q_\sigma}{z_\sigma}\sigma'^2\right] = -\frac{H}{\dot\varphi^2 + \dot\sigma^2}(\dot\varphi q_\varphi + \dot\sigma q_\sigma). \qquad (11.177)$$

In Eq. (11.177) the second equality arises just by using the cosmic time coordinate and by taking into account that[e]

$$z_\varphi = \frac{a\varphi'}{\mathcal{H}}, \qquad z_\sigma = \frac{a\sigma'}{\mathcal{H}}. \qquad (11.178)$$

If the system consists of two interacting scalar fields (as it can happen in some classes of inflationary models) the procedure is very simple: we have to solve Eqs. (11.174) and (11.175) and then compute the interesting quantities, for instance, the curvature perturbations. Notice, once again, another example of the general observation that has been already discussed in the present chapter: $\mathcal{R}$ is the fluctuation of the scalar curvature in the comoving orthogonal gauge. In the off-diagonal gauge it *does not* have the same interpretation.

We conclude this section by pointing out that, in the limit where σ is absent, q_φ is the canonical normal mode of the system that has been already

---
[e]In chapter 10 we had only a single scalar degree of freedom and, therefore, there was no ambiguity in the definition of z (see, for instance, Eq. (10.27)). In the present exercise we are dealing with two minimally coupled scalars. We are therefore obliged to specify which z we discuss by introducing a subscript, as in Eq. (11.178).

discussed in Eq. (10.29) and in Appendix C. In the case where σ is absent we have $z_\sigma \to 0$ and $\mathcal{A}_{\varphi\sigma} \to 0$. In this case Eq. (11.174) implies that

$$\mathcal{A}_{\varphi\varphi} - \frac{a''}{a} = -\frac{z_\varphi''}{z_\varphi}. \tag{11.179}$$

The result (11.179) follows by using Eqs. (11.157), (11.158) and (11.159). In particular, the following two background relations are rather useful:

$$\frac{\mathcal{H}''}{\mathcal{H}} = 2\mathcal{H}' - 2\frac{\varphi''}{\varphi'}\left(1 - \frac{\mathcal{H}'}{\mathcal{H}^2}\right), \tag{11.180}$$

$$a^2 \frac{\partial^2 W}{\partial \varphi^2} = 4\mathcal{H}^2 - 2\mathcal{H}' - \frac{\varphi'''}{\varphi'}. \tag{11.181}$$

11.6 Mixed gauge-invariant treatments

Based on the developments discussed in the present chapter, it is possible to infer that there exists rather useful gauge-invariant descriptions of the cosmological perturbations that mix different gauge-invariant variables, i.e. gauge-invariant variables whose physical interpretation is particularly simple in a specific coordinate system. As in the selection of the coordinate system there is no gauge with a particular status, also in the selection of gauge-invariant quantities there is not a single unique and ultimate choice that is intrinsically better than the other possible choices. So, for instance, the gauge-invariant quantities defined in Eqs. (11.35)–(11.39) have the agreeable property of being the direct gauge-invariant generalization of the longitudinal variables. However there exists other sets of gauge-invariant combinations having equally nice dynamical properties. This observation leads us directly to what we will call mixed gauge-invariant descriptions.

Consider, to be more specific, the system defined by $\mathcal{R}$, the Bardeen potentials Ψ and Φ and the density contrast on comoving orthogonal hypersurfaces ϵ_m. It will now be shown how it is possible to have a closed system of equations involving the four mentioned variables. The first observation we want to make is that Eq. (11.143) is the relativistic generalization of the Poisson equation. Namely, by writing Eq. (11.143) as

$$\nabla^2 \Psi = 4\pi G a^2 \rho_t \epsilon_m, \tag{11.182}$$

it is apparent that this equation has the same form of the Poisson equation where, however, Ψ and ϵ_m hold in the relativistic limit. We then need the evolution equation for ϵ_m. The quickest way to derive this equation is to write the evolution of ζ and $\mathcal{R}$; then take the difference of the obtained equations and finally use, in the obtained expression, Eq. (11.141) which gives ϵ_m as the difference between ζ and $\mathcal{R}$. Let us follow this program. The evolution equation for ζ can be written as

$$\zeta' = -\frac{\mathcal{H}}{\rho_t(1+w_t)}\delta p_{\text{nad}} - \frac{\Theta_g}{3}, \qquad (11.183)$$

and this has been already discussed in chapter 7. The evolution equation for $\mathcal{R}$ is instead

$$\mathcal{R}' = -\frac{\mathcal{H}}{\rho_t(1+w_t)}\delta p_{\text{nad}}$$
$$+ \frac{\mathcal{H}}{12\pi G a^2(p_t+\rho_t)}\nabla^2(\Phi-\Psi) - \frac{\mathcal{H}c_{\text{st}}^2}{4\pi G a^2(p_t+\rho_t)}\nabla^2\Psi. \qquad (11.184)$$

In Eqs. (11.183) and (11.184), w_t and c_{st}^2 are the total barotropic index and the total sound speed, i.e.

$$w_t = \frac{p_t}{\rho_t}, \qquad c_{\text{st}}^2 = w_t - \frac{w_t'}{3\mathcal{H}(w_t+1)} = w_t - \frac{1}{3}\frac{d\ln(w_t+1)}{d\ln a}. \qquad (11.185)$$

Taking then the difference of Eqs. (11.184) and (11.185) and using repeatedly Eq. (11.141), the evolution equation for ϵ_m turns out to be

$$\epsilon_m' - 3\mathcal{H}w_t\epsilon_m = -(1+w_t)\Theta_g - 3\mathcal{H}(1+w_t)\sigma_t \qquad (11.186)$$

where the notation

$$\partial_i\partial^j\Pi_j^i = (p_t+\rho_t)\nabla^2\sigma_t. \qquad (11.187)$$

This set of gauge-invariant variables is rather useful in several applications like the ones dealing with the semi-analytical evaluation of temperature autocorrelations. Alternatively, the evolution equation of ϵ_m can be inferred from Eq. (11.182). The idea is to take the first derivative of both sides of Eq. (11.182) and then use the momentum constraint of Eq. (11.41) to eliminate the dependence upon $\nabla^2\Psi'$. Then Eq. (11.42) can be used to trade Φ (arising as a result of the use of the momentum constraint) for Ψ and the anisotropic stress σ_t. Finally, the remaining $\nabla^2\Psi$ can be expressed in terms of ϵ_m through the generalized Poisson equation (see Eq. (11.182)).

The last relevant equation we could derive involves Θ_g directly. From Eq. (11.46) it is possible to derive the following expresssion:

$$\Theta_g' + \mathcal{H}\Theta_g + \frac{c_{\text{st}}^2}{w_t+1}\nabla^2\epsilon_m + \nabla^2\Phi - \nabla^2\sigma_t = 0. \qquad (11.188)$$

Now the obtained system of equations can be used in different ways. To illustrate the usefulness of the described approach let us consider a simple problem, namely, the physical situation provided by the radiation-matter transition which can be described by the following exact form of the barotropic index and of the total sound speed:

$$w_t = \frac{1}{3(\alpha + 1)}, \quad c_{st}^2 = \frac{4}{3(3\alpha + 4)}, \tag{11.189}$$

which have already been derived in Eq. (7.97). Recall that, in Eq. (11.189), $\alpha = a/a_{eq} = x^2 + 2x$ where $x = \tau/\tau_1$.

The evolution equations can then be solved in a kind of iterative approach where, to zeroth order, the anisotropic stress is neglected, i.e. $\sigma_t = 0$. The skilled reader will recognize that this is essentially the same physical situation already encountered in chapter 7. Equation (11.43) stipulates that:

$$\nabla^2(\Phi - \Psi) = 12\pi G a^2 (p_t + \rho_t)\sigma_t, \tag{11.190}$$

so, if $\sigma_t = 0$, $\Phi = \Psi$. In this approximation Eqs. (11.182), (11.186) and (11.188) can be written, respectively, as

$$\nabla^2 \Psi = 4\pi G a^2 \rho_t \epsilon_m, \tag{11.191}$$

$$\epsilon_m' - 3\mathcal{H} w_t \epsilon_m + (1 + w_t)\Theta_g = 0, \tag{11.192}$$

$$\Theta_g' + \mathcal{H}\Theta_g + \nabla^2 \Psi = 0. \tag{11.193}$$

In Eq. (11.193) we neglected a term going as $\nabla^2 \epsilon_m$. This is justified since we are working for typical scales larger than the Hubble radius. However, we have to be careful. It would be indeed very sloppy to say that, since we are outside the Hubble radius, *all* the Laplacians should be neglected. In chapter 7 we used to neglect terms containing the Laplacians. Here, on the contrary, one of the dynamical variables *is* effectively the Laplacian of Ψ. Consequently, since $\epsilon_m \simeq \mathcal{O}(\nabla^2 \Psi)$ the Laplacian of Ψ must be kept in Eq. (11.193) and (11.191). Similarly, in chapter 7 we neglected the peculiar velocity. But the peculiar velocity was of higher order in that treatment: it would be wrong to neglect it with the variables employed in the present example. The logic of the calculation will be to start with the zeroth order solution where $\sigma_t = 0$. Under this approximation, as we will see, both ϵ_m and Θ_g can be computed. Then, using Eq. (11.191) it will be possible to compute Ψ. The obtained solution will then be inserted back into Eq. (11.190) and the anisotropic stress will be switched on and the contribution of σ_t can be perturbatively included.

To achieve this program the first step is to find the zeroth order solution and, for this purpose it is useful to recall that $\mathcal{H}$ can be written directly in terms of α (see Eq. (11.189)):

$$\mathcal{H}(\alpha) = \frac{2}{\tau_1} \frac{\sqrt{\alpha+1}}{\alpha}. \tag{11.194}$$

Equations (11.192) and (11.193) can then be rearranged by using as the integration variable α directly instead of τ. Eliminating $\nabla^2 \Psi$ in Eq. (11.12) with help of Eq. (11.191), Eqs. (11.192) and (11.193) can be written, respectively, as

$$\frac{d\epsilon_m}{d\alpha} - \frac{\epsilon}{\alpha(\alpha+1)} + \frac{\tau_1(3\alpha+4)}{6(\alpha+1)^{3/2}} \Theta_g = 0, \tag{11.195}$$

$$\frac{d\Theta_g}{d\alpha} + \frac{\Theta_g}{\alpha} + \frac{3}{\tau_1} \frac{\sqrt{\alpha+1}}{\alpha^2} \epsilon_m = 0. \tag{11.196}$$

Equations (11.195) and (11.196) can be clearly decoupled by taking the first derivative of Eq. (11.195) with respect to α and by using Eq. (11.196) to eliminate the derivative of Θ_g. The bottom line of this straightforward manipulation will be the following second-order equation for ϵ_m:

$$\frac{d^2 \epsilon_m}{d\alpha^2} + \mathcal{J}_1(\alpha) \frac{d\epsilon_m}{d\alpha} + \mathcal{J}_2(\alpha) \epsilon_m = 0, \tag{11.197}$$

$$\mathcal{J}_1(\alpha) = \frac{9\alpha + 14}{2(\alpha+1)(3\alpha+4)},$$

$$\mathcal{J}_2(\alpha) = -\frac{(3\alpha+4)(3\alpha^2+3\alpha+2) + 9\alpha^2 + 20\alpha + 8}{2\alpha^2(\alpha+1)^2(3\alpha+4)}.$$

Equation (11.197) can be solved with standard methods and the solution will be

$$\epsilon_m(k,\tau) = c_1(k) f_1(\alpha) + c_2(k) f_2(\alpha) \tag{11.198}$$

$$f_1(\alpha) = \frac{1}{\alpha\sqrt{\alpha+1}}, \tag{11.199}$$

$$f_2(\alpha) = \frac{9\alpha^3 + 2\alpha^2 - 8\alpha - 16 + 16\sqrt{\alpha+1}}{9\alpha(\alpha+1)}. \tag{11.200}$$

In Eq. (11.198), $c_1(k)$ and $c_2(k)$ are two integration constants that depend, in Fourier space, on the comoving momentum k. The solution $f_1(\alpha)$ is decaying while the solution $f_2(\alpha)$ is growing. The growing mode of the total density contrast ϵ_m is responsible for the decay of the Bardeen potential across the radiation-matter transition. Indeed, from Eq. (11.191) the Bardeen potential will be, in Fourier space,

$$\Psi(k,\alpha) = -\frac{6c_2(k)}{k^2 \tau_1^2} \frac{9\alpha^3 + 2\alpha^2 - 8\alpha + 16(\sqrt{\alpha}-1)}{\alpha^3}, \tag{11.201}$$

where the background identity $4\pi G a^2 \rho_t = 3\mathcal{H}^2/2 = (6/\tau_1^2)(\alpha+1)/\alpha^2$ has been used to extract $\Psi(k,\alpha)$ from Eq. (11.191). The normalization implied by Eq. (11.201) is not ideal. We then have to normalize the Bardeen potential to its value in the radiation epoch. After simple manipulations we then have

$$\Psi(k,\alpha) = \frac{\Psi_r(k)}{10\alpha^3}[9\alpha^3 + 2\alpha^2 - 8\alpha + 16(\sqrt{\alpha+1}-1)], \qquad (11.202)$$

where $\Psi_r(k) = \Psi(k,0)$. Equation (11.202) has the same content of Eq. (7.108) that has been obtained through a completely different procedure and, in particular, within the longitudinal gauge. Within the normalization implied by Eq. (11.202), the total density contrast and the total velocity field are:

$$\epsilon_m(k,\tau) = -\frac{k^2\tau_1^2\Psi_r(k)}{60\alpha(\alpha+1)}[9\alpha^3 + 2\alpha^2 - 8\alpha$$
$$+ 16(\sqrt{\alpha+1}-1)], \qquad (11.203)$$

$$\Theta_g(k,\tau) = \frac{k^2\tau_1}{10\alpha^2}\Psi_r(k)[\sqrt{\alpha+1}(3\alpha^2 - 4\alpha + 8) - 8]. \qquad (11.204)$$

The remaining part of the calculation consists in including the neutrino anisotropic stress into account perturbatively. For typical scales larger than the Hubble radius, Θ_g is a good estimate of the neutrino velocity. Therefore this allows us to compute σ_ν which can be inserted back into Eq. (11.190), allowing, ultimately, an estimate of Φ. We refer the interested reader to a pair of interesting papers where, among other things, this formalism has been exploited [276, 279].

Chapter 12

Interacting Fluids

Entropy perturbations are constant for typical wavelengths larger than the Hubble radius. This statement is not a theorem even if this kind of standard lore has been safely assumed with various derivations discussed in the previous chapters. For the introduction of entropy perturbations see, in particular, chapter 7 and, more formally, chapter 11.

In this chapter we are going to prove that the conservation of entropy perturbations is likely across the radiation-matter transition but it can be very well violated along the early stages of evolution of our Universe.[a] In particular, this will be the case if the fluids of the plasma are allowed to interact in different ways. And in connection with the discussion of chapter 2, the following situations will simultaneously be examined:

- the case of energy-momentum exchange between the fluids of the mixture;
- the case where one (or more) fluids of the mixture have nonvanishing bulk viscous stresses.

The discussion reported here follows the logic reported in [97, 98].

To introduce the problem consider the case where the plasma has two fluids with constant barotropic indices[b] w_a and w_b. Suppose, for simplicity, that we are in the longitudinal gauge. With the experience gained along the previous chapters and with the results of Appendix D it is immediate for us to write down the evolution equations, in particular, for the velocity and

---

[a] In this sense the topics covered in the present chapter go beyond the standard ΛCDM model.

[b] Since the barotropic index of each fluid is constant the sound speed will coincide, by definition, with the corresponding barotropic index, i.e. $c_{sa} \equiv w_a$ and $c_{sb} \equiv w_b$. Later on, in this chapter, we will be interested in avoiding this assumption.

for the density contrast. So, for the a-fluid we will have[c]:

$$\delta'_a = 3(w_a + 1) - (w_a + 1)\theta_a,\tag{12.1}$$

$$\theta'_a + \mathcal{H}(1 - 3w_a)\theta_a = -\frac{w_a}{w_a + 1}\nabla^2\delta_a - \nabla^2\phi,\tag{12.2}$$

while for the b-fluid

$$\delta'_b = 3(w_b + 1) - (w_b + 1)\theta_b,\tag{12.3}$$

$$\theta'_b + \mathcal{H}(1 - 3w_b)\theta_b = -\frac{w_b}{w_b + 1}\nabla^2\delta_b - \nabla^2\phi.\tag{12.4}$$

The entropy fluctuations for this fluid are defined, according to chapter 11, as

$$\mathcal{S}_{ab} = -3(\zeta_a - \zeta_b) \equiv \frac{\delta_a}{w_a + 1} - \frac{\delta_b}{w_b + 1}.\tag{12.5}$$

Note that in the second equality of Eq. (12.5) the longitudinal gauge definition of ζ_a and ζ_b:

$$\zeta_a = -\psi + \frac{\delta_a}{w_a + 1}, \qquad \zeta_b = -\psi + \frac{\delta_b}{w_b + 1},\tag{12.6}$$

have been used in order to express the entropy fluctuation in terms of the density contrast. Using now Eq. (12.6) in Eqs. (12.1) and (12.3) it is easy to show that

$$\zeta'_a = -\frac{\theta_a}{3}, \qquad \zeta'_b = -\frac{\theta_b}{3}.\tag{12.7}$$

Finally, using Eq. (12.7) together with the definition of $\mathcal{S}_{ab}$ given in Eq. (12.5) we immediately obtain:

$$\mathcal{S}'_{ab} = (\theta_a - \theta_b).\tag{12.8}$$

Concerning the result of Eq. (12.8) the following three remarks are in order:

- from Eq. (12.8) it is apparent that $\mathcal{S}_{ab}$ is constant, to a good approximation, for typical wavelengths larger than the Hubble radius since θ_a and θ_b are in this limit suppressed;
- indeed, Eqs. (12.2) and (12.4) show that both θ_a and θ_b are, in Fourier space, $\mathcal{O}(k^2\tau)$, i.e. negligible in the super-Hubble limit;
- this occurrence helped our study of the CDM-radiation system (see chapter 7) where it has been shown that the entropy fluctuations are constant along the matter-radiation transition.

---

[c]Along this derivation we will assume, without loss of generality, that the anisotropic stress vanishes. Indeed, as it will become clear in the following sections, we are interested in the evolution of the mixture for typical wavelengths larger than the Hubble radius. Notice that, under this approximation, $\phi = \psi$.

Interacting Fluids

In the following chapter we are going to address the following question: is the conclusion implied by Eq. (12.8) still true when the fluids are allowed to interact?

Having spelled out the main theme of the chapter let us remark on the more technical ground that:

- the analysis will be conducted in the off-diagonal (or uniform curvature) gauge which has been thoroughly introduced in chapter 11;
- we will be working in natural Planck units where $8\pi G = 1$;
- the situation of two generic fluids will be considered and, towards the end of the chapter, some more specific applications will be outlined.

12.1 Interacting fluids with bulk viscous stresses

As already discussed in chapter 2 the two dissipative effects that enter the description of a mixture of different fluids are the energy-momentum exchange between the fluids of the mixture and the possible presence of bulk viscous stresses. Since the background is isotropic the momentum transfer will be evident only to first order (i.e. when the peculiar velocities are present). Furthermore, the presence of bulk viscous stresses can be viewed, on a physical ground, as the relativistic generalization of the finite size effects characterizing real as opposed to perfect gases in thermodynamics.

Therefore, consider a mixture of two relativistic fluids (the a-fluid and the b-fluid) obeying a set of generally covariant evolution equations formed by the Einstein equations[d]

$$R^\nu_\mu - \frac{1}{2}\delta^\nu_\mu R = \frac{1}{2}T^\nu_\mu \qquad (12.9)$$

and by the evolution equations of the energy-momentum tensors of each fluid of the mixture, i.e.

$$\nabla_\mu T_a^{\mu\nu} = -\Gamma g^{\nu\beta} u_\beta (p_a + \rho_a), \qquad (12.10)$$

$$\nabla_\mu T_b^{\mu\nu} = \Gamma g^{\nu\beta} u_\beta (p_a + \rho_a), \qquad (12.11)$$

---

[d]As already mentioned, units $8\pi G = 1$ will be used throughout the chapter. The Latin (lower-case roman) subscripts a, b, c d, ... will denote, in the present chapter (and in analogy to what already done in chapter 2), different fluids present in the relativistic plasma. Greek (lower-case) subscripts will denote tensor indices. Latin (lower-case italic) subscripts $i, j, k, \ldots$ will denote the spatial components of a tensor.

where u_β is the total velocity field of the mixture. Equations (12.10) and (12.11) describe the situation where the a-fluid decays into the b-fluid with decay rate Γ. It is evident from the form of Eqs. (12.10) and (12.11) that the total energy-momentum tensor of the mixture, i.e.

$$T^{\mu\nu} = T_a^{\mu\nu} + T_b^{\mu\nu}, \qquad (12.12)$$

is covariantly conserved, i.e. $\nabla_\mu T^{\mu\nu} = 0$. The total energy-momentum tensor of each species is given by the sum of an inviscid contribution, denoted by $T_{a,b}^{\mu\nu}$ and by a viscous contribution, denoted by $\tilde{T}_{a,b}^{\mu\nu}$, i.e.

$$\mathcal{T}_{a,b}^{\mu\nu} = T_{a,b}^{\mu\nu} + \tilde{T}_{a,b}^{\mu\nu}, \qquad (12.13)$$

$$T_{a,b}^{\mu\nu} = (p_{a,b} + \rho_{a,b}) u_{a,b}^\mu u_{a,b}^\nu - p_{a,b} g^{\mu\nu}, \qquad (12.14)$$

$$\tilde{T}_{a,b}^{\mu\nu} = \xi_{a,b} \left(g^{\mu\nu} - u_{a,b}^\mu u_{a,b}^\nu \right) \nabla_\alpha u_{a,b}^\alpha, \qquad (12.15)$$

where the subscript in the various fluid quantities simply means that Eqs. (12.13), (12.14) and (12.15) hold, independently, for the a- and b-fluids. So, for instance, in Eqs. (12.14) and (12.15), u_a^μ and u_b^μ denote the peculiar velocities of each fluid of the mixture.

In a spatially flat metric of Friedmann–Robertson–Walker (FRW) type characterized by a background line element

$$ds^2 = \overline{g}_{\mu\nu} dx^\mu dx^\nu = a^2(\tau)[d\tau^2 - d\vec{x}^2], \qquad (12.16)$$

Eqs. (12.10) and (12.11) imply

$$\rho_a' + 3\mathcal{H}(\rho_a + \mathcal{P}_a) + a\overline{\Gamma}(\rho_a + p_a) = 0, \qquad (12.17)$$

$$\rho_b' + 3\mathcal{H}(\rho_b + \mathcal{P}_b) - a\overline{\Gamma}(\rho_a + p_a) = 0, \qquad (12.18)$$

where the usual notations established in chapter 2 have been followed. In Eqs. (12.17) and (12.18), $\mathcal{P}_{a,b}$ denote the total effective pressure of each species, i.e.

$$\mathcal{P}_{a,b} = p_{a,b} - 3\frac{\mathcal{H}}{a} \overline{\xi}_{a,b}, \qquad (12.19)$$

while $p_{a,b}$ denote the inviscid pressures of each species. In Eq. (12.19), $\overline{\xi}_{a,b}$ denote the bulk viscosity coefficient evaluated on the background geometry. As will be discussed later, the bulk viscosity coefficient may depend on both ρ_a and ρ_b. Equations (12.17) and (12.18) lead to the evolution of the total energy and pressure densities

$$\rho' + 3\mathcal{H}(\rho + \mathcal{P}) = 0, \qquad (12.20)$$

where $\rho = \rho_a + \rho_b$ and $\mathcal{P} = \mathcal{P}_a + \mathcal{P}_b$. Equations (12.17) and (12.18) must be supplemented by the explicit background form of Eq. (12.9), i.e.

$$3\mathcal{H}^2 = a^2\rho, \tag{12.21}$$

$$2(\mathcal{H}^2 - \mathcal{H}') = a^2(\rho + \mathcal{P}), \tag{12.22}$$

where, again, $(\rho + \mathcal{P})$ is the total effective enthalpy that contains the background viscosity coefficient of the mixture $\overline{\xi} = \overline{\xi}_a + \overline{\xi}_b$.

12.2 Evolution equations for the entropy fluctuations

We are now interested in deriving the evolution of the entropy and total-curvature fluctuations of the system. Both the entropy perturbations and the perturbations in the total spatial curvature can be written in terms of ζ_a and ζ_b which are related in the off-diagonal gauge [370] (see also [371, 372]) to the density contrasts of the individual fluids of the mixture [373]:

$$\mathcal{S} = -3(\zeta_a - \zeta_b), \tag{12.23}$$

$$\zeta = \frac{\rho'_a}{\rho'}\zeta_a + \frac{\rho'_b}{\rho'}\zeta_b. \tag{12.24}$$

In the following we are going to exploit the off-diagonal gauge [370, 371] which is particularly convenient for the problem at hand. The results will be exactly the same of those obtainable in the framework of gauge-independent descriptions (see [98]). In fact, the quantities $\mathcal{S}$ and ζ, defined in terms of ζ_a and ζ_b, are invariant under infinitesimal coordinate transformations. Consequently they can be computed in any suitable (non-singular) coordinate system.

In the off-diagonal gauge the spatial components of the perturbed metric vanish and, hence, the only components of the perturbed line element are:

$$\delta g_{00} = 2a^2\phi, \quad \delta g_{0i} = -a^2\partial_i B. \tag{12.25}$$

As we learned in chapter 11 (see, in particular, Eq. (11.176) and recall that, in the long wavelength limit $\mathcal{R} \simeq \zeta$), in the long-wavelength limit,

$$(\mathcal{H}' - \mathcal{H}^2)\phi = \mathcal{H}^2\zeta. \tag{12.26}$$

Thus, in the off-diagonal gauge, δg_{00} is connected to ζ. As anticipated, ζ_a and ζ_b can be expressed in terms of the fluctuations of the density contrasts of the individual fluids, i.e.[e]

$$\zeta_a = -\frac{\mathcal{H}\rho_a}{\rho_a'}\delta_a, \qquad \zeta_b = -\frac{\mathcal{H}\rho_b}{\rho_b'}\delta_b. \qquad (12.27)$$

The evolution equations obeyed by the density contrasts $\delta_a = \delta\rho_a/\rho_a$ and $\delta_b = \delta\rho_b/\rho_b$ are derived by perturbing Eqs. (12.10) and (12.11) to first order in the amplitude of the metric and hydrodynamical fluctuations:

$$\delta_a' + (3\mathcal{H} + a\overline{\Gamma})(c_{sa}^2 - w_a)\delta_a + \frac{9\mathcal{H}^2}{a\rho_a}[\overline{\xi}_a(\phi + \delta_a) - \delta\xi_a] + a(1+w_a)\overline{\Gamma}(\delta_\Gamma + \phi)$$

$$+ \left[(1+w_a) - 6\frac{\mathcal{H}\overline{\xi}_a}{a\rho_a}\right]\theta_a = 0, \qquad (12.28)$$

$$\delta_b' + 3\mathcal{H}(c_{sb}^2 - w_b)\delta_b + a\overline{\Gamma}\frac{\rho_a}{\rho_b}[(1+w_a)(\delta_b - \delta_\Gamma - \phi) - (1+c_{sa}^2)\delta_a]$$

$$+ \frac{9\mathcal{H}^2}{a\rho_b}[\overline{\xi}_b(\phi + \delta_b) - \delta\xi_b] + \left[(1+w_b) - 6\frac{\mathcal{H}\overline{\xi}_b}{a\rho_b}\right]\theta_b = 0. \qquad (12.29)$$

For notational convenience the barotropic indices (i.e. w_a, w_b) and the sound speeds (i.e. c_{sa}^2 and c_{sb}^2) have been introduced for the inviscid component of each species of the plasma. If the inviscid component is parametrized in terms of a perfect relativistic fluid, $c_{sa,b}^2 \equiv w_{a,b}$. In Eqs. (12.28) and (12.29):

- $\delta_\Gamma = \delta\Gamma/\overline{\Gamma}$ denotes the fractional fluctuation of the decay rate computed in the off-diagonal gauge;
- $\delta\xi_a$ and $\delta\xi_b$ denote the fluctuations of the bulk viscosity coefficients.

The fluctuation in the total viscosity is defined as

$$\delta\xi = \delta\xi_a + \delta\xi_b. \qquad (12.30)$$

Finally, in Eqs. (12.28) and (12.29)

$$\theta_a = \partial_i v_a^i = \nabla^2 v_a, \qquad \theta_b = \partial_i v_b^i = \nabla^2 v_b. \qquad (12.31)$$

The global velocity $\theta = \partial_i v^i$ field (with $\delta T_0^i = (\rho + \mathcal{P})v^i$) is recovered from θ_a and θ_b by recalling that $(p + \rho)\theta = (p_a + \rho_a)\theta_a + (p_b + \rho_b)\theta_b$.

Equations (12.28) and (12.29) must be supplemented by the perturbed components of Eq. (12.9); in particular by the Hamiltonian and momentum

---

[e]Notice that Eq. (12.27) holds in the off-diagonal gauge and *not* in the longitudinal gauge where, for instance, Eq. (12.6) has been written. This is another example of how it is possible to exploit the freedom of surfing through the gauges which has been the guiding theme of chapter 11.

constraints:
$$\mathcal{H}\nabla^2 B + 3\mathcal{H}^2\phi + \frac{a^2}{2}\delta\rho = 0, \qquad (12.32)$$

$$\nabla^2[\mathcal{H}\phi + (\mathcal{H}^2 - \mathcal{H}')B] + \frac{a^2}{2}(\rho + \mathcal{P})\theta = 0, \qquad (12.33)$$

and by the other two equations stemming from the spatial components (i.e. $(i=j)$ and $(i \neq j)$) of Eq. (12.9):

$$\phi' + \left(\mathcal{H} + 2\frac{\mathcal{H}'}{\mathcal{H}}\right) - \frac{a^2}{2\mathcal{H}}\left[\delta p - 3\frac{\mathcal{H}}{a}\delta\xi - \frac{\overline{\xi}}{a}(\theta - 3\mathcal{H}\phi)\right] = 0. \qquad (12.34)$$

$$B' + 2\mathcal{H}B + \phi = 0. \qquad (12.35)$$

In Eq. (12.32) the global energy and pressure density fluctuations (i.e. $\delta\rho$ and δp) have been introduced. As it is clear from Eqs. (12.32)–(12.35), that one of the advantages of the off-diagonal formulation is the absence of second time derivatives of the metric fluctuations. Strictly speaking the evolution equations of θ_a and θ_b should be added to the system. However, they are only relevant for typical length scales smaller than the Hubble radius at a given time. Since we are interested in the opposite regime, these equations will be omitted, but they will be discussed elsewhere in their full generality (see [98]).

To derive Eqs. (12.32), (12.33), (12.34) and (12.35) it is relevant to recall the first-order form of the total energy-momentum tensor. The scalar fluctuation of the effective energy-momentum tensor of the fluid sources is, in general coordinates,[f]

$$\delta_s T^\nu_\mu = \overline{u}_\mu \overline{u}^\nu[(\delta p + \delta\rho) - \delta\xi\nabla_\lambda \overline{u}^\lambda - \overline{\xi}\nabla_\lambda \delta u^\lambda]$$
$$+ \delta^\nu_\mu[\delta\xi\nabla_\lambda \overline{u}^\lambda - \delta p + \overline{\xi}\nabla_\lambda \delta u^\lambda]$$
$$+ (\delta u_\mu \overline{u}^\nu + \overline{u}_\mu \delta u^\nu)[(p + \rho) - \overline{\xi}\nabla_\lambda \overline{u}^\lambda], \qquad (12.36)$$

where the velocity field satisfies $g^{\mu\nu}u_\mu u_\nu = 1$ and

$$\delta_s(\nabla_\lambda u^\lambda) = \partial_\lambda \delta u^\lambda + \delta\Gamma^\beta_{\alpha\beta}u^\alpha \overline{u}^\alpha + \overline{\Gamma}^\beta_{\alpha\beta}\delta u^\alpha, \qquad (12.37)$$

where $\overline{\Gamma}^\beta_{\beta\alpha}$ and $\delta\Gamma^\beta_{\beta\alpha}$ are, respectively, the background and the perturbed Christoffel connections. The various components of Eq. (12.37) can be written, without choosing a specific gauge, as:

$$\delta T^0_0 = \delta\rho, \qquad (12.38)$$

$$\delta T^i_0 = (\rho + \mathcal{P})v^i, \qquad (12.39)$$

$$\delta T^j_i = -\delta^j_i\left\{\delta p - 3\frac{\mathcal{H}}{a}\delta\xi - \frac{\overline{\xi}}{a}[\theta - 3(\psi' + \mathcal{H}\phi) + \nabla^2 E']\right\}. \qquad (12.40)$$

---

[f]As in Appendices C and D δ_s denotes the scalar fluctuation of the corresponding quantity. Equation (12.36) can be simply obtained by perturbing, to first order, Eq. (12.12).

Recalling that, in the off-diagonal gauge, $E = \psi = 0$ (see chapter 11) the specific form of Eqs. (12.32), (12.33), (12.34) and (12.35) can be recovered.

By combining Eqs. (12.32) and (12.34) the evolution equation for ζ can be easily obtained and it is given by

$$\dot\zeta = -\frac{3}{2}\frac{H}{\dot H}[(\dot{\overline\xi}_a + \dot{\overline\xi}_b)\zeta + H(\delta\xi_a + \delta\xi_b)]$$
$$+ \frac{\dot\rho_a}{2\dot H}(c_{sb}^2 - c_{sa}^2)(\zeta_a - \zeta), \qquad (12.41)$$

where we passed, for later convenience, from the conformal time coordinate τ to the cosmic time coordinate t (i.e. $dt = a(\tau)d\tau$). Note that, in Eq. (12.41) the higher order terms (involving spatial gradients) have been neglected since we are directly interested in the super-Hubble evolution of the perturbations.

Equations (12.28) and (12.29) lead to the evolution equations of ζ_a and ζ_b whose explicit form is given by

$$\dot\zeta_a + \left[\frac{\dot q_a}{q_a} + (3H + \overline\Gamma)(1 + c_{sa}^2)\right]\zeta_a + \frac{9}{q_a}[H^2\delta\xi_a - \overline\xi_a \dot H \zeta]$$
$$= \frac{p_a + \rho_a}{q_a}\overline\Gamma\left[\delta_\Gamma + \frac{\dot H}{H^2}\zeta\right], \qquad (12.42)$$

and by

$$\dot\zeta_b + \left[\frac{\dot q_b}{q_b} + 3H(1 + c_{sb}^2)\right]\zeta_b - \overline\Gamma\frac{\dot q_a}{q_b}(1 + c_{sa}^2)\zeta_a + \frac{9}{q_b}[H^2\delta\xi_b - \dot H \overline\xi_b \zeta]$$
$$= -\frac{p_a + \rho_a}{q_b}\overline\Gamma\left[\delta_\Gamma + \frac{\dot H}{H^2}\zeta\right], \qquad (12.43)$$

where

$$q_a = \frac{\dot\rho_a}{H}, \qquad q_b = \frac{\dot\rho_b}{H}. \qquad (12.44)$$

Various identities can be used to bring Eqs. (12.41), (12.42) and (12.43) to slightly different (but equivalent) forms. In particular:

- using Eq. (12.24), we can always trade the combinations $(\zeta_a - \zeta)$ and $(\zeta_b - \zeta)$ for $\dot\rho_a/\dot\rho(\zeta_a - \zeta_b)$ and $\dot\rho_b/\dot\rho(\zeta_a - \zeta_b)$;
- according to Eq. (12.23), $(\zeta_a - \zeta_b) = -\mathcal{S}/3$;
- by virtue of the background equations (12.21) and (12.22), $\dot H/H = \dot\rho/(2\rho)$;
- if the inviscid component of each fluid of the mixture is a perfect fluid, then $c_{s,a}^2 = w_a$ and $c_{s,b}^2 = w_b$;
- finally the background evolution of each fluid, i.e. Eqs. (12.17) and (12.18), may always be employed to obtain equivalent forms of the above equations.

12.3 Specific physical limits

Specific limits of Eqs. (12.41)–(12.43) will now be reproduced. In the limit $\bar{\xi}_a = \bar{\xi}_b = 0$, with $\delta_\Gamma = \delta\xi_a = \delta\xi_b = 0$ and $\dot{\bar{\Gamma}} = 0$, Eqs. (12.41)–(12.43) read

$$\dot{\zeta} = -\frac{H\dot{\rho}_a\dot{\rho}_b}{\dot{\rho}^2}(w_b - w_a)\mathcal{S}, \tag{12.45}$$

$$\dot{\zeta}_a = \frac{\bar{\Gamma}}{6}(w_a + 1)\frac{\dot{\rho}_b \rho_a}{\dot{\rho}_a}\mathcal{S}, \tag{12.46}$$

$$\dot{\zeta}_b = 0. \tag{12.47}$$

In the case $w_a = 0$ and $w_b = 1/3$, Eqs. (12.45)–(12.47) coincide with the set of equations used in Ref. [374] to describe the radiative decay of a massive curvaton whose effective pressure, at the oscillatory stage, reproduces that of dusty matter, i.e. $w_a = 0$. It is then clear, taking the difference between Eqs. (12.47) and (12.46), that the evolution equation of entropy perturbations

$$\dot{\mathcal{S}} = \frac{\bar{\Gamma}}{2}(w_a + 1)\frac{\dot{\rho}_a \rho_a}{\rho\dot{\rho}_b}\left(1 - \frac{\dot{\rho}_b^2}{\dot{\rho}_a^2} - 2\frac{\rho}{\rho_a}\right)\mathcal{S} \tag{12.48}$$

is homogeneous and does not contain any ζ-dependent source term.

Sticking to the case of the radiative decay of a dusty fluid, but including the fluctuations of the decay rate, the following system of evolution equations

$$\dot{\zeta} = \frac{\dot{\rho}}{6\dot{H}}(\zeta_a - \zeta), \tag{12.49}$$

$$\dot{\zeta}_a - \frac{g_a}{g_a}\zeta_a = -g_a\left(\delta_\Gamma + \frac{\dot{H}}{H^2}\zeta\right), \tag{12.50}$$

can be derived from Eqs. (12.41)–(12.43) when $w_b = 1/3$ and $w_a = 0$. In Eqs. (12.49) and (12.50), $g_a = -H\rho_a/\dot{\rho}_a$. Equations (12.49) and (12.50) describe the situation discussed in Ref. [375], where the dynamics of the inflaton with inhomogeneous decay rate has been discussed (see, for instance, also [376, 377, 379, 378]). If the spatial fluctuations of the decay rate are not a function of the local energy density of the mixture, curvature fluctuations may be generated for length scales larger than the Hubble radius.

Consider now the case where the $\bar{\Gamma}$ is constant, the decay is homogeneous (i.e. $\delta_\Gamma = 0$), but $\xi_a = \xi_a(\rho_a)$ and $\xi_b = \xi_b(\rho_b)$. This occurrence implies that

$$\delta\xi_a = -\frac{\dot{\bar{\xi}}_a}{H}\zeta_a, \qquad \delta\xi_b = -\frac{\dot{\bar{\xi}}_b}{H}\zeta_b. \tag{12.51}$$

Hence, from Eqs. (12.41)–(12.43), we obtain, respectively

$$\dot{\zeta} = -\frac{\dot{\rho}_b}{\rho}\left[\frac{H\rho_a}{\dot{\rho}}(w_b - w_a) + \frac{\dot{\rho}}{4\rho}\left(\dot{\bar{\xi}}_a - \frac{\rho_a}{\dot{\rho}_b}\dot{\bar{\xi}}_b\right)\right]\mathcal{S}, \qquad (12.52)$$

$$\dot{\zeta}_a = \frac{\dot{\rho}_b}{6\rho\dot{\rho}_a}[\overline{\Gamma}(w_a + 1)\rho_a + 9H^2\bar{\xi}_a]\,\mathcal{S}, \qquad (12.53)$$

$$\dot{\zeta}_b = -\frac{\dot{\rho}_a}{3\dot{\rho}_b}\left[\overline{\Gamma}(w_a+1)\left(1 - \frac{\rho_a}{2\rho}\right) + \frac{3}{2}\bar{\xi}_b\right]\mathcal{S}. \qquad (12.54)$$

Again, in this case, it can be easily argued that the evolution of entropy fluctuations obeys a homogeneous equation in $\mathcal{S}$. In fact, combining Eqs. (12.53) and (12.54) it is possible to obtain:

$$\dot{\mathcal{S}} = -\left[\frac{\dot{\rho}_a\rho_a}{2\rho\dot{\rho}_b}\overline{\Gamma}(w_a+1)\left(1 - \frac{\dot{\rho}_b^2}{\dot{\rho}_a^2} - 2\frac{\rho}{\rho_a}\right) + \frac{3}{2}\left(\dot{\bar{\xi}}_a + \frac{\dot{\rho}_b^2}{\dot{\rho}_a^2}\dot{\bar{\xi}}_b\right)\right]\mathcal{S}. \qquad (12.55)$$

This conclusion can be, however, evaded if ξ_a and ξ_b are functions of both ρ_a and ρ_b, i.e. $\xi_a = \xi_a(\rho_a, \rho_b)$ and $\xi_b = \xi_b(\rho_a, \rho_b)$. In this case

$$\delta\xi_a = -\frac{\dot{\bar{\xi}}_a}{H}(\zeta_a + \zeta_b), \qquad \delta\xi_b = -\frac{\dot{\bar{\xi}}_b}{H}(\zeta_a + \zeta_b). \qquad (12.56)$$

Thus, in the situation described by Eq. (12.56), Eq. (12.55) will inherit two extra terms at the right-hand side, i.e.

$$-\frac{9H^2}{\dot{\rho}_b\dot{\rho}_a}\left(\dot{\rho}_a\dot{\bar{\xi}}_b\zeta_a - \dot{\rho}_b\dot{\bar{\xi}}_a\zeta_b\right), \qquad (12.57)$$

which cannot be recast, for generic ξ_a and ξ_b, in a single term proportional to $\mathcal{S}$.

12.4 Mixing between entropy and curvature perturbations

A relevant issue to be addressed concerns the phenomenological viability of interacting viscous mixtures. Consider, for instance, a model where the decay rate is constant but inhomogeneous (i.e. $\delta_\Gamma \neq 0$) and $\xi_a = \epsilon\sqrt{\rho_a}$ (where ϵ is constant). The viscosity coefficient of the b-fluid vanishes, i.e. $\xi_b = 0$. This model describes the situation where the a-fluid is initially dominant and characterized by a viscosity proportional to ϵ. Furthermore, if we want the Universe to be expanding, we must also require $\epsilon < (w_a + 1)/\sqrt{3}$. The a-fluid will start its decay for a typical cosmic time $t_\Gamma \sim \overline{\Gamma}^{-1}$, and then the background will be dominated by the b-fluid while the energy density of the a-fluid, i.e. ρ_a, will decay exponentially. Also the background viscosity will decay exponentially, since $\bar{\xi}_a = \epsilon\sqrt{\rho_a}$. These aspects are

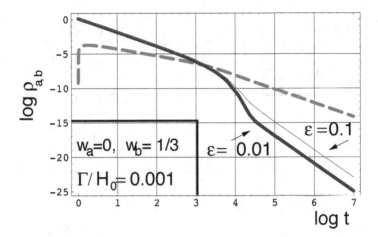

Fig. 12.1 The evolution of the background is illustrated. The parameters of the mixture are fixed in such a way that $w_a = 0$, $w_b = 1/3$, $\Gamma/H_0 \sim 10^{-3}$, $\xi_a = \epsilon\sqrt{\rho_a}$, $\xi_b = 0$; H_0 denotes the value of the Hubble parameter at the initial integration time. The dashed curve represents the evolution of the energy density of the decay products (radiation) while the full lines represent the evolution of the energy density of the decaying component for different values of ϵ. This figure is adapted from [97]. Reproduced by permission of Elsevier.

illustrated in Fig. 12.1 where, for two different values of ϵ, the common logarithm (i.e. the logarithm to base 10) of ρ_a and ρ_b are reported. From the point of view of the background, this model is perfectly viable and it leads to a final stage of expansion dominated by the b-fluid. To make the example even more explicit, one can think of the situation where the a-fluid is given by dust (i.e. $w_a = 0$) or stiff matter (i.e. $w_a = 1$). The b-fluid may be taken, for instance, to coincide with radiation (i.e. $w_b = 1/3$).

The dynamics of curvature fluctuations may be described, for practical reasons, by expressing the evolution equations in terms of ζ and δ_a, i.e. the curvature fluctuations and the density contrast of the a-fluid. Given the relations (12.24) and (12.27) ζ_b, ζ_a and δ_b can always be obtained as linear combinations (with background-dependent coefficients) of ζ and δ_a. From Eqs. (12.28) and (12.41) the relevant evolution equations can be written as

$$\dot{\zeta} = -\frac{1}{4\dot{H}}\left[3\frac{H\epsilon}{\sqrt{\rho_a}} + 2(w_b - w_a)\right](\dot{\rho}_a\zeta + H\rho_a\delta_a), \tag{12.58}$$

$$\dot{\delta}_a + \frac{9\epsilon H^2}{2\sqrt{\rho_a}}\delta_a + \frac{\dot{H}}{H^2}\left[\frac{9\epsilon H^2}{\sqrt{\rho_a}} + \overline{\Gamma}(w_a + 1)\right]\zeta = -\overline{\Gamma}(w_a + 1)\delta_\Gamma. \tag{12.59}$$

Equations (12.58) and (12.59) describe the evolution of ζ and δ_a for typical wavelengths larger than the Hubble radius. Initial conditions of the system

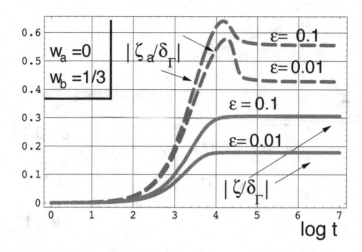

Fig. 12.2 The evolution of the fluctuations induced by the background of Fig. 12.1 is illustrated. The dashed curves illustrate the behaviour of $|\zeta_a|$ while the full lines represent the evolution of $|\zeta|$. Both $|\zeta|$ and $|\zeta_a|$ are given in units of δ_Γ. This figure is adapted from [97]. Reproduced by permission of Elsevier.

are then set by requiring $\zeta(t_0) = 0$ and $\delta_a(t_0) = \delta_b = 0$, where t_0 is the initial integration time. From Fig. 12.2 the evolution is such that curvature fluctuations grow from 0 to a value proportional to δ_Γ, i.e. proportional to the fluctuations of the decay rate over length scales larger than the Hubble radius. The final asymptotic value of ζ can be determined analytically and it turns out to be

$$|\zeta_{\text{final}}| \simeq \frac{1}{6}\left(\frac{1 + 3\sqrt{3}\,\epsilon}{1 - \sqrt{3}\,\epsilon}\right). \tag{12.60}$$

In the limit $\epsilon \to 0$, the results reproduce the findings of Ref. [375] leading to a Bardeen potential $|\Psi_{\text{final}}| \simeq \delta_\Gamma/9$, which implies $|\zeta_{\text{final}}| \simeq \delta_\Gamma/6$ by using the well-known relation of ζ and Ψ in a radiation-dominated phase. In the example discussed so far the values of δ_Γ have been taken in the ranges 10^{-6}–10^{-9}.

The class of examples reported so far can be generalized in various ways. Different barotropic indices for the fluids of the mixture can be studied. Equation (12.60) can then be generalized to the cases of generic w_a and w_b. Furthermore, the functional dependence of the viscosity coefficients can be chosen to be different. These generalizations can be found, for instance, in [98]. We would like to point out that the simple examples presented here may be made more realistic by thinking that a dusty fluid is an effective description of a scalar field oscillating in a quadratic potential [380]. Thus, the simple fluid model of a dust fluid decaying into radiation has been used [375] (with some caveats [378]) to infer some properties of the inflaton decay when the inflaton decay rate is not homogeneous. If the inhomogeneous decay occurs after an inflationary phase at low curvature (i.e. $H_{\rm inf} \ll 10^{-6} M_{\rm P}$), it is plausible to argue that the spectrum of δ_Γ may be converted into the spectrum of ζ for typical frequencies smaller than the Hubble rate. We are not interested here in supporting a specific model of inhomogeneous reheating. The purpose of the examples discussed so far is purely illustrative. However, the lesson to be drawn is that bulk viscous stresses may play a relevant role.

In the present chapter various results have been achieved. First of all, the concept of interacting viscous mixtures has been introduced, i.e. a mixture of interacting fluids with viscous corrections. In this framework, the coupled evolution of curvature and entropy fluctuations have been derived in the case where both the decay rate and the bulk viscosity coefficients are allowed to fluctuate over typical length scales larger than the Hubble radius. Different situations have been systematically discussed. If the decay rate is constant and homogeneous, with bulk viscosities that depend separately on the energy density of each fluid of the mixture, the evolution of entropy fluctuations obeys a source-free evolution equation. If, on the contrary, the bulk viscosity has a more general dependence on the energy densities of the fluids composing the mixture, the evolution equations of the entropy perturbations may inherit a source term that involves, in one way or another, curvature fluctuations. In similar terms, if the decay rate is allowed to fluctuate without being a function of the local density of the fluid, entropy fluctuations will not obey a source-free equation. These findings show that the conservation of entropy fluctuations, safely assumed across the radiation-matter equality, may not be the most generic situation in the early Universe.

Chapter 13

Spectator Fields

It is rather plausible that the early Universe is not described by a single scalar degree of freedom. On the contrary it seems conceivable that various fields are simultaneously present. In this situation various interesting phenomena can take place. One of them will be specifically discussed in the present chapter and it has to do with the possibility that non-adiabatic fluctuations in the early stages of the life of the Universe lead to fully adiabatic initial conditions for the CMB anisotropies.

In chapters 7 and 8 (and also in chapter 11) the distinction between adiabatic and non-adiabatic initial conditions has been made clear and, according to our presentation, these two possibilities represented physically distinct initial conditions for the evolution of the lowest multipoles of the Boltzmann hierarchy of the brightness perturbations. It is important to appreciate, however, that CMB initial conditions are set after neutrino decoupling. This is mainly due to the fact that we do have some control on the thermodynamic history of the early Universe only below temperatures of few MeV mainly thanks to the success of BBN.

As we learned in chapter 10 there is a different way of setting initial conditions for CMB anisotropies and it is related to the assumption that we do know pretty well the full thermodynamic history of the Universe. In its simplest realization this assumption amounts to state that there was an inflationary phase suddenly (or almost suddenly) followed by a radiation-dominated phase of expansion. This assumption is consistent with the data but it is not tested directly. In a complementary perspective it could happen that

- inflation is not of slow-roll type;
- inflation is not driven by a single scalar degree of freedom;

- the post-inflationary history is more complicated than what is normally assumed.

A complete discussion of the mentioned list of topics would require more than one chapter. In spite of this it is useful to point out that there are situations where the curvature perturbations induced by the master field that drives inflation might be negligibly small. Later on, thanks to the dynamical evolution, these fluctuations may grow and inherit the fluctuation spectra of other spectator fields which are present during inflation. By *spectator field* we mean a field whose energy density does not determine (or even affect) the evolution of the background. Still, its fluctuations must be appropriately taken into account when computing the evolution of curvature perturbations.

There are different (but related) concrete examples of such dynamics. The first obvious case is the one where the dynamics of the background is driven by a scalar degree of freedom (for instance, the dilaton in pre-big bang models [104, 105]). While the Universe inflates, some other spectator field (that is not source of the background geometry) gets amplified with a quasi-flat spectrum. The role of spectator field is played, in pre-big bang models, by the Kalb-Ramond axion [105]. When inflation terminates, the large-scale fluctuations in the curvature are vanishingly small owing to the steepness of the dilatonic spectrum. After the dilaton/inflaton decay, the Universe will presumably be dominated by radiation. However, owing to the presence of the spectator field, the dynamics of the inhomogeneities will be much richer than the one of a radiation-dominated Universe. Since the radiation peculiar velocity and density contrast are present together with the fluctuations of the spectator field, a non-adiabatic fluctuation of the pressure density, $\delta p_{\rm nad}$ can be expected. Since $\delta p_{\rm nad}$ is the source term for the evolution of the curvature perturbations, the quasi-flat spectrum of the spectator/axion field may be converted, under some dynamical requirements, into curvature fluctuations after the axion's decay.

Another possible example is the one where inflation occurs at relatively low curvature scales, i.e. $H \ll 10^{-6} \overline{M}_{\rm P}$. In this case one can imagine the situation where the fluctuations of a scalar field (that is light during inflation and later on decays), can be efficiently converted into adiabatic curvature perturbations.

In [381] the possible conversion of isocurvature fluctuations into adiabatic modes was investigated in a simple set up where on top of the inflaton field there is only an extra spectator field. The chief objective of Ref. [381] was not the possibility of converting isocurvature into adiabatic modes;

on the contrary, there was the hope that fluctuations of a pseudo-scalar spectator could indeed give rise to isocurvature modes [382] after the decay (taking place after baryogenesis in the model of [381]). The possibility of conversion was also briefly mentioned in Ref. [383]; the main goal of [383] was however the analysis of possible non-Gaussian effects arising in isocurvature models.

In [384] (see also [385]) the role of spectator field was played by some super-string moduli that could be amplified during a conventional inflationary phase. The authors correctly pointed out the possibility of obtaining, after modulus decay, a correlated mixture of adiabatic and isocurvature fluctuations. In [386] the conversion of isocurvature modes has been analyzed in the context of pre-big bang models. In fact, while the spectrum the dilaton and of the graviton is rather steep, the spectrum of axionic fluctuations may be rather flat or even red. The axionic fluctuation amplified during the pre-big bang phase can then be converted into adiabatic modes. In Ref. [387] the authors called the spectator field curvaton and provided further support to this idea.

On a more technical ground, in the present chapter, we will be led to analyze a rather interesting system of equations combining fluid perturbations and scalar field fluctuations. This system is rather interesting per se and it arises since it will be crucial, in what follows, to study the evolution of a scalar field in a fluid background. In a different perspective (and with rather different physical assumptions) this system also describes the fluctuations of a quintessential field after equality. This problem, however, will not be explicitly treated in the present script.

13.1 Spectator fields in a fluid background

Let us suppose that a scalar degree of freedom[a] σ, characterized by a potential $W(\sigma)$, evolves together with a fluid background with constant barotropic index w. The Friedmann-Lemaître equations can be written, in this case, as

---

[a]In this sccript the variable σ has also been employed to denote the cross section. In the present chapter the two variables will never appear together and no confusion is possible.

$$\overline{M}_{\rm P}^2 \mathcal{H}^2 = \frac{a^2}{3}(\rho_\sigma + \rho), \qquad (13.1)$$

$$\overline{M}_{\rm P}\mathcal{H}' = -\frac{a^2}{6}[(\rho_\sigma + 3p_\sigma) + (\rho + 3p)], \qquad (13.2)$$

$$\rho' + 3\mathcal{H}(\rho + p) = 0, \qquad (13.3)$$

$$\sigma'' + 2\mathcal{H}\sigma' + \frac{\partial W}{\partial \sigma}a^2 = 0, \qquad (13.4)$$

where $\overline{M}_{\rm P}$ denotes, as usual, the reduced Planck mass defined in Eq. (5.59); the prime denotes, as usual, a derivation with respect to the conformal time coordinate. Furthermore, in Eqs. (13.1) and (13.2),

$$\rho_\sigma = \frac{\sigma'^2}{2a^2} + W, \qquad p_\sigma = \frac{\sigma'^2}{2a^2} - W. \qquad (13.5)$$

In more explicit notations Eqs. (13.1) and (13.2) can also be written as:

$$\overline{M}_{\rm P}\mathcal{H}^2 = \frac{1}{3}\left[\frac{\sigma'^2}{2} + Wa^2 + \rho a^2\right], \qquad (13.6)$$

$$\overline{M}_{\rm P}\mathcal{H}' = -\frac{1}{3}\left[\sigma'^2 - Wa^2 + \left(\frac{3w+1}{2}\right)\rho a^2\right]. \qquad (13.7)$$

Consider now, as a warm-up, the situation where the scalar field σ is not a source of the background geometry but rather it evolves in a background which is dominated by the fluid sources. In this situation we will have the following conditions verified:

$$\overline{M}_{\rm P}\mathcal{H}^2 \gg a^2 \rho_\sigma, \qquad \rho \gg \rho_\sigma. \qquad (13.8)$$

Under this approximation, the evolution of σ can be studied in two complementary regimes:

- the kinetic regime where $\sigma' \gg \sqrt{2W}a$;
- the slow-roll regime where $\sigma' \ll \sqrt{2W}a$.

In the kinetic regime the evolution depends upon the specific evolution of the background geometry. For instance, in the case of conventional inflationary models σ may relax to the minimum of $W(\sigma)$ where the kinetic approximation eventually breaks down. In different models the evolution might be slightly different. Once the slow-roll regime takes place the dynamics of σ obeys the following approximate equation

$$\sigma' = -\frac{1}{\alpha \mathcal{H}}\frac{\partial W}{\partial \sigma}a^2. \qquad (13.9)$$

The parameter α depends upon the specific barotropic index of the dominant component of the fluid background. Suppose, for simplicity, that only a single fluid is present in the game. Then inserting Eq. (13.9) into Eq. (13.4) we obtain

$$\frac{1}{\alpha}\frac{\mathcal{H}'}{\mathcal{H}^2}\frac{\partial W}{\partial \sigma}a^2 - \frac{4}{\alpha}\frac{\partial W}{\partial \sigma}a^2 + a^2\frac{\partial W}{\partial \sigma} - \frac{1}{\alpha \mathcal{H}}\frac{\partial^2 W}{\partial \sigma^2}a^2 \sigma' = 0 \qquad (13.10)$$

Now, the last term at the left hand side of Eq. (13.10) contains three derivatives of the potential since, according to Eq. (13.9), σ' is proportional to the first derivative of W with respect to σ. Consequently, this term can be neglected and α is determined to be:

$$\alpha = \frac{1}{4 - \frac{\mathcal{H}'}{\mathcal{H}^2}} = \frac{6 + 3(w+1)}{2}, \qquad (13.11)$$

where the second equality follows since the fluid-dominated background implies that $\mathcal{H}'/\mathcal{H}^2 = -(3w+1)/2$. Thus, from Eqs. (13.9) and (13.11)

$$\sigma' = -\frac{2}{\mathcal{H}[6 + 3(w+1)]}\frac{\partial W}{\partial \sigma}a^2. \qquad (13.12)$$

Equation (13.12) can also be easily written in the cosmic time parametrization:

$$\dot{\sigma} = -\frac{2}{H[6 + 3(w+1)]}\frac{\partial W}{\partial \sigma} \qquad (13.13)$$

The discussion can be developed in reasonably general terms, however, for sake of simplicity it is easier to focus on the case when the background is dominated by radiation and the potential term of σ is quadratic, i.e.

$$p_{\rm r} = \frac{\rho_{\rm r}}{3}, \qquad W(\sigma) = \frac{m^2}{2}\sigma^2. \qquad (13.14)$$

Under these assumptions Eqs. (13.3)–(13.4) and (13.6)–(13.7) can be written as:

$$\overline{M}_{\rm P}^2 \mathcal{H}^2 = \frac{a^2}{3}\left[\rho_{\rm r} + \frac{\sigma'^2}{2a^2} + W(\sigma)\right], \qquad (13.15)$$

$$\overline{M}_{\rm P}\mathcal{H}' = -\frac{a^2}{3}\left[\rho_{\rm r} + \frac{\sigma'^2}{a^2} - W(\sigma)\right], \qquad (13.16)$$

$$\sigma'' + 2\mathcal{H}\sigma' + a^2\frac{\partial W}{\partial \sigma} = 0, \qquad (13.17)$$

$$\rho_{\rm r}' + 4\mathcal{H}\rho_{\rm r} = 0. \qquad (13.18)$$

Furthermore Eq. (13.12) implies

$$\sigma' = -\frac{m^2}{5\mathcal{H}}\sigma a^2. \qquad (13.19)$$

Since, during a radiation, furthermore, $\mathcal{H}a$ is constant, Eq. (13.19) can be solved and it gives:

$$\sigma(\tau) = \sigma_i \exp\left[-\frac{m^2 \tau_i^2}{20}\left(\frac{\tau^4}{\tau_i^4} - 1\right)\right].\tag{13.20}$$

Equation (13.20) tells that, basically, $\sigma(\tau)$ remains fixed at a constant value σ_i as long as $m\tau_i \simeq m/H_i \ll 1$. At H_i, Eq. (13.8) implies that the following hierarchy of scales holds[b]

$$\overline{M}_{\rm P}^2 H_i^2 \gg \frac{m^2\sigma^2}{2} \gg \frac{\dot{\sigma}^2}{2}.\tag{13.21}$$

As time goes by the Hubble rate $H = \mathcal{H}/a$ decreases. There are, therefore, two important moments:

- the moment at which $H \simeq m$ (denoted in what follows by H_m);
- the moment at which $\overline{M}_{\rm P}H \simeq m\sigma$ (denoted in what follows by H_σ).

For $H > H_m$, grossly speaking, $\sigma \simeq \sigma_i$. For $H < H_m$, σ will undergo a phase of coherent oscillations where $\dot{\sigma}^2 \propto a^{-3}$. This statement can be easily appreciated by writing the evolution of σ according to Eq. (13.17) in cosmic time:

$$\ddot{\sigma} + 3H\dot{\sigma} + m^2\sigma = 0.\tag{13.22}$$

By eliminating the first derivative we have that

$$\ddot{\Sigma} + \left[m^2 - \left(\frac{9}{4}H^2 + \frac{3}{2}\dot{H}\right)\right]\Sigma = 0, \qquad \Sigma = a^{3/2}\sigma.\tag{13.23}$$

Since $\dot{H} = -2H^2$, it is clear that for $m^2 \gg H^2$ the rescaled variable Σ will have a solution going as $\sin mt$ and the other solution evolving as $\cos mt$. Furthermore, σ will also oscillate but with amplitude decreasing as $a^{-3/2}$. It is important to stress, at this point, that the simple trigonometric oscillations are a direct consequence of the quadratic nature of $W(\sigma)$. This choice implies that, on average, $\langle p_\sigma \rangle \simeq 0$ which explains why σ^2 decreases as a^{-3}, i.e. as dusty matter. For instance in the case of a quartic potential, $\langle \rho_\sigma \rangle \simeq 3\langle p_\sigma \rangle$ [388, 391] and the oscillatory regime will have a different analytical description.

For $H > H_\sigma$ the background is dominated by radiation (or, more generally, by the fluid content of the model). For $H < H_\sigma$ the potential energy of

---

[b]In what follows we will often switch from the cosmic to the conformal time parametrization (and vice-versa) without any specific comment.

σ dominates against the background sources. When $H < H_\sigma$ the spectator field changes its status and becomes the dominant source of the evolution of the geometry.

Finally, at $H \simeq H_d \simeq m^3/M_P^2$ the field σ will decay. For an order of magnitude estimate we can impose, for instance, that $H_d > H_{BBN}$. The latter requirement imposes that the decay should occur prior to the moment at which light nuclei are formed (i.e. the moment of big-bang nucleosynthesis). If $H_d < H_{BBN}$ the production of light elements might be jeopardized. If $H_d > H_{BBN}$, $m > 10$ TeV taking the BBN temperature of the order of 0.1 MeV. This bound can be made more restrictive by requiring that the decay takes place prior the electroweak epoch (for temperatures of the order of 0.1 TeV).

With all the ingredients mentioned so far it is possible to sketch the evolution of the system recalling that, for accurate estimates, numerical methods are preferable especially in the light of the evolution of the corresponding fluctuations which will be addressed in the following section. For practical reasons we define the ratio between the energy density of the spectator field and the energy density of the radiation background, i.e.

$$r(a) = \frac{m^2 \sigma_i^2}{H_i^2 \overline{M}_P^2} \left(\frac{a}{a_i}\right)^4 \quad (13.24)$$

As indicated by Eq. (13.24) $r(a)$ is a sharply increasing function of the scale factor for $H > H_m$. Now we have two possibilities:

- σ can oscillate before becoming dominant, i.e. $H_m > H_\sigma$;
- σ can oscillate after being dominant against the radiation background, i.e. $H_m < H_\sigma$.

Let us discuss separately these two possibilities. If $H_m > H_\sigma$, then

$$m^2 \sigma^2 \simeq m^2 \sigma_i^2 \left(\frac{a_i}{a}\right)^3, \quad H < m. \quad (13.25)$$

Consequently, Eq. (13.24) will become, for $H < m$,

$$r(a) = \frac{\sigma_i^2}{\overline{M}_P^2} \left(\frac{a}{a_m}\right), \quad H < m. \quad (13.26)$$

By comparing Eqs. (13.25) and (13.26) it emerges that $r(a)$, for $H < m$, is always increasing but at a slower rate. According to Eq. (13.26) $r(a_\sigma) \simeq 1$ which implies that the duration of the oscillatory will be given by

$$\left(\frac{a_\sigma}{a_m}\right) \simeq \left(\frac{\sigma_i}{\overline{M}_P}\right)^{-2}. \quad (13.27)$$

Consequently, if $\sigma_i \gg \overline{M}_P$, the duration of the oscillating phase can be rather long and the moment at which σ starts dominating can be delayed. From Eq. (13.27) it is also clear that

$$H_\sigma \simeq m\left(\frac{\sigma_i}{\overline{M}_P}\right)\left(\frac{a_m}{a_\sigma}\right)^{3/2} \simeq m\left(\frac{\sigma_i}{\overline{M}_P}\right)^4. \qquad (13.28)$$

Let us now see what happens with decay. Here, again, there are two different possibilities which are particularly relevant for the fate of the perturbations. If the decay takes place *before* the spectator field is dominant, i.e. $H_\sigma < H_d$, then we get to the conclusion that $\sigma_i/\overline{M}_P$ should indeed satisfy:

$$\left(\frac{\sigma_i}{\overline{M}_P}\right) < \sqrt{\frac{m}{\overline{M}_P}}. \qquad (13.29)$$

In the opposite case, i.e. when σ decays *after* being dominant, the opposite inequality will apply and Eq. (13.29) will be

$$\left(\frac{\sigma_i}{\overline{M}_P}\right) > \sqrt{\frac{m}{\overline{M}_P}}. \qquad (13.30)$$

Owing to the absolute lower bound on the mass m (stemming from BBN) both inequalities of Eqs. (13.29) and (13.30) can be satisfied when $\sigma_i < \overline{M}_P$.

Let us briefly examine, finally, the case when $\sigma_i > \overline{M}_P$. Consider, then Eq. (13.24) and suppose that, indeed, $\sigma_i > \overline{M}_P$. The scale of oscillations will still be given by m and the scale of dominance will still be given by $H_\sigma \simeq m(\sigma_i/\overline{M}_P)^4$. Therefore, if $\sigma_i > \overline{M}_P$ we will necessarily have $H_\sigma > H_m$, i.e. the spectator field will become dominant without experiencing any oscillatory dynamics. From Eq. (13.24) it is then rather easy to estimate when the spectator field will eventually dominate:

$$\left(\frac{a_\sigma}{a_i}\right) \simeq \sqrt{\frac{\overline{M}_P H_i}{\sigma_i m}}. \qquad (13.31)$$

After the spectator field becomes dominant there are two possibilities:

- if the dominance takes place before the oscillatory regime, then a secondary (i.e. low curvature) inflationary phase will take place with $H_\sigma \overline{M}_P \simeq \sqrt{W(\sigma)}$;
- if the dominance takes place after oscillations have started (i.e. in the case $\sigma_i < \overline{M}_P$) then there will be an intermediate phase dominated by a dusty phase induced by the coherent oscillations of σ.

Notice, as remarked before, that the second statement depends heavily upon the quadratic nature of the potential and can be partially evaded if the potential has a different analytic form. In the limiting case of purely quartic self-interactions the intermediate phase will still be, effectively, dominated by radiation.

In connection with Eq. (13.31) it is important to appreciate that the spectator field must actually be effectively massless at the beginning, i.e.

$$\frac{m}{H_\mathrm{i}} \ll 1. \tag{13.32}$$

This condition stems from the requirement that we ought to have the fluctuations of σ amplified with nearly scale-invariant spectrum. Last but not least, to have a predominantly Gaussian contribution to the Sachs-Wolfe plateau we must also require that

$$\frac{H_\mathrm{i}}{2\pi} < \sigma_\mathrm{i}. \tag{13.33}$$

All the constraints and the possibilities listed in the present section define the parameter space of the model and few specific examples (the related exclusion plots can be consulted in Refs. [389] and [407]).

As a last remark we wish to stress that the initial radiation-dominated phase can be replaced by a different kind of phase. For instance, in the case of quintessential inflationary models [407], the initial phase is dominated by the kinetic energy of the inflaton/quintessence field. The discussion of the evolution of the background then follows similar lines but the nature and implications of the constraints is clearly different.

Before the spectator fields becomes dominant, an interesting phenomenon can take place, i.e. the conversion of an initial isocurvature mode into an adiabatic mode. There are two independent motivations to discuss in some detail for this possibility:

- the first motivation is that the fluctuations of the field which drives the evolution of the background before τ_i are not amplified with nearly scale-invariant spectrum, as required by experimental data;
- the second motivation is that inflation takes place at a curvature scale that is much smaller than $10^{-6}\overline{M}_\mathrm{P}$.

Two examples of these two situations will now be provided.

13.2 Unconventional inflationary models

During the pre-big bang evolution the Kalb-Ramond axion is amplified with quasi-flat spectrum [105], i.e.

$$\mathcal{P}_\sigma \simeq \left(\frac{H_1}{\overline{M}_P}\right)\left(\frac{k}{k_1}\right)^{n_\sigma-1} \qquad (13.34)$$

where the spectral index n_σ depends on the dynamics of the internal dimensions [388, 389]. On the contrary, the spectrum of curvature fluctuations is characterized by a rather steep spectral index. More precisely, the considerations introduced earlier in this section allow us to compute the scalar and tensor spectral indices [370]:

$$n_s = 4, \qquad n_T = 3. \qquad (13.35)$$

At the beginning of the post-big bang evolution the background is characterized by a "maximal" curvature scale H_1, whose finite value regularizes the big bang singularity of the standard cosmological scenario, and provides a natural cutoff for the spectrum of quantum fluctuations amplified by the phase of pre-big bang inflation. In string cosmology models such an initial curvature scale is at most of the order of the string mass scale, i.e. $H_1 \lesssim M_s \sim 10^{17}$ GeV.

The evolution of σ is different depending upon the initial value of σ in Planck units, i.e. $\sigma_i/\overline{M}_P$. If $\sigma_i > \overline{M}_P$ the axion dominates earlier since the condition of dominance is saturated faster. If $\sigma_i < \overline{M}_P$ the axion dominates later, i.e. its starts oscillating coherently at H_m and it will dominate at a redshift $(a_m/a_\sigma) \sim (\sigma_i/\overline{M}_P)^2$. For a more specific discussion we refer to [386, 388–390].

Consider then, as an illustration, the case of $\sigma_i < \overline{M}_P$. Initially $6H_{\text{inf}}^2 \overline{M}_P^2 \gg m^2 \sigma_i$ since σ is not dominant. Later on σ oscillates at $H_m \sim m$ and, around this scale the approximation of Eq. (13.12) breaks down and $\sigma(\tau) \simeq \sigma_i (a_m/a)^{3/2}$. At a typical scale $H_\sigma \simeq (\sigma_i/\overline{M}_P)^4 m/36$, the field starts dominating the background inducing, possibly, a short period of inflationary expansion. The axion can decay either before being dominant or after. The decay of the curvaton has been discussed both analytically and numerically by different authors [389, 392, 393]. In the following, for the sake of simplicity the approximation of sudden decay will be adopted. If the decay follows the dominance, i.e. $H_d < H_\sigma$, then

$$\left(\frac{\sigma_i}{\overline{M}_P}\right) \gg \sqrt{6}\sqrt{\frac{m}{\overline{M}_P}}. \qquad (13.36)$$

The opposite inequality holds if $H_d > H_\sigma$.

The evolution equations for the fluctuations are obtained by considering the simultaneous presence of the the fluctuations of σ and of the fluctuations of the radiation background. The Hamiltonian constraint then becomes, in the longitudinal gauge,

$$-k^2\psi - 3\mathcal{H}(\mathcal{H}\psi + \psi') = 4\pi G a^2(\rho_r \delta_r + \delta\rho_\sigma), \qquad (13.37)$$

while the $i = j$ component of the perturbed Einstein equations gives

$$\psi'' + 3\mathcal{H}\psi' + (\mathcal{H}^2 + 2\mathcal{H}')\psi = \frac{4\pi G}{3} a^2 \rho_r \delta_r + 4\pi G a^2 \delta p_\sigma, \qquad (13.38)$$

where

$$\delta\rho_\sigma = \frac{1}{a^2}\left[-\psi\sigma'^2 + \sigma'\chi' + \frac{\partial W}{\partial \sigma}a^2\chi\right], \qquad (13.39)$$

$$\delta p_\sigma = \frac{1}{a^2}\left[-\psi\sigma'^2 + \sigma'\chi' - \frac{\partial W}{\partial \sigma}a^2\chi\right]. \qquad (13.40)$$

In all the perturbation equations it has been assumed that anisotropic stresses are absent (i.e. $\psi = \phi$). Notice that we denoted with χ the fluctuation of σ. Finally the evolution of χ and of the fluid variables is given by

$$\chi'' + 2\mathcal{H}\chi' - \nabla^2\chi + \frac{\partial^2 W}{\partial \sigma^2}a^2\chi - 4\sigma'\Phi' + 2\frac{\partial W}{\partial \sigma}a^2\Phi = 0, \qquad (13.41)$$

$$\delta_r' = -\frac{4}{3}\theta_r + 4\psi', \qquad (13.42)$$

$$\theta_r' = \frac{k^2}{4}\delta_r + k^2\psi. \qquad (13.43)$$

We are interested in the solution of the system of Eqs. (13.37)–(13.38) and (13.41)–(13.43) with non-adiabatic initial conditions, i.e. $\psi(k, \tau_i) = \delta_r(k, \tau_i) = 0$ but $\chi(k, \tau_i) = \chi_i(k)$.

Since for large-scale inhomogeneities $\theta_r \ll 1$, Eq. (13.42) implies

$$\delta_r(k, \tau) \simeq 4\psi(k, \tau), \qquad (13.44)$$

because initially $\psi(\tau_i) \to 0$. Using Eq. (13.39) and (13.44), the Hamiltonian constraint (13.37) can be written, for $k \to 0$ as[c]

$$\frac{d(a^3\psi)}{d\ln a} = \frac{4\pi G}{3\mathcal{H}^2}a^3[\sigma'\chi' + W_{,\sigma}a^2\chi]. \qquad (13.45)$$

---

[c] In what follows the stenographic notation $W_{,\sigma}$ will be used to indicate the derivative of the potential with respect to σ.

In order to derive Eq. (13.45) we used the fact that $(p_\sigma + \rho_\sigma) \ll \rho_r$. Similarly to Eq. (13.12), Eq. (13.41) implies, with isocurvature initial conditions, that

$$\chi' \simeq -\frac{1}{5\mathcal{H}} W_{,\sigma\sigma} a^2 \chi. \tag{13.46}$$

Hence, from Eq. (13.45), direct integration leads to

$$\psi(k,\tau) = -\frac{1}{42\overline{M}_P} \frac{a^2}{\mathcal{H}^2} W_{,\sigma} \chi_i(k) + \mathcal{O}(W_{,\sigma}^2) \simeq -\frac{1}{14} \frac{1}{\rho_r} W_{,\sigma} \chi_i(k) + \mathcal{O}(W_{,\sigma}^2). \tag{13.47}$$

Inserting Eq. (13.47) into Eqs. (13.42) and (13.43)

$$\delta_r(k,\tau) = -\frac{2}{21\overline{M}_P^2} \frac{a^2}{\mathcal{H}^2} W_{,\sigma} \chi_i + \mathcal{O}(W_{,\sigma}^2),$$

$$\theta_r(\tau) = -\frac{1}{105\overline{M}_P^2} \frac{a^2}{\mathcal{H}^3} W_{,\sigma} \chi_i + \mathcal{O}(W_{,\sigma}^2). \tag{13.48}$$

The time evolution of $\mathcal{R}$ in the radiation-dominated, slow-roll regime can finally be determined from the definition of $\mathcal{R}$ in terms of ψ (see, for instance, Eq. (7.101) or the more complete discussion in chapter 11):

$$\mathcal{R}(k,\tau) \simeq \frac{1}{4\rho_r} \frac{\partial W}{\partial \sigma} \chi_i(k) + \mathcal{O}(W_{,\sigma}^2), \tag{13.49}$$

so that ψ and $\mathcal{R}$ obey the approximate relation

$$\psi(k,\tau) \simeq -(2/7)\mathcal{R}(k,\tau) + \mathcal{O}(W_{,\sigma}^2). \tag{13.50}$$

The same result can be obtained directly by integrating the evolution equation for $\mathcal{R}$ i.e.

$$\frac{d\mathcal{R}}{d\ln a} \simeq -\frac{\delta p_{\text{nad}}}{p+\rho}, \tag{13.51}$$

having approximately determined the form of δp_{nad}. The exact form of δp_{nad} and of the sound speed is, in our system,

$$\delta p_{\text{nad}} = \rho_r \left(\frac{1}{3} - c_s^2\right) \delta_r + \psi(c_s^2 - 1)(p_\sigma + \rho_\sigma) + \frac{\sigma'\chi'}{a^2}(1 - c_s^2)$$
$$- W_{,\sigma} \chi (1 + c_s^2), \tag{13.52}$$

$$c_s^2 = \frac{1}{3}\left[\frac{4\rho_r + 9(p_\sigma + \rho_\sigma) + 6(\sigma'/\mathcal{H})W_{,\sigma}}{4\rho_r + 3(p_\sigma + \rho_\sigma)}\right]. \tag{13.53}$$

When the curvaton is subdominant $c_s^2 \simeq 1/3$ and the leading term in Eq. (7.91) is the last one. Hence, to leading order, $\delta p_{\text{nad}} \simeq -4V_{,\sigma}\chi_i/3$;

inserting the approximate form of $\delta p_{\rm nad}$ into Eq. (13.51) we obtain again, by direct integration Eq. (13.49) since

$$\frac{d\mathcal{R}}{d\ln a} \simeq \frac{V_{,\sigma}}{\rho_{\rm r}}\chi. \qquad (13.54)$$

After the initial phase, when the oscillation starts, we have to assume a specific form of the potential which we take to be quadratic. For $H \leq m$, $\chi \sim \chi_{\rm i}(a_m/a)^{3/2}$, $\sigma \sim \sigma_{\rm i}(a_m/a)^{3/2}$ and Eq. (13.54) then leads to

$$\mathcal{R}(k,\tau) \simeq \left(\frac{\sigma_{\rm i}}{\overline{M}_{\rm P}}\right)\left(\frac{\chi_{\rm i}(k)}{\overline{M}_{\rm P}}\right)\left(\frac{a}{a_m}\right) = r(a)\frac{\chi_{\rm i}}{\sigma_{\rm i}}, \qquad (13.55)$$

$$r(a) = \left(\frac{\sigma_{\rm i}}{\overline{M}_{\rm P}}\right)^2 \left(\frac{a}{a_{\rm m}}\right). \qquad (13.56)$$

where for $H \leq m$ we used $\rho_{\rm r} \sim m^2 \overline{M}_{\rm P}^2 (a_m/a)^4$. The second equality follows from the definition of $r(a)$, i.e. the time-dependent ratio of the axion energy density over the energy density of the dominant component of the background. Recalling now that $\chi_i(k)$ has a quasi-flat power spectrum, Eq. (13.55) expresses the conversion of the initial isocurvature mode into the wanted adiabatic mode. The expression of $\mathcal{R}(a)$ will be frozen at the moment of sudden decay taking place at the scale $H_{\rm d}$. In the context of pre-big bang models the power spectrum of the axionic fluctuations depends on various parameters connected with the specific model of pre-big bang evolution. By comparing the generated curvature fluctuations with the value experimentally measured various constraints on the parameter of the model can be obtained [388, 389]. These constraints seem to suggest that the scale H_1 should be of the order of $10^{-1}M_{\rm s}$, where $M_{\rm s} \sim 0.01\overline{M}_{\rm P}$ is the string mass scale. Notice, furthermore, that we have just treated the case where the axion decays before becoming dominant. However, in the context of pre-big bang models the opposite case is probably the most relevant. If the axion decays after becoming dominant, numerical calculations have to back the analytical approach (especially in the case $\sigma_{\rm i} > \overline{M}_{\rm P}$).

13.3 Conventional inflationary models

It is conceivable that the curvaton could also play a rôle in the context of conventional inflationary models, provided the inflationary scale is sufficiently low (i.e. $H_{\rm inf} < 10^{-6}\overline{M}_{\rm P}$). In this case, it is argued [394], the adiabatic perturbations generated directly by the fluctuations of the inflaton will be too small. In [394] a simple model has been discussed where the potentials of the curvaton and of the inflaton are both quadratic. During inflation the curvaton should be effectively massless, i.e. $m \ll H_{\rm inf}$. There are some classes of inflationary models leading naturally to a low curvature scale [395] (see also [396]). As far as the evolution of the curvaton in a radiation dominated epoch is concerned, the results obtained in the case of the pre-big bang are also valid provided the scale H_1 is now interpreted as the curvature scale at the end of inflation.

To identify a particle physics candidate acting as a curvaton, various proposals have been made. Different authors [397–400] suggest that the right-handed sneutrinos (the supersymmetric partners of the righ-handed neutrinos) may act as curvatons. Models have also been proposed where the curvaton corresponds to a MSSM flat direction [401, 402] and to a MSSM Higgs [403].

In the following we would like to concentrate on one of the besic questions arising in low-scale inflationary models: is it possible to lower the inflationary scale *ad libitum*?

In the standard curvaton scenario the energy density of the curvaton increases with time with respect to the energy density of the radiation background. From this aspect of the theoretical construction, a number of constraints can be derived; these include an important aspect of the inflationary dynamics occurring prior to the curvaton oscillations, namely, the minimal curvature scale at the end of inflation compatible with the curvaton idea. Suppose, for simplicity, that the curvaton field σ has a massive potential and that its evolution, after the end of inflation, occurs during a radiation dominated stage of expansion. As previously discussed the ratio $r(a)$ increases with time during the radiation-dominated oscillations

$$r(a) \simeq \left(\frac{\sigma_{\rm i}}{\overline{M}_{\rm P}}\right)^2 \left(\frac{a}{a_{\rm m}}\right) \qquad H < H_{\rm m}, \tag{13.57}$$

where $\sigma_{\rm i}$ is the nearly constant value of the curvaton throughout the later stages of inflation. When σ decays the ratio r gets frozen to its value at decay, i.e. $r(t) \simeq r(t_{\rm d}) = r_{\rm d}$ for $t > t_{\rm d}$. Equation (13.57) then implies

$$m = \frac{\sigma_{\rm i}^2}{r_{\rm d}\overline{M}_{\rm P}}. \tag{13.58}$$

The energy density of the background fluid just before decay has to be larger than the energy density of the decay products, i.e. $\rho_{\rm r}(t_{\rm d}) \geq T_{\rm d}^4$. Since

$$\rho_{\rm r}(t_{\rm d}) \simeq m^2 \overline{M}_{\rm P}^2 \left(\frac{a_{\rm m}}{a_{\rm d}}\right)^4, \tag{13.59}$$

the mentioned condition implies

$$\frac{m}{T_{\rm d}}\sqrt{\frac{m}{\overline{M}_{\rm P}}} > 1, \qquad (13.60)$$

which can also be written, using Eq. (13.58), as

$$\left(\frac{\sigma_{\rm i}}{\overline{M}_{\rm P}}\right)^3 \geq r_{\rm d}^{3/2}\left(\frac{T_{\rm d}}{\overline{M}_{\rm P}}\right). \qquad (13.61)$$

Equation (13.61) has to be compared with the restrictions coming from the amplitude of the adiabatic perturbations, which should be consistent with observations. If σ decays before becoming dominant the curvature perturbations at the time of decay are

$$\mathcal{R}(t_{\rm d}) \simeq \frac{1}{\rho_{\rm r}}\frac{\partial V}{\partial \sigma}\chi\bigg|_{t_{\rm d}} \simeq r_{\rm d}\frac{\chi_k^i}{\sigma_{\rm i}}. \qquad (13.62)$$

Recalling that $\chi_k^i \sim H_{\rm inf}/(2\pi)$ the power spectrum of curvature perturbations

$$\mathcal{P}_\mathcal{R}^{1/2} \simeq \frac{r_{\rm d} H_{\rm inf}}{4\pi\sigma_{\rm i}} \simeq 5\times 10^{-5} \qquad (13.63)$$

implies, using Eq. (13.61) together with Eq. (13.58),

$$\left(\frac{H_{\rm inf}}{\overline{M}_{\rm P}}\right) \geq 10^{-4}\times r_{\rm d}^{-1/2}\left(\frac{T_{\rm d}}{\overline{M}_{\rm P}}\right)^{1/3}. \qquad (13.64)$$

Recalling now that $r_{\rm d} \leq 1$ and $m < H_{\rm inf}$, the above inequality implies that, at most, $H_{\rm inf} \geq 10^{-12}\,\overline{M}_{\rm P}$ if $T_{\rm d} \sim 1$ MeV is selected. This estimate is, in a sense, general since the specific relation between $T_{\rm d}$ and $H_{\rm d}$ is not fixed. The bound (13.64) can be even more constraining, for certain regions of parameter space, if the condition $T_{\rm d} \geq \sqrt{H_{\rm d}\,\overline{M}_{\rm P}}$ is imposed with $H_{\rm d} \simeq m^3/\overline{M}_{\rm P}^2$. In this case, Eq. (13.64) implies $H_{\rm inf}^2 \geq 10^{-8}m\,\overline{M}_{\rm P}$, which is more constraining than the previous bound for sufficiently large values of the mass, i.e. $m \geq 10^{-4}H_{\rm inf}$. Thus, in the present context, the inflationary curvature scale is bound to be in the interval [395, 404]

$$10^{-12}\overline{M}_{\rm P} \leq H_{\rm inf} \leq 10^{-6}\overline{M}_{\rm P}. \qquad (13.65)$$

This bound can be relaxed by a bit, if the evolution of the curvaton takes place after inflation in a background dominated by the kinetic energy of the inflaton field itself. Consider now the case where the inflationary epoch is not immediately followed by radiation. Different models of this kind may be constructed. For instance, if the inflaton field is identified with the quintessence field, a long kinetic phase occurred prior to the usual radiation-dominated stage of expansion. The evolution of a massive curvaton field in quintessential inflationary models has been recently studied [407] in the simplest scenario where the curvaton field is decoupled from the quintessence field and it is minimally coupled to the metric. In order to be specific, suppose that, the inflaton potential, $V(\varphi)$ is chosen to be a typical power law during inflation and an *inverse* power during the quintessential regime:

$$V(\varphi) = \lambda(\varphi^4 + M^4), \quad \varphi < 0,$$
$$V(\varphi) = \frac{\lambda M^8}{\varphi^4 + M^4}, \quad \varphi \geq 0, \tag{13.66}$$

where λ is the inflaton self-coupling and M is the typical scale of quintessential evolution, i.e. by appropriately fixing M, the field φ will be dominant about today. The potential of the curvaton may be taken to be, for simplicity, quadratic. In this model the curvaton will evolve, right after the end of inflation, in an environnment dominated by the kinetic energy of φ. The curvaton starts oscillating at $H_m \sim m$ and becomes dominant at a typical curvature scale $H_\sigma \sim m(\sigma_i/\overline{M}_P)^2$. Due to the different evolution of the background geometry, the ratio $r(t)$ will take the form

$$r(t) \simeq m^2 \left(\frac{\sigma_i}{\overline{M}_P}\right)^2 \left(\frac{a}{a_m}\right)^3, \quad H < H_m. \tag{13.67}$$

to be compared with Eq. (13.57), valid in the standard case. For $t > t_d$, $r(t)$ gets frozen to the value r_d whose relation to σ_i is different from the one obtained previously (see Eq. (13.58)) and valid in the case when σ relaxes in a radiation dominated environnment. In fact, from Eq. (13.67),

$$m \simeq \frac{\sigma_i}{\sqrt{r_d}}. \tag{13.68}$$

From the requirement

$$\rho_k(t_d) \simeq m^2 \overline{M}_P^2 \left(\frac{a_m}{a_d}\right)^6 \geq T_d^4, \tag{13.69}$$

it can be inferred, using (13.68), that

$$\left(\frac{\sigma_i}{\overline{M}_P}\right)^{3/2} \geq r_d^{3/4}\left(\frac{T_d}{\overline{M}_P}\right) \tag{13.70}$$

Following the analysis reported in [407], the amount of produced fluctuations can be computed.

The calculation exploits Eqs. (11.174) and (11.175) which are valid for a two-field model. Solving Eqs. (11.174) and (11.175) with the appropriate initial condition, and using $a^3\dot\varphi^2/H$ that is constant during the kinetic phase, Eq. (11.177) can be written as [407]

$$\mathcal{R}(t) \simeq \frac{H}{\rho_k}\dot\sigma q_\sigma \simeq \frac{\chi_\sigma}{\rho_k}\frac{\partial W}{\partial \sigma} \simeq r_d\left(\frac{\chi_k^{(i)}}{\sigma_i}\right). \tag{13.71}$$

Recalling that $\chi_k^{(i)} \sim H_{\text{inf}}/(2\pi)$, the observed value of the power spectrum, i.e. $\mathcal{P}_\mathcal{R}^{1/2} \sim 5 \times 10^{-5}$, implies

$$\left(\frac{H_{\text{inf}}}{\overline{M}_P}\right) \geq 10^{-4} r_d^{-3/2}\left(\frac{T_d}{\overline{M}_P}\right)^{2/3}. \tag{13.72}$$

The same approximations discussed in the standard case can now be applied to the case of quintessential inflation. They imply the absolute lower bound of $H_{\text{inf}} > 100$ GeV [404, 405]

Up to now we have always discussed the case of quadratic potentials. However, the potential of σ should be allowed to have different profiles. There is only one danger. If the potential leads to some attractor solution for the fluctuations, it may happen that the fluctuations of the curvaton are erased [404, 391]. If, for some reason, the curvaton does not decay completely, it is possible to have residual isocurvature fluctuations which will be completely correlated with the adiabatic component (see, for instance [406]).

Appendix A

The Concept of Distance in Cosmology

In cosmology there are different concepts of distance. Galaxies emit electromagnetic radiation. Therefore we can ask what the distance light travelled from the observed galaxy to the receiver is (that we may take, for instance, located at the origin of the coordinate system). But we can also ask what the actual distance of the galaxy was at the time the signal was emitted. Or we can ask what the distance of the galaxy is now. Furthermore, distances in FRW models depend on the matter content. If we would know precisely the matter content we would also know accurately the distance. However, in practice, the distance of an object is used to infer a likely value of the cosmological parameters. Different distance concepts can therefore be introduced such as:

- the proper coordinate distance;
- the redshift;
- the distance measure;
- the angular diameter distance;
- the luminosity distance.

In what follows these different concepts will be swiftly introduced and physically motivated.

A.1 The proper coordinate distance

The first idea coming to mind to define a distance is to look at a radial part of the FRW line element and define the coordinate distance to r_e as

$$s(r_e) = a(t) \int_0^{r_e} \frac{dr}{\sqrt{1-kr^2}}, \tag{A.1}$$

where, conventionally, the origin of the coordinate system coincides with the position of the observer while r_e defines the position of the emitter. Since k can be either positive, or negative or even zero, the integral at the right hand side of Eq. (A.1) will change accordingly so that we have:

$$s(r_e) = a(t) r_e, \qquad k = 0,$$

$$s(r_e) = \frac{a(t)}{\sqrt{k}} \arcsin[\sqrt{k}\, r_e], \qquad k > 0, \qquad (A.2)$$

$$s(r_e) = \frac{a(t)}{\sqrt{|k|}} \operatorname{arcsinh}[\sqrt{|k|}\, r_e], \qquad k < 0,$$

The only problem with the definition of Eq. (A.1) is that it involves geometrical quantities that are not directly accessible through observations. A directly observable quantity, at least in principle, is the redshift and, therefore it will be important to substantiate the dependence of the distance upon the redshift and this will lead to complementary (and commonly used) definitions of distance.

Before plunging into the discussion a lexical remark is in order. It should be clarified that the distance r_e is *fixed* and it is an example of a *comoving coordinate* system. On the contrary, the distance $s(r_e) = a(t) r_e$ (we consider, for simplicity, the spatially flat case) is called *physical* distance since it gets stretched as the scale factor expands. It is a matter of convenience to use either comoving or physical systems of units. For instance, in the treatment of the inhomogeneities, it is practical to use comoving wavelengths and wavenumbers. In different situations physical frequencies are more appropriate. It is clear that physical distances make sense at a given moment in the life of the Universe. For instance, the Hubble radius at the electroweak epoch will be of the order of the cm. The same quantity, evaluated today, will be much larger (and of the order of the astronomical unit, i.e. 10^{13} cm) since the Universe expanded a lot between a temperature of 100 GeV and the present temperature (of the order of 10^{-13} GeV). In contrast, comoving distances are the same at any time in the life of the Universe. A possible convention (which is, though, not strictly necessary) is to normalize to 1 the present value of the scale factor so that, today, physical and comoving distances would coincide.

A.2 The redshift

Suppose that a galaxy or a cloud of gas emits electromagnetic radiation of a given wavelength λ_e. The wavelength received by the observer will be denoted by λ_r. If the Universe does not expand we would simply have $\lambda_e = \lambda_r$. However since the Universe expands (i.e. $\dot{a} > 0$), the observed wavelength will be more red (i.e. redshifted) in comparison with the emitted frequency, i.e.

$$\lambda_r = \frac{a(t_r)}{a(t_e)}\lambda_e, \qquad 1 + z = \frac{a_r}{a_e}. \tag{A.3}$$

where z is the redshift. If the wavelength of the emitted radiation is precisely known (for instance a known emission line of the hydrogen atom or of some other molecule or chemical compound) the amount of expansion (i.e. the redshift) between the emission and the observation could be accurately determined. Equation (A.3) can be directly justified from the FRW line element. Light rays follow null geodesics, i.e. $ds^2 = 0$ in Eq. (2.2). Suppose then that a signal is emitted at the time t_e (at a radial position r_e) and received at the time t_r (at a radial position $r_r = 0$). Then from Eq. (2.2) with $ds^2 = 0$ we will have[a]

$$\int_{t_e}^{t_r} \frac{dt}{a(t)} = \int_0^{r_e} \frac{dr}{\sqrt{1-kr^2}}. \tag{A.4}$$

Suppose then that a second signal is emitted at a time $t_e + \delta t_e$ and received at a time $t_r + \delta t_r$. It will take a time $\delta t_r = \lambda_r/c$ to receive the signal and a time $\delta t_e = \lambda_e/c$ to emit the signal:

$$\int_{t_e+\delta t_e}^{t_r+\delta t_r} \frac{dt}{a(t)} = \int_0^{r_e} \frac{dr}{\sqrt{1-kr^2}}. \tag{A.5}$$

By subtracting Eqs. (A.4) and (A.5) and by rearranging the limits of integration we get:

$$\int_{t_e}^{t_e+\delta t_e} \frac{dt}{a(t)} = \int_{t_r}^{t_r+\delta t_r} \frac{dt}{a(t)}, \tag{A.6}$$

implying

$$\frac{\delta t_r}{\delta t_e} = \frac{\lambda_r}{\lambda_e} = \frac{a(t_r)}{a(t_e)} = \frac{\nu_e}{\nu_r}. \tag{A.7}$$

As far as the conventions are concerned we will remark that often (even if not always) the emission time will be generically denoted by t while the

---
[a]The position of the emitter is *fixed* in the comoving coordinate system.

observation time will be the present time t_0. Thus, with this convention $z + 1 = a_0/a(t)$.

Let us then suppose (incorrectly, as we shall see in a moment) that $t_r = t_e + r_e$. Then, recalling the definition of the Hubble parameter evaluated at $t_r = t_0$, i.e. $H_r = H_0 = \dot{a}_r/a_r$ we have:

$$\lambda_r = \frac{a(t_r)}{a(t_e)} \simeq \lambda_e[1 + H_0 r_e], \tag{A.8}$$

where the second equality in Eq. (A.8) follows by expanding in Taylor series around t_r and by assuming that $H_0 r_e < 1$. Recalling the definition of Eqs. (A.3) and (A.7), we obtain an approximate form of the Hubble law, i.e.

$$H_0 r_e \simeq z, \qquad z = \frac{\lambda_r - \lambda_e}{\lambda_r}. \tag{A.9}$$

This form of the Hubble law is approximate since it holds for small redshifts, i.e. $z < 1$. Indeed the assumption that $t_r \simeq t_e + r_e$ is not strictly correct since it would imply, for a flat Universe, that the scale factor is approximately constant. To improve the situation it is natural to expand systematically the scale factor and the redshift. Using such a strategy we will have:

$$\frac{a(t)}{a_0} = 1 + H_0(t - t_0) - \frac{q_0}{2} H_0^2 (t - t_0)^2 + ...$$
$$\frac{a_0}{a(t)} = 1 - H_0(t - t_0) + \left(\frac{q_0}{2} + 1\right) H_0^2 (t - t_0)^2 + ... \tag{A.10}$$

where q_0 is the deceleration parameter introduced in Eq. (2.63). From the definition of redshift (see Eq. (A.3)) it is easy to obtain

$$z = H_0(t_0 - t) + \left(\frac{q_0}{2} + 1\right) H_0^2 (t_0 - t)^2, \tag{A.11}$$

i.e.

$$(t_0 - t) = \frac{1}{H_0}\left[z - \left(\frac{q_0}{2} + 1\right)^2 z^2\right]. \tag{A.12}$$

Using then Eq. (A.10) to express the integrand appearing at the left hand side of Eq. (A.4) we obtain, in the limit $|kr_e^2| < 1$,

$$r_e = \int_t^{t_0} \frac{dt'}{a(t')} = \frac{1}{H_0 a_0}\left[z - \frac{1}{2}(q_0 + 1)z^2\right], \tag{A.13}$$

where, after performing the integral over t', Eq. (A.12) has been used to eliminate t in favour of z. Notice that while the leading result reproduces the one previously obtained in Eq. (A.9), the correction involves the deceleration parameter.

A.3 The distance measure

As we saw in the previous subsection the term *distance* in cosmology can have different meanings. Not all the distances we can imagine can actually be measured. A meaningful question can be, for instance, the following: we see an object at a given redshift z; how far is the object we see at redshift z? Consider, again, Eq. (A.4) where, now the integrand at the left hand side can be expressed as

$$\frac{dt}{a} = \frac{da}{Ha^2} = -\frac{1}{a_0}\frac{dz}{H(z)}. \qquad (A.14)$$

From Eq. (2.56), $H(z)$ can be expressed as

$$H(z) = H_0\sqrt{\Omega_{M0}(1+z)^3 + \Omega_{\Lambda 0} + \Omega_k(1+z)^2}, \qquad (A.15)$$

where

$$\Omega_{M0} + \Omega_{\Lambda 0} + \Omega_k = 1, \qquad \Omega_k = -\frac{k}{a_0^2 H_0^2}. \qquad (A.16)$$

In Eq. (A.16), Ω_k accounts for the curvature contribution to the total density. This parametrization may be confusing but will be adopted to match the existing notations in the literature. Notice that if $k > 0$ then $\Omega_k < 0$ and vice-versa. Defining now the integral[b]

$$D_0(z) = \int_0^z \frac{dx}{\sqrt{\Omega_{M0}(1+x)^3 + \Omega_{\Lambda 0} + \Omega_k(1+x)^2}}, \qquad (A.17)$$

we have, from Eq. (A.4), that

$$r_e(z) = \frac{D_0(z)}{a_0 H_0}, \qquad k = 0,$$

$$r_e(z) = \frac{1}{a_0 H_0 \sqrt{|\Omega_k|}} \sin\left[\sqrt{|\Omega_k|} D_0(z)\right], \qquad k > 0,$$

$$r_e(z) = \frac{1}{a_0 H_0 \sqrt{\Omega_k}} \sinh\left[\sqrt{\Omega_k} D_0(z)\right], \qquad k < 0. \qquad (A.18)$$

The quantity $r_e(z)$ of Eq. (A.18) is sometimes also called *distance measure* and it is denoted by $d_M(z)$. In the limit $|kr_e^2| < 1$ the open and closed expressions reproduce the flat case $k = 0$.

---

[b]It should be borne in mind that the inclusion of the radiation term (scaling, inside the squared root of the integrand, as $\Omega_{R0}(1+x)^4$) is unimportant for moderate redshifts (i.e. up to $z \simeq 20$). For even larger redshifts (i.e. of the order of z_{eq} or z_{dec}) the proper inclusion of the radiation contribution is clearly mandatory. See Eqs. (2.89) and (2.117) for the derivation of z_{eq} and z_{dec}. See also the applications at the end of this Appendix (in particular, Eqs. (A.34), (A.35) and (A.36)).

Some specific examples will now be given. Consider first the case $\Omega_{M0} = 1$, $\Omega_{\Lambda 0} = 0$ and $\Omega_k = 0$ (i.e. flat, matter-dominated Universe). In such a case Eq. (A.18) gives

$$r_e(z) = \frac{1}{a_0 H_0}\left[1 - \frac{1}{\sqrt{z+1}}\right] \simeq \frac{1}{a_0 H_0}\left[z - \frac{3}{4}z^2\right], \qquad (A.19)$$

where the second equality has been obtained by expanding the exact result for $z, 1$. Notice that the second equality of Eq. (A.19) reproduces Eq. (A.13) since, for a flat matter-dominated Universe, $q_0 = 1/2$. The other example we wish to recall is the one where $\Omega_k = 1$, $\Omega_{M0} = 0$, and $\Omega_{\Lambda 0} = 0$ (i.e. open Universe dominated by the spatial curvature). In this case Eqs. (A.17) and (A.18) give, respectively,

$$D_0(z) = \ln(z+1), \qquad r_e(z) = \frac{1}{2a_0 H_0}\left[(z+1) - \frac{1}{z+1}\right]. \qquad (A.20)$$

The considerations developed so far suggest the following chain of observations:

- in cosmology the distance measure to redshift z depend upon the cosmological parameters;
- while for $z < 1$, $r_e H_0 \simeq z$, as soon as $z \simeq 1$ quadratic and cubic terms start contributing to the distance measure;
- in particular quadratic terms contain the deceleration parameter and are thus sensitive to the total matter content of the Universe.

If we are able to scrutinize objects at high redshifts we may be able to get important clues not only on the expansion rate of the present Universe (i.e. H_0) but also on its present acceleration (i.e. $q_0 < 0$) or deceleration (i.e. $q_0 > 0$). This type of reasoning is the main rationale for the intensive study of type Ia supernovae. It has indeed been observed that type Ia supernovae are dimmer than expected. This experimental study suggests that the present Universe is accelerating rather than decelerating, i.e. $q_0 < 0$ rather than $q_0 > 0$. Since

$$q_0 = -\frac{\ddot{a}_0 a_0}{\dot{a}_0^2} = \frac{\Omega_{M0}}{2} + (1 + 3w_\Lambda)\Omega_{\Lambda 0}, \qquad (A.21)$$

where the dark energy contribution is $p_\Lambda = w_\Lambda \rho_\Lambda$ with $w_\Lambda \simeq -1$. So, to summarize, suppose we know an object of given redshift z. Then we also know precisely the matter content of the Universe. With these information we can compute what is the distance to redshift z by computing $r_e(z)$. The distance measure of Eq. (A.18) depends both on the redshift and on the precise value of the cosmological parameters. But cosmological parameters are, in some sense, exactly what we would like to measure. Astronomers are therefore interested in introducing more operational notions of distance like the angular diameter distance and the luminosity distance.

A.4 Angular diameter distance

Suppose we are to be in Eucledian (non-expanding) geometry. Then we do know that the arc of a curve s is related to the diameter d as $s = d\vartheta$ where ϑ is the angle subtended by s. Of course, this is true in the situation where $\vartheta < 1$. Suppose that s is known, somehow. Then $d \simeq s/\vartheta$ can be determined by determining ϑ, i.e. the angular size of the object. In FRW space-times the angular diameter distance can be defined from the angular part of the line element. Since $ds_\vartheta^2 = a^2(t)r^2 d\vartheta^2$, the angular diameter distance to redshift z will be

$$D_A(z) = \frac{s_\vartheta}{\vartheta} = a(t)r_e = \frac{a_0 r_e(z)}{1+z}, \qquad (A.22)$$

where $r_e(z)$ is exactly the one given, in the different cases, by Eq. (A.18). The quantity introduced in Eq. (A.22) is the physical angular diameter distance. We can also introduce the comoving angular diameter distance $\overline{D}_A(z)$:

$$D_A(z) = a(t)\overline{D}_A = \frac{a_0}{1+z}\overline{D}_A(z), \qquad (A.23)$$

which implies that the comoving angular diameter distance coincides with the distance measure defined in Eq. (A.18).

To determine the angular diameter distance we need, in practice, a set of standard rulers, i.e. objects that have the same size for different redshifts. Then the observed angular sizes will give us the physical angular diameter distance. Using the results of Eqs. (A.19) and (A.20) in Eq. (A.22) we obtain, respectively,

$$D_A(z) = \frac{1}{H_0} \frac{\sqrt{z+1}-1}{(z+1)^{3/2}}, \qquad (A.24)$$

$$D_A(z) = \frac{1}{2H_0}\left[1 - \frac{1}{(z+1)^2}\right], \qquad (A.25)$$

where Eq. (A.24) applies in the case of a matter-dominated Universe and Eq. (A.25) applies in the case of an open Universe dominated by the spatial curvature.[c] It is interesting to notice that the angular diameter distance in a flat (matter-dominated) Universe decreases for $z > 1$: objects that are further away appear larger in the sky. In such a case D_A given by Eq. (A.24) has a maximum and then decreases.

A.5 Luminosity distance

Suppose we are to be in Eucledian (transparent) space. Then, if the *absolute* luminosity (i.e. radiated energy per unit time) of an object at distance D_L is $\mathcal{L}_{\text{abs}}$, the *apparent* luminosity $\mathcal{L}_{\text{app}}$ will be

$$\mathcal{L}_{\text{app}} = \frac{\mathcal{L}_{\text{abs}}}{4\pi D_L^2}, \qquad D_L = \sqrt{\frac{\mathcal{L}_{\text{abs}}}{4\pi \mathcal{L}_{\text{app}}}}. \qquad (A.26)$$

If the observer is located at a position r_e from the source, the detected photons will be spread over an area $\mathcal{A} = 4\pi a_0^2 r_e^2$. Thus, from the emission time t_e to the observation time t_0 we will have the energy density of radiation evolve as:

$$\rho(t_0) = \rho(t_e)\left(\frac{a_e}{a_0}\right)^4, \qquad \sqrt{\frac{\rho_e}{\rho_0}} = (1+z)^{-2}. \qquad (A.27)$$

But now,

$$\mathcal{L}_{\text{abs}} \propto \sqrt{\rho_e}, \qquad \mathcal{L}_{\text{app}} \propto \frac{\sqrt{\rho_0}}{4\pi a_0^2 r_e^2}. \qquad (A.28)$$

Thus the luminosity distance as a function of the redshift becomes:

$$D_L(z) = (1+z)a_0 r_e(z), \qquad (A.29)$$

By comparing Eq. (A.22) to Eq. (A.29) we also have

$$D_A(z) = \frac{D_L(z)}{(1+z)^2}. \qquad (A.30)$$

To give two examples consider, as usual, the cases of a (flat) matter-dominated Universe and the case of an open (curvature-dominated) Universe. In these two cases, Eq. (A.29) in combination with Eqs. (A.19) and (A.20) leads to

$$D_L(z) = \frac{1}{H_0}\sqrt{z+1}[\sqrt{z+1} - 1], \qquad D_L(z) = \frac{1}{2H_0}[(z+1)^2 - 1]. \qquad (A.31)$$

---

[c] It is clear that, in this case, from Eqs. (2.56) and (2.57) the scale factor expands linearly. In fact, suppose we take Eqs. (2.56) and (2.57) in the limit $\rho = 0$ and $p = 0$. In this case the spatial curvature must be negative for consistency with Eq. (2.56) where the right hand side is positive semi-definite. By summing up Eqs. (2.56) and (2.57) we get to the condition $H^2 + \dot{H} = 0$, i.e. $\ddot{a} = 0$ which means $a(t) \sim t$. In this case, by definition, the deceleration parameter introduced in Eq. (2.63) vanishes.

As in the case of the angular diameter distance where it is mandatory to have a set of standard rulers, in the case of the luminosity distance there is the need of a set of standard candles, i.e. a set of objects that are known to have all the same absolute luminosity $\mathcal{L}_{\rm abs}$. Then by measuring the apparent luminosity $\mathcal{L}_{\rm app}$, the luminosity distance can be obtained at a given redshift. The observed $D_{\rm L}(z)$ can then be compared with various theoretical models and precious information on the underlying cosmological parameters can be obtained.

A.6 Horizon distances

In the discussion of the kinematical features of FRW models, a key rôle is played by the concept of *event horizon* and of *particle horizon*. The physical distance of the event horizon is defined as

$$d_{\rm e} = a(t) \int_t^{t_{\rm max}} \frac{dt'}{a(t')}. \tag{A.32}$$

The quantity defined in Eq. (A.32) measures the maximal distance over which we can admit, even in the future, a causal connection. If $d_{\rm e}(t)$ is finite in the limit $t_{\rm max} \to \infty$ (for finite t) we can conclude that the event horizon exists. In the opposite case, i.e. $d_{\rm e}(t) \to \infty$ for $t_{\rm max} \to \infty$ the event horizon does not exist.

The physical distance of the particle horizon is defined as

$$d_{\rm p}(t) = a(t) \int_{t_{\rm min}}^{t} \frac{dt'}{a(t')}. \tag{A.33}$$

Equation (A.33) measures the extension of the regions admitting a causal connection at time t. If the integral converges in the limit $t_{\rm min} \to 0$ we say that there exists a particle horizon. In connection with the concept of horizon in inflationary cosmology the Ref. [408] can be usefully consulted.

A.7 Few simple applications

In CMB studies it is often useful to compute the (comoving) angular diameter distance to decoupling or to equality (see Eq. (A.23) for a definition of the comoving angular diameter distance). As already discussed, the (comoving) angular diameter distance coincides with the distance measure. So suppose, for instance, we are interested in the model where $\Omega_{\rm M0} = 1$. In this case we have to compute

$$\overline{D}_A(z) = \frac{D_A(z)}{a(t)} = \frac{a_0 r_e(z)}{a(1+z)} = r_e(z), \quad (A.34)$$

where the last equality follows from the definition of redshift and where $\overline{D}_A$ denotes the (comoving) angular diameter distance. If $\Omega_{M0} = 1$, the comoving angular diameter distance to decoupling and to equality is given, respectively, by:

$$\begin{aligned}\overline{D}_A(z_{\text{dec}}) &= \frac{2}{H_0}\left(1 - \frac{1}{\sqrt{1+z_{\text{dec}}}}\right) \simeq \frac{1.939}{H_0}, \\ \overline{D}_A(z_{\text{eq}}) &= \frac{2}{H_0}\left(1 - \frac{1}{\sqrt{1+z_{\text{eq}}}}\right) \simeq \frac{2}{H_0},\end{aligned} \quad (A.35)$$

where we assume that the Universe was always matter-dominated from equality onwards (which is not an extremely good approximation since radiation may modify the estimate a bit). The comoving angular diameter distance enters crucially in the determination of the multipole number on the last scattering sphere. Suppose indeed we are interested in the following questions:

- what is the multipole number corresponding to a wavelength comparable with the Hubble radius at decoupling (or at equality)?
- what is the angle subtended by such a wavelength?

Before answering these questions let us just remark that when the spatial curvature vanishes (and when $\Omega_{M0} + \Omega_{\Lambda 0} = 1$), the comoving angular diameter distance must be computed numerically but a useful approximate expression is

$$\overline{D}_A(z_{\text{dec}}) = \frac{2}{\Omega_{M0}^{0.4}} H_0^{-1}. \quad (A.36)$$

When the (comoving) wave-number k_{dec} is comparable with the Hubble radius, the following chain of equalities holds:

$$k_{\text{dec}} = \mathcal{H}_{\text{dec}} = a_{\text{dec}} H_{\text{dec}} = \sqrt{\Omega_{M0}} H_0 \sqrt{1+z_{\text{dec}}}, \quad (A.37)$$

where the second equality follows by assuming that the Hubble radius at decoupling is (predominantly) determined by the matter contribution, i.e.

$$H_{\text{dec}}^2 \simeq \frac{8\pi G}{3}\rho_M = H_0^2 \Omega_{M0}(1+z_{\text{dec}})^3. \quad (A.38)$$

The corresponding multipole number on the last scattering sphere will then be given by

$$\ell_{\text{dec}} = k_{\text{dec}} \overline{D}_A(z_{\text{dec}}) = 2\sqrt{1+z_{\text{dec}}}\,\Omega_{M0}^{0.1} \simeq 66.3\,\Omega_{M0}^{0.1}, \quad (A.39)$$

where $z_{\rm dec} \simeq 1100$. The angle subtended by $\pi/k_{\rm dec}$ will then be

$$\theta_{\rm dec} = \frac{180^0}{\ell_{\rm dec}} = 2.7^0\, \Omega_{\rm M0}^{-0.1}. \qquad (A.40)$$

Following analog steps, it is possible to show that

$$k_{\rm eq} = \frac{h_0^2 \Omega_{\rm M0}}{14\ {\rm Mpc}}, \qquad \ell_{\rm eq} = k_{\rm eq} \overline{D}_{\rm A}(z_{\rm eq}) = 430\, h_0 \Omega_{\rm M0}^{0.6}. \qquad (A.41)$$

Equation (A.38) holds under the assumption that the Hubble radius at decoupling is predominantly given by the matter contribution. Equation (A.41) holds under the same assumption, namely, that at equality the main contribution to ρ_t comes from the dusty matter. This is reasonable within the ΛCDM paradigm since the dark energy will become dominant much later, i.e. at a much smaller redshift. So, effectively, for the above estimates it is as if $\Omega_{\rm M0} = \Omega_{\rm t0}$.

Appendix B

Kinetic Description of Hot Plasmas

B.1 Generalities on thermodynamic systems

In thermodynamics we distinguish, usually, intensive variables (like the pressure and the temperature) which do not depend upon the total matter content of the system and extensive variables (like internal energy, volume, entropy, number of particles). The first principle of thermodynamics tells us that

$$d\mathcal{E} = TdS - pdV + \mu dN, \tag{B.1}$$

where S is the entropy, p is the pressure, V is the volume, μ is the chemical potential, N the number of particles and $\mathcal{E}$ the internal energy. From Eq. (B.1) it can be easily deduced that

$$T = \left(\frac{\partial \mathcal{E}}{\partial S}\right)_{V,N}, \quad p = -\left(\frac{\partial \mathcal{E}}{\partial V}\right)_{S,N}, \quad \mu = \left(\frac{\partial \mathcal{E}}{\partial N}\right)_{V,S}, \tag{B.2}$$

where the subscripts indicate that each partial derivation is done by holding fixed the remaining two variables. Suppose now that the system is described only by an appropriate set of extensive variables. In this situation we can think that, for instance, the internal energy is a function of the remaining extensive variables, i.e. $\mathcal{E} = \mathcal{E}(S, V, N)$. Let us then perform a scale transformation of all the variables, i.e.

$$\mathcal{E} \to \sigma \mathcal{E}, \quad S \to \sigma S, \quad V \to \sigma V, \quad N \to \sigma N. \tag{B.3}$$

We consequently have $\sigma \mathcal{E} = \mathcal{E}(\sigma S, , \sigma V, \sigma N)$. By taking the derivative of the latter relation with respect to σ and then by fixing $\sigma = 1$ we get the following relation:

$$\mathcal{E} = \left(\frac{\partial \mathcal{E}}{\partial S}\right)_{V,N} S + \left(\frac{\partial \mathcal{E}}{\partial V}\right)_{S,N} V + \left(\frac{\partial \mathcal{E}}{\partial N}\right)_{V,S} N. \tag{B.4}$$

Using now Eqs. (B.2) in Eq. (B.4) we get the following important relation sometimes called *fundamental thermodynamic identity*:
$$\mathcal{E} = TS - pV + \mu N. \tag{B.5}$$
If the system is formed by different particle species, a chemical potential for each species is introduced and, consequently, $\mu N = \sum_i \mu_i N_i$. Equation (B.5) tells us that if the chemical potential vanishes (as in the case of a gas of photons) the entropy will be simply given by $S = (\mathcal{E} + pV)/T$. In statistical mechanics it is sometimes useful to introduce different *potentials* such as the free energy F, the Gibbs free energy G and the so-called thermodynamic potential Ω:
$$F = \mathcal{E} - TS, \qquad G = F + pV, \qquad \Omega = F - \mu N. \tag{B.6}$$
The free energies F and G or the thermodynamic potential Ω allow us to reduce the number of extensive variables employed for the description of a given system in favour of one or more intensive variables.[d] So the description provided via the potentials is always semi-extensive in the sense that it includes always one or more intensive variables. Notice, for instance, that $\Omega(T, V, \mu)$ and that, using Eqs. (B.6) and (B.1)
$$d\Omega = -pdV - SdT - Nd\mu. \tag{B.7}$$
Equation (B.7) implies that $\Omega = \Omega(V, T, \mu)$ and
$$S = -\left(\frac{\partial \Omega}{\partial T}\right)_{V,\mu}, \qquad p = -\left(\frac{\partial \Omega}{\partial V}\right)_{T,\mu}, \qquad N = -\left(\frac{\partial \Omega}{\partial \mu}\right)_{T,V}. \tag{B.8}$$
In the case of a gas of photons $\mu = 0$, and $\Omega = F$. This implies, using Eq. (B.6) and (B.1) that $dF = (\mu dN - pdV - SdT)$. Hence, the condition of equilibrium of a photon gas is given by
$$\mu = \left(\frac{\partial F}{\partial N}\right)_{V,T} = 0. \tag{B.9}$$
For a boson gas and for a fermion gas we have Ω that can be written, respectively, as
$$\Omega^{B} = \sum_{\vec{k}} \Omega^{B}_{\vec{k}}, \qquad \Omega^{F} = \sum_{\vec{k}} \Omega^{F}_{\vec{k}}, \tag{B.10}$$
$$\Omega^{B}_{\vec{k}} = T \sum_{\vec{k}} \ln\left[1 - e^{\frac{\mu - E_k}{T}}\right], \tag{B.11}$$
$$\Omega^{F}_{\vec{k}} = -T \sum_{\vec{k}} \ln\left[1 + e^{\frac{\mu - E_k}{T}}\right]. \tag{B.12}$$

---

[d]To avoid ambiguities in the notation we did not mention the enthalpy, customarily defined as $H = \mathcal{E} + pV$ (this nomenclature may clash with the notation employed for the Hubble parameter H). Note, however, that the enthalpy density is exactly what appears in the second of the Friedmann-Lemaître equations (see Eq. (2.57)).

Recalling that the Bose-Einstein and Fermi-Dirac occupation numbers are defined as
$$N^{\rm B} = \sum_{\vec{k}} \overline{n}_k^{\rm B}, \qquad N^{\rm F} = \sum_{\vec{k}} \overline{n}_k^{\rm F}, \tag{B.13}$$
the third relation reported in Eq. (B.8) allows us to determine $\overline{n}_k^{\rm B}$ and $\overline{n}_k^{\rm F}$:
$$\overline{n}_k^{\rm B} = -\left(\frac{\partial \Omega_k^{\rm B}}{\partial \mu}\right)_{T,V} = \frac{1}{e^{(E_k-\mu)/T} - 1}, \tag{B.14}$$
$$\overline{n}_k^{\rm F} = -\left(\frac{\partial \Omega_k^{\rm F}}{\partial \mu}\right)_{T,V} = \frac{1}{e^{(E_k-\mu)/T} + 1}. \tag{B.15}$$

B.2 Fermions and bosons

To determine the concentration, the energy density, the pressure and the entropy we can now follow two complementary procedures. For instance, the entropy and the pressure can be deduced from Eq. (B.8). Then Eq. (B.5) allows us to determine the internal energy $\mathcal{E}$. It is also possible to write the energy density, the pressure and the concentration in terms of the occupation numbers:
$$n^{\rm B/F} = \frac{g}{(2\pi)^3} \int d^3k \; \overline{n}_k^{\rm B/F}, \tag{B.16}$$
$$\rho^{\rm B/F} = \frac{g}{(2\pi)^3} \int d^3k \; E_k \, \overline{n}_k^{\rm B/F}, \tag{B.17}$$
$$p^{\rm B/F} = \frac{g}{(2\pi)^3} \int d^3k \; \frac{|\vec{k}|^2}{3E_k} \, \overline{n}_k^{\rm B/F}, \tag{B.18}$$
where g denotes the effective number of relativistic degrees of freedom and where the superscripts indicate that each relation holds independently for the Bose-Einstein or Fermi-Dirac occupation number. Then, Eq. (B.5) can be used to determine the entropy or the entropy density.

Consider, for instance, the ultra-relativistic case when the temperature is much larger than both the masses and the chemical potential:
$$T \gg m, \qquad T \gg |\mu|, \qquad E_k = \sqrt{k^2 + m^2} \simeq k. \tag{B.19}$$
In this case we will have
$$n^{\rm B} = \frac{\zeta(3)}{\pi^2} g T^3, \qquad n^{\rm F} = \frac{3}{4} n^{\rm B}, \tag{B.20}$$
$$\rho^{\rm B} = \frac{\pi^2}{30} g T^4, \qquad \rho^{\rm F} = \frac{7}{8} \rho^{\rm B}, \tag{B.21}$$
$$p^{\rm B} = \frac{\pi^2}{90} g T^4, \qquad p^{\rm F} = \frac{7}{8} p^{\rm B}, \tag{B.22}$$

From Eq. (B.5) the entropy density will then be quickly determined as

$$s^{\text{B}} = \frac{2\pi^2}{45} g T^3, \qquad s^{\text{F}} = \frac{7}{8} s^{\text{B}}. \tag{B.23}$$

To perform the integrations implied by the above results it is useful to recall that

$$\mathcal{I}_{\text{B}} = \int_0^\infty \frac{x^3 dx}{e^x - 1} = \frac{\pi^4}{15}, \qquad \mathcal{I}_{\text{F}} = \int_0^\infty \frac{x^3 dx}{e^x + 1} = \frac{7}{8} \mathcal{I}_{\text{B}} = \frac{7\pi^4}{120}. \tag{B.24}$$

Notice that the value of $\mathcal{I}_{\text{F}}$ can quickly be determined once the value of $\mathcal{I}_{\text{B}}$ is known. In fact

$$\mathcal{I}_{\text{F}} - \mathcal{I}_{\text{B}} = -2 \int_0^\infty \frac{x^3 dx}{e^{2x} - 1} = -\frac{1}{8} \mathcal{I}_{\text{B}}, \tag{B.25}$$

where the second equality follows by changing the integration variable from x to $y = 2x$. Furthermore, the following pair of integrals is useful to obtain the explicit expressions for the concentrations:

$$\int_0^\infty dx \frac{x^{s-1}}{e^{ax} - 1} = \frac{1}{a^s} \Gamma(s) \zeta(s), \qquad \int_0^\infty dx \frac{x^{s-1}}{e^{ax} + 1} = \frac{1}{a^s} \Gamma(s)(1 - 2^{1-s}) \zeta(s), \tag{B.26}$$

where $\Gamma(s)$ is the Euler Gamma function and $\zeta(s)$ is the Riemann ζ function (recall $\zeta(3) = 1.20206$).

As an example of the first procedure described consider the calculation of the entropy density of a boson gas directly from the first relation of Eq. (B.8):

$$S^{\text{B}} = \frac{gV}{2\pi^2} \left\{ \mathcal{I}_{\text{B}} - \int_0^\infty x^2 dx \ln[1 - e^{-x}] \right\} = \frac{2gV}{45\pi^2} T^3, \tag{B.27}$$

where the second integral can be evaluated by parts and where the sum has been transformed into an integral according to

$$\sum_{\vec{k}} \to \frac{g}{(2\pi)^3} V \int d^3k. \tag{B.28}$$

The result of Eq. (B.27) clearly coincides with the one of Eq. (B.23), recalling that, by definition, $s^{\text{B}} = S^{\text{B}}/V$. In similar terms, the wanted thermodynamic variables can be obtained directly from the thermodynamic potentials.

In the ultra-relativistic limit the boson and fermion gases have a radiative equation of state, i.e.

$$p^{\text{B}} = \frac{\rho^{\text{B}}}{3}, \qquad p^{\text{F}} = \frac{\rho^{\text{F}}}{3}. \tag{B.29}$$

In the nonrelativistic limit the equation of state is instead $p = 0$ for both bosons and fermions. In fact, in the nonrelativistic limit,

$$|m - \mu| > T, \qquad E_k = \sqrt{k^2 + m^2} \simeq m + \frac{k^2}{2m}. \qquad (B.30)$$

Then, in this limit

$$\overline{n}_k^{B/F} = \overline{n}_k = e^{(\mu - E_k)/T}. \qquad (B.31)$$

Then, from the definitions (B.16), (B.17) and (B.18) it can be easily obtained, after Gaussian integration

$$n = \frac{g}{(2\pi)^{3/2}} (mT)^{3/2} e^{(\mu-m)/T},$$

$$\rho = mn, \qquad p = nT = \rho\left(\frac{T}{m}\right) \ll \rho, \qquad (B.32)$$

which shows that, indeed $p = 0$ in the nonrelativistic limit. Note that to derive Eq. (B.32) the well known result is

$$\mathcal{I}(\alpha) = \int_0^\infty dk\, e^{-\alpha k^2} = \frac{1}{2}\sqrt{\frac{\pi}{\alpha}},$$

$$\int_0^\infty dk\, k^2 e^{-\alpha k^2} = -\frac{\partial}{\partial \alpha} \mathcal{I}(\alpha) = \frac{\sqrt{\pi}}{4} \alpha^{-3/2}. \qquad (B.33)$$

The energy density in the ultra-relativistic limit (i.e. Eq. (B.21)) goes as T^4. The energy density in the nonrelativistic limit (i.e. Eq. (B.32)) is exponentially suppressed as $e^{-m/T}$. Therefore, as soon as the temperature drops below the threshold of pair production, the energy density and the concentration are exponentially suppressed. This is the result of particle-antiparticle annihilations. At very high temperatures $T \gg m$ particles annihilate with anti-particles but the energy-momentum supply of the thermal bath balances the annihilations with the production of particles-antiparticles pairs. At lower temperatures (i.e. $T < m$) the thermal energy of the particles is not sufficient for a copious production of pairs.

B.3 Thermal, kinetic and chemical equilibrium

Let us now suppose that the primordial plasma is formed by a mixture of N_b bosons and N_f fermions. Suppose also that the ultra-relativistic limit holds so that the masses and the chemical potentials of the different species can be safely neglected. Suppose finally that, in general, each bosonic species carries g_b degrees of freedom at a temperature and that each fermionic species carries g_f degrees of freedom. Each of the bosonic degrees of freedom will be in thermal equilibrium at a temperature T_b and; similarly each of the fermionic degrees of freedom will be in thermal equilibrium at a temperature T_f. Under the aforementioned assumptions, Eqs. (B.21) and (B.23) imply that the total energy density and the total entropy density of the system are given by

$$\rho(T) = g_\rho(T) \frac{\pi^2}{30} T^4, \qquad s(T) = g_s(T) \frac{2\pi^2}{45} T^3, \qquad (B.34)$$

where

$$g_\rho(T) = \sum_{b=1}^{N_b} g_b \left(\frac{T_b}{T}\right)^4 + \frac{7}{8} \sum_{f=1}^{N_f} g_f \left(\frac{T_f}{T}\right)^4, \qquad (B.35)$$

$$g_s(T) = \sum_{b=1}^{N_b} g_b \left(\frac{T_b}{T}\right)^3 + \frac{7}{8} \sum_{f=1}^{N_f} g_f \left(\frac{T_f}{T}\right)^3, \qquad (B.36)$$

Equations (B.35) and (B.36) are clearly different. If all the fermionic and bosonic species are in thermal equilibrium at a common temperature T, then

$$T_b = T_f = T, \qquad g_\rho = g_s. \qquad (B.37)$$

However, if at least one of the various species has a different temperature, then $g_\rho \neq g_s$. In more general terms we can say that:

- if all the species are in equilibrium at a common temperature T, then the system is in thermodynamic equlibrium;
- if some species are in equilibrium at a temperature different from T, then the system is said to be in kinetic equilibrium.

There is a third important notion of equilibrium, i.e. the chemical equilibrium. Consider, indeed, the situation where $2H + 0 \to H_2 0$. In chemical equilibrium the latter reaction is balanced by $H_2 0 \to 2H + 0$. We can attribute a coefficient to each of the reactants of a chemical process. For instance, in the aforementioned naive example we will have $\alpha_H = 1$, $\alpha_O = 1$, $\alpha_{H_2O} = -1$ satisfying the sum rule $\sum_R \alpha_R R = 0$ where R denotes each of the reactants. Such a sum rule simply means that the disappearance of a water molecule implies the appearance of two atoms of hydrogen and one atom of oxygen and vice versa. This concept of chemical equilibrium can be generalized to more general reactions, like $e^+ + e^- \to \gamma$ or $e + p \to H + \gamma$ and so on. By always bearing in mind the chemical analogy, let us suppose we conduct a chemical reaction at fixed temperature and fixed pressure (as it is sometimes the case for practical applications). Then the Gibbs free energy is the appropriate quantity to use since

$$dG = \sum_R \mu_R dN_R - SdT + Vdp. \tag{B.38}$$

If $dT = 0$ and $dp = 0$ the condition of chemical equilibrium is expressed by $dG = 0$ which can be expressed as

$$\sum_R \mu_R \left(\frac{\partial N_R}{\partial \lambda}\right) d\lambda \equiv \sum_R \alpha_R \mu_R = 0, \tag{B.39}$$

where the second equality follows from the first one since dN_R are not independent and are all connected by the fact that dN_R/α_R must have the same value $d\lambda$ for all the reactants. Thus, in the case of $e^+ + e^- \to \gamma$ (and vice versa) the condition of chemical equilibrium implies $\mu_{e^+} + \mu_{e^-} = \mu_\gamma$, i.e. $\mu_{e^+} = -\mu_{e^-}$ since $\mu_\gamma = 0$. Similarly, for the reactions $p + e \to H + \gamma$ (hydrogen formation) and $\gamma + H \to p + e$ (hydrogen photo-dissociation) the condition of chemical equilibrium implies that $\mu_e + \mu_p = \mu_H$.

B.4 An example of primordial plasma

The considerations presented in this Appendix will now be applied to a few specific examples that are useful in the treatment of the hot Universe. Consider first the case when the primordial soup is formed by all the degrees of freedom of the Glashow-Weinberg-Salam (GWS) model $SU_L(2) \otimes U_Y(1) \otimes SU_c(3)$. Suppose that the plasma is at a temperature T larger than 175 GeV, i.e. a temperature larger than the top mass which is the most massive species of the model (the Higgs mass, still unknown, such that $m_H > 115$ GeV). In this situation all the species of the GWS model are in thermodynamic equilibrium at the temperature T. In this situation g_ρ and g_s are simply given by

$$g_\rho = g_s = \sum_b g_b + \frac{7}{8} \sum_f g_f, \tag{B.40}$$

where the sum now extends over all the fermionic and bosonic species of the GWS model. In the GWS the quarks come in six flavours

$$(m_t, m_b, m_c, m_s, m_u, m_d) = (175, 4, 1, 0.1, 5 \times 10^{-3}, 1.5 \times 10^{-3}) \text{GeV} \tag{B.41}$$

where the quark masses have been listed in GeV. The lepton masses are, roughly,

$$(m_\tau, m_\mu, m_e) = (1.7, 0.105, 0.0005) \text{ GeV}. \tag{B.42}$$

Finally, the $W^\pm$ and Z^0 are, respectively, 80.42 and 91.18 GeV.

Let us then compute, separately, the bosonic and the fermionic contributions. In the GWS model the fermionic species are constituted by the six quarks, by the three massive leptons and by the neutrinos (that we will take as massless). For the quarks the number of relativistic degrees of freedom is given by $(6 \times 2 \times 2 \times 3) = 72$, i.e. 6 particles times a factor 2 (for the corresponding antiparticles) times another factor 2 (the spin) times a factor 3 (since each quark may come in three different colors). Leptons do not carry color so the effective number of relativistic degrees of freedom of e, μ and τ (and of the corresponding neutrinos) is 18. Globally, the fermions carry 90 degrees of freedom. There are eight massless gluons each of them with two physical polarizations amounting to $8 \times 2 = 16$ bosonic degrees of freedom. The $SU_L(2)$ (massless) gauge bosons and the $U_Y(1)$ (massless) gauge boson lead to $3 \times 2 + 2 = 8$ bosonic degrees of freedom. Finally, the Higgs field (an $SU_L(2)$ complex doublet) carries 4 degrees of freedom. Globally, the bosons carry 28 degrees of freedom. Therefore, we will have, overall in the GWS model

$$g_\rho = g_s = 28 + \frac{7}{8} \times 90 = 106.75. \tag{B.43}$$

Notice as a side remark that the counting given above assumes indeed that the electroweak symmetry is restored (as it is probably the case for $T > 175$ GeV). When the electroweak symmetry is broken down to the $U_{\text{em}}(1)$, the vector bosons acquire a mass and the Higgs field loses three of its degrees of freedom. It is therefore easy to understand that the gauge bosons of the electroweak sector lead to the same number of degrees of freedom obtained in the symmetric phase: the three massive gauge bosons will carry 3×3 degrees of freedom, the photons 2 degrees of freedom and the Higgs field (a real scalar, after symmetry breaking) 1 degree of freedom.

When the temperature drops below 170 GeV, the top quarks start annihilating and $g_\rho \to 96.25$. When the temperature drops below 80 GeV the gauge bosons annihilate and $g_\rho \to 86.25$. When the temperature drops below 4 GeV, the bottom quarks start annihilating and $g_\rho \to 75.75$ and so on. While the electroweak phase transition takes place around 100 GeV the quark-hadron phase transition takes place around 150 MeV. At the quark-hadron phase transition the quarks are not free anymore and start combining forming colorless hadrons. In particular, bound states of three quarks are called baryons while bound states of quark-antiquark pairs are mesons. The massless degrees of freedom around the quark-hadron phase transition are π^0, $\pi^\pm$, $e^\pm$, $\mu^\pm$ and the neutrinos. This implies that $g_\rho = 17.25$. When the temperature drops even more (i.e. $T < 20\,\text{MeV}$) muons and also pions annihilate and for $T \sim \mathcal{O}(\text{MeV})$, $g_\rho = 10.75$.

It is relevant to notice that the information on the temperature and on the number of relativistic degrees of freedom can be usefully converted into information on the Hubble expansion rate and on the Hubble radius at each corresponding epoch. In fact, using Eq. (2.56) the temperature of the Universe determines directly the Hubble rate and, consequently, the Hubble radius. According to Eq. (2.56) and taking into account Eq. (B.34) we will have,

$$H = 1.66\sqrt{g_\rho}\frac{T^2}{M_{\rm P}}, \qquad (B.44)$$

or in Planck units:

$$\left(\frac{H}{M_{\rm P}}\right) = 1.15 \times 10^{-37} \left(\frac{g_\rho}{106.75}\right)^{1/2} \left(\frac{T}{\text{GeV}}\right)^2. \qquad (B.45)$$

Equation (B.45) measures the curvature scale at each cosmological epoch. For instance, at the time of the electroweak phase transition, $H \simeq 10^{-34} M_{\rm P}$ while for $T \sim \text{MeV}$, $H \simeq 10^{-44}\,M_{\rm P}$. To get an idea of the size of the Hubble radius we we can express H^{-1} in centimeters:

$$r_H = 1.4 \left(\frac{106.75}{g_\rho}\right)^{1/2} \left(\frac{100\,\text{GeV}}{T}\right)^2 \text{ cm}, \qquad (B.46)$$

which shows that the Hubble radius is of the order of the centimeter at the electroweak epoch while it is of the order of 10^4 Mpc at the present epoch. Finally, recalling that during the radiation phase $H^{-1} = 2t$,

$$t_H = 23 \left(\frac{106.75}{g_\rho}\right)^{1/2} \left(\frac{100\,\text{GeV}}{T}\right)^2 \text{ psec}, \qquad (B.47)$$

which shows that the Hubble time is of the order of 20 psec at the electroweak time while it is $t \simeq 0.73\,\text{s}$ right before electron positron annihilation and for $T \simeq \text{MeV}$.

B.5 Electron-positron annihilation and neutrino decoupling

As soon as the Universe is one second old two important phenomena take place: the annihilation of electrons and positrons and the decoupling of neutrinos. For sufficiently high temperatures the weak interactions are in equilibrium and the reactions

$$e^- + p \to n + \nu_e, \qquad e^+ + n \to p + \bar{\nu}_e, \qquad (B.48)$$

are balanced by their inverse. However, at some point, the rate of the weak interactions equals the Hubble expansion rate and, eventually, it becomes smaller than H. Recalling that, roughly,

$$\Gamma_{\text{weak}} \simeq \sigma_{\text{F}} T^3, \qquad \sigma_{\text{F}} = G_{\text{F}}^2 T^2, \qquad G_{\text{F}} = 1.16 \times 10^{-5} \text{ GeV}^{-2}, \qquad (\text{B.49})$$

we immediately show, neglecting numerical factors, that

$$\frac{\Gamma_{\text{weak}}}{H} \simeq \left(\frac{T}{\text{MeV}}\right)^3, \qquad H \simeq \frac{T^2}{M_{\text{P}}}. \qquad (\text{B.50})$$

Thus, as soon as T drops below a temperature of the order of the MeV the weak interactions are not in equilibrium anymore. This temperature scale is also of upmost importance for the formation of the light nuclei since, below the MeV, the neutron to proton ratio is depleted via free neutron decay.

It is important to appreciate, at this point, that the neutrino distribution is preserved by the expansion and, therefore, we may still assume that $a(t)T$ is constant. However, around the MeV the electrons and positrons start annihilating. This occurrence entails the sudden heating of the photons since the annihilations of electrons and positrons end up in photons. The net result of this observation is an increase of the temperature of the photons with respect to the kinetic temperature of the neutrinos. In equivalent terms, after electron-positron annihilation, the temperature of the neutrinos will be systematically smaller than the temperature of the photons. To describe this phenomenon let us consider an initial time t_i before e^+-e^- annihilation and a final time t_f after e^+-e^- annihilation. Using the conservation of the entropy density we can say that

$$g_s(t_i) \, a^3(t_i) \, T_\gamma^3(t_i) = g_s(t_f) \, a^3(t_f) \, T_\gamma^3(t_f), \qquad (\text{B.51})$$

$$g_s(t_i) = 10.75, \qquad g_s(t_f) = 2 + 5.25 \left[\frac{T_\nu(t_f)}{T_\gamma(t_f)}\right]^3, \qquad (\text{B.52})$$

where it has been taken into account that while before e^+-e^- annihilation the temperature of the photons coincides with the temperature of the neutrinos, it may not be the case after e^+-e^- annihilation. Using the same observation, we can also say that

$$a^3(t_i) \, T_\nu^3(t_i) = a^3(t_f) \, T_\nu^3(t_f). \qquad (\text{B.53})$$

Dividing then Eq. (B.52) by Eq. (B.53) the anticipated result is that

$$T_\gamma(t_f) = \left(\frac{11}{4}\right)^{1/3} T_\nu(t_f), \qquad (\text{B.54})$$

i.e. the temperature of the photons gets larger than the kinetic temperature of the neutrinos. This result also implies that the energy density of a massless neutrino background would be today

$$\rho_{\nu 0} = \frac{21}{8}\left(\frac{4}{11}\right)^{4/3}\rho_{\gamma 0}, \qquad h_0^2 \Omega_{\nu 0} = 1.68 \times 10^{-5}. \tag{B.55}$$

It is worth noticing that, according to the considerations related to the phenomenon of neutrino decoupling, the effective number of relativistic degrees of freedom around the eV is given by

$$g_\rho = 2 + \frac{7}{8} \times 6 \times \left(\frac{4}{11}\right)^{4/3} = 3.36, \tag{B.56}$$

where two refers to the photon and the neutrinos contribute weighted by their kinetic temperature.

B.6 Big-bang nucleosynthesis (BBN)

In this subsection the salient aspects of BBN will be summarized. BBN is the process where light nuclei are formed. Approximately one quarter of the baryonic matter in the Universe is in the form of ^{4}He. The remaining part is made, predominantly, by hydrogen in its different incarnations (atomic, molecular and ionized). During BBN the protons and neutrons (formed during the quark-hadron phase transition) combine to form nuclei. According to Eq. (B.47) the quark-hadron phase transition takes place when the Universe is 20 μsec old to be compared with $t_{\rm BBN} \simeq$ sec. During BBN only light nuclei are formed and, more specifically, ^{4}He, ^{3}He, D, ^{7}Li. The reason why the ^{4}He is the most abundant element has to do with the fact that ^{4}He has the largest binding energy for nuclei with atomic number $A < 12$ (corresponding to carbon). Light nuclei provide stars with the initial set of reactions necessary to turn on the synthesis of heavier elements (iron, cobalt and so on).

In short, the logic of BBN is the following:

- after the quark-hadron phase transition the antinucleons annihilate with the nucleons, thus the total baryon concentration will be given by $n_{\rm B} = n_{\rm n} + n_{\rm p}$;
- if the baryon number is conserved (as it is rather plausible) the baryon concentration will stay constant and, in particular, it will be $n_{\rm B} \simeq 10^{-10} n_\gamma$;

- for temperatures lower than $T \simeq \mathcal{O}(\text{MeV})$ weak interactions fall out of equilibrium; at this stage the concentrations of neutrons and protons are determined from their equilibrium values and, approximately,

$$\frac{n_n}{n_p} = e^{-\frac{Q}{T}} \simeq \frac{1}{6}, \qquad Q = m_n - m_p; \qquad (B.57)$$

for $T \simeq 0.73$ MeV;
- if nothing else would happen, the neutron concentration would be progressively depleted by free neutron decay, i.e. $n \to p + e^- + \bar{\nu}_e$;
- however, when $T \simeq 0.1$ MeV the reactions for the formation of the deuterium (D) are in equilibrium, i.e. $p + n \to D + \gamma$ and $D + \gamma \to p + n$;
- the reactions for the formation of deuterium fall out of thermal equilibrium only at a much lower temperature (i.e. $T_D \simeq 0.06$ MeV);
- as soon as deuterium is formed, ^{4}He and ^{3}He can arise according to the following chain of reactions:

$$D + n \to T + \gamma, \qquad T + p \to\,^4\text{He} + \gamma, \qquad (B.58)$$
$$D + p \to\,^3\text{He} + \gamma, \qquad ^3\text{He} + n \to\,^4\text{He} + \gamma; \qquad (B.59)$$

As soon as the helium is formed a little miracle happens: since the temperature of equilibration of the helium is rather large (i.e. $T \simeq 0.3$ MeV), the helium is not in equilibrium at the moment when it is formed. In fact the helium can only be formed when deuterium is already present. As a consequence the reactions (B.58) and (B.59) only take place from right to left. When the deuterium starts being formed, free neutron decay has already depleted by a bit the neutron to proton ratio which is equal to 1/7. Since each ^{4}He has two neutrons, we have $n_n/2$ nuclei of ^{4}He that can be formed per unit volume. Therefore, the ^{4}He mass fraction will be

$$Y_p = \frac{4(n_n/2)}{n_n + n_p} \simeq \frac{2(1/7)}{1 + 1/7} \simeq 0.25. \qquad (B.60)$$

The abundances of the other light elements are comparatively smaller than the one of ^{4}He and, in particular:

$$D/H \simeq 10^{-5}, \qquad ^3\text{He}/H \simeq 10^{-3}, \qquad ^7\text{Li}/H \simeq 10^{-10}. \qquad (B.61)$$

The abundances of the light elements computed from BBN calculations agree with the observations. The simplest BBN scenario implies that the only two free parameters are the temperature and what has been called η_b

i.e. the baryon to photon ratio which must be of the order of 10^{-10} to agree with experimental data. BBN represents, therefore, one of the successes of the Standard Cosmological Model.

Appendix C

Scalar Modes of the Geometry

In this Appendix we are going to derive the results that are the starting point for the study of the evolution of the scalar modes both around equality and during the primeval inflationary phase. As exemplified in section 10 it will always be possible to pass from a gauge-dependent description to a fully gauge-invariant set of evolution equations. In this Appendix the longitudinal gauge will be consistently followed.

C.1 Fluctuations of the Einstein tensor

The scalar fluctuations of the geometry with covariant and controvariant indices can be written, to first order and in the longitudinal gauge, as

$$\delta_s g_{00} = 2a^2 \phi, \qquad \delta_s g^{00} = -\frac{2}{a^2}\phi,$$
$$\delta_s g_{ij} = 2a^2 \psi \delta_{ij}, \qquad \delta_s g^{ij} = -\frac{2}{a^2}\psi \delta^{ij}, \qquad \text{(C.1)}$$

where the notations of section 6 have been followed. Since the fluctuations of the Christoffel connections can be expressed as

$$\delta_s \Gamma^\mu_{\alpha\beta} = \frac{1}{2}\overline{g}^{\mu\nu}(-\partial_\nu \delta_s g_{\alpha\beta} + \partial_\beta \delta_s g_{\nu\alpha} + \partial_\alpha \delta_s g_{\beta\nu})$$
$$+ \frac{1}{2}\delta_s g^{\mu\nu}(-\partial_\nu \overline{g}_{\alpha\beta} + \partial_\beta \overline{g}_{\nu\alpha} + \partial_\alpha \overline{g}_{\beta\nu}), \qquad \text{(C.2)}$$

where $\overline{g}_{\mu\nu} = a^2(\tau)\eta_{\mu\nu}$ denotes the background metric and $\delta_s g_{\mu\nu}$ the first-order fluctuations in the longitudinal gauge which are given, in explicit terms, by Eq. (C.1). Inserting Eq. (C.1) into Eq. (C.2) the explicit form

of the fluctuations of the Christoffel connections can be obtained:
$$\delta_s \Gamma^0_{00} = \phi',$$
$$\delta_s \Gamma^0_{ij} = -[\psi' + 2\mathcal{H}(\phi + \psi)]\delta_{ij}$$
$$\delta_s \Gamma^0_{i0} = \delta_s \Gamma^0_{0i} = \partial_i \phi,$$
$$\delta_s \Gamma^i_{00} = \partial^i \phi, \qquad (C.3)$$
$$\delta_s \Gamma^k_{ij} = (\partial^k \psi \delta_{ij} - \partial_i \psi \delta^k_j - \partial_j \psi \delta^k_i),$$
$$\delta_s \Gamma^j_{0i} = -\psi' \delta^j_i.$$

We remark, incidentally, that the fluctuations of the Christoffel connections in an inhomogeneous Minkowski metric (used in section 7 for the derivation of the scalar Sachs-Wolfe effect) are simply obtained from Eqs. (C.3) by setting $\mathcal{H} = 0$.

The fluctuations of the Ricci tensor can be now expressed, as
$$\delta_s R_{\mu\nu} = \partial_\alpha \delta_s \Gamma^\alpha_{\mu\nu} - \partial_\nu \delta_s \Gamma^\beta_{\mu\beta} + \delta_s \Gamma^\alpha_{\mu\nu} \overline{\Gamma}^\beta_{\alpha\beta}$$
$$+ \overline{\Gamma}^\alpha_{\mu\nu} \delta_s \Gamma^\beta_{\alpha\beta} - \delta_s \Gamma^\beta_{\alpha\mu} \overline{\Gamma}^\alpha_{\beta\nu} - \overline{\Gamma}^\beta_{\alpha\mu} \delta_s \Gamma^\alpha_{\beta\nu}. \qquad (C.4)$$

where, as usual, $\overline{\Gamma}^\mu_{\alpha\beta}$ are the background values of the Christoffel connections, i.e., as already obtained:
$$\overline{\Gamma}^0_{00} = \mathcal{H}, \quad \overline{\Gamma}^0_{ij} = \mathcal{H}\delta_{ij}, \quad \overline{\Gamma}^j_{0i} = \mathcal{H}\delta^j_i. \qquad (C.5)$$

Using Eqs. (C.3) in Eq. (C.4) and taking into account Eq. (C.5) the explicit form of the components of the (perturbed) Ricci tensors can be easily obtained and they are:
$$\delta_s R_{00} = 3[\psi'' + \mathcal{H}(\phi' + \psi')],$$
$$\delta_s R_{0i} = 2\partial_i(\psi' + \mathcal{H}\phi),$$
$$\delta_s R_{ij} = -\delta_{ij}\{[\psi'' + 2(\mathcal{H}' + 2\mathcal{H}^2)(\psi + \phi) + \mathcal{H}(\phi' + 5\psi') - \nabla^2\psi]\}$$
$$+ \partial_i\partial_j(\psi - \phi) \qquad (C.6)$$

The fluctuations of the Ricci tensor with mixed (i.e. one covariant the other controvariant) indices can be easily obtained since
$$\delta_s R^\beta_\alpha = \delta_s(g^{\alpha\mu} R_{\beta\mu}) = \delta_s g^{\alpha\mu} \overline{R}_{\beta\mu} + \overline{g}^{\alpha\mu} \delta_s R_{\beta\mu}, \qquad (C.7)$$

where $\overline{R}_{\alpha\beta}$ are the Ricci tensors evaluated on the background, i.e.
$$\overline{R}_{00} = -3\mathcal{H}', \quad \overline{R}^0_0 = -\frac{3}{a^2}\mathcal{H}',$$
$$\overline{R}_{ij} = (\mathcal{H}' + 2\mathcal{H}^2)\delta_{ij}, \quad \overline{R}^j_i = -\frac{1}{a^2}(\mathcal{H}' + 2\mathcal{H}^2)\delta^j_i, \qquad (C.8)$$
$$\overline{R} = -\frac{6}{a^2}(\mathcal{H}^2 + \mathcal{H}').$$

Using Eqs. (C.6) into Eq. (C.7) and recalling Eq. (C.8) we get

$$\delta_s R_0^0 = \frac{1}{a^2}\{\nabla^2\phi + 3[\psi'' + \mathcal{H}(\phi' + \psi') + 2\mathcal{H}'\phi]\},$$

$$\delta_s R_i^j = \frac{1}{a^2}[\psi'' + 2(\mathcal{H}' + 2\mathcal{H}^2)\phi + \mathcal{H}(\phi' + 5\psi') - \nabla^2\psi]\delta_i^j$$
$$- \frac{1}{a^2}\partial_i\partial^j(\psi - \phi), \qquad (C.9)$$

$$\delta_s R_i^0 = \frac{2}{a^2}\partial_i(\psi' + \mathcal{H}\phi),$$

$$\delta_s R_0^i = -\frac{2}{a^2}\partial^i(\psi' + \mathcal{H}\phi).$$

Finally the fluctuations of the components of the Einstein tensor with mixed indices are computed to be

$$\delta_s \mathcal{G}_0^0 = \frac{2}{a^2}\{\nabla^2\psi - 3\mathcal{H}(\psi' + \mathcal{H}\phi)\}, \qquad (C.10)$$

$$\delta_s \mathcal{G}_i^j = \frac{1}{a^2}\{[-2\psi'' - 2(\mathcal{H}^2 + 2\mathcal{H}')\phi - 2\mathcal{H}\phi' - 4\mathcal{H}\psi' - \nabla^2(\phi - \psi)]\delta_i^j$$
$$- \partial_i\partial^j(\psi - \phi)\}, \qquad (C.11)$$

$$\delta_s \mathcal{G}_i^0 = \delta_s R_i^0. \qquad (C.12)$$

Equations (C.10), (C.11) and (C.12) are extensively used in chapters 7 and 10.

C.2 Fluctuations of the energy-momentum tensor(s)

All along in this book two relevant energy-momentum tensors have been extensively used, namely the energy-momentum tensor of a minimally coupled scalar field and the energy-momentum tensor of a mixture of perfect fluids. For the applications discussed in chapters 7, 8, 9 and 10 it is relevant to derive the first-order form of the energy-momentum tensor. Needless to say that since the inverse metric appears in several places in the explicit form of the energy-momentum tensor(s), an explicit dependence upon the scalar fluctuations of the metric may enter the various (perturbed) components of T_μ^ν.

Let us start with the case of a fluid source. The energy-momentum tensor of a perfect fluid is

$$T_{\mu\nu} = (p + \rho)u_\mu u_\nu - p g_{\mu\nu}. \qquad (C.13)$$

By perturbing to first-order the normalization condition $g_{\mu\nu}u^\mu u^\nu = 1$ we have

$$\delta_s g^{\mu\nu}\overline{u}_\mu\overline{u}_\nu + \overline{g}^{\mu\nu}(\overline{u}_\mu\delta_s u_\nu + \delta_s u_\mu\overline{u}_\nu) = 0, \qquad (C.14)$$

implying, together with Eq. (C.1)

$$\bar{u}_0 = a, \qquad \delta_s u^0 = -\phi/a. \qquad (C.15)$$

It is important to stress that, on the background, the spatial component of $\bar{u}_\mu$, i.e. $\bar{u}_i$, vanish exactly. The contribution to the peculiar velocity arises instead to first-order since, in the longitudinal gauge, $\delta_s u_i \neq 0$. By taking the first-order fluctuation of Eq. (C.13) the result is

$$\delta_s T_{\mu\nu} = (\delta p + \delta\rho)\bar{u}_\mu \bar{u}_\nu + (p+\rho)(\bar{u}_\mu \delta_s u_\nu + \delta_s u_\mu \bar{u}_\nu) - \delta p \bar{g}_{\mu\nu} - p\delta_s g_{\mu\nu}. \qquad (C.16)$$

The perturbed components of energy-momentum tensor with mixed indices can of course be obtained from the expression

$$\delta_s T_\alpha^\beta = \delta_s(g^{\alpha\mu}T_{\beta\mu}) = \delta_s g^{\alpha\mu}\bar{T}_{\beta\mu} + \bar{g}^{\alpha\mu}\delta_s T_{\beta\mu} \qquad (C.17)$$

where $\bar{T}_{\mu\nu}$ denote the background components of the energy-momentum tensor, i.e.

$$\bar{T}_{00} = a^2\rho, \qquad \bar{T}_{ij} = a^2 p. \qquad (C.18)$$

Inserting Eqs. (C.1) into Eq. (C.16) and taking into account Eqs. (C.17) and (C.18) we obtain:

$$\delta_s T_0^0 = \delta\rho, \qquad \delta_s T^{00} = \frac{1}{a^2}(\delta\rho - 2\rho\phi), \qquad (C.19)$$

and

$$\delta_s T_i^j = -\delta p \delta_i^j, \qquad (C.20)$$

$$\delta_s T^{ij} = \frac{1}{a^2}[\delta p \delta^{ij} + 2p\psi\delta^{ij}], \qquad (C.21)$$

$$\delta_s T_0^i = (p+\rho)v^i, \qquad (C.22)$$

$$\delta_s T^{0i} = \frac{1}{a^2}(p+\rho)v^i. \qquad (C.23)$$

where we have chosen to define $\delta u^i = v^i/a$.

Let us now consider the fluctuations of the energy-momentum tensor of a scalar field φ characterized by a potential $V(\varphi)$:

$$T_{\mu\nu} = \partial_\mu\varphi\partial_\nu\varphi - g_{\mu\nu}\left[\frac{1}{2}g^{\alpha\beta}\partial_\alpha\varphi\partial_\beta\varphi - V(\varphi)\right]. \qquad (C.24)$$

Denoting with χ the first-order fluctuation of the scalar field φ we will have

$$\delta_s T_{\mu\nu} = \partial_\mu\chi\partial_\nu\varphi + \partial_\mu\varphi\partial_\nu\chi - \delta_s g_{\mu\nu}\left[\frac{1}{2}g^{\alpha\beta}\partial_\alpha\varphi\partial_\beta\varphi - V\right]$$

$$- g_{\mu\nu}\left[\frac{1}{2}\delta_s g^{\alpha\beta}\partial_\alpha\varphi\partial_\beta\varphi + g^{\alpha\beta}\partial_\alpha\chi\partial_\beta\varphi - \frac{\partial V}{\partial\varphi}\chi\right]. \qquad (C.25)$$

Inserting Eqs. (C.1) into Eq. (C.25) we obtain, in explicit terms:
$$\delta_s T_{00} = \chi'\varphi' + 2a^2\phi V + a^2\frac{\partial V}{\partial \varphi}\chi,$$
$$\delta_s T_{0i} = \varphi'\partial_i\chi, \qquad (C.26)$$
$$\delta_s T_{ij} = \delta_{ij}\left[\varphi'\chi' - \frac{\partial V}{\partial \varphi}\chi a^2 - (\phi+\psi)\varphi'^2 + 2a^2 V\psi\right].$$

Recalling that
$$\overline{T}_{00} = \frac{\varphi'^2}{2} + a^2 V \equiv a^2\rho_\varphi,$$
$$\overline{T}_{ij} = \left[\frac{\varphi'^2}{2} - a^2 V\right]\delta_{ij} \equiv a^2 p_\varphi \delta_{ij}, \qquad (C.27)$$

the perturbed components of the scalar field energy-momentum tensor with mixed (one covariant the other controvariant) indices can be written, following Eq. (C.17), as

$$\delta_s T^0_0 = \frac{1}{a^2}\left(-\phi\varphi'^2 + \frac{\partial V}{\partial \varphi}a^2\chi + \chi'\varphi'\right), \qquad (C.28)$$

$$\delta_s T^j_i = \frac{1}{a^2}\left(\phi\varphi'^2 + \frac{\partial V}{\partial \varphi}a^2\chi - \chi'\varphi'\right)\delta^j_i, \qquad (C.29)$$

$$\delta_s T^i_0 = -\frac{1}{a^2}\varphi'\partial^i\chi. \qquad (C.30)$$

These equations have been extensively used in chapter 10.

C.3 Fluctuations of the covariant conservation equations

The perturbed Einstein equations are sufficient to determine the evolution of the perturbations. However, for practical purposes, it is often useful to employ the equations stemming from the first-order counterpart of the covariant conservation equations. Consider, first, the case of a fluid, then the perturbation of the covariant conservation equation can be written as:

$$\partial_\mu \delta_s T^{\mu\nu} + \overline{\Gamma}^\mu_{\mu\alpha}\delta_s T^{\alpha\nu} + \delta_s\Gamma^\mu_{\mu\alpha}\overline{T}^{\alpha\nu} + \overline{\Gamma}^\nu_{\alpha\beta}\delta_s T^{\alpha\beta} + \delta_s\Gamma^\nu_{\alpha\beta}\overline{T}^{\alpha\beta} = 0, \qquad (C.31)$$

where the definition of the covariant derivative has been made explicit. Recalling Eqs. (C.1) and (C.20)–(C.22) the (0) and (i) components of Eq. (C.31) can be written as

$$\partial_0 \delta_s T^{00} + \partial_j \delta_s T^{j0} + (2\delta_s\Gamma^0_{00} + \delta_s\Gamma^k_{k0})\overline{T}^{00} + (2\overline{\Gamma}^0_{00} + \overline{\Gamma}^k_{k0})\delta_s T^{00}$$
$$+ \overline{\Gamma}^0_{ij}\delta_s T^{ij} + \delta_s\Gamma^0_{ij}\overline{T}^{ij} = 0, \qquad (C.32)$$

$$\partial_0 \delta_s T^{0j} + \partial_k \delta_s T^{kj} + (\delta_s\Gamma^0_{0k} + \delta_s\Gamma^m_{mk})\overline{T}^{kj}$$
$$+ (\overline{\Gamma}^0_{00} + \overline{\Gamma}^k_{k0})\delta_s T^{0j} + \delta_s\Gamma^j_{00}\overline{T}^{00} + \delta_s\Gamma^j_{km}\overline{T}^{km} + 2\overline{\Gamma}^j_{0k}\delta_s T^{0k} = 0. \qquad (C.33)$$

Inserting now the specific form of the perturbed connections of Eqs. (C.3) into Eqs. (C.32) and (C.33) the following result can be obtained respectively:

$$\delta\rho' - 3\psi'(p+\rho) + (p+\rho)\theta + 3\mathcal{H}(\delta\rho + \delta p) = 0, \qquad \text{(C.34)}$$

for the (0) component, and

$$(p+\rho)\theta' + \theta[(p'+\rho') + 4\mathcal{H}(p+\rho)] + \nabla^2\delta p + (p+\rho)\nabla^2\phi = 0 \qquad \text{(C.35)}$$

for the (i) component. In the above equations, as explained in the text, the divergence of the velocity field, i.e. $\theta = \partial_i v^i = \partial_i \partial^i v$, has been directly introduced. Notice that the possible anisotropic stress (arising, for instance, in the case of neutrinos) has been neglected. Its inclusion modifies the left hand side of Eq. (C.35) by the term $-(p+\rho)\nabla^2\sigma$. In section 7, Eqs. (C.34) and (C.35) have been written in the case of a mixture of fluids.

Finally, to conclude this Appendix, it is relevant to compute the fluctuation of the Klein-Gordon equation which is equivalent to the covariant conservation of the energy-momentum tensor of the the (minimally coupled) scalar degree of freedom that has already been extensively discussed. The Klein-Gordon equation in curved spaces can be written as (see section 5)

$$g^{\alpha\beta}\nabla_\alpha\nabla_\beta\varphi + \frac{\partial V}{\partial\varphi}a^2 = 0. \qquad \text{(C.36)}$$

From Eq. (C.36) the perturbed Klein-Gordon equation takes the form

$$\delta_s g^{\alpha\beta}[\partial_\alpha\partial_\beta\varphi - \overline{\Gamma}^\sigma_{\alpha\beta}\partial_\sigma\varphi] + \overline{g}^{\alpha\beta}[\partial_\alpha\partial_\beta\chi - \delta_s\Gamma^\sigma_{\alpha\beta}\partial_\sigma\varphi - \overline{\Gamma}^\sigma_{\alpha\beta}\partial_\sigma\chi] + \frac{\partial^2 V}{\partial\varphi^2}\chi$$
$$= 0. \qquad \text{(C.37)}$$

Using Eqs. (C.1) and (C.3) we have:

$$\delta_s g^{00}[\varphi'' - \mathcal{H}\varphi'] - \delta_s g^{ij}\overline{\Gamma}^0_{ij}\varphi' + \overline{g}^{00}[\chi'' - \delta_s\Gamma^0_{00}\varphi' - \overline{\Gamma}^0_{00}\chi']$$
$$+ \overline{g}^{ij}[\partial_i\partial_j\chi - \delta_s\Gamma^0_{ij}\varphi' - \overline{\Gamma}^0_{ij}\chi'] + \frac{\partial V}{\partial\varphi^2}\chi = 0. \qquad \text{(C.38)}$$

Finally, recalling the explicit forms of the Christoffel connections and of the metric fluctuations the perturbed Klein-Gordon equation becomes:

$$\chi'' + 2\mathcal{H}\chi' - \nabla^2\chi + \frac{\partial^2 V}{\partial\varphi^2}a^2\chi + 2\phi\frac{\partial V}{\partial\varphi}a^2 - \varphi'(\phi' + 3\psi') = 0. \qquad \text{(C.39)}$$

It should be appreciated that the perturbed Klein-Gordon equation also contains a contribution arising from the metric fluctuations and it is not only sensitive to the fluctuations of the scalar field.

C.4 Some algebra with the scalar modes

We will now develop some algebra that is rather useful when dealing with the scalar modes induced by a minimally coupled scalar degree of freedom. We will assume that Eqs. (10.4), (10.5) and (10.6) are valid in the longitudinal gauge. Subtracting Eq. (10.4) from Eq. (10.5), the following equation can be obtained (recall that $\phi = \psi$ since the scalar field to first-order does not produce an anisotropic stress):

$$\psi'' + 6\mathcal{H}\psi' + 2(\mathcal{H}' + 2\mathcal{H}^2)\psi - \nabla^2\psi = -8\pi G \frac{\partial V}{\partial \varphi} a^2 \chi. \quad (C.40)$$

From Eq. (10.6) it follows easily that

$$\chi = \frac{\psi' + \mathcal{H}\psi}{4\pi G \varphi'}. \quad (C.41)$$

Using then Eq. (C.41) in Eq. (C.40) (to eliminate χ) and recalling Eq. (5.58) (to eliminate the derivative of the scalar potential with respect to φ) we obtain the following decoupled equation:

$$\psi'' + 2\left[\mathcal{H} - \frac{\varphi''}{\varphi'}\right]\psi' + 2\left[\mathcal{H}' - \mathcal{H}\frac{\varphi''}{\varphi'}\right]\psi - \nabla^2\psi = 0. \quad (C.42)$$

Note that Eq. (5.58) is written in the cosmic time coordinate. Here we need its conformal time counterpart which is easily obtained:

$$\varphi'' + 2\mathcal{H}\varphi' + a^2 \frac{\partial V}{\partial \varphi} = 0. \quad (C.43)$$

It is appropriate to mention, incidentally, that Eq. (C.42) can also be written in a slightly simpler form that may be of some use in specific applications, namely:

$$f'' - \nabla^2 f - z\left(\frac{1}{z}\right)'' f = 0, \quad f = \frac{a}{\varphi'}\psi, \quad z = \frac{a\varphi'}{\mathcal{H}}. \quad (C.44)$$

It could be naively thought that the variable defined in Eq. (C.44) is the canonical normal mode of the system. This is not correct since, as we see, Eq. (C.44) does not contain any information on the fluctuation of the scalar field. The correct normal mode of the system will now be derived.

Recall now the definition of the curvature perturbations introduced in section 7 (see Eq. (7.101)):

$$\mathcal{R} = -\psi - \frac{\mathcal{H}}{\mathcal{H}^2 - \mathcal{H}'}(\psi' + \mathcal{H}\phi). \quad (C.45)$$

By setting $\phi = \psi$ in Eq. (C.45) we can express the first (conformal) time derivative of $\mathcal{R}$ as:

$$\frac{\partial \mathcal{R}}{\partial \tau} = -\psi' - \frac{\mathcal{H}}{\mathcal{H}^2 - \mathcal{H}'}[\psi'' + \mathcal{H}'\psi + \mathcal{H}\psi] - [\psi' + \mathcal{H}\psi]\frac{\partial}{\partial \tau}\left(\frac{\mathcal{H}}{\mathcal{H}^2 - \mathcal{H}'}\right). \quad (C.46)$$

Recalling the conformal time analog of Eq. (5.57), i.e.

$$\mathcal{H}^2 - \mathcal{H}' = 4\pi G\,\varphi'^2, \quad (C.47)$$

the derivation appearing in the second term of Eq. (C.46) can be made explicit. Using then Eq. (C.42) inside the obtained expression, all the terms can be eliminated except the one containing the Laplacian. The final result will be:

$$\mathcal{R}' = -\frac{\mathcal{H}}{4\pi G\,\varphi'^2}\nabla^2\psi. \quad (C.48)$$

Equation (C.48) can be used to obtain a decoupled equation for $\mathcal{R}$ that has been quoted and used in chapter 10 (see, in particular, Eq. (10.28)). From Eqs. (C.45) and (C.47) we can write:

$$\frac{\mathcal{H}}{4\pi G\,\varphi'^2}(\psi' + \mathcal{H}\psi) = -(\mathcal{R} + \psi). \quad (C.49)$$

Taking the Laplacian of both sides of Eq. (C.49) and recalling Eq. (C.48) the following relation can be derived:

$$\frac{\mathcal{H}}{4\pi G\,\varphi'^2}\nabla^2\psi' = -\nabla^2\mathcal{R} + \left[2\mathcal{H} - \frac{\mathcal{H}'}{\mathcal{H}}\right]\mathcal{R}'. \quad (C.50)$$

By now taking the derivative of Eq. (C.48) we obtain

$$\mathcal{R}'' = -\frac{\mathcal{H}}{4\pi G\,\varphi'^2}\nabla^2\psi' - \frac{\mathcal{H}}{4\pi G\,\varphi'^2}\left[\frac{\mathcal{H}'}{\mathcal{H}} - 2\frac{\varphi''}{\varphi'}\right]\nabla^2\psi. \quad (C.51)$$

Using now Eqs. (C.48) and (C.50) in Eq. (C.51) (to eliminate, respectively, $\nabla^2\psi$ and $\nabla^2\psi'$) the following equation is readily derived:

$$\mathcal{R}'' + 2\left[\mathcal{H} - \frac{\mathcal{H}'}{\mathcal{H}} + \frac{\varphi''}{\varphi'}\right]\mathcal{R}' - \nabla^2\mathcal{R} = 0. \quad (C.52)$$

Equation (C.52) can finally be rewritten as

$$\mathcal{R}'' + 2\frac{z'}{z}\mathcal{R}' - \nabla^2\mathcal{R} = 0, \qquad z = \frac{a\varphi'}{\mathcal{H}}, \quad (C.53)$$

which is exactly Eq. (10.28). As discussed in Eq. (10.26), $\mathcal{R}$ is related to the scalar normal mode as $q = -\mathcal{R}z$. Recalling Eq. (C.42) and Eq. (C.47), we have

$$q = a\chi + z\psi. \quad (C.54)$$

The derivation presented in this Appendix is gauge-dependent. However, since $\mathcal{R}$ and q are both gauge-invariant, their equations will also be gauge-invariant. Finally, it should be stressed that the same result obtained here by working with the evolution equations of the fluctuations can be obtained by perturbing (to second-order) the (non-gauge-fixed) scalar tensor action. This procedure is rather lengthy and the final results (already quoted in the bulk of this book) are the ones of Eqs. (10.26) and (10.29).

Appendix D

Metric Fluctuations: Gauge-Independent Treatment

In the previous Appendix the scalar modes of the geometry have been treated by exploiting the longitudinal gauge. In practical problems related to the evolution of the various modes of the geometry, it is often profitable to exploit not only one gauge but many different gauges. As already exemplified in this book, some equations may acquire a more tractable form in a specific gauge (see chapter 11). Moreover, certain gauge-invariant observables may readily be expressed in a gauge that is different from the longitudinal one. Last but not least, there are perfectly physical solutions of the lowest multipoles of the Einstein-Boltzmann hierarchy that are divergent in the longitudinal description but that are completely regular in the synchronous gauge. The bottom line of these remarks could be that there is nothing deep in choosing a coordinate system but if this choice is made according to physical considerations, the resulting expressions may appear much easier to interpret and to evaluate.

To facilitate the task of employing different gauges in the course of a physical calculation it is therefore important to perturb the relevant geometric quantities *without assuming* a specific gauge. This is the main theme of the present Appendix. In short the logic will be to take the decomposition expressed by Eqs. (6.3), (6.4), (6.5) and (6.6) and to repeat all the steps outlined in Appendix C. Furthermore, for sake of completeness, we will discuss, in a unified perspective, the scalar, vector and tensor modes. According to this perspective it is practical do denote by δ_s, δ_v and δ_t the scalar, vector and tensor fluctuations of a given four-dimensional tensor. We recall, indeed, that in Eqs. (6.3), (6.4) and (6.5) the scalar, vector and tensor modes are classified according to three-dimensional rotations.

D.1 The scalar problem

The general expression reported in Eq. (C.2) is independent of the specific coordinate system since it only involves the (formal) fluctuations of the four-dimensional metric tensor. Using now the decomposition given in Eqs. (6.3), (6.4) and (6.5), different perturbed components of the connections can be found to be, after some algebra:

$$
\begin{aligned}
&\delta_s \Gamma^0_{00} = \phi', \\
&\delta_s \Gamma^0_{ij} = -[\psi' + 2\mathcal{H}(\phi + \psi)]\delta_{ij} + \partial_i \partial_j [(E' + 2\mathcal{H}E) - B], \\
&\delta_s \Gamma^0_{i0} = \delta_s \Gamma^0_{0i} = \partial_i(\phi + \mathcal{H}B), \\
&\delta_s \Gamma^i_{00} = \partial^i(\phi + B' + \mathcal{H}B), \\
&\delta_s \Gamma^k_{ij} = (\partial^k \psi \delta_{ij} - \partial_i \psi \delta^k_j - \partial_j \psi \delta^k_i) + \partial_i \partial_j \partial^k E - \mathcal{H} \partial^k B \delta_{ij}, \\
&\delta_s \Gamma^j_{0i} - \psi' \delta^j_i + \partial_i \partial^j E'.
\end{aligned} \quad \text{(D.1)}
$$

To deduce Eqs. (D.1) it is practical to recall that, for scalar fluctuations:

$$
\begin{aligned}
&\delta_s g_{00} = 2a^2 \phi, \qquad &&\delta_s g^{00} = -\frac{2}{a^2}\phi \\
&\delta_s g_{ij} = 2a^2(\psi \delta_{ij} - \partial_i \partial_j E), \qquad &&\delta_s g^{ij} = -\frac{2}{a^2}(\psi \delta^{ij} - \partial^i \partial^j E), \\
&\delta_s g_{0i} = -a^2 \partial_i B, \qquad &&\delta_s g^{0i} = -\frac{1}{a^2}\partial^i B.
\end{aligned} \quad \text{(D.2)}
$$

In similar terms, using Eqs. (D.1) and (D.2) in Eqs. (C.4), the components of the Ricci tensor perturbed to first-order are:

$$
\begin{aligned}
\delta_s R_{00} &= \nabla^2[\phi + (B - E')' + \mathcal{H}(B - E')] + 3[\psi'' + \mathcal{H}(\phi' + \psi')], \\
\delta_s R_{0i} &= \partial_i[(\mathcal{H}' + 2\mathcal{H}^2)B + 2(\psi' + \mathcal{H}\phi)], \\
\delta_s R_{ij} &= -\delta_{ij}\{\psi'' + 2(\mathcal{H}' + 2\mathcal{H}^2)(\psi + \phi) \\
&\quad + \mathcal{H}(\phi' + 5\psi') - \nabla^2 \psi] + \mathcal{H}\nabla^2(B - E')\} \\
&\quad + \partial_i \partial_j [(E' - B)' + 2(\mathcal{H}' + 2\mathcal{H}^2)E + 2\mathcal{H}(E' - B) + (\psi - \phi)].
\end{aligned}
$$
(D.3)

From Eqs. (D.3) the first-order fluctuation of the Ricci scalar can be found easily:

$$
\begin{aligned}
\delta_s R = \frac{2}{a^2}\{&3\psi'' + 6(\mathcal{H}' + \mathcal{H}^2)\phi + 3\mathcal{H}(\phi' + 3\psi') \\
&+ \nabla^2[(\phi - 2\psi) + (B - E')' + 3\mathcal{H}(B - E')]\}.
\end{aligned} \quad \text{(D.4)}
$$

Using Eqs. (D.3) into Eqs. (C.7), the fluctuations of the components of the Ricci tensor with mixed indices will then be:

$$\delta_s R^0_0 = \frac{1}{a^2}\{\nabla^2[\phi + (B - E')' + \mathcal{H}(B - E')] + 3[\psi'' + \mathcal{H}(\phi' + \psi') + 2\mathcal{H}'\phi]\},$$

$$\delta_s R^j_i = \frac{1}{a^2}[\psi'' + 2(\mathcal{H}' + 2\mathcal{H}^2)\phi + \mathcal{H}(\phi' + 5\psi') - \nabla^2\psi + \mathcal{H}\nabla^2(B - E')]\delta^j_i$$

$$- \frac{1}{a^2}\partial_i\partial^j[(E' - B)' + 2\mathcal{H}(E' - B) + (\psi - \phi)],$$

$$\delta_s R^0_i = \frac{2}{a^2}\partial_i[\psi' + \mathcal{H}\phi],$$

$$\delta_s R^i_0 = \frac{2}{a^2}\partial^i[-(\psi' + \mathcal{H}\phi) + (\mathcal{H}' - \mathcal{H}^2)B].$$

(D.5)

The fluctuations of the components of the Einstein tensor with mixed indices are computed to be

$$\delta_s \mathcal{G}^0_0 = \frac{2}{a^2}\{\nabla^2\psi - \mathcal{H}\nabla^2(B - E') - 3\mathcal{H}(\psi' + \mathcal{H}\phi)\}, \quad \text{(D.6)}$$

$$\delta_s \mathcal{G}^j_i = \frac{1}{a^2}\{[-2\psi'' - 2(\mathcal{H}^2 + 2\mathcal{H}')\phi - 2\mathcal{H}\phi' - 4\mathcal{H}\psi']$$

$$- \nabla^2[(\phi - \psi) + (B - E')' + 2\mathcal{H}(B - E')]\}\delta^j_i$$

$$- \frac{1}{a^2}\partial_i\partial^j[(E' - B)' + 2\mathcal{H}(E' - B) + (\psi - \phi)], \quad \text{(D.7)}$$

$$\delta_s \mathcal{G}^0_i = \delta R^0_i. \quad \text{(D.8)}$$

D.2 The vector problem

Recalling the decomposition of Eqs. (6.3), (6.4), (6.5) and (6.6) the vector fluctuations of the geometry can be written as:

$$\delta_v g_{0i} = -a^2 Q_i, \quad \delta_v g^{0i} = -\frac{Q^i}{a^2},$$

$$\delta_v g_{ij} = a^2(\partial_i W_j + \partial_j W_i), \quad \delta_v g^{ij} = -\frac{1}{a^2}(\partial^i W^j + \partial^j W^i). \quad \text{(D.9)}$$

The general form of the vector fluctuations of the Christoffel connections (as well as of the Ricci tensor) can be easily obtained from Eqs. (C.2),

(C.4) and (C.7) by replacing $\delta_s \to \delta_v$. Consequently the result is:
$$\delta_v \Gamma^0_{i0} = \mathcal{H} Q_i,$$
$$\delta_v \Gamma^0_{ij} = -\frac{1}{2}(\partial_i Q_j + \partial_j Q_i) - \mathcal{H}(\partial_i W_j + \partial_j W_i) - \frac{1}{2}(\partial_i W'_j + \partial_j W'_i),$$
$$\delta_v \Gamma^i_{00} = Q^{i\prime} + \mathcal{H} Q^i,$$
$$\delta_v \Gamma^k_{ij} = -\mathcal{H} Q^k \delta_{ij} + \frac{1}{2}\partial^k(\partial_i W_j + \partial_j W_i)$$
$$- \frac{1}{2}\partial_j(\partial^k W_i + \partial_i W^k) - \frac{1}{2}\partial_i(\partial_j W^k + \partial^k W_j),$$
$$\delta_v \Gamma^j_{i0} = \frac{1}{2}(\partial_i Q^j - \partial^j Q_i) - \frac{1}{2}(\partial^j W'_i + \partial_i W^{j\prime}),$$
(D.10)

where the conditions of Eq. (6.6) on Q_i and W_i have been extensively used. Following the same steps discussed above in the scalar case the fluctuations of the Ricci tensors are:

$$\delta_v R_{0i} = [\mathcal{H}' + 2\mathcal{H}^2]Q_i - \frac{1}{2}\nabla^2 Q_i - \frac{1}{2}\nabla^2 W'_i$$
$$\delta_v R_{ij} = -\frac{1}{2}\left\{(\partial_i Q_j + \partial_j Q_i)' + 2\mathcal{H}(\partial_i Q_j + \partial_j Q_i)\right\}$$
$$- \frac{1}{2}(\partial_i W''_j + \partial_j W''_i) - \mathcal{H}(\partial_i W'_j + \partial_j W'_i)$$
$$- \frac{1}{2}(\partial_i W_j + \partial_j W_i)(2\mathcal{H}' + 4\mathcal{H}^2),$$
(D.11)

D.3 The tensor problem

Consider now the case of the tensor modes of the geometry, i.e. according to Eq. (6.6), the two polarizations of the graviton:

$$\delta_t g_{ij} = -a^2 h_{ij}, \qquad \delta_t g^{ij} = \frac{h^{ij}}{a^2}. \qquad (D.12)$$

Again, the tensor fluctuations of the Christoffel connections can be easily deduced from Eq. (C.2) by replacing $\delta_s \to \delta_t$. The result of this manipulation will then be:

$$\delta_t \Gamma^0_{ij} = \frac{1}{2}(h'_{ij} + 2\mathcal{H} h_{ij}),$$
$$\delta_t \Gamma^j_{i0} = \frac{1}{2} h_i^{j\prime}, \qquad (D.13)$$
$$\delta_t \Gamma^k_{ij} = \frac{1}{2}[\partial_j h_i^k + \partial_i h_j^k - \partial^k h_{ij}].$$

Inserting these results into the perturbed expressions of the Ricci tensors it is easy to obtain:

$$\delta_t R_{ij} = \frac{1}{2}[h''_{ij} + 2\mathcal{H}h'_{ij} + 2(\mathcal{H}' + 2\mathcal{H}^2)h_{ij} - \nabla^2 h_{ij}], \quad \text{(D.14)}$$

$$\delta_t R_i^j = -\frac{1}{2a^2}[h_i^{j''} + 2\mathcal{H}h_i^{j'} - \nabla^2 h_i^j]. \quad \text{(D.15)}$$

D.4 Inhomogeneities of the sources

The background energy-momentum tensor of a perfect fluid is

$$T_{\mu\nu} = (p + \rho)u_\mu u_\nu - p g_{\mu\nu}. \quad \text{(D.16)}$$

From the normalization condition $g_{\mu\nu}u^\mu u^\nu = 1$, it can be concluded that $u_0 = a$ and $\delta u^0 = -\phi/a$. Hence,

$$\delta_s T_0^0 = \delta\rho, \qquad \delta_s T^{00} = \frac{1}{a^2}(\delta\rho - 2\rho\phi), \quad \text{(D.17)}$$

and

$$\delta_s T_i^j = -\delta p \,\delta_i^j, \quad \text{(D.18)}$$

$$\delta_s T^{ij} = \frac{1}{a^2}[\delta p \,\delta^{ij} + 2p(\psi\delta^{ij} - \partial^i\partial^j E)], \quad \text{(D.19)}$$

$$\delta_s T_0^i = (p + \rho)v^i, \quad \text{(D.20)}$$

$$\delta_s T^{0i} = \frac{1}{a^2}[(p + \rho)v^i + \partial^i B]. \quad \text{(D.21)}$$

where we have chosen to define $\delta u^i = v^i/a$. Notice that the velocity field can be split into divergenceless and divergencefull parts, i.e.

$$v^i = \partial^i v + \mathcal{V}^i, \quad \partial_i \mathcal{V}^i = 0 \quad \text{(D.22)}$$

Notice that in the case of neutrinos (and possibly also in the case of other collisionless species) the anisotroopic stress has to be considered both in the Einstein equations and in the covariant conservation equations. The anisotropic stress is the introduced as

$$\delta_s T_i^j = -\delta p \,\delta_i^j + \Pi_i^j, \quad \text{(D.23)}$$

where $\Pi_i^j = T_i^j - \delta_i^j T_k^k/3$. In the case of scalar fluctuations, it is practical to adopt the following notation

$$\partial_j \partial^i \Pi_i^j = (p + \rho)\nabla^2 \sigma, \quad \text{(D.24)}$$

which is equivalent, in Fourier space, to the following identity

$$(p + \rho)\sigma = \left(\hat{k}_j \hat{k}^i - \frac{1}{3}\delta_j^i\right)\Pi_i^j. \quad \text{(D.25)}$$

There could be a potential confusion in the definition of the perturbed velocity field. It is plausible to define the peculiar velocity in two slightly different ways, namely

$$\delta_s T^i_0 = (p + \rho)u_0 \delta u^i \equiv (p + \rho)v^i \equiv (p + \rho)\partial^i v \quad (D.26)$$
$$\delta_s T^0_i = (p + \rho)u^0 \delta u_i = (p + \rho)\bar{v}_i \equiv (p + \rho)\partial_i \bar{v}, \quad (D.27)$$

where from the normalization condition $g_{\mu\nu} u^\mu u^\nu = 1$, $u_0 = a$ and $u^0 = 1/a$. The velocity fields $\bar{v}$ and v defined in Eqs. (2.8) and (2.16) are not equivalent. In fact $\delta u_i = a\bar{v}_i$ and $\delta u^i = v^i/a$. Recall that

$$\delta u^i = \delta(g^{i\alpha} u_\alpha) \equiv \delta g^{i0} u_0 + g^{ik} \delta u_k. \quad (D.28)$$

Inserting now the explicit definitions of δu^i and δu_k in terms of v^i and $\bar{v}_k$ and recalling that $\delta g^{i0} = -\partial^i B/a^2$, we have

$$v_i = -\bar{v}_i - \partial_i B. \quad (D.29)$$

This difference in the definition of the velocity field reflects in the gauge transformations which are different for v_i and $\bar{v}_i$, namely

$$v_i \to \tilde{v}_i = v_i + \partial_i \epsilon',$$
$$\bar{v}_i \to \tilde{\bar{v}}_i = \bar{v}_i - \partial_i \epsilon_0. \quad (D.30)$$

In this book we always define the velocity field as in Eq. (2.8). However, this remark should be borne in mind since different authors use different definitions which may only be equivalent, up to a sign, in specific gauges (like the class of gauges where $B = 0$).

Let us now consider the fluctuations of the energy-momentum tensor of a scalar field φ characterized by a potential $V(\varphi)$:

$$T_{\mu\nu} = \partial_\mu \varphi \partial_\nu \varphi - g_{\mu\nu} \left[\frac{1}{2} g^{\alpha\beta} \partial_\alpha \varphi \partial_\beta \varphi - V(\varphi) \right]. \quad (D.31)$$

Denoting with χ the first-order fluctuation of the scalar field φ we will have

$$\delta_s T_{\mu\nu} = \partial_\mu \chi \partial_\nu \varphi + \partial_\mu \varphi \partial_\nu \chi - \delta_s g_{\mu\nu} \left[\frac{1}{2} g^{\alpha\beta} \partial_\alpha \varphi \partial_\beta \varphi - V \right]$$
$$- g_{\mu\nu} \left[\frac{1}{2} \delta_s g^{\alpha\beta} \partial_\alpha \varphi \partial_\beta \varphi + g^{\alpha\beta} \partial_\alpha \chi \partial_\beta \varphi - \frac{\partial V}{\partial \varphi} \chi \right], \quad (D.32)$$

and, in explicit terms,

$$\delta_s T_{00} = \chi' \varphi' + 2a^2 \phi V + a^2 \frac{\partial V}{\partial \varphi} \chi,$$

$$\delta_s T_{0i} = \varphi' \partial_i \chi + a^2 \partial_i B \left[\frac{\varphi'^2}{2a^2} - V \right],$$

$$\delta_s T_{ij} = \delta_{ij} \left[\varphi' \chi' - \frac{\partial V}{\partial \varphi} \chi a^2 - (\phi + \psi)\varphi'^2 + 2a^2 V \psi \right]$$
$$+ 2a^2 \left[\frac{\varphi'^2}{2a^2} - V \right] \partial_i \partial_j E. \quad (D.33)$$

Recalling that

$$\overline{T}_{00} = \frac{\varphi'^2}{2} + a^2 V, \qquad \overline{T}_{ij} = \left[\frac{\varphi'^2}{2} - a^2 V\right]\delta_{ij}, \qquad (D.34)$$

the perturbed components of the scalar field energy-momentum tensor with mixed (one covariant the other controvariant) indices can be written as

$$\delta_s T^\nu_\mu = \delta_s T_{\alpha\mu} \overline{g}^{\alpha\nu} + \overline{T}_{\alpha\mu} \delta_s g^{\alpha\nu}, \qquad (D.35)$$

i.e. in explicit terms,

$$\delta_s T^0_0 = \frac{1}{a^2}\left(-\phi\varphi'^2 + \frac{\partial V}{\partial \varphi} a^2 \chi + \chi'\varphi'\right), \qquad (D.36)$$

$$\delta_s T^j_i = \frac{1}{a^2}\left(\phi\varphi'^2 + \frac{\partial V}{\partial \varphi} a^2 \chi - \chi'\varphi'\right)\delta^j_i, \qquad (D.37)$$

$$\delta_s T^i_0 = -\frac{1}{a^2}\varphi'\partial^i\chi - \frac{\varphi'^2}{a^2}\partial^i B. \qquad (D.38)$$

The covariant conservation of the energy-momentum tensor implies, in the scalar field case, the validity of the Klein-Gordon equation which can be written as

$$g^{\alpha\beta}\nabla_\alpha\nabla_\beta\varphi + \frac{\partial V}{\partial \varphi} a^2 = 0. \qquad (D.39)$$

From Eq. (D.39) the perturbed Klein-Gordon equation can be easily deduced:

$$\delta_s g^{\alpha\beta}[\partial_\alpha\partial_\beta\varphi - \overline{\Gamma}^\sigma_{\alpha\beta}\partial_\sigma\varphi] + \frac{\partial^2 V}{\partial \varphi^2}\chi$$
$$+ \overline{g}^{\alpha\beta}[\partial_\alpha\partial_\beta\chi - \delta_s\Gamma^\sigma_{\alpha\beta}\partial_\sigma\varphi - \overline{\Gamma}^\sigma_{\alpha\beta}\partial_\sigma\chi] = 0, \qquad (D.40)$$

which becomes, in explicit form,

$$\chi'' + 2\mathcal{H}\chi' - \nabla^2\chi + \frac{\partial^2 V}{\partial \varphi^2}a^2\chi + 2\phi\frac{\partial V}{\partial \varphi}a^2$$
$$-\varphi'(\phi' + 3\psi') + \varphi'\nabla^2(E' - B) = 0. \qquad (D.41)$$

Different gauges are suitable for different problems and this has been the guiding theme of the discussion developed in chapter 11. Hence, it is practical to collect here the generalized evolution equations of the scalar fluctuations obtained without fixing a particular coordinate system. The sources will be assumed to be barotropic fluids but this is not a severe limitation since, with the appropriate identifications, this set of equations can even describe more general situations.

From Eqs. (D.6),(D.17) and from Eqs. (D.8), (D.20) the general expressions for the Hamiltonian and momentum constraints reads

$$\nabla^2 \psi - \mathcal{H}\nabla^2(B - E') - 3\mathcal{H}(\psi' + \mathcal{H}\phi) = 4\pi G a^2 \delta\rho, \quad (D.42)$$

$$\partial^i[(\psi' + \mathcal{H}\phi) + (\mathcal{H}^2 - \mathcal{H}')B] = -4\pi G a^2 (p + \rho)v^i. \quad (D.43)$$

From Eqs. (D.7) and (D.19) the generalized expression of the (ij) component of the perturbed equations is:

$$[\psi'' + (\mathcal{H}^2 + 2\mathcal{H}')\phi + \mathcal{H}(\phi' + 2\psi')]\delta_i^j$$
$$+ \frac{1}{2}\delta_i^j \nabla^2[(\phi - \psi) + (B - E')' + 2\mathcal{H}(B - E')]$$
$$- \frac{1}{2}\partial_i \partial^j[(\phi - \psi) + (B - E')' + 2\mathcal{H}(B - E')]$$
$$= 4\pi G a^2 [\delta p \delta_i^j - \Pi_i^j]. \quad (D.44)$$

Separating, in Eq. (D.44), the tracefull from the traceless part, we obtain, respectively,

$$\psi'' + \mathcal{H}(\phi' + 2\psi') + (\mathcal{H}^2 + 2\mathcal{H}')\phi$$
$$+ \frac{1}{3}\nabla^2[(\phi - \psi) + (B - E')' + 2\mathcal{H}(B - E')] = 4\pi G a^2 \delta p \quad (D.45)$$

and

$$\nabla^2[(\phi - \psi) + (B - E')' + 2\mathcal{H}(B - E')] = 12\pi G a^2 (p + \rho)\sigma. \quad (D.46)$$

In general terms the the (0) and (i) components of

$$\partial_\mu \delta T^{\mu\nu} + \overline{\Gamma}^\mu_{\mu\alpha}\delta T^{\alpha\nu} + \delta\Gamma^\mu_{\mu\alpha}\overline{T}^{\alpha\nu} + \overline{\Gamma}^\nu_{\alpha\beta}\delta T^{\alpha\beta} + \delta\Gamma^\nu_{\alpha\beta}\overline{T}^{\alpha\beta} = 0 \quad (D.47)$$

can be written, for scalar fluctuations as

$$\partial_0 \delta_s T^{00} + \partial_j \delta_s T^{j0} + (2\delta_s\Gamma^0_{00} + \delta_s\Gamma^k_{k0})\overline{T}^{00}$$
$$+ (2\overline{\Gamma}^0_{00} + \overline{\Gamma}^k_{k0})\delta_s T^{00} + \overline{\Gamma}^0_{ij}\delta_s T^{ij} + \delta_s\Gamma^0_{ij}\overline{T}^{ij} = 0, \quad (D.48)$$

$$\partial_0 \delta_s T^{0j} + \partial_k \delta_s T^{kj} + (\delta_s\Gamma^0_{0k} + \delta_s\Gamma^m_{mk})\overline{T}^{kj}$$
$$+ (\overline{\Gamma}^0_{00} + \overline{\Gamma}^k_{k0})\delta_s T^{0j} + \delta_s\Gamma^j_{00}\overline{T}^{00}$$
$$+ \delta_s\Gamma^j_{km}\overline{T}^{km} + 2\overline{\Gamma}^j_{0k}\delta_s T^{0k} = 0. \quad (D.49)$$

Inserting now the specific form of the perturbed connections of Eqs. (D.1) into Eqs. (D.48) and (D.49) the following result can be obtained respectively:

$$\delta\rho' - 3\psi'(p + \rho) + (p + \rho)\theta + 3\mathcal{H}(\delta\rho + \delta p) + (p + \rho)\nabla^2 E' = 0, \quad (D.50)$$

for the (0) component, and

$$(p+\rho)\theta' + \theta[(p'+\rho') + 4\mathcal{H}(p+\rho)] + (p+\rho)\nabla^2 B',$$
$$+[p' + \mathcal{H}(p+\rho)]\nabla^2 B + \nabla^2 \delta p + (p+\rho)\nabla^2 \phi$$
$$-(p+\rho)\nabla^2 \sigma = 0, \qquad (D.51)$$

for the (i) component. In the above equations, as explained in the text, the divergence of the velocity field, i.e. $\theta = \partial_i v^i = \partial_i \partial^i v$, has been directly introduced. In the text, several gauge-dependent discussions are present. The longitudinal gauge equations are obtained, for instance, by setting everywhere in the above equations $E = B = 0$. The off-diagonal gauge equations are obtained from the above equations by setting everywhere $E = \psi = 0$. The sychronous gauge equations can be derived by setting everywhere in the above equations $\phi = B = 0$ and so on for the gauge that is most suitable in a given calculation.

Bibliography

[1] A. A. Penzias and R. Wilson, *Astrophys. J.* **142**, 419 (1965).
[2] R. H. Dicke, P. J. E. Peebles, P. G. Roll, and D. T. Wilkinson, *Astrophys. J.* **142**, 414 (1965).
[3] R. A. Alpher and R. C. Herman, *Phys. Today* **41**, 24 (1988).
[4] F. Govoni, L. Feretti, *Int. J. Mod. Phys.* D **13**, 1549 (2004).
[5] A. G. Lyne, F. G. Smith, *Pulsar Astronomy*, (Cambridge University Press, Cambridge, UK, 1998).
[6] B. M. Gaensler, R. Beck, L. Feretti, *New Astron. Rev.* **48**, 1003 (2004).
[7] J. Abraham et al. [Pierre Auger Collaboration], *Nucl. Instrum. Meth.* A **523**, 50 (2004).
[8] K. S. Capelle, J. W. Cronin, G. Parente and E. Zas, *Astropart. Phys.* **8**, 321 (1998).
[9] T. Yamamoto [Pierre Auger Collaboration], *The UHECR Spectrum Measured at the Pierre Auger Observatory and its Astrophysical Implications*, arXiv:0707.2638 [astro-ph].
[10] M. Aglietta et al. [Pierre Auger Collaboration], *Astropart. Phys.* **27**, 244 (2007).
[11] J. Abraham et al. [Pierre Auger Collaboration], *Science* **318**, 938 (2007).
[12] A. Dar and A. De Rujula, *A Theory of Cosmic Rays,* arXiv:hep-ph/0606199.
[13] A. Dar and A. De Rujula, *Mon. Not. Roy. Astron. Soc.* **323** 391 (2001).
[14] A. Dar, S. Dado and A. De Rujula, *On the Origin of the Diffuse Gamma-Ray Background Radiation*, arXiv:astro-ph/0607479.
[15] M. T. Ressell and M. S. Turner, *Bull. Am. Astron. Soc.* **22**, 753 (1990).
[16] G. Sironi, M. Limon, G. Marcellino et al. *Astrophys. J.* **357**, 301 (1990)
[17] T. Howell and J. Shakeshaft, *Nature* **216**, 753 (1966).
[18] R. B. Patridge, *3 K: The Cosmic Microwave Background Radiation* (Cambridge University Press, Cambridge, UK, 1995).
[19] J. C. Mather et al., *Astrophys. J.* **354** (1990) L37.
[20] E. L. Wright et al., *Astrophys. J.* **396** (1992) L13.
[21] G. F. Smoot et al., *Astrophys. J.* **396**, L1 (1992).
[22] A. Kogut et al., *Astrophys. J.* **419**, 1 (1993).

[23] C. L. Bennett et al., Astrophys. J. **436**, 423 (1994)
[24] J. C. Mather et al., Astrophys. J. **420** (1994) 439.
[25] D. J. Fixsen et al., Astrophys. J. **473**, 576 (1996).
[26] C. L. Bennett et al., Astrophys. J. **464**, L1 (1996)
[27] de Bernardis et al., Astrophys. J. **564**, 559 (2002).
[28] C. B. Netterfield et al., Astrophys. J. **571**, 604 (2002).
[29] N. W. Halverson et al., Astrophys. J. **568**, 38 (2002).
[30] A. T. Lee et al., Astrophys. J. **561**, L1 (2001).
[31] A. Benoît et al., Astron. Astrophys. **399**, L19 (2003).
[32] E. Gawiser and J. Silk, Phys. Rept. **333**, 245 (2000).
[33] M. Giovannini, Int. J. Mod. Phys. D **13**, 391 (2004).
[34] C. Dickinson et al., Mon. Not. Roy. Astron. Soc. **353**, 732 (2004)
[35] C. L. Bennett et al., Astrophys. J. Suppl. **148**, 1 (2003).
[36] L. Page et al, Astrophys. J. Suppl. **148**, 223 (2003).
[37] D. N. Spergel et al. [WMAP Collaboration], Astrophys. J. Suppl. **148**, 175 (2003).
[38] H. V. Peiris et al., Astrophys. J. Suppl. **148**, 213 (2003).
[39] D. N. Spergel et al. [WMAP Collaboration], Astrophys. J. Suppl. **170**, 377 (2007).
[40] L. Page et al. [WMAP Collaboration], Astrophys. J. Suppl. **170**, 335 (2007).
[41] B. S. Mason et al., Astrophys. J. **591**, 540 (2003).
[42] T. J. Pearson et al., Astrophys. J. **591**, 556 (2003).
[43] C. L. Kuo et al., Astrophys. J. **600**, 32 (2004).
[44] M. J. White, D. Scott and J. Silk, Ann. Rev. Astron. Astrophys. **32**, 319 (1994).
[45] W. Hu and S. Dodelson, Ann. Rev. Astron. Astrophys. **40**, 171 (2002).
[46] J. R. Bond, *Cosmology and Large Scale Structure* (Les Houches, Session LX, 1993), eds. R. Schaeffer, J. Silk, M. Spiro and J. Zinn-Justin, p. 469.
[47] A. Songaila et al., Nature **371**, 43 (1994).
[48] P. A. Thaddeus, Rev. Astr. Astrophys. **10**, 305 (1972).
[49] R. A. Sunyaev and Y. B. Zeldovich, Astron. Astrophys. **20**, 189 (1972).
[50] R. A. Sunyaev and Y. B. Zeldovich, Ann. Rev. Astron. Astrophys. **18**, 537 (1980).
[51] R. A. Sunyaev and Y. B. Zeldovich, Mon. Not. Roy. Astron. Soc. **190**, 413 (1980).
[52] H. Hebeling et al., Mon. Not. Astron. Soc. **281**, 799 (1996).
[53] Y. Rephaeli, Ann. Rev. Astron. Astrophys. **33**, 541 (1995)
[54] S. Sadeh and Y. Rephaeli, New Astron. **9**, 373 (2004)
[55] A. S. Kompaneets, Sov. Phys. JETP **4**, 730 (1957).
[56] P. J. Peebles, *Principles of Physical Cosmology*, (Princeton, Princeton University Press 1993).
[57] M. Tegmark et al., Astrophys. J. **606**, 702 (2004).
[58] N. Bachall et al., Astrophys. J. **585**, 182 (2003).
[59] A. G. Riess et al. [Supernova Search Team Collaboration], Astrophys. J. **607**, 665 (2004).
[60] B. Fields and S. Sarkar, Phys. Lett. B **592**, 202 (2004).

Bibliography

[61] J. Tonry et al., *Astrophys. J.* **594**, 1 (2003).
[62] S. Cole et al. [The 2dFGRS Collaboration], *Mon. Not. Roy. Astron. Soc.* **362**, 505 (2005).
[63] T. E. Montroy et al., *Astrophys. J.* **647**, 813 (2006).
[64] C. l. Kuo et al. [ACBAR collaboration], *Astrophys. J.* **600**, 32 (2004).
[65] A. C. S. Readhead et al., *Astrophys. J.* **609**, 498 (2004).
[66] C. Dickinson et al., *Mon. Not. Roy. Astron. Soc.* **353**, 732 (2004).
[67] W. L. Freedman et al., *Astrophys. J.* **553**, 47 (2001).
[68] D. J. Eisenstein et al. [SDSS Collaboration], *Astrophys. J.* **633**, 560 (2005).
[69] M. Tegmark et al. [SDSS Collaboration], *Astrophys. J.* **606**, 702 (2004).
[70] E. Semboloni et al., arXiv:astro-ph/0511090.
[71] H. Hoekstra et al., *Astrophys. J.* **647**, 116 (2006).
[72] P. Astier et al. [The SNLS Collaboration], *Astron. Astrophys.* **447**, 31 (2006).
[73] A. G. Riess et al. [Supernova Search Team Collaboration], *Astrophys. J.* **607**, 665 (2004).
[74] B. J. Barris et al., *Astrophys. J.* **602**, 571 (2004).
[75] S. Weinberg, *Gravitation and Cosmology*, (John Wiley & Sons, New York, US, 1972).
[76] P. Peebles, *The Large Scale Structure of the Universe*, (Princeton University Press, Princeton, New Jersey 1980).
[77] P. Peebles, *Principles of Physical Cosmology*, (Princeton University Press, Princeton, New Jersey 1993).
[78] E. W. Kolb and M. S. Turner, *The Early Universe*, (Addison-Wesley, US, 1990).
[79] Ya. B. Zeldovich, *Sov. Phys. Usp.* **6**, 475 (1964) [*Usp. Fiz. Nauk.* **80**, 357 (1963)].
[80] E. M. Lifshitz and I. M. Khalatnikov, *Sov. Phys. Usp.* **6**, 495 (1964) [*Usp. Fiz. Nauk.* **80**, 391 (1964)].
[81] A. Friedmann, *Z. Phys.* **10**, 377 (1922); **21**, 326 (1924).
[82] G. Lemaître, *Ann. Soc. Sci.* Bruxelles **47A**, 49 (1927).
[83] G. Lemaître, *Rev. Questions Sci.* **129**, 129 (1958).
[84] G. Lemaître, *Nature* **128**, 700 (1931).
[85] A. S. Eddington, *Month. Not. R. Astron. Soc.* **90**, 672 (1930).
[86] R. Tolman, *Relativity, Thermodynamics and Cosmology* (Dover Press), p. 364.
[87] M. Ryan and L. Shepley, *Homogeneous Relativistic Cosmologies*, (Princeton University Press, Princeton 1978).
[88] L. Landau and E. Lifshitz, *Fluid Mechanics* (Pergamon Press, Oxford, 1989).
[89] Ya. B. Zeldovich and I. D. Novikov, *The Structure and the Evolution of the Universe*, Vol. 2 (Chicago University Press, Chicago 1971).
[90] L. P. Grishchuk, *Ann. N. Y. Acad. Sci.* **302**, 439 (1977).
[91] C. Eckart, *Phys. Rev.* **58**, 267 (1940).
[92] C. Eckart, *Phys. Rev.* **58**, 269 (1940).
[93] C. Eckart, *Phys. Rev.* **58**, 919 (1940).

[94]   W. Israel, *Ann. Phys.* **100**, 310 (1976).
[95]   W. Israel and J. M. Stewart, *Ann. Phys.* **118**, 341 (1979).
[96]   R. Maartens, arXiv:astro-ph/9609119.
[97]   M. Giovannini, *Phys. Lett.* B **622**, 349 (2005).
[98]   M. Giovannini, *Class. Quant. Grav.* **22**, 5243 (2005).
[99]   V. A. Belinskii and I. M. Khalatnikov, *Sov. Phys. JETP* **42**, 205 (1976) [Zh. Eksp. Teor. Fiz. **69**, 401 (1975)].
[100] G. L. Murphy, *Phys. Rev.* D **8**, 4231 (1973).
[101] J. Barrow, *Nucl. Phys.* B **310**, 743 (1988).
[102] L. D. Landau, E. M. Lifshitz, *The classical theory of fields*, (Elsevier, Amsterdaam, 1980).
[103] P. A. M. Dirac, *General theory of relativity*, (Princeton University Press, Princeton, New Jersey, 1996).
[104] G. Veneziano, *Phys. Lett.* B **265**, 287 (1991).
[105] M. Gasperini and G. Veneziano, *Phys. Rept.* **373**, 1 (2003).
[106] M. Gasperini and G. Veneziano, *Astropart. Phys.* **1**, 317 (1993).
[107] M. Gasperini, M. Giovannini and G. Veneziano, *Nucl. Phys.* B **694**, 206 (2004).
[108] M. Gasperini, M. Giovannini and G. Veneziano, *Phys. Lett.* B **569**, 113 (2003).
[109] M. Gasperini, J. Maharana and G. Veneziano, *Nucl. Phys.* B **472**, 349 (1996).
[110] S. W. Hawking and R. Penrose, *Proc. R. Soc. London* **A314**, 529 (1970)
[111] S. W. Hawking and G. F. R. Ellis, *The Large Scale Structure of the Space-time* (Cambridge University Press, Cambridge, England, 1973).
[112] C.-I. Kuo and L. H. Ford, *Phys. Rev.* D **47**, 4510 (1992).
[113] J. Bernstein *et al.*, *Rev. Mod. Phys.* **61**, 25 (1989).
[114] P. J. E. Peebles and B. Ratra, *Rev. Mod. Phys.* **75**, 559 (2003).
[115] J. L. Tonry *et al.* [Supernova Search Team Collaboration], *Astrophys. J.* **594**, 1 (2003).
[116] B. J. Barris *et al.*, *Astrophys. J.* **602**, 571 (2004).
[117] T. Padmanabhan and T. R. Choudhury, *Mon. Not. Roy. Astron. Soc.* **344**, 823 (2003).
[118] J. Rehm and K. Jedamzik, *Phys. Rev. Lett.* **81**, 3307 (1998).
[119] H. Kurki-Suonio and E. Sihvola, *Phys. Rev. Lett.* **84**, 3756 (2000).
[120] M. Giovannini, H. Kurki-Suonio, E. Sihvola, *Phys. Rev.* D **66**, 043504 (2003).
[121] K. A. Olive, *Phys. Rept.* **190**, 307 (1990).
[122] A. D. Linde, *Particle Physics and Inflationary Cosmology* (Chur, Switzerland, Harwood, 1990).
[123] A. Guth, *Phys. Rev.* D **23**, 347 (1981).
[124] A. Linde, *Phys. Lett.* **108B** (1982).
[125] A. Albrecht and P. J. Steinhardt, *Phys. Rev. Lett.* **48**, 1220 (1982).
[126] A. Linde, *Phys. Lett.* **129B**, 177 (1983).
[127] A. R. Liddle and S. M. Leach, *Phys. Rev.* D **68**, 103503 (2003).
[128] S. Dodelson and L. Hui, *Phys. Rev. Lett.* **91**, 131301 (2003).

[129] A. Berera, *Phys. Rev. Lett.* **75**, 3218 (1995).
[130] M. Bastero-Gil and A. Berera, *Phys. Rev.* D **76**, 043515 (2007).
[131] A. Berera, *Nucl. Phys.* B **585**, 666 (2000).
[132] A. Berera and L. Z. Fang, *Phys. Rev. Lett.* **74**, 1912 (1995).
[133] J. Yokoyama and A. D. Linde, *Phys. Rev.* D **60**, 083509 (1999).
[134] R. Penrose, *General Relativity: An Einstein Centenary*, eds. S. Hawking and W. Israel (Cambridge University Press, Cambridge 1979).
[135] K. P. Tod, *Class. Quantum Grav.* **20**, 521 (2003).
[136] P. C. W. Davies, *Class. Quantum Grav.* **4**, L225 (1987).
[137] P. C. W. Davies, *Class. Quantum Grav.* **5**, 1349 (1988).
[138] M. Giovannini, *Phys. Rev.* D **59** 121301 (1999).
[139] W. Zimdahl and D. Pavon, *Phys. Rev.* D **61**, 108301 (2000).
[140] M. Giovannini, *Phys. Rev.* D **61**, 108302 (2000).
[141] M. Cataldo and P. Mella, *Phys. Lett.* B **642**, 5 (2006).
[142] M. Gasperini and M. Giovannini, *Phys. Lett.* B **301**, 334 (1993).
[143] M. Gasperini and M. Giovannini, *Class. Quant. Grav.* **10**, L133 (1993).
[144] G. Smoot, *CMB Observations and the Standard Model of the Universe*, talk given at the 11th Paris Cosmology Colloquium, Paris, August 16-18 2007.
[145] A. Borde and A. Vilenkin, *Phys. Rev. Lett.* **72**, 3305 (1994).
[146] A. Borde and A. Vilenkin, *Int. J. Mod. Phys.* D **5**, 813 (1996).
[147] A. Borde and A. Vilenkin, *Phys. Rev.* D **56**, 717 (1997).
[148] A. Borde, A. H. Guth and A. Vilenkin, *Phys. Rev. Lett.* **90**, 151301 (2003).
[149] E. M. Lifshitz and I. M. Khalatnikov, *Sov. Phys. JETP* **12**, 108 (1960).
[150] E. M. Lifshitz and I. M. Khalatnikov, *Sov. Phys. JETP* **12**, 558 (1961).
[151] I. M. Khalatnikov and E. M. Lifshitz, *Phys. Rev. Lett.* **24**, 76 (1970).
[152] V. A. Belinskii and I. M. Khalatnikov, *Sov. Phys. JETP* **30**, 1174 (1970).
[153] V. A. Belinskii and I. M. Khalatnikov, *Sov. Phys. JETP* **36**, 591 (1973).
[154] V. A. Belinskii, E. M. Lifshitz and I. M. Khalatnikov, *Sov. Phys. Usp.* **13**, 745 (1971).
[155] N. Deruelle and D. Goldwirth, *Phys. Rev.* D **51**, 1563 (1995).
[156] N. Deruelle and K. Tomita, *Phys. Rev.* D **50**, 7216 (1994).
[157] G. Comer, N. Deruelle, D. Langlois, and J. Parry, *Phys. Rev.* D **49**, 2759 (1994).
[158] I. M. Khalatnikov, A. Y. Kamenshchik, M. Martellini and A. A. Starobinsky, *JCAP* **0303**, 001 (2003).
[159] M. Giovannini, *Phys. Lett.* B **634**, 1 (2006).
[160] M. Giovannini, *JCAP* **0509**, 009 (2005).
[161] M. Giovannini, *Phys. Rev.* D **72**, 083508 (2005).
[162] E. Komatsu and T. Futamase, *Phys. Rev.* D **58**, 023004 (1998).
[163] E. Komatsu and T. Futamase, *Phys. Rev.* D **59**, 064029 (1999).
[164] L. Kofman and S. Mukohyama, arXiv:0709.1952 [hep-th].
[165] M. Gasperini, N. G. Sanchez and G. Veneziano, *Nucl. Phys.* B **364**, 365 (1991).
[166] M. Gasperini, N. G. Sanchez and G. Veneziano, *Int. J. Mod. Phys.* A **6**, 3853 (1991).
[167] H. J. de Vega and N. G. Sanchez, *Phys. Lett.* B **197**, 320 (1987).

[168] L. F. Abbott and M. B. Wise, *Nucl. Phys. B* **244**, 541 (1984).
[169] F. Lucchin and S. Matarrese, *Phys. Rev. D* **32**, 1316 (1985).
[170] F. Lucchin and S. Matarrese, *Phys. Lett. B* **164**, 282 (1985).
[171] D. Lyth and E. Stewart, *Phys. Lett. B* **274**, 168 (1992).
[172] G. F. R. Ellis and M. S. Madsen, *Class. Quant. Grav.* **8**, 667 (1991).
[173] G. F. R. Ellis and R. Maartens, *Class. Quant. Grav.* **21**, 223 (2004).
[174] G. F. R. Ellis, *Class. Quant. Grav.* **5**, 891 (1988).
[175] B. Ratra and P. J. E. Peebles, *Phys. Rev. D* **52**, 1837 (1995).
[176] L. Kofman, A. D. Linde and A. A. Starobinsky, *Phys. Rev. Lett.* **73**, 3195 (1994).
[177] L. Kofman, A. D. Linde and A. A. Starobinsky, *Phys. Rev. D* **56**, 3258 (1997).
[178] P. B. Greene, L. Kofman, A. D. Linde and A. A. Starobinsky, *Phys. Rev. D* **56**, 6175 (1997).
[179] G. N. Felder, L. Kofman and A. D. Linde, *Phys. Rev. D* **59**, 123523 (1999).
[180] G. N. Felder, L. Kofman and A. D. Linde, *Phys. Rev. D* **60**, 103505 (1999).
[181] M. Giovannini, *Phys. Rev. D* **58**, 083504 (1998).
[182] P. J. E. Peebles and A. Vilenkin, *Phys. Rev. D* **59**, 063505 (1999).
[183] B. Spokoiny, *Phys. Lett. B* **315**, 40 (1993).
[184] P. J. E. Peebles and B. Ratra, *Astrophys. J.* **325**, L17 (1988).
[185] L. H. Ford, *Phys. Rev. D* **35**, 2955 (1987).
[186] D. Cirigliano, H. J. de Vega and N. G. Sanchez, *Phys. Rev. D* **71**, 103518 (2005).
[187] D. Boyanovsky, H. J. de Vega and N. G. Sanchez, *Phys. Rev. D* **73**, 023008 (2006).
[188] H. J. de Vega and N. G. Sanchez, *Phys. Rev. D* **74**, 063519 (2006).
[189] D. Boyanovsky, H. J. de Vega and N. G. Sanchez, *Phys. Rev. D* **74**, 123006 (2006).
[190] D. Boyanovsky, H. J. de Vega and N. G. Sanchez, *Phys. Rev. D* **74**, 123007 (2006).
[191] C. Destri, H. J. de Vega and N. G. Sanchez, arXiv:astro-ph/0703417.
[192] A. D. Sakharov, *Sov. Phys. JETP* **22**, 241 (1966) [*Zh.Éksp. Teor. Fiz.* **49**, 345 (1965)].
[193] A. D. Sakharov, *JETP Lett.* **5**, 24 (1967) [*Pisma Zh. Eksp. Teor. Fiz.* **5**, 32 (1967)].
[194] J. M. Bardeen, *Phys. Rev. D* **22**, 1882 (1980).
[195] W. Press and E. Vishniac, *Astrophys. J.* **239**, 1 (1980).
[196] W. Press and E. Vishniac, *Astrophys. J.* **236**, 323 (1980).
[197] C.-P. Ma and E. Bertschinger, *Astrophys. J.* **455**, 7 (1995).
[198] J. Bardeen, P. Steinhardt, M. Turner, *Phys. Rev. D* **28**, 679 (1983).
[199] R. Brandenberger, R. Kahn, and W. Press, *Phys. Rev. D* **28**, 1809 (1983).
[200] D. H. Lyth, *Phys. Rev. D* **31**, 1792 (1985).
[201] E. M. Lifshitz and I. M. Khalatnikov, Sov. Phys. Usp. **6**, 495 (1964) [Usp. Fiz. Nauk. **80**, 391 (1964)].
[202] M. Giovannini, *Class. Quant. Grav.* **22**, 363 (2005).
[203] M. Giovannini, *Phys. Rev. D* **70**, 103509 (2004).

[204] L. Parker and S. A. Fulling, *Phys. Rev.* D **9**, 341 (1974).
[205] L. Parker and S. A. Fulling, *Phys. Rev.* D **7**, 2357 (1973).
[206] B. Hu and L. Parker, *Phys. Lett.* A **63**, 217 (1977).
[207] L. P. Grishchuk, *Sov. Phys. JETP* **40**, 409 (1975) [Zh. Éksp. Teor. Fiz. **67**, 825 (1974)].
[208] L. P. Grishchuk, *JETP Lett.* **23**, 293 (1976) [*Pis'ma Zh. Eksp. Teor. Fiz.* **23**, 326 (1976)].
[209] B. Allen, *Phys. Rev.* D **37**, 2078 (1988).
[210] V. Sahni, *Phys. Rev.* D **42**, 453 (1990).
[211] L. P. Grishchuk and M. Solokhin, *Phys. Rev.* D **43**, 2566 (1991).
[212] M. Gasperini and M. Giovannini, *Phys. Lett.* B **282**, 36 (1992).
[213] V. Bozza, M. Giovannini and G. Veneziano, *JCAP* **0305**, 001 (2003).
[214] M. Giovannini, *Class. Quant. Grav.* **20**, 5455 (2003).
[215] M. Abramowitz and I. A. Stegun, *Handbook of Mathematical Functions* (Dover, New York, 1972).
[216] A. Erdelyi, W. Magnus, F. Obehettinger, and F. Tricomi, *Higher Trascendental Functions* (Mc Graw-Hill, New York, 1953).
[217] N. D. Birrel and P. C. W. Daview, *Quantum Fields in Curved Space*, (Cambridge University Press, Cambridge England, 1982).
[218] J. garriga and E. Verdaguer, *Phys. Rev.* D **39**, 1072 (1991).
[219] B. Allen, in *Proceedings of the Les Houches School on Astrophysical Sources of Gravitational Waves*, edited by J. Marck and J.P. Lasota (Cambridge University Press, Cambridge England, 1996).
[220] B. Schutz, *Class. Quantum Grav.* **16**, A131 (1999).
[221] V. Kaspi, J. Taylor, and M. Ryba, *Astrophy. J.* **428**, 713 (1994).
[222] V. F. Schwartzman, *JETP Lett.* **9**, 184 (1969) [*Pis'ma Zh. Éksp. Teor. Fiz*, **9**, 315 (1969)].
[223] M. Giovannini, *Stochastic GW backgrounds and ground based detectors*, gr-qc/0009101.
[224] M. Giovannini, *Phys. Rev.* D **60**, 123511 (1999).
[225] M. Giovannini, *Class. Quant. Grav.* **16**, 2905 (1999).
[226] D. Babusci and M. Giovannini, *Phys. Rev.* D **60**, 083511 (1999).
[227] M. Gasperini and M. Giovannini, *Phys. Rev.* D **47**, 1519 (1993).
[228] R. Brustein, M. Gasperini, M. Giovannini and G. Veneziano, *Phys. Lett.* B **361**, 45 (1995).
[229] D. Babusci and M. Giovannini, *Int. J. Mod. Phys.* D **10**, 477 (2001).
[230] P. Bernard, G. Gemme, R. Parodi and E. Picasso, *Rev. Sci. Instrum.* **72**, 2428 (2001).
[231] R. Ballantini *et al.*, *Class. Quant. Grav.* **20**, 3505 (2003).
[232] R. Ballantini, P. Bernard, A. Chincarini, G. Gemme, R. Parodi and E. Picasso, *Class. Quant. Grav.* **21** (2004) S1241.
[233] A. M. Cruise and R. M. J. Ingley, *Class. Quant. Grav.* **23**, 6185 (2006).
[234] A. M. Cruise and R. M. J. Ingley, *Class. Quant. Grav.* **22**, S479 (2005).
[235] A. O. Barut and L. Girardello, *Commun. Math. Phys.* **21**, 41 (1971).
[236] D. Stoler, *Phys. Rev.* D **1**, 3217 (1970).
[237] H. P. Yuen, *Phys. Rev.* A **13**, 2226 (1976).

[238] R. Loudon, *The Quantum Theory of Light*, (Oxford University Press, 1991).
[239] L. Mandel and E. Wolf, *Optical Coherence and Quantum Optics*, (Cambridge University Press, Cambridge, England, 1995).
[240] B. L. Shumaker, *Phys. Rep.* **135**, 317 (1986).
[241] J. Grochmalicki and M. Lewenstein, *Phys. Rep.* **208**, 189 (1991).
[242] L. P. Grishchuk and Yu. V. Sidorov, *Phys. Rev. D* **42**, 3413 (1990).
[243] M. Giovannini, *Phys. Rev. D* **61**, 087306 (2000).
[244] M. Gasperini, M. Giovannini and G. Veneziano, *Phys. Rev. D* **48**, 439 (1993).
[245] M. Kruczenski, L. E. Oxman and M. Zaldarriaga, *Class. Quant. Grav.* **11**, 2317 (1994).
[246] D. Koks, A. Matacz and B. L. Hu, *Phys. Rev. D* **55**, 5917 (1997) [Erratum-ibid. D **56**, 5281 (1997)].
[247] D. R. Truax, *Phys. Rev. D* **13**, 1988 (1985).
[248] R. A. Fisher, M. M. Nieto, and V. Sandberg, *Phys. Rev. D* **29**, 1107 (1984).
[249] M. Giovannini, *Phys. Rev. D* **70**, 103509 (2004).
[250] M. Giovannini, *Class. Quant. Grav.* **22**, 363 (2005).
[251] M. Giovannini, *Class. Quant. Grav.* **23**, R1 (2006).
[252] M. Giovannini, *Int. J. Mod. Phys. D* **13**, 391 (2004).
[253] M. Giovannini, *Phys. Rev. D* **73**, 101302 (2006)
[254] M. Giovannini, *Class. Quant. Grav.* **23**, 4991 (2006).
[255] M. Giovannini, *Phys. Rev. D* **74**, 063002 (2006).
[256] A. L. Matacz, *Phys. Rev. D* **49**, 788 (1994).
[257] M. Giovannini, *Class. Quant. Grav.* **21**, 4209 (2004).
[258] M. Giovannini, *Phys. Rev. D* **73**, 083505 (2006).
[259] K. Enqvist, H. Kurki-Suonio and J. Valiviita, *Phys. Rev. D* **62**, 103003 (2000).
[260] K. Enqvist and H. Kurki-Suonio, *Phys. Rev. D* **61**, 043002 (2000).
[261] K. Enqvist, H. Kurki-Suonio and J. Valiviita, *Phys. Rev. D* **65**, 043002 (2002).
[262] M. Bucher, K. Moodley and N. Turok, *Phys. Rev. Lett.* **87**, 191301 (2001).
[263] M. Bucher, K. Moodley and N. Turok, *Phys. Rev. D* **66**, 023528 (2002).
[264] K. Moodley, M. Bucher, J. Dunkley, P. G. Ferreira and C. Skordis, *Phys. Rev. D* **70**, 103520 (2004).
[265] M. Bucher, J. Dunkley, P. G. Ferreira, K. Moodley and C. Skordis, *Phys. Rev. Lett.* **93**, 081301 (2004).
[266] H. Kurki-Suonio, V. Muhonen and J. Valiviita, *Phys. Rev. D* **71**, 063005 (2005).
[267] J. Valiviita and V. Muhonen, *Phys. Rev. Lett.* **91**, 131302 (2003).
[268] C. Gordon and A. Lewis, *Phys. Rev. D* **67**, 123513 (2003).
[269] C. Gordon and K. A. Malik, *Phys. Rev. D* **69**, 063508 (2004).
[270] F. Ferrer, S. Rasanen and J. Valiviita, *JCAP* **0410**, 010 (2004).
[271] J. M. Bardeen, J. R. Bond, N. Kaiser and A. S. Szalay, *Astrophys. J.* **304**, 15 (1986).
[272] M. Bucher, K. Moodley and N. Turok, *Phys. Rev. D* **62**, 083508 (2000).
[273] A. Challinor and A. Lasenby, *Astrophys. J.* **513**, 1 (1999).

[274] S. Chandrasekar, *Radiative Transfer*, (Dover, New York, US, 1966).
[275] A. Peraiah, *An Introduction to Radiative Transfer*, (Cambridge University Press, Cambridge, UK, 2001).
[276] W. Hu and N. Sugiyama, *Astrophys. J.* **444**, 489 (1995).
[277] W. Hu and N. Sugiyama, *Phys. Rev. D* **51**, 2599 (1995).
[278] W. Hu and N. Sugiyama, *Phys. Rev. D* **50**, 627 (1994).
[279] W. Hu and N. Sugiyama, *Astrophys. J.* **471**, 542 (1996).
[280] W. Hu, D. Scott, N. Sugiyama and M. J. White, *Phys. Rev. D* **52**, 5498 (1995).
[281] D. D. Harari and M. Zaldarriaga, *Phys. Lett.* B **319**, 96 (1993).
[282] M. Zaldarriaga and D. D. Harari, *Phys. Rev. D* **52**, 3276 (1995).
[283] D. D. Harari, J. D. Hayward and M. Zaldarriaga, *Phys. Rev. D* **55**, 1841 (1997).
[284] A. Kosowsky, *Annals Phys.* **246**, 49 (1996).
[285] A. Kosowsky and A. Loeb, *Astrophys. J.* **469**, 1 (1996).
[286] J. R. Bond and A. S. Szalay, *Astrophys. J.* **274**, 443 (1983).
[287] J. D. Jackson, *Classical Electrodynamics*, (Wiley, New York, US, 1975).
[288] A. Kosowsky, *New Astron. Rev.* **43**, 157 (1999).
[289] See, for instance, http://www.cmbfast.org.
[290] U. Seljak and M. Zaldarriaga, *Astrophys. J.* **469**, 437 (1996).
[291] W. Hu, *Ann. Phys.* **303**, 203 (2003).
[292] M. Zaldarriaga, *Astrophys. J.* **503**, 1 (1998).
[293] W. Hu and M. J. White, *New Astron.* **2**, 323 (1997).
[294] W. Hu and M. White, *Phys. Rev. D* **56**, 596 (1997).
[295] J. R. Bond and G. Efstathiou, *Mon. Not. Roy. Astron. Soc.* **226**, 655 (1987).
[296] J. R. Bond and G. Efstathiou, *Astrophys. J.* **285**, L45 (1984).
[297] B. Jones and R. Wyse, *Astron. Astrophys.* **149**, 144 (1985).
[298] P. D. Naselsky, D. I. Novikov, and I. D. Novikov *The Physics of the Cosmic Microwave Background*, (Cambridge University press, Cambridge UK, 2006).
[299] L. Verde *et al.*, *Astrophy. J. Suppl.* **148**, 195 (2003).
[300] G. Esftathiou and J. Bond, *Mon. Not. Roy. Astron. Soc.* **304**, 75 (1999).
[301] S. Weinberg, *Phys. Rev. D* **62**, 127302 (2000).
[302] S. Weinberg, *Astrophys. J.* **581**, 810 (2002).
[303] S. Weinberg, *Phys. Rev. D* **64**, 123512 (2001).
[304] S. Weinberg, *Phys. Rev. D* **64**, 123511 (2001).
[305] F. Atrio-Barandela, A. Doroshkevich, and A. Klypin, *Astrophys. J.* **378**, 1 (1991).
[306] P. Naselsky and I. Novikov, *Astrophys. J.* **413**, 14 (1993).
[307] H. Jorgensen, E. Kotok, P. Naselsky, and I Novikov, *Astron. Astrophys.* **294**, 639 (1995).
[308] A. G. Doroshkevich, I. P. Naselsky, P. D. Naselsky and I. D. Novikov, *Astrophys. J.* **586**, 709 (2003).
[309] A. G. Doroshkevich and P. D. Naselsky, *Phys. Rev. D* **65** (2002) 123517.
[310] V. F. Mukhanov, *Int. J. Theor. Phys.* **43**, 623 (2004).
[311] U. Seljak and M. Zaldarriaga, *Phys. Rev. Lett.* **78**, 2054 (1997).

[312] M. Zaldarriaga and U. Seljak, *Phys. Rev. D* **55**, 1830 (1997).
[313] W. Hu and M. White, *Phys. Rev. D* **56**, 596 (1997).
[314] J. J. Sakurai, *Modern Quantum Mechanics*, (Addison-Wesley, New York, 1985).
[315] M. Kamionkowski, A. Kosowsky and A. Stebbins, *Phys. Rev. D* **55** , 7368 (1997).
[316] A. Polnarev, *Sov. Astron.* **29**, 607 (1985).
[317] R. Crittenden, J. Bond, R. Davis, G. Efsthathiou, and P. Steinhardt, *Phys. Rev. Lett.* **71**, 324 (1993).
[318] L. Knox, Y.-S. Song, *Phys. Rev. Lett.* **89**, 011303 (2002).
[319] W. Hu, U. Seljak, M. White, and M. Zaldarriaga, *Phys. Rev. D* **57**, 3290 (1998).
[320] U. Seljak and M. Zaldarriaga, *Astrophys. J.* **469**, 437 (1996).
[321] V. F. Mukhanov, H. A. Feldman and R. H. Brandenberger, *Phys. Rept.* **215**, 203 (1992).
[322] V. N. Lukash, *Sov. Phys. JETP* **52**, 807 (1980) [*Zh. Eksp. Teor. Fiz.* **79**, 807 (1980)]
[323] V. F. Mukhanov, *Sov. Phys. JETP* **67**, 1297 (1988) [*Zh. Eksp. Teor. Fiz.* **94**, 1 (1988)].
[324] E. Stewart and D. Lyth, *Phys. Lett. B* **302**, 171 (1992).
[325] J. Martin and R. H. Brandenberger, *Phys. Rev. D* **63**, 123501 (2001).
[326] R. H. Brandenberger and J. Martin, *Mod. Phys. Lett. A* **16**, 999 (2001).
[327] J. Martin and R. H. Brandenberger, *Phys. Rev. D* **65**, 103514 (2002).
[328] J. C. Niemeyer, *Phys. Rev. D* **63**, 123502 (2001).
[329] J. C. Niemeyer and R. Parentani, *Phys. Rev. D* **64**, 101301 (2001).
[330] J. Kowalski-Glikman, *Phys. Lett. B* **499**, 1 (2001).
[331] T. Jacobson and D. Mattingly, *Phys. Rev. D* **63**, 041502 (2001).
[332] C. S. Chu, B. R. Greene and G. Shiu, *Mod. Phys. Lett. A* **16**, 2231 (2001).
[333] S. Tsujikawa, R. Maartens and R. Brandenberger, *Phys. Lett. B* **574**, 141 (2003).
[334] M. Bastero-Gil, P. H. Frampton and L. Mersini, *Phys. Rev. D* **65**, 106002 (2002).
[335] D. Amati, M. Ciafaloni and G. Veneziano, *Phys. Lett. B* **197**, 81 (1987).
[336] D. Amati, M. Ciafaloni and G. Veneziano, *Phys. Lett. B* **216**, 41 (1989).
[337] D. J. Gross and P. F. Mende, *Nucl. Phys. B* **303**, 407 (1988).
[338] D. J. Gross and P. F. Mende, *Phys. Lett. B* **197**, 129 (1987).
[339] A. Kempf, *Phys. Rev. D* **63**, 083514 (2001).
[340] A. Kempf and J. C. Niemeyer, *Phys. Rev. D* **64**, 103501 (2001).
[341] S. F. Hassan and M. S. Sloth, *Nucl. Phys. B* **674**, 434 (2003).
[342] R. Brandenberger and P. M. Ho, *Phys. Rev. D* **66**, 023517 (2002).
[343] U. H. Danielsson, *Phys. Rev. D* **66**, 023511 (2002).
[344] U. H. Danielsson, *JHEP* **0207**, 040 (2002).
[345] A. A. Starobinsky, *JETP Lett.* **73**, 371 (2001) [*Pisma Zh. Eksp. Teor. Fiz.* **73**, 415 (2001)].
[346] N. Kaloper, M. Kleban, A. E. Lawrence and S. Shenker, *Phys. Rev. D* **66**, 123510 (2002).

[347] N. Kaloper, M. Kleban, A. E. Lawrence and S. Shenker, *Phys. Rev.* D **66**, 123510 (2002).
[348] R. Easther, B. R. Greene, W. H. Kinney and G. Shiu, *Phys. Rev.* D **64**, 103502 (2001).
[349] R. Easther, B. R. Greene, W. H. Kinney and G. Shiu, *Phys. Rev.* D **67**, 063508 (2003).
[350] R. Easther, B. R. Greene, W. H. Kinney and G. Shiu, *Phys. Rev.* D **66**, 023518 (2002).
[351] N. A. Chernikov and E. A. Tagirov, *Annales Poincare Phys. Theor.* A **9**, 109 (1968).
[352] E. A. Tagirov, *Ann. Phys.* **76**, 561 (1973).
[353] K. Goldstein and D. A. Lowe, *Nucl. Phys.* B **669**, 325 (2003).
[354] H. Collins and R. Holman, *Phys. Rev.* D **70**, 084019 (2004).
[355] H. Collins, R. Holman and M. R. Martin, *Phys. Rev.* D **68**, 124012 (2003).
[356] V. F. Mukhanov, L. R. W. Abramo and R. H. Brandenberger, *Phys. Rev. Lett.* **78**, 1624 (1997).
[357] L. R. W. Abramo, R. H. Brandenberger and V. F. Mukhanov, *Phys. Rev.* D **56**, 3248 (1997).
[358] T. Tanaka, arXiv:astro-ph/0012431.
[359] M. Porrati, *Phys. Lett.* B **596**, 306 (2004).
[360] A. A. Starobinsky and I. I. Tkachev, *JETP Lett.* **76**, 235 (2002) [*Pisma Zh. Eksp. Teor. Fiz.* **76**, 291 (2002)].
[361] B. R. Greene, K. Schalm, G. Shiu and J. P. van der Schaar, hep-th/0411217.
[362] R. Brandenberger and J. Martin, hep-th/0410223.
[363] D. Wands et al., *Phys. Rev.* D **62**, 043527 (2000).
[364] S. Weinberg, *Phys. Rev.* D **67**, 123504 (2003).
[365] S. Weinberg, *Phys. Rev.* D **70**, 043541 (2004).
[366] S. Weinberg, *Phys. Rev.* D **70**, 083522 (2004).
[367] S. Weinberg, *Phys. Rev.* D **69**, 023503 (2004).
[368] T.-P. Cheng and L.-F. Lee, *Gauge theory of elementary particle physics*, (Clarendon Press, Oxford, UK, 1984).
[369] J. Goldstone, A. Salam and S. Weinberg, *Phys. Rev.* **127** (1962).
[370] R. Brustein, M. Gasperini, M. Giovannini, V. F. Mukhanov and G. Veneziano, *Phys. Rev.* D **51**, 6744 (1995).
[371] J. Hwang, *Astrophys. J.* **375**, 443 (1990).
[372] J. c. Hwang and H. Noh, *Class. Quant. Grav.* **19**, 527 (2002).
[373] K. A. Malik and D. Wands, *JCAP* **0502**, 007 (2005).
[374] K. A. Malik, D. Wands and C. Ungarelli, *Phys. Rev.* D **67**, 063516 (2003).
[375] G. Dvali, A. Gruzinov and M. Zaldarriaga, *Phys. Rev.* D **69**, 023505 (2004).
[376] L. Kofman, arXiv:astro-ph/0303614.
[377] M. Postma, *JCAP* **0403**, 006 (2004).
[378] A. Mazumdar and M. Postma, *Phys. Lett.* B **573**, 5 (2003), [Erratum-ibid. **585**, 295 (2004)]
[379] R. Allahverdi, *Phys. Rev.* D **70**, 043507 (2004).
[380] M. S. Turner, *Phys. Rev.* D **28**, 1243 (1983).
[381] S. Mollerach, *Phys. Rev.* D **42**, 313 (1990).

[382] L. Kofman and A. Linde, *Nucl. Phys. B* **282**, 555 (1987).
[383] A. Linde and V. Mukhanov, *Phys. Rev. D* **56**, 535 (1997).
[384] T. Moroi and T. Takahashi, *Phys. Lett. B* **522**, 215 (2001) [Erratum, *ibid* **539**, 303 (2002)].
[385] T. Moroi and T. Takahashi, *Phys. Rev. D* **66**, 063501 (2002).
[386] K. Enqvist and M. S. Sloth, *Nucl. Phys. B* **626**, 395 (2002).
[387] D. H. Lyth and D. Wands, *Phys. Lett. B* **524**, 5 (2002).
[388] V. Bozza, M. Gasperini, M. Giovannini and G. Veneziano, *Phys. Lett. B* **543**, 14 (2002).
[389] V. Bozza, M. Gasperini, M. Giovannini and G. Veneziano, *Phys. Rev. D* **67**, 063514 (2003).
[390] M. Sloth, *Nucl. Phys. B* **656**, 239 (2003).
[391] K. Dimopoulos et al., *JHEP* **05**, 057 (2003).
[392] K. A. Malik, D. Wands and C. Ungarelli, *Phys. Rev. D* **67**, 063516 (2003).
[393] S. Gupta, K. A. Malik and D. Wands, *Phys. Rev. D* **69**, 063513 (2004).
[394] N. Bartolo and A. Liddle, *Phys. Rev. D* **65**, 121301 (2002).
[395] D. H. Lyth, *Phys. Lett. B* **579**, 239 (2004).
[396] K. Dimopoulos and D. H. Lyth, *Phys. Rev. D* **69**, 123509 (2004).
[397] K. Enqvist and A. Mazumdar, *Phys. Rept.* **380**, 99 (2003).
[398] J. McDonald, *Phys. Rev. D* **68**, 043505 (2003).
[399] T. Moroi and H. Murayama, *Phys. Lett. B* **553**, 126 (2003).
[400] M. Postma, *Phys. Rev. D* **67**, 063518 (2003).
[401] K. Enqvist, A. Jokinen, S. Kasuya, A. Mazumdar, *Phys. Rev. D* **68**, 103507 (2003).
[402] K. Enqvist, *Mod. Phys. Lett. A* **19**, 1421 (2004)
[403] M. Bastero-Gil, V. Di Clemente, and S. F. King, *Phys. Rev. D* **67**, 103516 (2003).
[404] M. Giovannini, *Phys. Rev. D* **69**, 083509 (2004).
[405] M. Postma, *JCAP* **0405**, 002 (2004).
[406] D. Wands, *New Astron. Rev.* **47**, 781 (2003).
[407] M. Giovannini, *Phys. Rev. D* **67**, 123512 (2003).
[408] G. F. R. Ellis and W. R. . Stoeger, *Class. Quant. Grav.* **5**, 207 (1988).

Index

ΛCDM model, 25, 64, 90, 92, 275
   basic introduction, 22
   extensions, 28, 403
γ-rays, 6

absolute luminosity, 416
ACBAR, 14, 24
accelerating Universe, 42, 82, 84
adiabatic
   evolution, 33, 69, 75, 95
   fluctuations, 13
   initial conditions, 66, 199, 275
   mode, 198, 205, 254, 267, 305, 357
   regime, 138
adiabaticity
   condition, 208, 210, 221, 358
   violation, 224
angular power spectrum, 12, 77, 297
angular resolution, 12
anisotropic stress, 110, 181, 253, 440
apparent luminosity, 416
Archeops, 13
AUGER, 7

barotropic index, 43, 334, 351, 377, 424
baryon asymmetry, 130
baryon-to-photon ratio, 218, 239, 291
baryons, 132, 183
BEPPO-SAX, 8
Bessel functions, 146, 185, 314, 316
   spherical, 184, 201, 265

big-bang nucleosynthesis, 45, 306, 431
BKL
   formalism, 102
blackbody, 3
   entropy, 14
   pressure, 14
Bogoliubov transformation, 146, 167
Boltzmann
   constant, 4
   suppression, 55
Boltzmann equation
   collisionless, 248
   for brightness perturbations, 258
   for massless particles, 251
   for neutrinos, 251
   for the tensor modes, 286
   in different gauges, 344
   in synchronous gauge, 354
   in the longitudinal gauge, 354
   in the synchronous gauge, 354
   perturbed, 247
BOOMERANG, 13, 24
Bose and Fermi gases
   entropy density, 424
   concentration, 423
   energy density, 423
   non-relativistic limit, 425
   pressure, 423
   ultrarelativistic limit, 423
Bose-Einstein
   distribution, 250, 255
   occupation number, 9, 423

statistics, 422
brightness perturbations
  definition, 257
  evolution, 258
  gauge transformation, 345
  in different gauges, 357
  solutions, 259
bulk viscous stresses, 379

canonical fluctuations
  scalar problem, 311, 441
  tensor problem, 142, 315
canonical quantization, 144, 315
CBI, 14, 24
CDM perturbations, 208, 216
CDM velocity and synchronous gauge, 357
CDM-radiation system, 183, 359
  evolution equations, 186
  non-adiabatic modes, 188
chemical potential, 11, 33, 421
Christoffel connections, 33, 133, 435
clusters of galaxies, 6, 18, 19
CMB
  anisotropies, 13, 77, 172, 201, 285
  history, 12
  initial conditions, 182, 222, 305
  temperature, 8, 17
CMB polarization
  analytical estimates, 278
  and first-order in tight-coupling, 278
  and Stokes parameters, 255
COBE, 12
coherent states, 154
collisionless species, 207, 251
comoving orthogonal gauge, 362
Comptonization parameter, 20
concentration
  of Bose and Fermi gas, 423
  of baryons, 56
  of electrons, 56
  of photons, 16, 56
conformal coupling, 108
conjugate momenta, 248
consistency condition, 125, 320

contraction, 42
correlation cosine, 201
cosine integral function, 210
cosmic rays, 6
cosmological
  constant, 22, 40, 94, 206
  parameters, 22, 45, 191
Coulomb scattering, 59
covariant conservation, 32, 47, 105
  of dusty matter, 45
covariant derivative, 33, 38, 103, 439
critical
  density, 9, 15, 43, 74
  parameter, 43
  units, 4, 10, 43
critical fraction
  of baryons, 23, 45
  of dark energy, 22, 45, 46
  of dark matter, 23, 45
  of dusty matter, 45
  of neutrinos, 23, 46
  of photons, 10, 46
  of radiation, 46
  of spatial curvature, 48
cross-correlation, 285
cross-corrrelation, 281
cross-over frequency, 20
curvature
  extrinsic, 44, 103
  fine tuning, 75
  intrinsic, 85, 103
  intrinsic(spatial), 44, 74
curvature perturbations
  as a technical tool, 191
  conservation, 192
  evolution, 191
  rigorous definition, 362
cyanogen (CN), 18

dark
  energy, 15, 41, 48, 102, 126, 206
  matter, 16, 45, 132, 171, 183
Dasi, 13
de Sitter
  inflation, 113
  space-time, 83, 326

deceleration, 42, 70
deceleration parameter
  homogeneous, 44
  inhomogeneous, 104
decoupling
  of neutrinos, 254, 305, 429
  of photons, 54, 55, 417
degree of linear polarization, 259
density contrast
  comoving orthogonal
      hypersurfaces, 365
  gauge transformation, 135
  longitudinal gauge, 178, 179
  on uniform curvature hypersurfaces
    as a practical tool, 197
  rigorous definition, 364
differential energy spectrum
  of gravitons, 148
  of photons, 9
diffusion damping, 242, 288
dipole, 247, 253, 345
dispersion measurement, 20
dispersion relations, 288, 289
distance
  angular diameter, 49, 415
  examples, 417
  luminosity, 49, 416
  measure, 49, 413
Doppler
  contribution, 235
  oscillations, 206, 216, 292
  peak, 13, 74, 188, 235, 267, 275
  region, 203
  term, 177, 264

Eddington, 29
Einstein
  equations, 39, 103, 109, 136, 179, 247, 308
  tensor, 40, 131, 179
  Universe, 29
Einstein-Boltzmann
  equations, 247
  hierarchy, 171, 266, 271, 305
Einstein-Hilbert
  action, 37, 151

gravity, 113, 334
energy conditions, 42
energy-momentum tensor
  fluctuations, 307
  of a scalar field, 109, 179
  of perfect fluids, 32, 103, 132, 135
energy-momentun tensor
  fluctuations, 438
enthalpy, 35, 422
entropy, 422
  and squeezed states, 156
  black-hole, 15
  density, 14, 424, 426, 430
  fluctuations, 223, 378
  generation in reheating, 94
  gravitational, 15, 96
  modes, 206, 359
  of the Universe, 15, 94
  per dark matter particle, 188
  problem, 75, 81
  thermodynamic, 15
entropy fluctuations
  in CDM-radiation system, 188
  evolution, 193
  in CDM-radiation system, 359
equilibrium
  chemical, 55, 426
  kinetic, 426
  thermal, 55, 426
equilibrium distribution, 251
Euler Γ function, 202, 316
Euler-Mascheroni constant, 210
evolution of the inhomogeneities
  during dusty matter, 186
  general discussion, 132
  radiation matter transition, 192
  scalar case, 180
  tensor case, 137
evolution of the inhomogeneities;
  during radiation, 184
expansion, 42
extensive variables, 97, 421

Faraday rotation, 6
Fermi-Dirac
  distribution, 250, 255

occupation number, 423
statistics, 422
first law of thermodynamics, 33, 421
flatness problem, 74, 85
fluctuations
  classical, 88
  covariant conservation, 181, 439
  of the Christoffel connections, 436, 448
  of the Einstein tensor, 180
  of the Ricci tensor, 436
  of the sources, 135, 180
  quantum mechanical, 88
fluids
  barotropic, 30
  baryon-photon, 216
  dusty, 34, 52
  interacting, 37, 377
  perfect, 32, 103
  stiff, 34, 110
  viscous, 34
free energy, 422
Friedmann, 29
Friedmann-Lemaître equations
  fully inhomogeneous, 105
Friedmann-Lemaître equations
  fully inhomogeneous, 101
  homogeneous, 41, 44
  scalar field case, 111
fundamental thermodynamic identity, 33

galaxy, 4
Gamow, 6
gauge
  comoving orthogonal, 347, 362
  longitudinal, 180, 341, 346
  parameters, 343
  synchronous, 349
  transformations, 133
  uniform density, 344
  uniform field, 344
  unifrom density, 364
gauge parameters
  scalar case, 133, 342
  vector case, 134

gauge transformation
  scalar case, 134, 342
  vector case, 134
gauge-independent fluctuations, 445
gauge-invariant
  entropy fluctuations, 360
gauge-invariant fluctuations
  scalar modes, 347
  simple example, 134
General Relativity, 31, 37
geodesic
  equation, 98, 173
  incompleteness, 97
geodesics
  incomplete, 97
  null, 98, 173
Gibbs free energy, 422, 427
GLAST, 8
Glauber operator, 154

Hamiltonian
  operator, 145
  scalar problem, 322
  tensor problem, 142–144
Hamiltonian constraint, 180, 197, 206, 225, 335, 352
Hankel functions, 146, 316, 326
Harrison-Zeldovich spectrum, 198, 233, 270, 293, 319
Heisenberg algebra, 155
Heisenberg representation, 141, 312
homogeneity, 31
horizon
  crossing, 90
  distances, 417
  event, 70, 82, 417
  particle, 70, 71, 81, 417
  problem, 70, 81
  reentry, 182
Hubble
  constant, 10, 15
  radius, 71
    crossing, 91
  rate, 22, 33
  space telescope, 24
  volume, 14, 33

Index

hypersurfaces
  comoving orthogonal, 363
  of uniform density, 364, 366
  of uniform scalar field, 368

inflation
  conventional models, 81, 108, 403
  unconventional models, 113, 400
inflationary backgrounds
  large and small field, 115
  dynamical classification, 114
  exact solutions, 116
  geometric classification, 113
inflaton
  decay, 81
inhomogeneities
  Bardeen formalism, 131
  in FRW backgrounds, 129
  scalar, 131
  tensor, 131
  vector, 131
integrated Sachs-Wolfe term, 177
intensive variables, 421
interacting fluids, 377
internal energy, 33
ionization fraction, 55, 56
isotropy, 31, 77

Klein-Gordon equation
  in arbitrary geometry, 109
  in FRW background, 112
  perturbed form, 308, 440
Kompaneets, 19

Legendre polynomials, 77, 248, 270
  orthonormality relation, 252
  recurrence relation, 252, 265
Lemaître, 29
leptons, 428
Lie derivative, 133
line of sight integrals, 261, 281, 293

mass-shell condition, 249
Maxima, 13
metric
  Bianchi, 32

FRW, 31
  fully inhomogeneous, 102
  inhomogeneity, 43
metric tensor, 31, 33
  determinant, 37
minimal coupling, 108
mixed gauge-invariant treatments, 372
mixing coefficients
  numerical evaluation, 164
  sudden approximation, 147
  unitarity, 147
mode functions, 145, 160, 313, 318
momentum constraint, 180, 207, 231, 358
monopole, 247, 253, 345

natural gravitational units, 379
natural units, 4, 9, 145
neutrinos, 132, 183
non-adiabatic
  contribution, 189, 191, 193
  fluctuation, 194
  fluctuations, 189
  mode, 193
  modes, 190, 195, 198, 205, 305, 306
non-adiabatic modes
  uncorrelated, 203
non-relativistic limit, 425
number of e-folds
  and slow-roll parameters, 124
  minimal, 83
number of efolds
  maximal, 91

objectivity, 3
Occhialini, 8
octupole, 208, 254, 279
off-diagonal gauge, 342, 367
ordinary Sachs-Wolfe term, 177

Palatini identity, 38
peculiar velocity, 104
Penzias, 6, 12
phase space distribution
  neutrinos, 180, 208

phase-space distribution
  neutrinos, 253
  photons, 257
photon-baryon
  synchronization, 270
  velocity, 218
photon-baryon velocity, 354
photons, 132, 183
pivot scale, 198, 268
Planck
  length, 83
  mass, 15
    reduced, 111, 141, 394
  time, 71, 74
  units, 15
Planck Explorer, 12
Poisson equation
  relativistic generalization, 372
polarization
  line of sight solutions, 265
  linear, 255, 278, 285
  power spectrum, 254, 281
potential barriers, 139
power spectrum
  scalar, 318
  tensor, 315
power-law inflation, 113
pre-big bang, 42, 151, 392
pre-decoupling
  phase, 179
  physics, 205
pre-deoupling
  physics, 171
pre-equality
  evolution equations, 205
pre-equality initial conditions, 223
preheating, 122
pressure
  of Bose and Fermi gas, 423
primordial plasma, 427
pulsar astronomy, 6
pump filed, 139

quadrupole, 253
quadrupole and anisotropic stress, 180

quantum field, 42, 141, 144
quantum mechanical
  normalization, 88
quantum mechanics, 3, 85
quantum state
  of perturbations, 130, 151
quantum-mechanical
  inhomogeneities, 92
  normalization, 130, 138
  operators, 132
  origin, 93
  treatment, 130, 141
quarks, 428
quasi de Sitter expansion, 113, 120
quintessential inflation, 122, 151, 399, 406

radiation-matter transition, 53, 191
radiative transfer, 247
Rayleigh-Jeans region, 21
rays
  gamma, 6
  x, 6
recombination, 54, 91
redshift, 17, 47
  definition, 411
  of equality, 52
  of recombination, 64
reheating, 94
  temperature, 95
relativistic species, 423, 429, 431
relic graviton background, 150
relic graviton spectrum, 148
  frequency range, 150
Ricci
  scalar, 38
  tensor, 38
Riemann tensor, 38
Riemann zeta function, 16, 424
ROSAT, 8, 18

Sachs-Wolfe effect
  tensor modes, 172
  integrated contribution, 177
  ordinary contribution, 177, 235
  scalar modes, 175

Sachs-Wolfe plateau
    analytical estimates, 196, 266
    scale-invariant limit, 203
Saha equation, 56
Sakharov, 129, 233
scalar field
    action in curved backgrounds, 108
    conformal coupling, 112
    first-order fluctuations, 307
    homogeneous evolution, 108
    minimal coupling, 108, 307
Schrödinger representation, 143, 152
Silk damping scale, 218, 235, 242
singularity problem, 79, 81, 107
SKA, 6
slow-roll
    algebra, 123, 312
    parameters, 120, 313
    regime, 120, 394
sound horizon, 235, 273, 276, 294
    at decoupling, 276
spatial gradients, 107
spectator field, 392
    fluctuations, 401
spectral distortion, 22
spectral index, 23
    scalar, 125, 319, 400
    tensor, 125, 317, 400
speed
    of light, 4
    of sound, 189, 211, 377, 402
        baryon-photon system, 273
        explicit form, 190
        in multi-fluid systems, 190, 224
        photon-baryon system, 220
spherical harmonics, 77, 282
    addition theorem, 77
    spin-s, 284
squeezed states
    harmonic oscillator, 153
    and unitary transformation, 153
    general discussion, 151
    non-compact groups, 155
squeezing operator, 155
Standard Cosmological Model

assumptions, 30
    entropy problem, 70, 75, 93
    flatness problem, 70, 75, 85
    horizon problem, 70, 81
    inhomogeneity problem, 70, 76
    problems, 69
    singularity problem, 70, 79
stiff phase, 91
Stokes parameters, 248, 255, 282
Sunyaev-Zeldovich effect, 18
super-adiabatic
    amplification, 136
        potential barrier, 138
super-adiabatic amplification, 311
    tensor modes, 139
superinflation, 113
supernovae, 22, 50
survey
    2dF, 24
    Sloan, 24
    supernova legacy, 24
    gold sample, 24
synchronous gauge, 349

tensor modes
    conjugate momentum, 143
    evolution, 136
    Hamiltonian, 142
    quantum description, 141
    second order action, 141
    two plarizations, 137
theory of angular momentum, 283
thermal history of the Universe, 90
thermodynamic potential, 422
thermodynamics
    second law, 15, 75, 96
Thompson
    cross section, 59, 61, 216
    mean free path, 62, 219
    scattering, 19, 58
three-momentum
    comoving, 249
    physical, 249
tight-coupling
    approximation, 232, 272
    baryon-lepton fluid, 63

expansion, 233, 267, 272
first order, 278
limit, 221, 278
photon-baryon fluid, 35, 216
second order, 287
zeroth order, 272
time
  conformal, 32, 44
  cosmic, 31
  of equality, 54
total angular momentum method, 286
trans-planckian effects, 144, 324
transfer function, 215
two-fluid treatment, 171

ultra-relativistic limit, 424
uniform density gauge, 344, 364
Universe
  spatially closed, 31, 43
  spatially flat, 31, 43
  spatially open, 31, 50
  spatiallyopen, 43

vacua in curved backgrounds, 156
viscosity
  bulk, 35
  shear, 218
visibility function, 64, 259, 264, 294
VSA, 14, 24

wave-numbers
  comoving, 148, 205
  physical, 17
wavelength
  of extra-galactic emissions, 5
  physical, 88, 90
Wien region, 21
Wien's law, 9
Wigner matrix elements, 284
Wilson, 6, 12
WKB approximation, 238, 275
WMAP, 12, 24
Wronskian, 212

x-rays, 5

zeldovich, 29